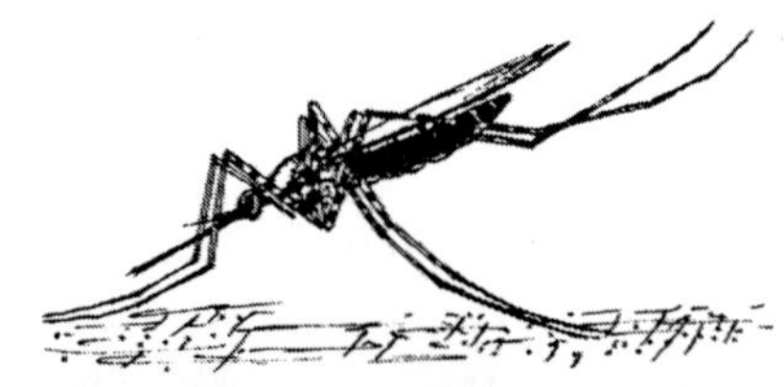

MOSQUITO HUNTERS

An Indian history of hostilities against man's deadliest foe – the Mosquito – since 1881

Professor Dr B.K. Tyagi

Emtd. Scientist 'G' (Director) & Director in-Charge,
ICMR - Centre for Research in Medical Entomology Madurai, TN,
India;

Advisor, VIT University, Vellore, Tamil Nadu, India;
Consultant, BCIL-UN FAO Sri Lanka Project on Biosafety
&
President, Society of Medical Arthropodology

2020

Published by

SCIENTIFIC PUBLISHERS (INDIA)
5 A, New Pali Road, P.O. Box 91
Jodhpur 342 001 (India)
E-mail: info@scientificpub.com
Website: www.scientificpub.com

ISBN : 978-93-89832-88-4
E-ISBN : 978-93-89832-90-7

Printed in India

Dedicated

to

all those medical arthropodologists and mosquito ambushers who have given their everything – time, energy and resources, for a better understanding of the man's deadliest foe – mosquito, for a happier world to live in.

डॉ कैलाश चन्द्र
निदेशक
Dr Kailash Chandra
Director

भारत सरकार
भारतीय प्राणि सर्वेक्षण
पर्यावरण, वन एवं जलवायु परिवर्तन मंत्रालय
Government of India
Zoological Survey of India
Ministry of Environment, Forest and Climate Change

FOREWORD

Mosquitoes (Order Insecta, Family Culicidae) are best known for their mediation of several deadly and/or debilitating vector-borne diseases of public health importance. They are indeed no small enemy of human being to whom they carry viruses, bacteria and parasites, infecting some 700 million people and killing more than one million every year. Thus, no creature on Earth has touched directly the lives of more men, women and children than the mosquito which may appear smaller in size but is a giant of a killer. Mosquito is possibly man's deadliest foe on the planet!

Mosquitoes have attracted human attention since the time immemorial and *Susruta Samhita*, India's medicinal classic of yore, has made extensive references to them. It is India where, in 1897, Dr Ronald Ross had made the greatest medical discovery of the inextricable link between mosquito and malaria, for which he won the Nobel Prize (1902). The country has also shown to the world how the deadly malaria could be controlled through a nation-wide vertical programme during the 1960s. Presently, of course, the nation is targeting to eliminate lymphatic filariasis, along with malaria, in next one decade. Truly, hats off to all those mosquito-borne disease specialists, particularly medical entomologists, who have given their entire life to make a better future to live in for all of us.

Mosquito Hunters of India, in my opinion, is a unique book, encompassing in one volume India's great history of hostilities against man's deadliest foe – the Mosquito, during past more than a century! Dr B.K. Tyagi, an eminent arthropodologist of India, has commensurately touched upon all the important characteristics of a mosquito's biology associated with the disease transmission, and, at the same time, taken the reader through a long and meandering journey of rich Indian history of victories and failures, discoveries and inventions, feuds and treacheries, and, above all the importance of comprehensive team work – all against one animal alone, the mosquito, who is still at large!

I wish Dr Tyagi and this book a humongous success among the professional and amateur researchers alike, and hope the book will make an invaluable reading material to the students and general masses for their own wisdom.

23th April, 2020

[Dr. Kailash Chandra]
Director

प्राणि विज्ञान भवन, 535, एम. ब्लॉक, न्यु अलीपुर, कोलकाता - 700 053, दूरभाष : +91 33 2400 6893, टेलीफैक्स : +91 33 2400 8595
Prani Vigyan Bhawan, 535, M-Block, New Alipore, Kolkata - 700 053, Phone : +91 33 2400 6893, Telefax : +91 33 2400 8595
E-mail : director@zsi.gov.in, kailash.chandra61@gov.in, Website : zsi.gov.in

PREFACE

I must have already written over three dozen books on varied medico-arthropodological/entomological subjects especially pertaining to mosquitoes, but one book which I always wanted to pen is this, "**Mosquito Hunters: An Indian history of hostilities against man's deadliest foe - the Mosquito, since 1881**", basically to highlight, on the one hand, the grand fables of valour and the 'Never Say Never' die-hard spirit of an extraordinary league of mosquito ambushers of India – the country which had once almost defeated malaria during the mid-1960s and is currently marching ahead to pin down another mosquito-borne disease, the lymphatic filariasis, in near future, by 2030, and, on the other hand, the superb evolutionary and adaptive brilliance of this creature, the mosquito, which appears painfully feeble-bodied measuring not more than the size of a barley seed, and yet has posed on the Earth the greatest threat to peoples' health, economy and intelligence, earning her a commensurate sobriquet "Man's deadliest foe". This book in its entirety propels the reader to respect these truly heroic and incredibly intelligent and daring men and women scientists who have given their everything – time, energy and resources, to better our understanding about the deadliest enemy, mosquito, for a happier world to live in! After all, only few creatures on earth can match the uncanny wisdom of a mosquito and are more worthy of scientific inquiry and human wonder.

This book, written from an Indian perspective, is all about a fascinating and terrifying vicious little killer, the mosquito, and the monumental efforts directed against his *bête noire* by the man himself. In this age old war between them, however, the mosquito has so far emerged invincible outwitting man with her (the female mosquito bites, feeds on blood and transmits disease pathogens, hence the focus for attraction and entire discussion) constantly evolving new strategies to stay alive beyond comprehension, though man also has a few splintered fables of success to celebrate. Filled with little-known facts and remarkable anecdotes that bring this tiny being into larger focus, the book offers fascinating, alarming, and convincing evidence that the sooner we get to know this pesky insect – the mosquito, the better off whole mankind will be.

The book also tells the story of some of the mosquito species that contribute to human diseases such as malaria, filariasis, dengue, Zika, chikungunya, and Japanese encephalitis. These diseases have played an important role in slowing down the national progress through depleted economy, health and intelligentsia. The country spends almost 50% of its health budget in fighting against these ailments. Therefore, it emerges that, besides the brutal facts of how the mosquito has insinuated itself into human history, from the earliest times of malaria that devastated invaders of ancient Rome (Alexander 'The Great' had reportedly died due to *Plasmodium falciparum* malaria while returning home after the battle in 326 BC against the Indian king Porus in the malaria infested Punjab region), and the story of man's struggle to live with the mosquito, from the early 19th century malaria-defeat in Mian Mir under direct charge of one of the greatest medical entomologists ever, Dr Samuel Rickard Christophers, who advocated to Dr Ronalad Ross's theory of 'environment sanitation', to the malaria-deaths of hundreds of rural inhabitants living in The Thar Desert's irrigated Command Area under the world famous Indira Gandhi Nahar Pariyojana (IGNP), in the early 1990s, and the recent panic over the chikungunya virus' in Kerala and many other States and Union Territories, crippling thousands of people, in 2006, and deaths from dengue all over the country during 2012-14, the stories of their numerous encounters need to be told to the modern generation of medical entomologists and vector-borne disease specialists to relive the moments of victories and defeats in this vicious age-old battle between man and mosquito. At the end we find that we have only ourselves to be blamed to a great extent for accelerating the spread of mosquitoes and the diseases they transmit; with climate change and increased international travel, mosquito-borne illnesses are flaring up all over the globe. Catastrophic failures of mosquito control have ensured that worldwide even now one person dies of malaria every twelve seconds.

India is blessed to boast of a large number of ingenious entomologists, both those who specialized in the discipline as a college or university subject in their Master's Degree or did doctoral research in the subject, and those who by vocation or profession were though basically a medico, engineer, geographer, statistician or historian etc. but acquired deeply the knowledge of insects of medical significance as a part of expansion and improvement in resources under establishment of the Britishers' colonial rule in India. The latter, in particular, varied from having been born and brought up in India to those who either served in India in some kind of vocation or even connected with Indian mosquitoes or the mosquito-borne diseases through research projects, travelogues and global writings, such as books and reviews, etc. etc. All these are bracketed as "Indian mosquito hunters" for convenience of completing the illustrated history of Indian mosquito scientists' combat against the humankind's deadliest foe as well as bringing into light the glory of medical entomology in India. I do not, however, claim in anyway that I would have covered 'all' medical entomologists, although I made fairly a well-judged decision to portray only those who in my opinion have made an impact, however trifle, in our understanding of the biology of mosquito, with focused attention on control.

Mosquito's biology seems quite simple and yet it is so extraordinary in its comprehensive portrayal of the villainous role in history and its threat to mankind. The famous culicidologist, Dr A.N. Clements (2011) has signified mosquito's life in human learning as: "The great importance of mosquitoes lies in their role as transmitters of pathogens and parasites, and in their use as experimental animals well suited to laboratory investigations into aspects of biochemistry, physiology and behaviour." This book describes, in a mosquito's-eye view, how mosquito breeds, rests, feeds, flies, mates, and dies, besides makes interaction with her natural enemies. The book also deals with the current constraints and future control prospects of mosquito control. In view of the increasing resistance to insecticides and chemotherapy, the book throws light on the subject of greatest promise to ending mosquitoes' deadly assault on man by rendering them impotent by genetic manipulation or replacing them through paratransgenesis involving micro-organisms such as bacteria and fungi.

In essaying this book I have felt both the thrill of recounting my own endless confrontations with mosquitoes of varied nature in far flung nooks and corners of the country, describing in course as new species *Anopheles pseudosundaicus*, after a gap of more than six decades for an Indian anopheline mosquito, and *Toxorhynchites darjeelingensis*, the first ever under the genus since early 20th Century, besides being honoured in having been given to a mosquito my name, *Tx. tyagii* (for authorities describing these species see Chapter 5), discovering a new malaria paradigm, 'Desert Malaria' in the Thar's xeric ecosystem, expounding *de novo* about *Aedes albopictus* transmitting dengue in Kerala as a 'primary' and 'major' vector without the presence of *Ae. aegypti*, inventing a 'mosquito sampler', and executing transfer of two technologies (ToT) to the Public Health Department of Tamil Nadu etc. etc. But after close to forty years now, my work on mosquito continues with double the energy and enthusiasm. Still I do not know the kind of challenges and surprises this little critter has in store to offer! My quest for mosquito-related knowledge will continue forever!

Prof. Dr B.K. Tyagi

August 20th , 2020

ACKNOWLEDGEMENT

It is unimaginable to encompass in a small composition like this a vast and diverse life attribution of one of the world's diverse group of animals which is also the man's deadliest foe, mosquito, without the immense help of the generous Librarians of various national institutional libraries, particularly personal courtesies of xerox copies and other material of the various different scientific materials by a large number of friends and colleagues all over country and beyond!

As always I sincerely thank my consort, Ajita, who was all supportive and motivating to me throughout the period of completing this rare composition.

Last, but not least, I owe much to Mr Pawan Kumar Sharma, CEO of the Scientific Publishers (India), Jodhpur, for his trust in me and for the patience and help extended throughout the process of writing this monograph.

Prologue

"No creature has touched directly the lives of more human beings than the mosquito.

She has been a nuisance, a pollinator of plants and an angel of death all over the globe.

And throughout history, much of our trouble with the mosquito has been caused by man himself."

- Andrew Spielman & Michael D'Antonio (2001)

CONTENTS

Chapter 1

MOSQUITO - *THE 'FEMME FATALE'*

1.1. Introduction

There are lots of reasons why people should be interested in mosquitoes. Not only mosquitoes trouble people, of both sexes and of all age groups, by their irritating and painful bites, but, in the process, they also mediate transmission of many life-threatening and debilitating diseases, besides a plethora of vexations of varied nature including psychological disturbance (Clements, 1992, 1999, 2011; Eldridge & Edman, 2012; Tyagi, 2003, 2004). A mosquito bite *per se* may not bring instant havoc but it is capable to cause disasters if it carried disease pathogens. Through their bites mosquitoes also levy unbearable uneasiness, ill temperament and loss of sleep, thereby affecting human psychology in attending to important jobs at hand. They are branded objects of deep abhorrence and dislike for all they offer to the mankind - the annoyance, the pain, the nuisance and the deadly infections. For centuries in India, like all other fast developing nations, mosquitoes, by mere virtue of their involvement in the transmission of several deadly and/or incapacitating infections like malaria, dengue, chikungunya, Zika, West Nile, filariasis and Japanese encephalitis, have been responsible in wasting humongous national resources, in particular, by impoverishing human health, reducing economy and emaciating intelligentsia.

Mosquitoes are some of the most deadly creatures on the planet. They belong to the Class Insecta of which more than one million species have been described to exist at present, although some estimates indicate there could be insects anywhere between 3-30 millions. What ever the total, a great deal of scientific information on life come mainly from insects. The dominance of insects among the world's animals is a fundamental scientific insight which means that in numbers of species beyond our comprehension these animals permeate diverse and essential natural processes in Earth's terrestrial, aerial and freshwater ecosystems, contributing to the function of the natural world as a self-sustaining biological system. They are, in fact, an integral and complex part of the terrestrial and freshwater ecosystems with which the future of human being is inextricably linked. Insects are generally understood to be the most significant creatures due largely to their remarkable structural and behavioural diversity as well as stupendous number of species. Among members of all insect orders and families fewer can match the popularity and importance that these tiny creatures, the mosquitoes, carry to the human life! In fact no animal on earth has affected human lives so directly and profusely as is brought about by the mosquito. A mosquito is small in size but is giant of a killer. It is indeed no small enemy of humans to whom they carry viruses, bacteria and parasites, which they transmit through bites, infecting some 700 million people and killing more than one million every year.

Mosquitoes are ubiquitous in distribution and are found in all regions of the world, except the regions near the two poles and altitudes beyond 2000 meters. There are some 3500 species of mosquitoes, with a worldwide population of more than 100 trillion!! Most of them exist in tropical climates, but there are mosquitoes in arctic and desert regions, ranging in distribution from sea level to altitudes as high as 10,800 feet (3,600 meters). References to their very being can be traced in the annals of yore of many countries; in India they were referred to in '*Susruta Samhita*', a Sanskrit classic in medicine dated fifth century B.C. where the mosquitoes were described as a cause of both nuisance and human disease. In India multiple names exist for mosquitoes, *e.g.*, Mashâ, Mashakâh (in Sanskrit), Mashâ, Mashak (in Assamese, Bengali, Oriya), Dās (in Gujarati), Machchar (in Hindi), Solle (Kannada), Moh (in Kashmiri), Kothuku (in Malayalam), Dās (in Marathi), Machhad (in Punjabi), Koshu, Koschu (in Tamil), Doma (in Telugu) and Machchad (in Urdu). The Indian medicine's *magnum opus* of yore 'Susruta Samhita' described mosquitoes as "*Their bite is painful as that of serpents, and causes disease resulting from the three humours together: the bite as if burnt with caustic or fire, is red, yellow, white and pink colour accompanied by fever, pain of limbs, hair standing on end, pain, vomiting, diarrhoea, thirst, heat, giddiness, yawning, shivering, hiccup, burning sensation, intense cold, vesicles or pustules increasing, swelling, knots under the skin, circles etc.*" In *Susruta* also five kinds of mosquitoes are named, *viz.*, (i) *Samudrāh* (marine or coastal mosquito), (ii) *Parimandalah* (very small mosquito), (iii) *Hastimashakah* (elephant mosquito, perhaps because of large size), (iv) *Krishnah* (Black or dark mosquito, and (v) Pārvatiyah (hill mosquito).

The famous adage "Mosquitoes are harmful" is generally true since they are often incriminated with some kind of deadly or debilitating pathogen. It is amazing that this tiny and feeble-looking organism has in fact assumed the reputation of the most formidable envy of man in modern times. A mosquito that hardly measures the size and weight of a grape seed, is ubiquitously abhorred for the nuisance, pain and death it hails every year on millions of humans. Rao (1984), a celebrated Indian medical entomologist, cautioned to not overwhelmingly indulge in self-aggrandization in our understanding about the biology of mosquito - her strengths and weaknesses, and warned to not underestimate the miniatured enemy who has so far outwitted humans in their fight against various mosquito-borne and life-threatening diseases. In past, they were responsible, amongst others, for (i) felling great leaders such as for example Alexander 'The Great' was suspected to have succumbed to the infection of malaria parasite *Plasmodium falciparum* while retreating after a decisive battle against the Indian king Porus in the mosquito-ridden marshy lands of the Sindhu river valley of Punjab, in 323 BC (*vide* SOVE Newsletter), and similarly the much dreaded and ruthless ruler, Sultan Muhammed bin Tughluk contracted malaria in 1351 AD while on a military campaign against rebels, and within a short time succumbed to the disease, (ii) decimating armies such as for example the deadly Yellow Fever was stated to have taken its toll on thousands of U.S. Army recruits during the brief Spanish-American War, and that death from disease among American troops was much more common than death from bullets, and (iii) deciding the fates of nations such as for example the slim Panama, which still belonged to Colombia until late nineteenth century, raged a war for independence, following which a bitter civil war ensued between 1900 and 1903, when it finally won its independence (Spielman and D'Antonio, 2001).

1.2. Mosquitoes are both ugly and beautiful

The impact of painful biting of mosquitoes is so long-lasting and deeply impactful in human psyche that any amount of arguments to appreciate other far less-known, yet quite astonishingly beneficial, traits of some of their cousins get simply underscored, giving an impression that there is no such thing like a good mosquito. The fact, however, is that there are several mosquito species that are indeed elegant in appearance and sometimes these stand out to compete with other natural beauties among arthropods.

Foremost, take the example of the Asian Tiger mosquito, *Aedes albopictus* whose overall beauty and serenity have few rivals among her siblings, near or far (Fig.1.1). A native of West Bengal, *Ae. albopictus* is a competent vector of many viruses including dengue fever, chikungunya and Eastern equine encephalitis virus etc. Its life cycle is closely associated with human habitat, and it breeds in containers with standing water, often tires or other containers. It is a daytime feeder and can be found in shady areas where it rests in shrubs near the ground. *Aedes albopictus* feeding peaks in the early morning and late afternoon; it is an opportunistic and aggressive biter with a wide host range including man, and domestic and wild animals. The mosquito is a much sturdier animal compared to its close ally, *Ae. aegypti.* Surprisingly, many a time *Ae. albopictus* males have been seen to sire the females of *Ae. aegypti,* but never the reverse. In spite of all such villainous attributes, the sheer beauty of its colour patterns, with silvery stripes on most parts of the body topped by a sparkling stripe along the middle of the head extending backward across the entire length of the thorax, and annules on legs, make *Ae. albopictus* the centre of attraction to many, including its avian predators.

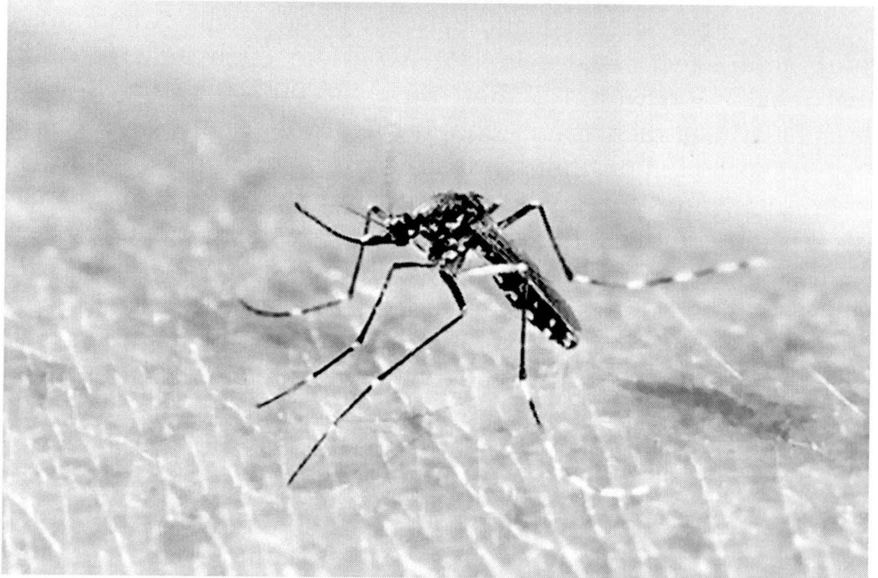

Fig. 1.1. The Asian Tiger mosquito, *Aedes albopictus.* (*Source:* Free Wikipedia).

In the whole realm of insects, and particularly mosquitoes, the paddle-legged beauty, *Sabethes cyaneus* is an exception as far as the discipline of courtship goes (Fig. 1.2). This tropical forest species lay their eggs in shaded natural and artificial containers such as tree-holes, bamboo hollows, and even fallen banana leaves filled with rainwater. *Sabethes* larvae develop in these small source pools and feed on bacteria, algae, and other organic material. The beautiful adult *Sabethes* mosquitoes are usually encountered in close proximity to the small habitats of their aquatic forms.

Fig. 1.2. The paddle-legged beauty, *Sabethes cyaneus* (Adapted from: Robert G. Hancock).

The mating or courtship behaviour of this mosquito is quite cool. The vast majority of species do not display any behavioural patterns complex enough to even qualify for the term, 'mating'. For most mosquitoes, mating is a fast and haphazard occurrence with swarms of males circling around and waiting for the sound of female wing-beats. Once detected, the male rushes after the female and immediately attempts copulation mid-air without so much as a "how do you do?" - this is not courtship! In contrast, *Sa. cyaneus* is one notable exception.

Few people would use the words "beautiful" and "mosquito" in the same sentence, but *Sabethes cyaneus* has few rivals when it comes to looks. Its colour is predominantly a stunning iridescent blue, marked with contrasting silvery-white spots - especially, on the abdomen! Bold colours are not their only unique feature; the feather-like paddles catch the eye as well. Formed by elongated scales, located on the tibia of each middle leg, both the male and female sport these strange paddles. The males must go through a number of displays and movements, in a specific order, and the female has multiple opportunities to accept or reject his advances. The wooing begins with the female resting on the underside of a horizontal surface, such as a stick or twig. The male, apparently attracted by her stunning beauty, hovers around the female with his leg paddles outstretched. He then places one of these middle legs (midlegs) over her wing, perhaps preventing her from just flying away, but also to be used as a pivot point as he swings around to face his date and simultaneously grasps the twig with his front legs. Once aligned, the true courtship begins. The male waves his free midleg back and forth, apparently showing off his lovely paddle, but eliciting no apparent response from the female. Next, the male will flex his forelegs to create a bout of rhythmic body "swinging" while simultaneously flicking his proboscis (tube-like mouthpart) downward. If successful, this display will induce the female to lower her abdomen sufficiently for copulation to take place. If unsuccessful, he would return to waving his free midleg, followed by another round of rhythmic body swinging, until she either lowers her abdomen or uses her hind legs to reject him by pushing him away. During copulation, the male gyrates his abdomen while waving his middle and hind legs alternately until mating is completed and the pair disengages!

1.3. Mosquitoes have some good siblings too

A thought about mosquitoes spontaneously bring memories of itchiness, red bumps, and a feeling of frustration. To see the whole diverse world of mosquitoes it is imminent to see 'beyond the parapet'! Ultimately, there seem to be few things that mosquitoes do that other organisms can't do just as well — except perhaps for one; they are lethally efficient at sucking blood from one individual and mainlining it into another, providing an ideal route for the spread of pathogenic microbes. For their eggs to successfully develop into larvae female mosquitoes must provide them with a source of protein in the form of blood. Although all warm blooded creatures (mammals and birds) are fair game, our thin and relatively hairless human skin is especially appealing as an easy meal. When mosquitoes aren't blood feeding to support the development of healthy eggs, the adults feed off of plant nectars to fuel their flight. In 1913, it was first determined that mosquitoes are pollinators too! In fact, many mosquitoes around the world are efficient pollinators of small flowers that live in wetter environments.

Some mosquitoes such as *Aedes sollicittans, Ae. provocans* and *Ae. tritaeniorhynchus* spp. are floral feeder, hence very potential pollinators of some selected plant species in certain countries, *e.g.*, Canada. Similarly, *Aedes communis,* known by some as a snow pool mosquito, is an important pollinator of *Platanthera obtusata,* the blunt-leaf orchid, in northern regions (Fig.1.3). It is most common in the state of New Jersey, although it can also be found throughout much of the Northern United States, Canada, and Alaska. It usually inhabits dense forests at high elevations. Its larvae are mostly found a few feet from the shore in deep snow pools, which reside at greater than 1,500 feet in elevation. This snow pool mosquito eats the nectar from the floral spur of the *Platanthera obtusata,* during which time its eye naturally comes into contact with the pollinium, a cluster of pollen. The pollinium sticks to the mosquito's eye even when it flies away. Thus, when it eats from another flower, the pollinium touches the stigma of that flower, and the flower is pollinated!

Fig. 1.3. Mosquito pollinating a flower (Source photographer: Beatriz Moisset); *Aedes communis* on a fleece (*Source photographer*: Bretwood Higman).

Even though this mosquito is an important part of the propagation of this beautiful flower, that does not make them any less annoying to us when the females are looking for unwilling blood donors. *Aedes commnis* is mostly a bother in New York state and Pennsylvania.

Mosquitoes are also delectable food material as they are easy to catch as well. As larvae, mosquitoes make up substantial biomass in aquatic ecosystems globally. They abound in bodies of water ranging from ephemeral ponds to tree holes to old tyres, and the density of larvae on flooded plains can be so high that their writhing sends out ripples across the surface. They feed on decaying leaves, organic detritus and microorganisms. In the absence of their larvae, hundreds of species of fish would have to change their diet to survive which is not easy since trait of feeding behaviour is deeply imprinted genetically in those fish such as the mosquitofish, *Gambusia affinis* which is a specialized predator — so effective at killing mosquitoes that it is stocked in rice fields and swimming pools as pest control — that could go extinct. And the loss of these or other fish could have major effects up and down the food chain. Many species of insect, spider, salamander, lizard and frog would also lose a primary food source, i.e., mosquitoes. On the contrary if the mosquito immature aquatic population was sprayed with a microbial mosquito-control agent, the birds otherwise depended on them produce on average two chicks per nest after spraying, compared with three for birds at control sites. Most mosquito-eating birds would probably switch to other insects that, post-mosquitoes, might emerge in large numbers to take their place. Other insectivores might not miss them at all: bats feed mostly on moths, and less than 2% of their gut content is mosquitoes.

Mosquito larvae are important members of the tight-knit communities in the 25–100-millilitre pools inside pitcher plants, *Sarracenia purpurea,* on the east coast of North America. Species of mosquito, *Wyeomyia smithii,* and midge, *Metriocnemus knabi,* are the only insects that live there, along with microorganisms such as rotifers, bacteria and protozoa. When other insects drown in the water, the midges chew up their carcasses and the mosquito larvae feed on the waste products, making nutrients such as nitrogen available for the plant. In this case, eliminating mosquitoes might affect plant growth. A stronger argument for keeping mosquitoes might be found if they provide 'ecosystem services' — the benefits that humans derive from nature. Without mosquitoes, thousands of plant species would lose a group of pollinators. Adults depend on nectar for energy (only females of some species need a meal of blood to get the proteins necessary to lay eggs).

Despite their being universally despised for creating nuisance, painful bites, loss of sleep and blood, and, above all, their potential to transmit a large number of fatal and/or debilitating infections, mosquitoes, or at least some of them, offer a wonderful blend of colours and sheen, in addition to much hyped biological control tool. *Toxorhynchites* spp. are a rare group of mosquitoes, essentially vegetarian (Fig.1.4). All members of this genus subsist on plant materials; sap, nectar etc. Elegance nonpareil these non-biting mosquitoes make a significant constituent of the culicid biome. The metallic shining of azure on thorax, comparatively gigantic size, elephantine trunk-like sap-sucking proboscis and spidery long legs are something unseen in any other animal on the planet. Immatures of Toxorhynchites predate on the immature stages of other co-inhabiting mosquito species including dengue vectors. Accordingly, Toxorhynchites larvae had been successfully deployed as biological control agents against phytotelmata breeding *Aedes aegypti.*

Interestingly, a few mosquito species particularly those belonging to the Sabethine group of genera such as Malaya exhibit a quaint feeding behaviour on the ants belonging to the genus Cremastogaster (Fig.1.5). The Malaya mosquitoes, of which two species occur in India, are totally innocuous to man and other animals as they never bite due to presence of a highly specialized dew seeking proboscis or mouthparts only; the hairy proboocis is peculiar in having a flexible joint and a very much enlarged tip carrying four long hairs. They are small mosquitoes ornamented with brilliant silvery, blue or violet scales; These mosquitoes imbibe the sugary volume in the oral cavity of ants.

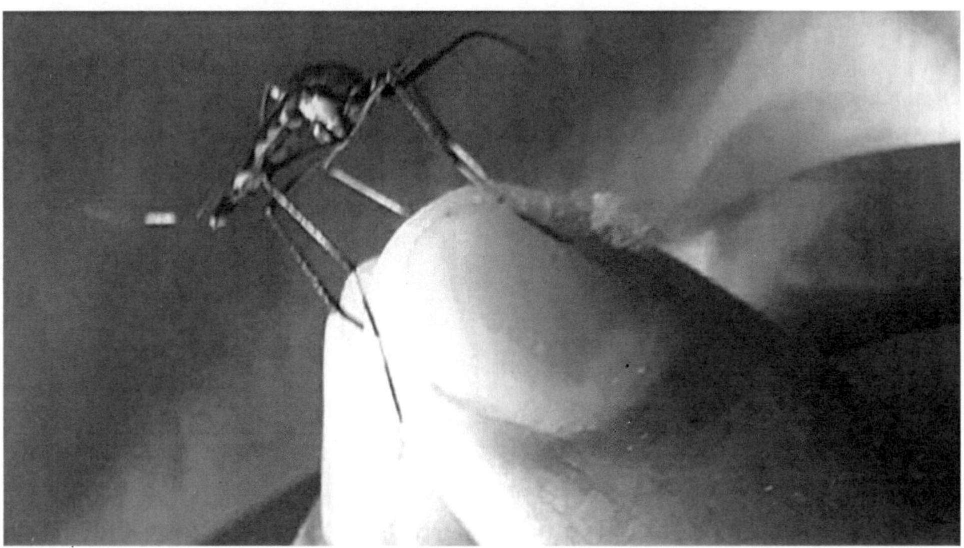

Fig.1.4. A non-biting plant sap sucking *Toxorhynchites* sp. (*Courtesy:* Dr Sreejith, *pers. comm.*)

Fig. 1.5. *Malaya* sp. taking food from mouth of an ant (*Cremastogaster* sp.) (*Source*: ZSI's free facsimile of Barraud, 1934)

The complex nature of saliva of mosquitoes has always been a subject of constant scientific exploration and its anticoagulatory properties are being highlighted as of great medicinal significance. In Guangzhou, China a huge mosquito producing factory is established at Sun Yat-sen University, primarily to use the cultures for developing genetically modified and/or *Wolbachia*-induced control of dengue vectors, but, at the same time as a by-product of huge production of mosquitoes the saliva from males (and possibly female mosquitoes, too) is extracted in great quantum for use in dug industry, likely in developing anaesthesia or specific proteins.

1.4. Why does a mosquito bite itch?

Mosquito's vicious bite often leaves behind an imprint, evanescent though, of a slowly developing itchy wheel. However, it is a first mosquito bite ever, the roundish wheel fails to appear. The immune system in human body begins thereafter, and develops antibodies to recognize and flag the foreign chemicals. But after just one dose of mosquito spit, the familiar itchy bump left behind by subsequent bites caused by an allergic reaction to all those chemicals the mosquitoes are squirting into the body. The immune system releases histamines, chemicals that make the cells in the lining of human blood vessels spread apart. That lets the fluid leak out into the skin. That fluid is carrying components of man's immune system that attacks and destroys the foreign chemicals from the mosquito. The extra fluid from the leaking blood vessels causes a bump, which irritates other recepters in the skin, leading to the itching. Children get bigger bumps from mosquito bites than adults because their immature immune systems respond with larger amounts of histamines. More leaking blood vessels equal bigger bumps, which mean more itching and scratching. However, following repeated mosquito biting, the human immune system also becomes more stronger and exhibits a different reaction to mosquito saliva. Instead of triggering histamines, another part of the immune system grabs onto the foreign chemicals and drags them away to macrophages, which destroy the chemicals in the mosquito's saliva.

1.5. Change in vector behaviour of mosquitoes due to anthropization inducing new disease paradigms:

Few case studies of conflagration of malaria, filariasis and dengue in the Thar Desert as well as origin of dengue in Kerala State are discussed below to highlight factors engendering alteration in the respective vector mosquito's behaviour. To understand how vectors change in their vectorial potential it is important to foremost have knowledge of the eco-physiographic alterations brought about in their environment under the impact anthropogenic activities. The Thar Desert for malaria conflagration during late 1980s onward and the Western Ghat region for dengue emergence during the mid-1990s are burning examples for their enhanced distribution, density, vector species, host preference for blood feeding, resting and breeding. The ecological and epidemiological complexities of the vectors involved in respective disease epidemics are dealt with *in extenso* here under.

1.5.1. MALARIA IN THE THAR DESERT

1.5.1.1. Thar Desert's physiography and ecology

In modern times, particularly since the first quarter of the 20th Century, malaria became almost invariably associated with the mega irrigation projects in the country, most notable of which is the Indira Gandhi Nahar Pariyojana (IGNP) (Tyagi, 2002, 2004). Indira Gandhi Nahar Pariyojana (IGNP) was among the first group of large surface water transfer projects taken up in the country aiming to transform the wastelands of Thar Desert in Rajasthan into agriculturally productive zones along with improvement of afforestation and environment, development and protection of livestock/animal health, human rehabilitation and settlement, and economic growth of the poor people of the desert (Gupta, 1975). The Thar Desert in the northwestern region of Rajasthan State, India, located at 24°–30° N and 68°–78° E and lying between irrigated lands of the Indus and Sutlej and the western edge of the Aravalli Hills, is the only desertic ecosystem in the country which support a unique malariologic paradigm, "The Desert Malaria" (Tyagi, 1995; Tyagi *et al.*, 1995). Organized in 12 desert districts, it is spread over 75,000 km^2 (62% of the total land of the State) harbouring a population of 12·8 million people (39% of the population in the State). The seasonal Luni river divides the desert into two parts, in the north a vast stretch of sand dunes or levees and in the south an expansive basin rimmed by flat topped hills of sandstone and limestone. The annual ambient temperature varies between –4 °C and 50 °C, while the average precipitation, which was until recently about 250 mm, has since 1990 been about 350–400 mm.

1.5.1.2. Construction of Indira Gandhi Nahar Pariyojana (IGNP) in The Thar Desert

Post 1920, three major canal systems have been introduced into the desert regions of Rajasthan, the Gang canal, the Bhakra-Sirhind feeder canal and the Indira Gandhi canal, running over 10,000 km in distribution. Consequently, the physiography of the land has changed a great deal in its soil composition, landscape, potential for water retention and climate. As a result, some areas, especially those in the vicinity of the canalized irrigation, have become perennial breeding sites for vectors and are regularly experiencing outbreaks of malaria, particularly that caused by *Plasmodium falciparum*. Periodic malaria epidemics, characterized by a high *P. falciparum* incidence, between 1983 and 1994 claiming hundreds of deaths are ample proof of the changing pattern of the disease epidemiology in the Thar Desert region.

Of the three major canal systems (Fig.1.6) existing in the Thar Desert region today, the first was the Gang canal project which became fully operative in 1928, with its 1251 km distribution system in Sri Ganganagar district only. Along with the Bikaner canal feeder, it provides water to about

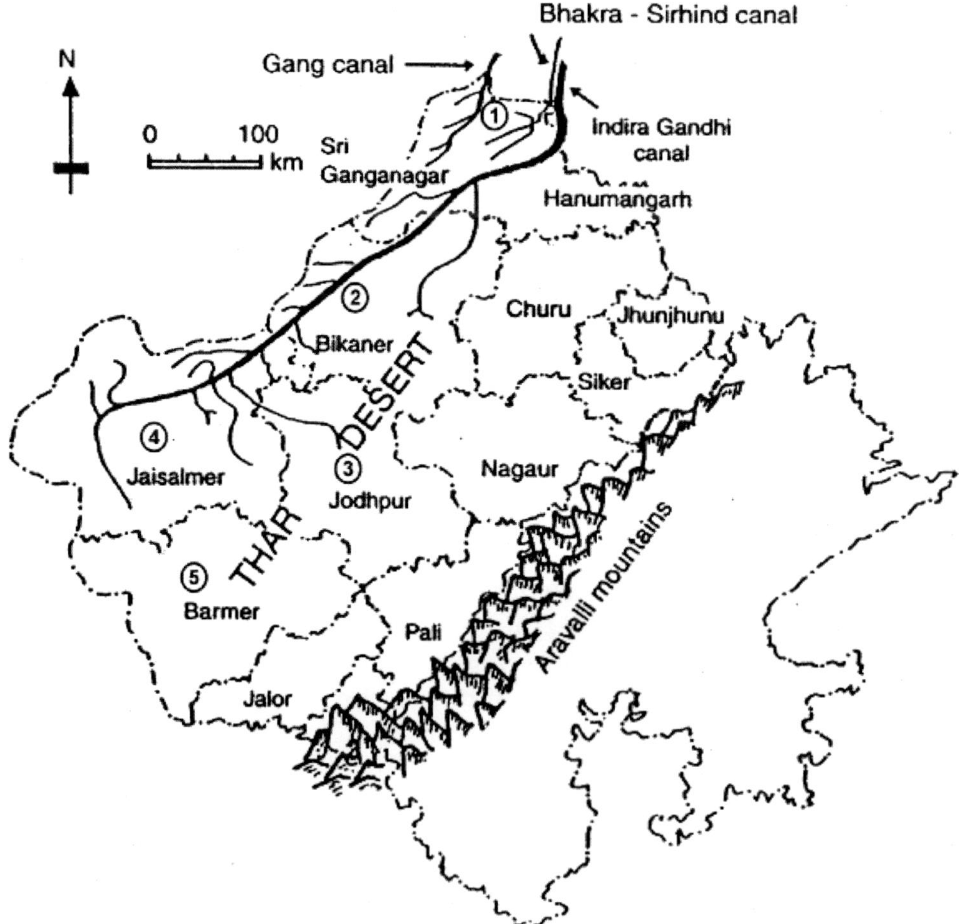

Fig. 1.6. Twelve desert districts (those encircled had severe epidemic history) in the Thar Desert region Rajasthan and the extent of the three major canal systems. (*Source:* Tyagi, 1995)

65% of the 300,000 ha of cultivable land. The Bhakra-Sirhind canal project, with its 1219 km long distribution system, was started in 1955 to utilize a further 300,000 ha of cultivable land. The most important of all the canal systems is the Indira Gandhi Nahar Pariyojana (IGNP) which is one of the world's largest projects of its kind in the desert. It started in 1958 and irrigation of part of the area began in 1961. The project, when completed, will include a 649-km long main canal (of which 445 km is in the Thar Desert) and over 9000 km of distributaries. Although only two of the three stages envisaged are completed by now, the main canal including feeder and 5500 km of distributaries is already irrigating 950,000 ha of land (approximately 50% of the irrigated area that is planned). At present the IGNP command area covers only 11% of the entire Thar Desert falling in Rajasthan, whereas the IGNP non-command area and the non-IGNP area are spread over 78·4% and 10·6%, respectively (Fig.1.7).

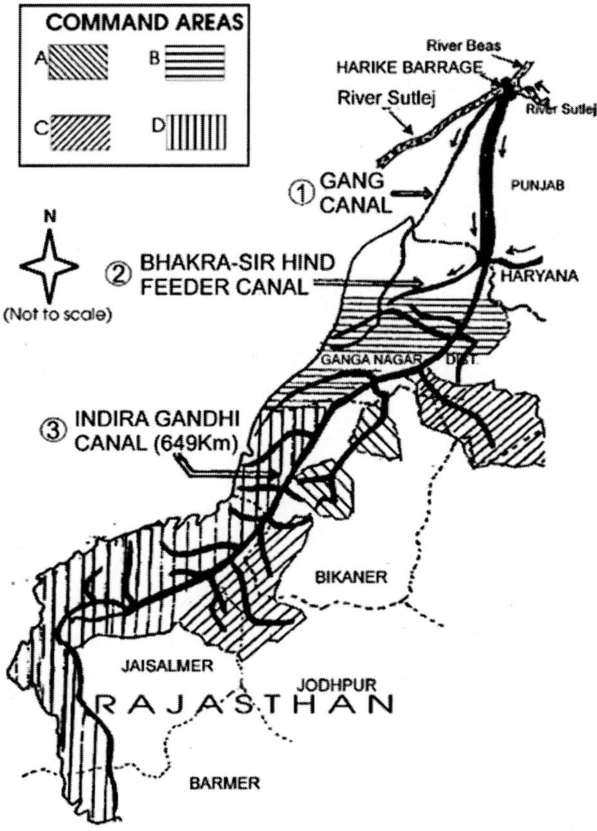

Fig. 1.7. Extent and types of flow of water under the Indira Gandhi Nahar Pariyojna (IGNP) in the Thar Desert, Western Rajasthan. Note that command areas comprise two types of lands which are either directly under flow (B and D) or under lift (A and C), depending on natural gradient of the land. (*Source:* Tyagi, 1995)

1.5.1.3. Agroeconomic and demographic changes in The Thar Desert due to IGNP-Gang-Bhakra driven canalized water irrigation system

This Herculean task, bringing Himalaya's water to the Thar Desert, aims at transforming vast stretches of western desert terrain into blossom. As a result, the annual agriculture production from the newly irrigated land now exceeds Rs. 6500 million. While several irrigation-intensive crops, such as sugarcane, mustard, groundnut, etc., have been established in some of the canalized desert districts; the production of rice in Sri Ganganagar district, the most irrigated among all desert districts, has topped the list in the whole State. Rice cultivation has often been associated with malaria transmission (Russell & Rao, 1940; Rao, 1984; Dossou-yovo *et al.*, 1994), although Sharma *et al.* (1994) have found a negative correlation between rice cultivation and the annual parasite incidence (API) in most States and Union Territories on a nationwide survey. Beside the surplus irrigation water that accumulates on irrigated land, vast areas along the course of the Indira Gandhi main canal have been engulfed by seepage water from the irrigation canals on a perennial basis. An estimated 8600 ha in the Nalibed area, Sri Ganganagar district, is permanently inundated under the IGNP Stage I by the Ghaggar river floodwater while 1000 ha is converted into marshy land due to waterlogging and excessive seepage from the canal in Stage II. Supply of canal water to the desert region over several decades has caused a rise of the water table and water-holding potential of the soil, and a plentiful growth of vegetation, particularly hydrophytic weeds. Such extensive breeding grounds are the preferred sites of many anophelines including *An. culicifacies*, the most serious vector of malaria in the whole Indian subcontinent.

1.5.1.4. Malaria scenario in The Thar Desert

In early and mid-twentieth century malaria in Rajasthan was known more characteristically from the mountainous and forested southern region where control programme was run regularly in accordance with the government policy. However, in the Thar desert of yore where malaria was only mild and unstable with potential for local and focal epidemics mostly along the border line the control programme was more event-based and far irregular. Contrary to the erstwhile desertic ecosystem without canalised irrigation avenues, the mountainous and sylvatic Mewar region in the southern Rajasthan continuously encountered high malaria endemicity with *An. fluviatilis* as the main vector. It was in the mid-eighties onwards when malaria epidemics began to occur regularly in the State, most of the which were reported from the hyperendemic hilly and forested terrain of

Udaipur in the southern part of Rajasthan. The malaria scenario thereafter catapulted and, ironically, more cases began to appear in the Thar Desert. Thus, while in 1986 the Udaipur zone had contributed 44·1% positive cases and 67·7% *P. falciparum* cases as against 24·4% positives and 11·6% *P. falciparum* cases in the desert, in 1994 only 14·1% positives (*P. falciparum* 26·6%) hailed from the Udaipur zone districts compared to 53·3% positives from the desert districts (Tyagi, 1995, 2002).

Malaria in the Thar Desert was less pernicious until the late seventies and eighties, with the sole vector *Anopheles stephensi* breeding characteristically insularly within the limits of underground man-made water reservoirs namely the '*Tanka*' and '*Beri*' (Tyagi, 1994a,1995; Tyagi & Yadav, 1996), exhibiting several hitherto unreported behavioural attributes (Tyagi, 1994b). However, the disease resurged with a vengeance in nearly the entire Thar Desert region during late 1980s following a radical change in the vector biota, particularly inveigling of the deadly *An. culicifacies*, and their interplay with the parasites in the changing desert environment. Malariologically, so far as Rajasthan State is concerned, most of the presently organized 30 districts including the 12 desert districts (Sri Ganganagar, Hanumangarh, Bikaner, Churu, Nagaur, Jodhpur, Barmer, Jaisalmer, Pali, Jhunjunu, Sikar and Jalore) have hitherto been considered to have low and variable endemicities of malaria with potential for sporadic outbreaks of the disease. Tyagi & Chaudhary (1997) have presented evidence that the outbreak has been caused by mismanagement of the widespread developmental activities of canal-based irrigation. The Thar Desert region in north-western India is conventionally regarded to be only mildly prone to the malaria disease. Preceding 1990s, a major part of this arid zone has undergone vast physiographic and climatic changes, resulting in the formation of extensive mosquito breeding habitats of varying nature, high vector densities and intrusion of new malaria mosquitoes earlier unknown for the region. As a result, the prevalence of disease in the desert, with several epidemics reported in the last 25 years, has changed a great deal for the worse. It is evident from data for the period 1961–1994 that not only has the slide positivity rate in the desert region increased 3.5-fold, but *Plasmodium falciparum* incidence also increased as much. Along with *Anopheles stephensi*, the traditional malaria vector in the xeric environment, another significant vector of the Indian mainland, *An. culicifacies*, has also established itself in the areas extensively irrigated through canals. It is evidenced (Tyagi, 2002) that the malaria situation in the State, particularly following extensive canalization in north-western desert areas, indicates the formation of new disease foci in the hitherto less malarious Thar Desert.

1.5.1.5. Malaria epidemics and their pathways in The Thar Desert

Documentable instances of outbreaks of malaria epidemics in the Thar Desert are available since 1975, although there is every possibility that mortality might have also occurred in the desert during the nation-wide epidemic in 1908, and perhaps thereafter as well on other such occasions! As a consequence, at least 13 episode-years of malaria epidemics are identifiable since 1975. Save for the massive outbreaks of malaria in 1975, 1976, 1990 and 1994, most other epidemics were focal in nature covering often a part of a district or two. While the epidemics of 1975 and 1976 were predominantly characterized by a resurgence in *P. vivax* cases, the outbreaks of 1983-1984 onwards saw a phenomenal rise in *P. falciparum* (Tyagi, 2002) (Table 1.1).

Table 1.1. Districts in the Thar Desert struck by epidemics

District	Years of epidemics	Total number of deaths
Barmer	1975, 1990, 1994	178
Bikaner	1975, 1976, 1990, 1992, 1993, 1994	113
Churu	1975, 1976	5
Sri Ganganagar	1983, 1984	Nil
Jaisalmer	1975, 1990, 1992, 1993, 1994, 1995, 1996, 1999, 2000, 2001	72
Jalor	1975,1976, 1990	20
Jhunjunu	1975, 1976	6
Jodhpur	1990, 1992, 1993, 1994	72
Nagaur	1975, 1976, 1990	2
Pali	1975, 1976, 1990, 1994	3
Sikar	1975, 1976	10

(i) *Malaria epidemic of 1975-77:* Even though no death was brought on record, it was crystal clear that, in view of the steep rise in the passive cases compared to the 1966 level of malaria cases, outbreaks of malaria had also struck the Thar Desert, along with other parts in the country.

(ii) *Malaria epidemic of 1983-1984*: Sri Ganganagar district experienced a focal *P. falciparum*-dominated malaria outbreak for the first time in 1983, with 4925 positive cases of which 43.4% were *P. falciparum* cases. In spite of the fact that in 1983 the annual parasite incidence (API) was but second lowest during the last few years, the malaria outbreak in the Nalibed area in the erstwhile Hanumangarh tehsil of Sri Ganganagar caused high morbidity. Two primary health centers (PHCs), Sangaria and Pilibanga, which were most severely affected, had jointly contributed 72.2% of the total positive cases in the district during the epidemic. It is noteworthy that *P. falciparum* had constantly increased for at least few years since 1987, while the API had kept a low ebb for the same period.

(iii) *Malaria epidemic of 1990*: In 1990, a widespread epidemic hit western Rajasthan districts, but particularly the Barmer district claiming 47 lives. A total of 19,322 positive cases (*P. falciparum* 66%) were recorded in Barmer district which was 2.5 times higher than what was reported in the preceding year. It was advocated that the epidemic in Barmer district occurred as a sequel to widespread floods in the region. It was demonstrated that the rise in the incidence of malaria was associated with the increase in the rainfall. During 1977, when annual rainfall was 13% above the normal, the annual parasite incidence (API) showed a conspicuous rise at 5.3 and annual *P. falciparum* incidence (AFI) between 0.03 and 0.1. However, 1985 was a year of exception when there was noticeable increase in cases, particularly of *P. falciparum*, despite low rainfall. This increase was, however, attributed to the migratory cases, since the cases did not proliferate locally due to non-conducive transmission conditions. This was substantiated by a significant decline in the incidence in the next 2-3 years

Both rural and urban areas were equally and simultaneously affected by the epidemic. Though all the eight PHCs presented a rising trend, the PHC Baitu was the worst affected with API touching a peak of 55.3 and AFI 19.6. The highest increase in AFI was in Mandli (4.5 times), Gudamalani (4 times), Ramsar (2.8 times), Ganga (2.3 times), Sindhari (1.75 times) and Chohtan (1.6 times). The Barmer town also recorded a 4-fold rise in API and 3-fold rise in AFI. The district overall API and AFI recorded 2.5 and 3-fold increases, respectively. The mass blood survey carried out during the epidemic showed that all age groups were affected. The slide positivity rate (SPR) was 61%. More than 93% of these had *P. falciparum* infection. A total of 122 infants were found positive of which less than 30% had *P. falciparum* infection. A total of 103 infant positives were from Baitu PHC only. About 88% of the positive infants were detected during September-November, 1990 only.

(iv) *Malaria epidemics of 1992-1993:* During late 1992 and early 1993 epidemics of malaria broke out simultaneously in several districts, *i.e.*, Bikaner (Bajju and Kolayat PHCs), Jodhpur (Baap and Phalodi PHCs) and Jaisalmer (Pokaran, Nachna and Sangad PHCs). In Jaisalmer district alone a total of 2128 positive cases (*P. falciparum* 73.6%) were reported during the epidemic as against a combined total of of 1809 cases (*P. falciparum* 27.6%) in the two preceding years, *i.e.*, 1990 and 1991. If the malariometric data from the three PHCs are reviewed it becomes clear that the malaria scenario has drastically changed over the years. A total of 2128 positive cases (*P. falciparum* 73.6%) were reported against a combined total for 1990, and 1991 of 1980 cases (*P. falciparum* 27.6%), clearly indicating a shift towards an ascending order.

During early 1993, the SPR was found to be as high as 85.0% and the *P. falciparum* proportion 88% in two desert villages near the Indira Gandhi canal, Madassar and Awai, (Tyagi *et al.*, 1994).

(v) *Malaria epidemic in 1994*: In mid-1994 a major epidemic of malaria broke out in whole of Rajasthan State taking a toll of 452 lives. The worst affected three districts included Barmer, Jaisalmer and Jodhpur, contributing nearly two-third of the total mortality. *Plasmodium falciparum* accounted for over 60% of cases in the desert and was considered as the main cause of death.

In a year-long pilot study during 1993-94, following a rapid mass blood survey immediately after the epidemic in the late autumn in 1992 in the two IGNP command villages (*P. vivax* 11.1%, *P. falciparum* 88.7%) and two unirrigated villages (*P. vivax* 17.1%, *P. falciparum* 82.7%), fever surveys conducted for one full year between April 1993 and March 1994 showed a slightly higher slide positivity rate in the IGNP villages (32.3%) than in the unirrigated villages (25.5%). The most important feature was, however, the higher proportion of *P. falciparum* in the IGNP villages (76.6%) compared to that in the unirrigated villages (16.6%). Influenced by the high percentage of *P. falciparum* in the study villages, the slide positivity rate maintained a high profile throughout. Mixed infections occurred in all villages, although more of them occurred in the unirrigated villages (11.5%) than the other two villages (3.2%). Interestingly, more gametocyte carriers were present in the IGNP non-command villages (16#, 17.8%) compared to the unirrigated villages (2#, 8.3%). Cumulatively, 73.7% of all the positive cases were found in children of less than 14 years of age. The male population (68.5%) suffered more than the females (31.5%).

(vi) *Malaria epidemic in 1995-1996*: A total of 45 lives (60.7% of the total deaths in Rajasthan) were reported in the Thar Desert region alone, even though there is some debate whether the year of 1995 to be regarded as an year of epidemic or not. However, Najera *et al.* (1998) have rightly characterized it, along with 1996, as an epidemic year. The malariometric data of 1995 are comparable with some of the past epidemic years.

(vii) *Malaria epidemic of 1999:* A focal outbreak of malaria occurred in the villages situated close to the main Indira Gandhi canal near Ramgarh in Jaisalmer district, western Rajasthan. Stagnation of water over a month's period in the main canal as well as long standing rain water in the form of large lake-like bodies near these villages formed vast breeding grounds for the vectors like *Anopheles culicifacies,* along with *A. stephensi* already breeding in the 'tanka' and 'beri' in the epidemic-hit villages. A mass blood survey was conducted covering 380 persons with a history of fever in last one month in the epidemic hit villages, namely Seuva, Raghava, Raimala and Sadhana, all within a distance of 2 km from the main Indira Gandhi Cana. Of these 49 (12%) were *P. vivax* and 40 (10.5%) were *P. falciparum,* while 8 (2.1%) had mixed parasitism with both parasites. Results of the mass blood survey point out towards a high slide positivity rate (25.5%) and *P. falciparum* proportion (49.5%) in the villages hit by the epidemic. While Raghwa village has presented the highest SPR (30.4%), Raimala showed the lowest SPR (13.1%). Interestingly, however, while on one hand Raimala singled out with the highest %*P. falciparum* (60%), Seuva offered the lowest *P. falciparum* proportion (41.0%), on the other. Maximum *P. falciparum* cases were nevertheless originated from Sadhna village.

1.5.1.6. Possible factors responsible for conflagration of malaria particularly P. falciparum-*dominated malaria in The Thar Desert*

It is an enigma so far as to how *P. falciparum* ushered in the interior of the Thar Desert and manifested so often in the form of epidemics with high morbidity and mortality. A few possible assumptions are offered here:

(i) With the establishment of extensive canal-based irrigation in the interior parts of the Thar Desert, a far more serious vector of malaria, *An. culicifacies,* ushered in the xeric environments. This species being highly adaptive to new physiographic situations built up its density very fast under given circumstances of a large variety of breeding places in the IGNP area particularly the inevitable and numerous seepage water collections from the canals. What *Anopheles culicifacies* lacks in its susceptibility to infection by *P. falciparum*, it compensates by its relatively much higher density, hence also vigorous and prodigious biting on human host;

(ii) *Anopheles stephensi*, chronologically the oldeset and most dominant vector of malaria in the typical sand-duned xeric environments of the Thar Desert, which traditionally bred only in the 'tanka', the only major source of reserved potable water invariably present in every desert village and but transmitted 'desert malaria' (Tyagi, 1996, 1999, 2002) at a low ebb, without epidemics, shifted its breeding habits with the availability of extensive and varied breeding sites due to extensive canalization to compete with that of *An. culicifacies;*

(iii) The massive Indira Gandhi Nahar Pariyojana, in continuation with the Gang canal system and the Bhakra-Sirhind feeder canal system, together spanning over nearly 70 years of initiation, had attracted large labour forces from distant and often hyperendemic States like Bihar, Uttar Pradesh, Punjab, Haryana, Gujarat and Madhya Pradesh. Such labour forces quite possibly brought along with them the parasite, possibly a resistant strain of, *P. falciparum*, which in the non-immune population had the potential to spread and manifest fast in suitable climatic situations of IGNP-induced environment in a part of the Thar Desert. It was argued that the 1994-utbreak of cerebral malaria in the Thar Desert had proved fatal because the affected population had been exposed to the parasite for the first time; and,

(iv) The resistant strain of *P. falciparum* inadvertently inveigled into the Thar interiors is uncontrollable by the normal regimen of the prescribed antimalarials.

The above hypothesis can be further aptly understood if the progress of pathways of malaria epidemics in the Thar Desert was traced out carefully (*cf.* Tyagi, 2002). These observations sufficiently indicate towards a shift in the abiotic characteristics of the Thar Desert under the IGNP command area as compared to that in the typical desertic environs.

1.5.1.7. Correlation between vector species predominance and malaria conflagration under the impact of IGNP

It is evident from the above interpretation that increase in vector taxa occurred in the Thar Desert with the availability of canalized water for irrigation. With this escalated the malaria incidence dynamically transforming to *Plasmodium falciparum*-dominated malaria outbreaks in the Thar Desert (Table 1.2).

Table 1.2. Overview of malaria epidemics in the Thar Desert districts from 1975 to 2001

District	Years of epidemics	Dominant malaria species at the epicenter of epidemic	Possible reason associated with epidemic
Sri Ganganagar (including Hanumangarh)	1983, 1984, 1985	*P. falciparum*	Increased vector activity with increase in varied breeding sites
Bikaner	1975, 1976, 1990, 1992, 1993, 1994	*P. falciparum*	Increased vector activity with increase in varied breeding sites
Churu	1975, 1976	*P. falciparum*	Inter-state population migration
Jhunjunu	1975, 1976	*P. falciparum*	Inter-state population migration
Nagaur	1975, 1976, 1990	*P. vivax*	Inter-state population migration
Jodhpur	1990, 1992, 1993, 1994, 2001, 2002	*P. vivax*, *P. falciparum*	Increased vector activity with increase in varied breeding sites; excessive quarry-mining brooding perennial vector breeding
Jaisalmer	1975, 1990, 1992-1996, 1999-2001, 2002	*P. vivax*, *P. falciparum*	Increased vector activity with increase in varied breeding sites; non-immune human population
Pali	1975, 1976, 1990	*P. vivax*, *P. falciparum*	Inundative vectorism
Barmer	1975, 1990, 1994	*P. vivax*, *P. falciparum*	Inundative vectorism
Jalor	1975, 1976, 1990	*P. falciparum*	Inundative vectorism
Sikar	1975, 1976	*P. falciparum*	Flooding

Malaria evolution in the Thar Desert was so obvious that a parallel could be drawn in the evolution of vector taxa and the malaria incidence by drawing a correlation between both the relative number of *Plasmodium vivax* and *P. falciparum* cases in both irrigated and unirrigated villages (note the high prevalence of *P. falciparum* in the irrigated villages), during 1993-1994 (Fig. 1.8).

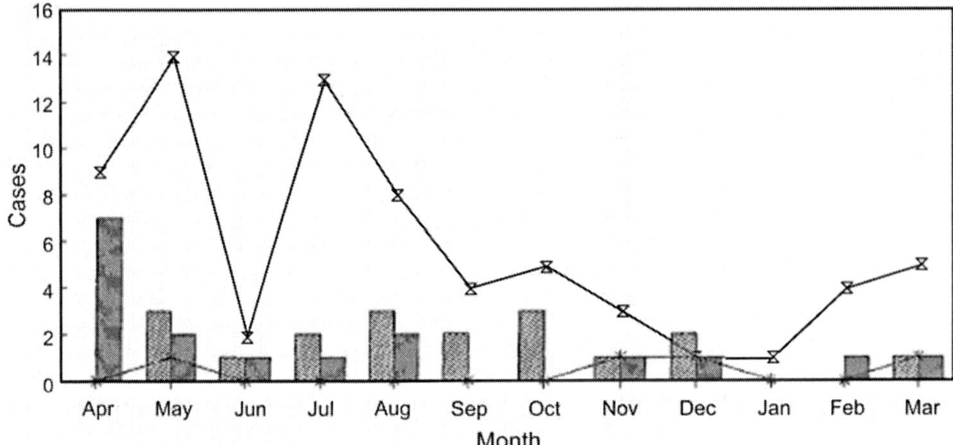

Fig. 1.8. Graph showing the relative number of *Plasmodium vivax* and *P. falciparum* cases in both irrigated and unirrigated villages (note the high prevalence of *P. falciparum* in the irrigated villages), during 1993-1994; *P. falciparum*: irrigated villages (X), unirrigated villages (+); *P. vivax*: irrigated villages (░), unirrigated villages (▓).

Contrarily, Russell (1938) had advocated that it was not irrigation *per se*, but defective and untidy irrigation which is responsible for vector proliferation and spread of malaria. Thus, by this understanding, it seems that the seepage of water from canals over vast areas is likely responsible for flare-up of malaria in districts like Sri Ganganagar, Bikaner, Jodhpur and Jaisalmer (IGNP command area). On the other hand, in the IGNP non-command area districts malaria transmission, though at a low ebb, is carried by *An. stephensi*, whereas in Barmer and Pali districts (non IGNP area) both *An. culicifacies* and *An. stephensi* play a role in transmission of the disease (Table 1.3).

Table 1.2. Overview of malaria epidemics in the Thar Desert districts from 1975 to 2001

Period (in years)	No. of anopheline species				Cases reported		Irrigation (in ha × 10^6)
		vectors		Non-vectors (including occasional or secondary vectors)	*P. vivax*	*P. falciparum* %	
Pre-1930	1	*An. stephensi*	-	-	-	-	-
1930-34	1	*An. stephensi*	-	-	-	-	-
1935-39	1	*An. stephensi, An. culicifacies*[a]	4	*An. subpictus, An. annularis, An. vagus, An. splendidus*	-	-	0.24
1940-44	1	*An. stephensi, An. culicifacies*	4	*An. subpictus, A n. annularis, An. vagus, An. splendidus*	-	-	0.24
1945-49	1	*An. stephensi, An. culicifacies*	5	*An. subpictus, An. annularis, An. vagus, An. splendidus, An. barbirostris*	-	-	0.24
1950-54	1	*An. stephensi, An. culicifacies*	5	*An. subpictus, An. annularis, An. vagus, An. splendidus, An. barbirostris*	-	-	0.24
1955-59	1	*An. stephensi, An. culicifacies*	5	*An. subpictus, An. annularis, An. vagus, An. splendidus, An. barbirostris*	-	-	0.24
1960-64	1	*An. stephensi, An. culicifacies*	5	*An.subpictus, An. annularis, An. vagus, An. splendidus, An. barbirostris*	-	-	0.32
1965-69	1	*An. stephensi, An. culicifacies*	5	*An. subpictus, An. annularis, An. vagus, An. splendidus, An. barbirostris*	-	-	0.37
1970-74	2	*An. stephensi, An. culicfacies*	5	*An. subpictus, An.annularis, An. vagus, An. splendidus, An. barbirostris*	-	-	0.71
1975-79	2	*An. stephensi, An. culicifacies*	5	*An. subpictus, An. annularis, An. vagus, An. splendidus, An. barbirostris*	639,754	21,649 (5%)	1.22

Period (in years)	No. of anopheline species				Cases reported		Irrigation (in ha × 10^6)
	vectors		Non-vectors (including occasional or secondary vectors)		*P. vivax*	*P. falciparum* %	
1980-84	2	*An. stephensi, An. culicifacies*	5	*An. subpictus, An.annularis, An. vagus, An. splendidus, An. barbirostris*	98,242	12,422 (13%)	1.23
1985-89	2	*An. stephensi, An. culicifacies*	5	*An.subpictus, An. annularis, An. vagus, An. splendidus, An. barbirostris*	11,746	16,533 (15%)	1.33
1990-94	3	*An. stephensi, An. culicifacies, An. fluviatilis*[b]	6	*An.annularis, An. subpictus*[c], *An. vagus, An. splendidus, An. barbirostris, An. nigerrimus*	386,208	106,963 (29%)	2.17
1995-2000	3	*An. stephensi, An. culcifacies, An. fluviatilis*	9	*An.annularis, An. subpictus*[c], *An. vagus, An. splendidus, An. barbirostris, An. nigerrimus, An. pulchrrimus, An. d'thali*	2,45,960	30,077 (11%)	2.17

[a] Until the late 1980s *An. culicifacies* was recorded only from the desert fringe urban areas.

[b] *An. fluviatilis* is yet to be incriminated with malaria parasites in the Thar Desert, though it is a very efficient vector in southern Rajasthan.

[c] *An. subpictus* is only rarely incriminated in the Thar Desert.

The Thar Desert region in western Rajasthan is currently encountering a radical change in % *P. falciparum* over the years have come about regardless of SPR. Therefore, in a region such as the Thar Desert, % *P. falciparum* is a rather significant factor in determining the severity of the magnitude of fatal infection or future epidemic. New foci of *P. falciparum* and its pathways in the Thar Desert Out of the three physiographic zones described earlier, the IGNP command area seems to correlate with the flare up of malaria in the desert. In Fig. 1.9 the annual parasite incidence and % *P. falciparum* are presented to relate the two parameters to each other and to the period of availability of canal water for irrigation in the four desert districts under the IGNP command area.

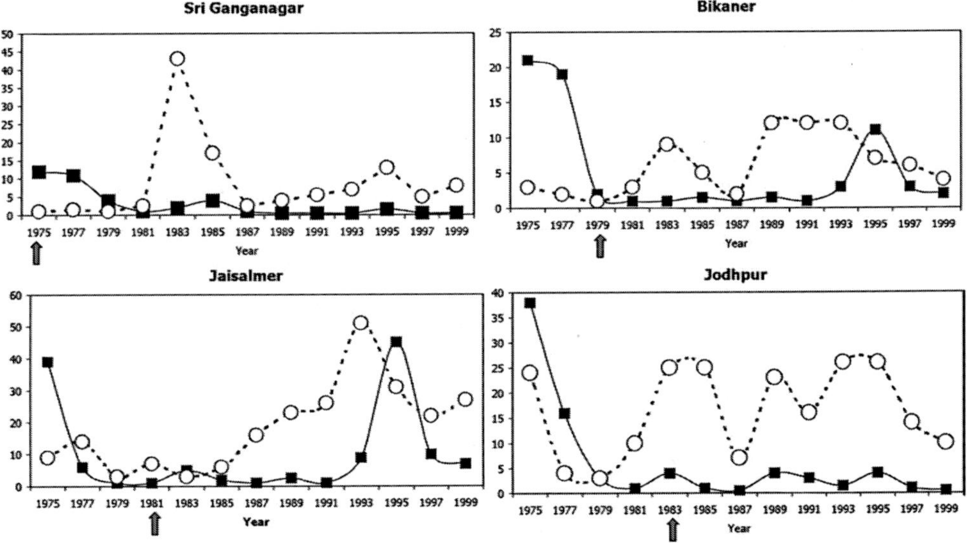

Fig. 1.9. Annual parasite incidence (API) (▪) and % *P. falciparum* (○) over time in four districts. The arrows indicate the year that canal water became available for irrigation.

Generally, it would appear that more malaria cases occurred in the early seventies. However, it is significant to note an appreciable decline in malaria cases since 1977 on one hand, and a significant rise in % *P. falciparum* with different magnitudes soon after that, on the other. Ironically, in Jaisalmer, and to some extent in Jodhpur, both annual parasite incidence and % *P. falciparum* are showing an increasing trend since the early nineties, clearly coinciding with the period of canalized irrigation and implying that a likely outbreak is foreseeable in the near future.

1.5.1.8. Anopheline fauna in The Thar Desert

Covell (1927) and Christophers (1933) were the first to put the State's anopheline fauna comprising four species on record. Puri (1955) described a total of 16 species of mosquitoes from the erstwhile State of Rajaputana. Most of these were restricted to the Udaipur zone in southern Rajasthan, and only seven (44%) were recorded from five of the eleven desert districts (Christophers, 1933; Puri, 1960). The Thar Desert anopheline fauna was updated by Tyagi (2004f) to a total of 12, as detailed below.

Foremost Tyagi & Verma (1991) and Tyagi (1994d) added three more anopheline species *viz., An. nigerrimus* from Sri Ganganagar district, and *An. splendidus* and *An. d'thali*, new to Rajasthan, from Jaisalmer district, raising the total to ten (56%). As far as the desert is concerned both *An. culicifacies* and *An. stephensi* have been incriminated for harbouring malaria parasites (Tyagi & Yadav, 2001). *Anopheles culicifacies,* a species widely recognized as a major vector of malaria in India, is rampant in the plains as well as the desert areas currently under extensive irrigation-based development. Tyagi & Verma (1991) have mapped out the distribution of *An. culicifacies* and *An. stephensi* in Sri Ganganagar district, of which two-thirds is highly irrigated (Sangaria Primary Health Centre) and the remaining area is covered with sand dunes (Pilibanga Primary Health Centre).

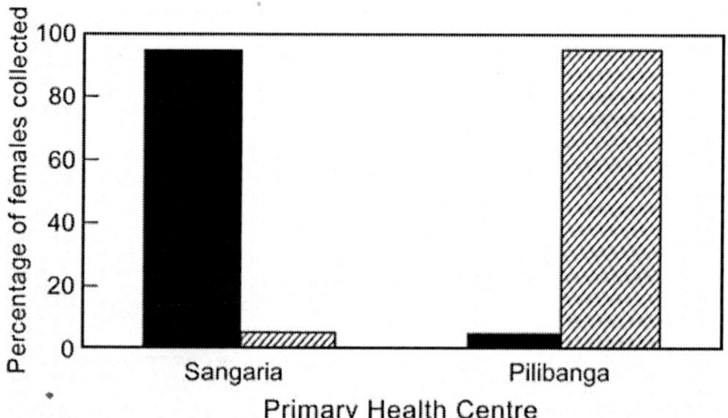

Fig. 1.10. Dominance of *An. culicifacies* (black pillar) and *An. stephensi* (lined pillar) in Sangaria Primary Health Centre (highly irrigated) and Pilibanga Primary Health Centre (xeric in nature) in Sri Ganganagar district.

It was found that *An. culicifacies,* breeding in a variety of habitats, formed the bulk (94·3%) of the two species in the Sangaria PHC area while exactly the reverse occurred in the Pilibanga PHC area with *An. stephensi* (95·2%) breeding more or less exclusively in 'tanka'. The predominance of *An. stephensi* in the desert environment and its breeding behaviour associated with 'tankas' were subsequently established by Tyagi & Yadav

(1996) in the north-western villages in Jodhpur district where *An. stephensi* (99%) was absolutely predominant (Figs. 1.10). In the Jaisalmer district where the IGNP is currently covering the maximum area, *An. culicifacies* occurs in more or less the same relative density as *An. stephensi.* Both these species occur above the critical density permitting malaria transmission (Rao, 1984).

1.5.1.9. 'Tanka' and 'beri' in The Thar Desert – their significance in sustaining Anopheles stephensi *and the evolution of a new malariological paradigm, "The Desert Malaria"*

It seems that *An. culicifacies* has entered the desert region with the canal irrigation rather recently (Tyagi & Verma, 1991). On the contrary, *An. stephensi,* which is believed to occur in two distinct forms, has been regarded as a desert species (Zahar, 1990). In the Thar Desert, *An. stephensi* breeds in well-like earthen pits locally called 'tanka' and 'beri' (Tyagi, 1992). Both 'tanka' and 'beri' are amongst the oldest type of man-made water reservoirs in the rural Thar region (Figs. 1.11, 1.12).

Fig. 1.11. An example of a *tanka* in the Thar Desert. (*Source*: Tyagi & Choudhary, 1995)

Fig. 1.12. Examples of *beri*: each mound of stones in the middle ground marks the entrance to a *beri.* (*Source*: Tyagi & Yadav, 1995).

1.5.1.10. Vector control in The Thar Desert.

Insecticide spraying in the whole of Rajasthan was initiated on a regular basis in 1989. As the insecticide of choice, DDT had been applied indoors at the rate of 1–2 g (a.i.)/m^2 twice a year, although HCH and deltamethrin have also been employed occasionally for emergent spraying in the border villages. In the Thar region, insecticide spraying with DDT has been very irregular and far from satisfactory. Tyagi (1994e) stressed the need for training the spray squads and discouraging ad hoc employment of unskilled labourers. Application of low doses of insecticides for a long time, coupled with faulty and inadequate coverage of sprayed areas often leads to development of resistance in the target vector species (Tyagi, 1992). Tyagi (1995) reported that both *An. stephensi* and *An. culicifacies* have exhibited an increasing tolerance towards DDT. The first report of DDT and dieldrin resistance in *An. culicifacies* is from Pali district (1962), followed by Sirohi (1963), Jodhpur (1964), Sri Ganganagar (1967) and Bikaner (1976). As for *An. stephensi,* the first instance of resistance against DDT was reported from the non-desert districts of Alwar and Kota in eastern Rajasthan. All vector species have been found to be absolutely susceptible to malathion since its first use in 1980.

Conclusively, the position regarding endemic malaria in Rajasthan, especially in the Thar Desert where *P. falciparum* is rampant across the entire region, has changed a great deal over the past few decades with the introduction and intensity of density build up of several different mosquito species, including *An. culicifacies* (Sharma, 1986a,b; Tyagi, 1994a). The overall number of positive cases in the State has increased 27 times, from 8494 cases in 1961 to 229,772 cases in 1994 due largely to the double impact of the *An. stephensi-An. culicifacies* duo, while the targeted population has increased only 2·5 times during the same period. While the slide positivity rate (SPR) (the number of blood slides found with Plasmodium parasites as a percentage of those examined) has significantly increased from 1·43 to 1961 to 5·07 in 1994 (3·5-fold increase), the slide falciparum rate (SFR) (the number of blood slides found with *Plasmodium falciparum* as a percentage of those examined) has also increased as much, from 0·55 to 1·95. However, the increase in either the SPR or the SFR was not steady and, therefore, no definite trend can be noticed. The maximum SPR was recorded in 1976, with several epidemics in non-desert parts of the State (Sharma, 1986a,b; Tyagi, 1994a). From the mid-eighties onwards, however, the trend has become clear and both the SPR and the SFR have steadily increased. In fact, the highest ever SFR has been recorded in 1994. The annual parasite incidence (API) (the number of parasite-positive blood slides per thousand population) also showed a steady increase from 0·48 to 1·76.

The IGNP which was initiated in 1957 and was to be completed in three stages, by early 1990s, could however never be completed with all its objectives achieved fully, and though many factors existed for its slow completion, nevertheless, the most compelling non-human reason was the development of extensive mosquitogenic situations in the entire IGNP-waterlogged region of several hectares. An additional Stage III also envisaged in the planning of extending objectives of the IGNP which, despite all claims, did not even see the Sun. Incessant malaria epidemics in the eighties and nineties as well as increased mosquito populations and bites, but also the inveigling into the Thar Desert a large number of new species, many of which more vicious biters and/or serious vectors of diseases, had obviously ceased the enthusiasm of completing the project that stalled for many years at different stages of development. The astronomical rising costs due to delay and withdrawal of funding resulted in restricting the IGNP development to only Stage I and II. Much of the labour forces, too, left the scene of work due largely to escalating malaria incidence together with vector and nuisance mosquitoes' unrelenting biting nuisance in the canal construction area.

1.5.2. NON-ENDEMIC FILARIASIS IN THE THAR DESERT

Lymphatic filariasis caused by *Wuchereria bancrofti* and transmitted through the common house mosquito, *Culex quinquefasciatus,* is endemic in most of peninsular and inland States and Union Territories; Rajasthan being a non-endemic State to the infection, albeit cases positive with microfilaraemia have been encountered now and then (Prasad & Nagda, 2018; Tyagi, 2018). Presence of sporadic incidences comprising both imported as well as indigenous cases have warranted a proper monitoring of emergence of positives especially in view of migration of labour forces from endemic neighbouring States of Uttar Pradesh and Gujarat, migration of cattle raisers in search of fodder, preponderance of the vector and likelihood of armed personnel and their associates who frequently move on duty or transfer amongst endemic and non-endemic States (Tyagi, 2018).

I, while serving as Officer in-Charge of Desert Medicine Research Centre (ICMR), Jodhpur, have myself worked on nonendemic lymphatic filariasis in the Thar Desert. I collaborated with Major Dr R. Srivastava, in-Charge of the Station Health Organization, Indian Army HQrs, Jodhpur during 1991-93 (*cf.* DMRC Ann. Rep., 1993-84). I had made night blood survey in the BJS Colony and the Army Residential Unit along Banad Road, and of a total of 2,894 samples collected/slides made, 24 were found positive with *W. bancrofti*. All positive cases from the army residential area were treated with DEC at the SHO clinic. I also sampled fortnightly from the above areas *Culex quinquefasciatus* and dissected 3,051 females but none was found parasitized with the filarial worm. In view of the above scenario of information Prasad & Nagda (2018) have propounded the following hypothesis represented graphically.

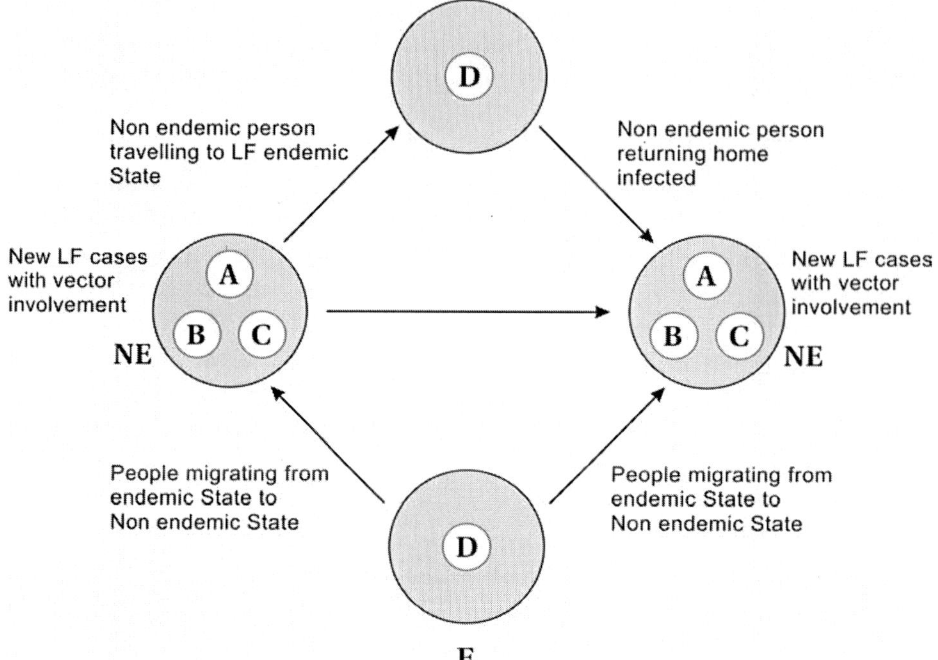

Fig. 1.13. Hypotheses of spread of 'lymphatic filariasis disease endemism' to a non-endemic state from neighbouring/endemic states (*Source*: Dr B.K. Tyagi, unpublished; A= Non-endemic population; B= Non-endemic infected population; C= Vector *Cx. quinquefasciatus*; D=Endemic population; NEN=Non-endemic State; EN Endemic State)

1.5.3. DENGUE FEVER IN THE THAR DESERT

The Thar Desert's physiography and ecology till 1980s did not support at all any survival of an Aedes mosquito being the vector for dengue and chikungunya infections. The annual precipitation was very low, rains were scanty, ground water was scarce, perennial river, if any, was absent and, above all, the local desert folks were very careful about keeping water resources, in whatever form available, tightly covered so as to conserve their lifeline for extended period of time. Any of these desert water resources bred, if at all any mosquito species, only *Anopheles stephensi,* the only original malaria vector in the Thar Desert (Tyagi, 1995). From 1980s onward, however, scenario seemed to change with respect to availability of drinking water as the conduit-facilitated supply was drawn to each doorstep in every village in the desert. This made the local folks desert their main source of conserving potable water in 'tanka' go largely unused, and the uncared source of water changed the quality from drinking water to somewhat permanently stagnant and uncovered infrequently traversed source. The urban vector of dengue, *Ae. aegypti*, which already encroached the peri-urban periphery in the towns now got inveigled into the middle of the desert with the availability of the 'tanka water', in hundreds (Tyagi & Hiriyan, 2004). The first dengue cases in the Thar desert thus began to appear since 1999/2000. The Thar Desert is now permanently under the impact of the disease.

1.5.4. INTENSIFICATION OF VECTORISM: EMERGENCE OF *AEDES ALBOPICTUS* AS A MAIN VECTOR OF DENGUE IN KERALA STATE (INDIA)

Dengue is one of the most serious and fast emerging tropical diseases which in certain socio-ecological settings exacts disease burden (465,000 DALYs across the globe) that can only be paralleled with that of malaria. One modelling estimate indicates 390 million dengue virus infections per year (95% credible interval 284–528 million), of which 96 million (67–136 million) manifest clinically (with any severity of disease). The number of dengue cases reported to WHO increased ~6 fold, from <0.5 million in 2010 to over 3.34 million in 2016. These numbers are from member States in only three WHO regions (SEARO, WPRO and PAHO). Despite a risk of infection existing in 128 countries, 70% of the actual burden is shouldered by Asia. Dengue, with its two severe clinical manifestations - dengue haemorrhagic fever (DHF) and dengue shock syndrome (DSS), posed until recently an increasingly perilous situation due to lack of specific antiviral drugs or vaccine. The first dengue vaccine, Dengvaxia (CYD-TDV) by Sanofi Pasteur, a live recombinant tetravalent dengue vaccine, was first registered in Mexico in December, 2015 and approved in the United States in May 2019, as the first vaccine approved for the prevention of dengue disease caused by all dengue virus serotypes (1, 2, 3 and 4) in people aged nine through 16 who have laboratory-confirmed previous dengue infection and who live in endemic areas.

Generally it is believed that in most parts of the world the dengue virus has been transmitted mainly through the bite of *Aedes aegypti* and that to a lesser extent by *Ae. albopictus*, popularly known as the Asian tiger mosquito, occurring predominantly in certain southeast Asian countries, but also in many Indian ocean island countries/entities, and already in the process of establishment in many European and western nations! The first confirmed report of dengue infection in India dates back to 1940s and thereafter several states began to report the disease which mostly struck in epidemic proportion often inflicting heavy morbidity and mortality, both in urban and rural environments. In Kerala, for the first time, cases of dengue with some deaths were reported in 1997 (Tyagi, 2003). albeit antibodies detected in human sera from Kozhikode, Cannur, Palakkad, Thrissur, Kottayam and Thiruvananthapuram districts as early as 1979. Only *Ae. aegypti* (L.) occurred in these seropositive districts. Later the author and his team isolated, besides the human sera, from the mosquito, *Ae albopictus*, both DEN-2 and DEN-3 serotypes which was a new discovery altogether and founded the theory that underlined the fact that *Ae. albopictus* was able to transmit infection without the aid from the primary vector *Ae. aegypti* and thus was a principal vector in as far as Kerala was concerned (Tyagi *et al.*, 2003, 2004, 2006). This discovery carried an important message which was bound to impact the national policy to control disease via suppression of the vector (*Ae. aegypti*). Therefore, it is very important to comprehend about the ecological and anthropization factors responsible to force this otherwise benign mosquito to come out of its sylvatic abode into the midst of the urban human congregation and turn up a vicious vector for the deadly dengue (Tyagi *et al.*, 2006)

1.5.4.1. Physiography and climate of the Kerala State in the lap of Western Ghats.

Kerala state had since 1950s undergone an enormous change in respect of both its physiography/climate. With a total area of 38,863 sq. km (urban 3365 sq. km and rural 35,498 sq. km), it is one of the smaller states in the country, ironically supporting a very high population density of 819 persons per sq. km (3,18,38,619; males 1,54,68,664, females 1,63,69,995; 2001 sensus) which is nearly three times the national average. Shaped as a high vertical slope with an average breadth of 50 km, it is endowed with diversified climatic richness, with a pinch of hydrological concern as 41 of its 44 rivers originating in the Western Ghat empty into the Arabian sea in less than 48 hours after a rain. The state's lowland region, which accounts for 10% of the total area, runs along the coastline and embodies beaches, swamps and lagoons, besides backwaters, paddy fields and coconut plantations. Kerala's midland (42% of the total landmass) is primarily made up of valleys with undulating small hills and meandering passages. A variety of seasonal, annual and perennial crops are grown in its mainly lateritic soil. Of all these crops, cocoa and pineapple plantations are of tremendous significance as dengue vectors have been found to breed there. The highland region (48% of the total state's area), with steep hills is rife with forests and small streams. Plantation crops like rubber, tea, coffee and cardamom are grown in this region. The diversity of its geographical features has resulted in a corresponding diversity in climate. While the high ranges along the eastern border have a cool and bracing climate throughout the year, the plains, invariably merging coastally into Arabian sea along the western flank, are hot and humid. The average annual rainfall (3000 mm) is quite high compared to other Indian states - almost three times higher than that in Karnataka while twice than in Tamil Nadu. The state basically enjoys four types of climate, *viz.*, winter, summer, southwest monsoon and northeast monsoon. The winter season sets in during the month of December and continues till the end of February. During this season comparatively there is less rainfall, especially in the northern parts of the state. Winter is followed by the summer season. It starts in February and continues till May. Temperature is very high during this period. Occasional showers is a characteristic of this season. Kottayam district tucked in the Western Ghat ranges receives the highest rainfall during this season. The southwest monsoon, accounting for 60-65% of the total rainfall, begins either in the end of May or in the beginning of June and fades out by September. The northeast monsoon, commencing in October, contributes to 25-30% of the total rainfall. The coastal areas record a maximum temperature of 32 °C. The interiors record a maximum of 37 °C during summer. In the coastal area it is hot and humid during April-May while cool during December-January. Since about mid-2001, I had carried out studies in Kottayam, Idukki and Ernakulam, besides the coastal Thrissur and Thiruvananthapuram districts - all of which have had epidemics during past few years (Fig. 1.14). This region is fairly humid and warm throughout the year with the relative humidity and temperature varying between 70-90% and 22-34.5 °C, respectively. The annual precipitation is high reaching up to 3000 mm, with the maximum number of rainy days (18 – 21 rainy days/ month) being in May to August.

A careful analysis of epidemics in Kerala illustrates that dengue not only seemed to have an epicentre in Kottayam before spreading over to entire Kerala, most of the cases had erupted in the mountainous and sylvan environs of the Western Ghat ranges on the south-western edge of the peninsular India facing Arabian sea coast in Kerala of which it covers nearly 90% of the total landmass. The Western Ghat ranges receive variable amount of rainfall from both the southwest (May - July) and the northeast monsoon (September - November), and is characterized with the Shola forest punctuated with man-grown extensive plantations of rubber (*Hevea brasiliensis*), cocoa (*Theobroma cacao*), coconut (*Cocoa nucifera*) and pineapple (*Ananas comosus*).

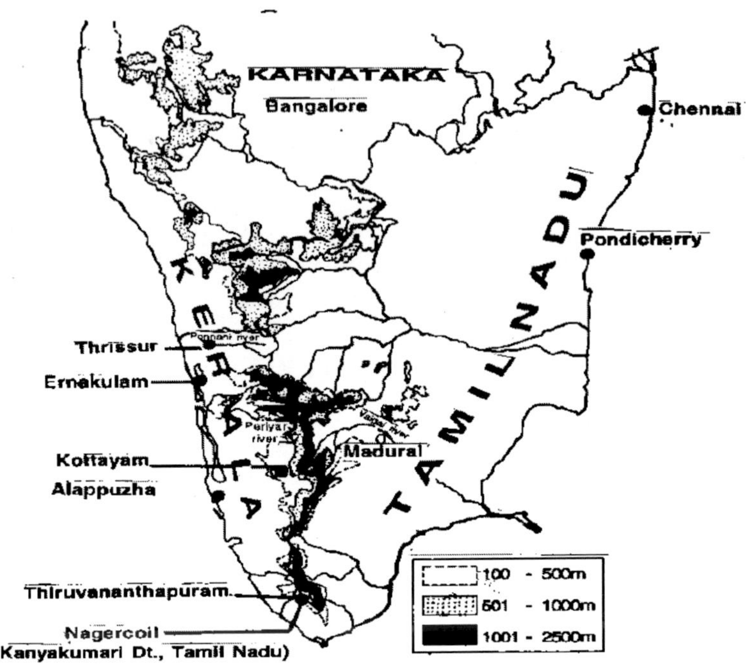

Fig. 1.14. Peninsular India exhibiting sites of investigation (*cf.* Tyagi *et al.*, 2006).

1.5.4.2. Impact of arthropization on Aedes albipictus behaviour and dengue transmission.

Anthropogenic impacts brought about a discernible vicissitude in Kerala's forest cover, agricultural practices, traditional water harvesting, demography, urbanization and human mobility. It would be hard to believe that Kerala experienced a drought in 2003, with the average annual rainfall dipping to an all-time low of 2000 mm! Although Kerala, which enjoys high rainfall, it still gets no more than 50-60 lit. of water for domestic use per person daily. Water allowance per capita depleted miserably in past few years. The Kerala state has an estimated 77.35 billion cubic meters (BCM) of fresh water, but nearly 40% of the resources are lost as run off, meaning only 42 BCM of water being available for the whole populace though the state requires a minimum of 49.70 BCM annually for various indispensably vital activities such as irrigation, domestic use, industries and others miscellaneous use. This shortage has led people to resort to exploiting ground water so severely that the water table in several districts has gone alarmingly low. The water scarcity also induces people to store water in a mélange of containers like pitchers, cement tanks, etc. as long as possible. A cadence in both ambient temperature and relative humidity is considered highly supportive to a wide dissemination of the dengue vector in time and space. There are evidences to prove a decline in rainfall, on one hand, and a marginal yet perceptible rise in ambient temperature, on the other. Since 1980 Kerala has suffered seven years of deficient rainfall and another five years below normal rainfall. An analysis of the southwest monsoon since 1998 shows that except Thiruvananthapuram in 2001, all districts have been receiving highly deficient rainfall in most years. Kerala had 2270 mm rainfall in 2003, still about 33% lesser than the average. It was amongst the few states in India declared drought-hit during 2003. It is noteworthy that since 1999 the annual rainfall has never touched the average value of 3107 mm; on an average, annual rainfall has dropped by 15-20%. The southwest monsoon has been particularly poor and in 2002 and 2003 it was deficient by as much as 33.3 and 26.4%, respectively. The year 2003 was further impoverished due to failure of northeast monsoon, with no area getting any rain during January-March. A close observation of rainfall data in various districts since 1900 shows that the number of rainy days in Kerala has dropped drastically. A discernible climatic cadence in Kerala is possibly linked with the atmospheric El Nino phenomenon which characterizes monsoon-related vagaries. Soil in Kerala can retain water only for a maximum of three months. Unless it is replenished periodically through precipitation many parts of the state are subject to acute water shortage. For decades the marshy lands and wetlands, which recharge ground water and help retaining soil moisture, have been reclaimed for constructing houses and setting up industrial units including coconut and other plantations. Forest, which help in orographic or convectional rainfall (i.e., rain in hill during evening), play an integral role in holding soil together and retaining rainwater, but large scale deforestation has been occurring in the state since 1970s, mostly with a view to planting cash crops like rubber plantation. Kerala had an estimated forest cover of 44.45% in 1900 which, according to the recent satellite images, had reduced to 14.7% in 1983 and to a pathetic 9% at present. Kerala originally had a total cropped area spanning over 2 million hectares, with paddy as the major crop covering about 35% of this area, followed by coconut (21%). By 1975 the paddy growing area reduced to 30% of the total cropped area, with only 0.881 million ha growing paddy. In 2003 the area under paddy cultivation nose-dived to a paltry 0.31 million ha. However, the coconut plantation sustained a good growth rate of 3% per annum through 1990s. It is noteworthy that, along with coconut, rubber plantation has *de novo* supported breeding of dengue vector, *Ae. albopictus*. At present while rubber plantations occupy the largest land area in Kerala, paddy fields have reduced and reclaimed largely for construction of human habitation.

1.5.4.3. Dengue outbreaks in Kerala State.

Cyclic dengue epidemics in Kerala state have been occurring since 2001, even though the first dengue report was brought on record from Kottayam district in 1997 with 14 cases and 4 deaths. This was followed by a more severe dengue outbreak implicating 67 cases (nearly 5-fold increase) and a toll of 13 human lives (3-fold increase) in 1998, again in the same district. The years of 1999 and 2000 were virtually free from any dengue cases. However, soon in 2001, epidemic dengue resurged mainly in Kottayam, Idukki and Ernakulam reporting 70 cases, followed by 219 cases in 2002 with some deaths. The year 2003 experienced the severest epidemic till date yielding as many as 3546 confirmed cases (253-fold increase) and a toll of 68 human lives, spread for the first time all over the Kerala's fourteen districts. The occurrence of DHF and DSS was also clinically and serologically documented. In spite of this history of dengue-related malady during past few years, no mosquito species was ever incriminated with a dengue virus in nature in Kerala, which created a void in comprehending the disease epidemiology. In view of a rather recent emergence of dengue infection in Kerala, the data on disease epidemiology are very scanty and inconclusive. During 2003, however, the dengue cases, which initially kept a low ebb, suddenly began to appear in large numbers from invariably all the fourteen districts soon after the southwest monsoon, reaching a maximum of 1337 in the month of July. It can be judged from these observations that 75.5% of the total dengue cases in Kerala in 2003 occurred only during the southwest monsoon period when *Ae. albopictus* also prevailed in higher density and after that nine of the fourteen districts continued to yield dengue cases till the end of the year, with Thiruvananthapuram reporting the highest incidence of 789/3546 (22.2%). It is also noteworthy that 9 of the 14 districts, of which four are landlocked in the high ranges of the Western Ghat mountains, had reported dengue cases for the first time. However, all those districts (*e.g.,*Kottayam, Idukki, Ernakulam, etc.) which had been reporting dengue earlier had shown an increase in the incidence. Clinically dengue patients presented a wide spectrum of signs and symptoms, often overlapping with flu and other infections, including leptospirosis. However, during an investigation carried out in 2003, major symptoms elicited by DF cases included fever (100%), headache (85.9%), retero-orbital pain (47.4%), myalgia (77.2%) and joint pain or arthralgia (47.4%). The haemorrhagic patterns of the DHF cases were manifested as rashes (86.3%), gum bleeding (21.1%) and purpural/echymosses (5.3%), etc. Children below 15 yr of age were quite severely affected (19.3%), although majority of infection occurred in the active adult age group of 16-60 yrs (77.2%) with the geriatric group (>60 yrs) people being also vulnerable (3.5%). While a total of 52.64% patients exhibited haemorrhagic patterns of DHF, only 47.36% manifested signs and symptoms of the classical dengue fever. Five DHF cases showed multiple haemorrhagic manifestations with skin rashes being the most common. Rarely though, Gullain-Barre syndrome and acute necrotising myelitis of unknown aetiology were also reported to be associated with dengue infection. Serologically, even though clinical cases had not appeared until late 1990s, the DEN-1 and DEN-4 viruses had been detected in 1970s from the human sera in a few districts of Kerala like Kozhikode, Kannur, Palakkad, Thrissur, Kottayam and Thiruvananthapuram. It was interesting that only *Ae. aegypti* without an instance of virus incrimination was recorded from all these districts covered under serological surveillance of local human population and *Ae. albopictus* was conspicuously absent. During the 2003 outbreak, a total of 112 finger-pricked blood samples were collected from dengue suspected and/or confirmed cases. Of these, 74 (66.0%) sera originating from Pattanamthitta (22.9%), Kottayam (22.2%), Alleppy (Alappuzha) (24.3%), Ernakulam (6.7%) and Thiruvananthapuram (25.6%) were found positive in antigen capture MAC-ELISA. Using dengue positive human sera, the biological TOXO-IFA system was employed and DEN-3 (67%) was isolated from Kerala cases, implying introduction of a new dengue virus serotype (later all the four serotypes were discovered from Kerala in *Aedes albopictus*). Additionally, a total of 37 human sera were tested by indigenously developed MAC-ELISA and 8 exhibited IgM antibodies against DEN-2 indicating recent dengue virus infection. These observations carry tremendous epidemiological significance from the viewpoint of evolving hyperendemicity.

1.5.4.4. Aedes albopictus *predominance in the Kerala State.*

A total of 9105 specimens belonging to 26 species under seven genera have been sampled in Kerala following various sampling methods. Among all the species, *Ae. albopictus* – the Asian tiger mosquito, was most dominant (69.7%), followed by *Armigeres subalbatus* (20.7%), Ae. aegypti (4.4%) and *Culex uniformis* (1.5%), with rest of the species accounting less than 1% individually. *Aedes aegypti* was generally conspicuous by either a total absence or a very low density. Sampled species exhibited good degree of flexibility in breeding adaptability to various kinds of habitats. *Aedes albopictus* surpassed all species by exhibiting a wide spectrum of breeding preferences with a clear cut predilection for coconut shells/plastic cups (79.7%) deployed in collecting latex from the rubber trees. *Aedes albopictus* breeding in cocoa pods (*Theobroma cacao*) has been widespread in Western Ghat forest fringe areas and its significance has been duly emphasized in preserving the vector density during high monsoon times when the main breeding habitats such as the latex-collecting cups tethered along the main trunk of the rubber tree are rendered unsuitable for breeding due to latter being turned upside down or covered with polythene sheets to avoid water collection in cups. *Ae. albopictus* maintained two peaks of high density correlating with the two monsoon seasons. This observation was considerably substantiated by both the container larval positivity and the adult landing index for *Ae. albopictus*. While the container larval positivity was as high as 64.1-73.8% in June and 80-84.3% in December, the adult landing index too was correspondingly high in June (7.2-11) and December (7.1-7.8), respectively. High density of *Ae. albopictus* (Breteau Index about 20) compared to relatively very low BI for *Ae. aegypti* seemed to bear a direct impact on dengue transmission and dissemination since the highest cases seemingly arose during these two periods only; the more conspicuous peak of both cases and deaths being during the southwest monsoon.

Anthropization is often associated with the spatial and temporal distribution of *Ae. albopictus* in Kerala which has a tendency to displace *Ae. aegypti* from its habitats. At least three human activity-related reasons can be attributed to the spread of *Ae. albopictus* far from its original sylvatic abode in the Western Ghat in yore to the coastal plains harbouring congested human settlements at present: (i) massive deforestation during past three decades that forced *Ae. albopictus* to come out of its natural abode, (ii) development of human settlements along forest fringe areas where mosquito frequently fed on human blood peridomestically, and (iii) its potential for virus transportation through different modes. Although, nothing is known at present about zoonosis in Kerala, it has been clearly demonstrated that dengue is a zoonotic disease of monkeys in Malaysia maintained by *Ae. pseudoniveus/subniveus* at canopy level. Therefore, it would be interesting to know if monkey-based zoonotic reservoir would be occurring in the Western Ghat region abounding both in *Ae. albopictus* and different types of monkeys?

Virus detection from a dengue vector in Kerala had been elusive until 2002. A long term investigation on both *Ae. aegypti* and *Ae. albopictus* in Kerala during 2002-2004 yielded a total of 240 pools (5173 adults) of *Ae. albopictus* comprising 112 pools (2456 adults) of males and 128 pools (2717 adult of which 413 were wild-caught) of females which were screened for the presence of dengue virus through ELISA. Of these, one pool consisting of 20 females, reared from the field collected immatures from Mundakayam in Kottayam district, was positive for dengue virus, indicating the transovarial mode of transmission. This phenomenon was further corroborated by a similar observation on *Ae. albopictus* in a nearby Nagercoil area (Kanyakumari district, Tamil Nadu) in the lower reaches of the Western Ghat ranges. Subsequently, during 2004, DEN-2 virus was finally isolated using the TOXO -IFA system and confirmed through ELISA from the two pools of wild caught *Ae. albopictus* female mosquitoes from Kottayam district, demonstrating with certainty the inevitably inextricable role of *Ae. albopictus* in the transmission of dengue as a major and principal, and not merely a conventionally believed secondary or bridge, vector in Kerala state.

In Kerala, *Ae. albopictus* exhibited a rather new biological dimension unknown hitherto particularly with reference to its potential to transmit dengue viruses. Because of its quaint biological attributes the mosquito differed almost completely from its main ally, *Aedes aegypti*, in its control strategies that was so far entirely based on the vector biology of *Ae. aegypti.* Significantly, soon thereafter many countries in Southeast Asia began reporting similar observations, substantiating strongly my viewpoint about both the disease transmission and control. The discovery turned out to be far more gigantic than I had possibly imagined!

1.7. Is it a good philosophy to annihilate mosquitoes?

Eradicating any organism *hook, line and sinker* would have serious consequences for ecosystems. Mosquito might be obnoxious and dangerous creatures to man and his economy, but they are having a role to play in our planet's ecosystem. There are currently over 3,500 species of mosquito, of which only a couple of hundred bite or bother humans, directly or indirectly. They live on almost every continent and habitat, and serve important functions in numerous ecosystems. They have been on Earth for more than 100 million years and have co-evolved with so many species along the way. Wiping out a species of mosquito could leave a predator without prey, or a plant without a pollinator, not to speak of immense artificial intelligence (AI) being derived out of their beiology, ecology and behaviour. Exploring a world without mosquitoes is more than an exercise in imagination, albeit intense efforts are under way to develop methods that might rid the world of the most pernicious, disease-carrying species. Even so the ecological scar left by a missing mosquito would be hard to heal quickly as the niche would take a long time to be filled by other organisms, if at all happens. Whether life would continue as before is difficult to foretell. A world without mosquitoes is out of question, though possibly workable sans a few notorious species. The human being must be prepared to realize the impact of the slightest of the collateral damage down their elimination.

A MOSQUITO'S-EYE VIEW

Mosquito, the deadliest foe of man on the planet, has many things good to offer as well! As an entomologist one must know both the ugly (*i.e.*, pestilent and vector) and beautiful (*i.e.*, pollinator, biocontrol agent, source of pharmacological and bioengineering tools etc.) sides of the life cycle of the mosquito.

Ugly side of the mosquito face:

- Deadly carriers of various different infections.
- They are reasons for varied nature of annoyance, psychopathy, allergie etc.

Beautiful side of the mosquito face:

- Larvae and pupae of mosquitoes live in water and provide food for fish and other wildlife, including larger larvae of other species such as dragonflies.
- Mosquito larvae help filtering detritus for plant life to thrive.
- Adults of some mosquitoes (*e.g., Aedes* spp.) help pollinate flowers. A few plants do exist that are preferentially mosquito-pollinated. All are orchids, namely cold-temperature ones. An example is *Platanthera obtusata,* the blunt-leaved orchid found across the Arctic, pollinated by mostly female Aedes mosquitoes as well as a few moths. It attracts mosquitoes by giving off a faint scent, detectable by mosquitoes, that is very similar to human body odour. The related *Platanthera flava* is also pollinated by Aedes primarily. Other Platanthera species are also occasionally pollinated by mosquitoes.
- The more than 90 species of one such harmless genus, Toxorhynchites, also known as the "elephant mosquito" because of their great size, are biocontrol agents in larval stage as they devour or predate larvae of other mosquitoes especially vectors.
- Like most blood-feeding organisms, proteins in mosquitoes' saliva are designed to prevent the host organism's blood from clotting, better allowing the mosquito to access its blood-meal (*e.g.*, the protein '**anophelin**' in Anopheles mosquitoes and '**aegyptin**' in *Aedes aegypti* mosquito). These protein target and bind to the central host blood coagulation enzyme thrombin, and therefore

prevents blood from clotting. Understanding of this mechanism has propelled pharmacological research in developing safe anticoagulants for the treatment of thromboembolic diseases such as stroke in the future. The world's largest mosquito factory in Guangzhou, China produces millions of mosquitoes daily (in connection with experiments involving *Wolbachia* treatment for control of *Aedes albopictus* - the vector for dengue, chikungunya and Zika) at the Sun Yat-Sen University-Michigan State University Joint Center of Vector Control for Tropical Disease and, as subsidiary activity, saliva is extracted in tons from Aedes mosquitoes for such and other purposes.

- Studying the human antibody response against the mosquito salivary proteins has provided new biomarkers for a direct and accurate evaluation of the human exposure to mosquito bites, at community and individual levels, and thus a possible development of a vaccine against a mosquito-borne disease in future.
- The purple pitcher-plant mosquito, *Wyeomyia smithii*, found in the Nearctic Region from the northeastern coast of the Gulf of Mexico to northeastern Saskatchewan is distributed in consonance with the range of distribution of the purple pitcher plant, *Sarracenia purpurea*, which is found from 30° N to 50° N in North America, implying its behavioural adaptation and shift in accordance with the climate change. *Wyeomyia smithii* thus becomes the first ever animal model to exhibit a shift in behavioural change according to its host plant.

♦ ♦ ♦ ♦ ♦ ♦

Chapter 2

MOSQUITO BODY - *MORPHOLOGY AND TAXONOMY*

2.1. Introduction

Of all the predators that stalk the planet, mosquito has taken more lives than any other organism. Consequently, other than mosquito, no animal on earth has touched so directly and profoundly the lives of so many human beings with respect to their health, economy, social activities and tranquillity. Each year some 500 million people are affected by mosquito-borne diseases and nearly 3 million die from these. With a tiny little bite this pest, with the skill of a phlebotomist, transmits more than a hundred diseases, from protozoan (*e.g.*, malaria) and helminthic (*e.g.*, lymphatic filariasis] to viral infections (dengue, yellow fever, several different kinds of encephalitides). There are over 3500 species of mosquitoes but only a small percentage of these phlebotomize and/or carry disease pathogens to humans. In fact there are more pestilent or nuisance mosquitoes than vectors.

Although, in public health arena, a pestilent mosquito is generally accorded much lesser importance, yet in view of multifarious repercussions associated with the pestilence, its importance can in no way be undermined. A pestilent mosquito causes harm to human being in several ways, for instance, by attacking for feeding blood, disturbing sleep through attacking for feeding and/ or making unpleasant whining sound near ears, causing painful welts or wheals at the site of bite, and, in certain situations like recreation parks near marshy beach areas, by causing disturbance and much loss of tranquillity through their efforts to enter into ears, eyes, nose and even mouth fearing induction of some kind of allergy due to their scales and hairs. Considering bare facts that a mosquito weighing 2.5 mg can take blood to the tune of 250% (i.e., approximately 6-7 mg) of its weight and there are over 100 trillion mosquitoes worldwide, then on a given day if all these are allowed to feed on human blood then the mankind loses about 625×10^{12} kg of the precious blood resource to mosquitoes! Realizing further, however, that around 100 mosquito species only incline to take human blood, and that a female will survive enough to take a minimum of four feeds in her lifetime within a period of two weeks, the total loss of human blood amounts to 196 × 10^5 kg in 15 days, 392 × 10^5 kg in 30 days, and an amazingly stupendous quantum of 4604 × 10^5 kg in an year. The significance of this enormous loss of the vital fluid can be further judged from the fact that this volume of blood equals to nearly 92 × 10^6 (Ninety two million) mortals at a given point of time. The enormity of this loss is further compounded by our awareness of the fact that the damage-causing process is not only recurrent but is also irreversible. Ironically, all mosquitoes do not like to feed human blood, and in fact less than 1% of all species prefer to essentially drink human blood. A larger proportion of mosquitoes, however, chance upon humans for blood feeding if the predisposed host kind is not available. This phenomenon in a way alludes towards the significance of other animals which provide larger share of blood feeding to mosquitoes. Providentially, in nature, realizing all kinds of biotic and abiotic stresses, less than 5% of the eggs laid by a female get to become adults, thereby reducing enormously the overall blood-feeding burden by the mosquito on vertebrate hosts, including man.

Mosquito's life presents many amazing features, but four of these, *viz.*, (i) General body plan, (ii) Life cycle, and (iii) Reproduction, and (iv) Blood feeding behaviour, which fit very well within the planned framework of discussion under this book, will be briefly summarized here under.

2.2. General body plan

Mosquitoes are insects belonging to the order Diptera, the true flies (Phylum Arthropoda, Class Insecta) (Eldridge & Edman, 2012; Tyagi, 2003). Like all true flies they have two wings but, unlike other flies, mosquito wings have scales. Female mosquito's mouthparts form a long piercing-sucking proboscis (Fig. 2.1a,b). Males differ from females by having feathery antennae, clasping genitalia and mouthparts not suitable for piercing skin (Fig. 2.1c). A mosquito's principal food is nectar or a similar sugar source; only fewer than 100 species (less than world's 3% mosquito species) transmit disease pathogens to humans. India is inhabited by approximately 410 mosquito species and subspecies, which is just over 10% of the global faunal wealth (Christophers, 1933, Barraud, 1934; Puri, 1949, 1955; Wattal & Kalra, 1961; Rao, 1984; Nagpal & Sharma, 1995; Tyagi *et al.*, 2012, 2014).

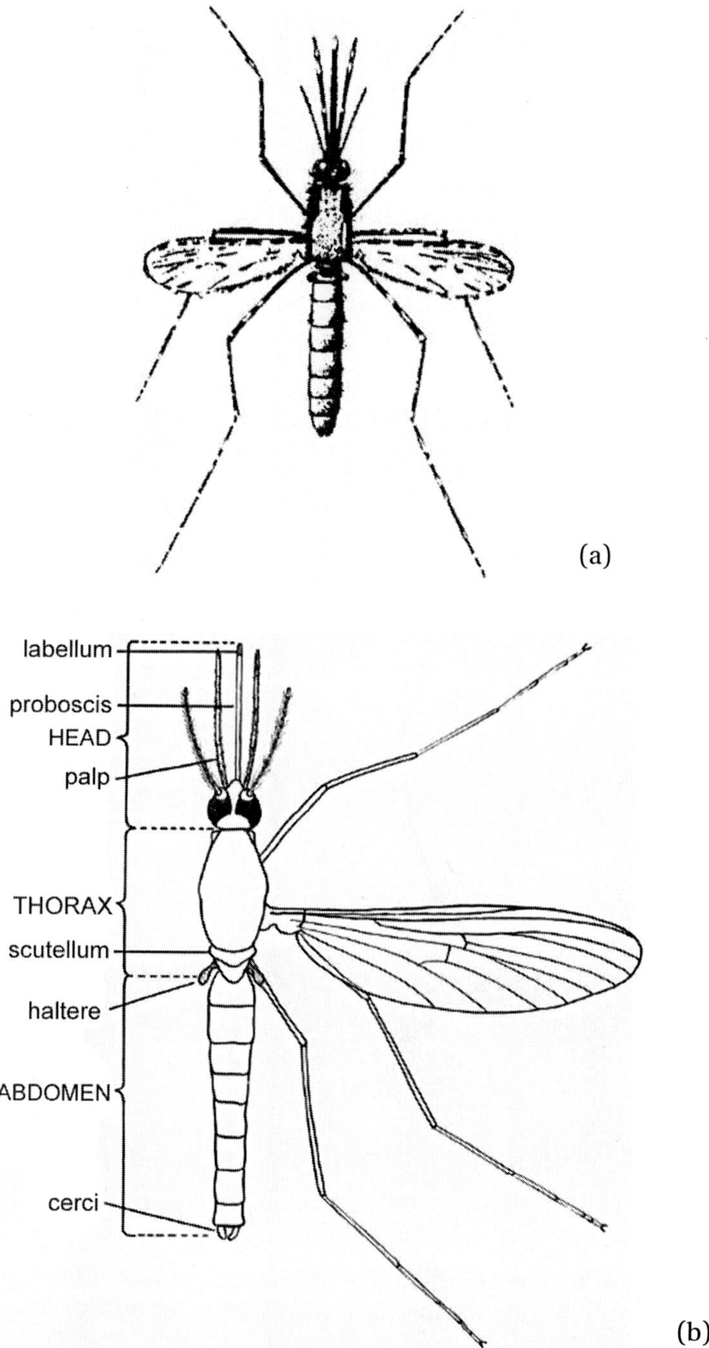

Fig. 2.1. (a,b): General morphology of an adult female anopheline mosquito (*Source:* Tyagi, 2004), and (b) Diagrammatic representation of an adult female anopheline mosquito body parts (*Source:* M.W. Service, 2008).

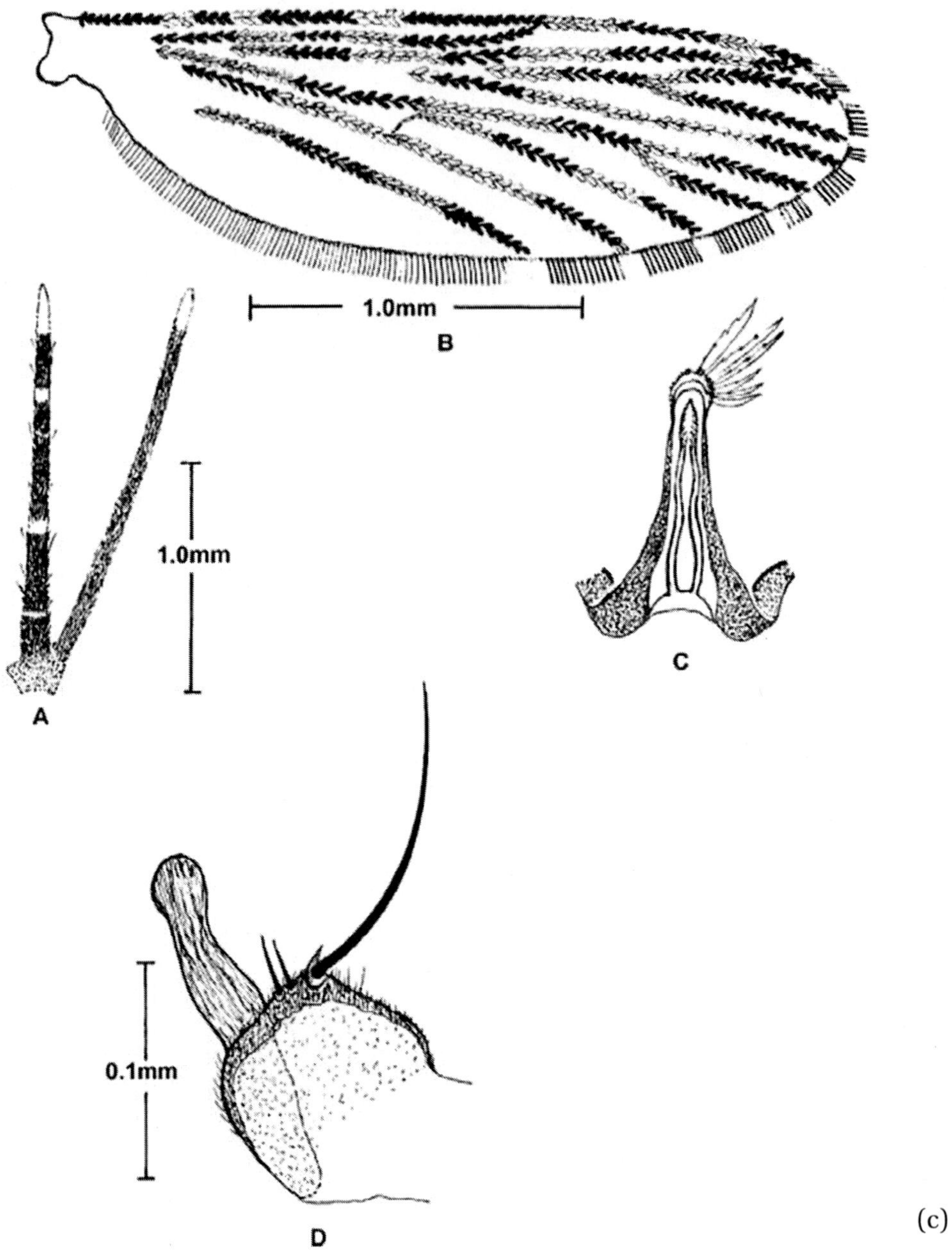

Fig. 2.1c. prominent features of identification in a mosquito; (a) Wing pattern, (b) Palpi, (c) Male genitalia, phallosome, and (d) Claspette. (*Source:* Tyagi *et al.*, 2014).

2.3. Classification

Every mosquito is assigned a Latin scientific name under binomial nomenclature system, such as *Anopheles culicifacies* or *Aedes aegypti*. The first word is the "generic" name of a group of closely related mosquitoes, while the second word is invariably the "species" name which denotes a group of individuals that are similar in structure and physiology, and capable of interbreeding (*vide infra*).

Phylum - Arthropoda
Class - Insecta
Order - Diptera
Family - Culicidae
Genus - *Anopheles / Aedes / Culex*
Species - *culicifacies / aegypti / quinquefasciatus*

Some mosquitoes, like *Anopheles culicifacies* (in India), *An. gambiae* (in Africa) and *An. maculipennis* (in Europe), have well-characterized sibling species-complexes whose individual species though morphologically indistinguishable from other members in the complex (both *sensu stricto* and *sensu lato*) are still good species as demonstrated by cytogenetical and immunological techniques. Such sibling species-complexes of certain vector mosquito species carry tremendous biomedical significance as there are some, and not all, species of the complex that are capable to transmit the human disease pathogens such as for example malaria parasites (Table 2.1).

Table 2.1. Indian anopheline species with complexes of species.

Species	# of Sibling Species in the Complex	Nomenclature assigned to sibling species
An. culicifacies	5	A, B, C, D and E
An. fluviatilis	3	S, T and U
An. subpictus	4	A, B, C and D
An. annularis	2	A and B

In some other species, however, albeit occurrence of distinct variants, occur forms or subspecies characterized by reasonably pronounced features in both morphology, behaviour and interaction with the parasite such as for example 'mysorensis' and an 'intermediate' form of *Anopheles stephensi* Liston in India (see Table 2.2) and 'domestic' (aegypti), 'sylvan' (formosus) and 'peridomestic' forms of *Aedes aegypti* Linn. in Africa.

Table 2.2. *Anopheles stephensi* with three variants, forms or sub-species characterized by distinctive features in both morphology, behaviour and interaction with malaria parasite (*Source:* Subbarao, 1987; Sweet & Rao, 1937).

S. No.	*Anopheles stephensi* forms	Ridges on the egg-floats	Aver. Number of ridges among the eggs laid	Spiracular length	Spiracular index (SI)
1	Type form (*An. stephensi* s.s.) **	14–22	16–19	Average spiracle length 0.11-0.12 mm	Average SI 8.09-9.23
2	Intermediate Subbarao, 1987	12–17	13-16	-	-
3	Mysorensis** Sweet & Rao, 1937	9–15	10–14	0.09-0.10 mm	6.82-7.60

** Rao *et al.* (1938) observed a certain degree of sterility in the reciprocal crosses between the two races, *i.e.*, Type and mysorensis.

The mosquito names are used in a descriptive manner so that the name tells something about each particular mosquito. For instance, in the name *Anopheles culicifacies* the generic word `Anopheles', taken from Greek, which in turn is itself derived from Sanskrit means 'hurtful' or 'harmful' [An = not; opheles = fruitful), while the specific word 'culicifacies' possibly refers to the Culex-like posture or resting habit, which the mosquito characteristically assumes soon after engorgement on blood. Some mosquitoes have what are called 'common names' as well as 'scientific names', such as *Aedes albopictus*, the 'Asian tiger mosquito', alluding to its possible cradle point - the Asia, or *Toxorhynchites darjeelingensis* discovered from Darjeeling area in West Bengal, India. Scientific investigators or taxonomists are continuously striving hard to discover new taxa of mosquitoes in all countries; the last anopheline mosquito species discovered from India was *Anopheles pseudosundaicus* Tyagi *et al.* (2009) from Kerala State after a gap of nearly six decades.

2.4. Life Cycle

Mosquitoes are very slender flies, generally with an elongated piercing proboscis and no ocelli (Fig. 2.2). The palpi are stiff and not pendulous. Legs are long, antennae densely plumose in the males, pilose in females. Wings are fringed with scales along the posterior margin and the veins. The larvae and pupae are aquatic and very active, though only the former feed. The larvae are metapneustic, with an elongated thoracic mass.

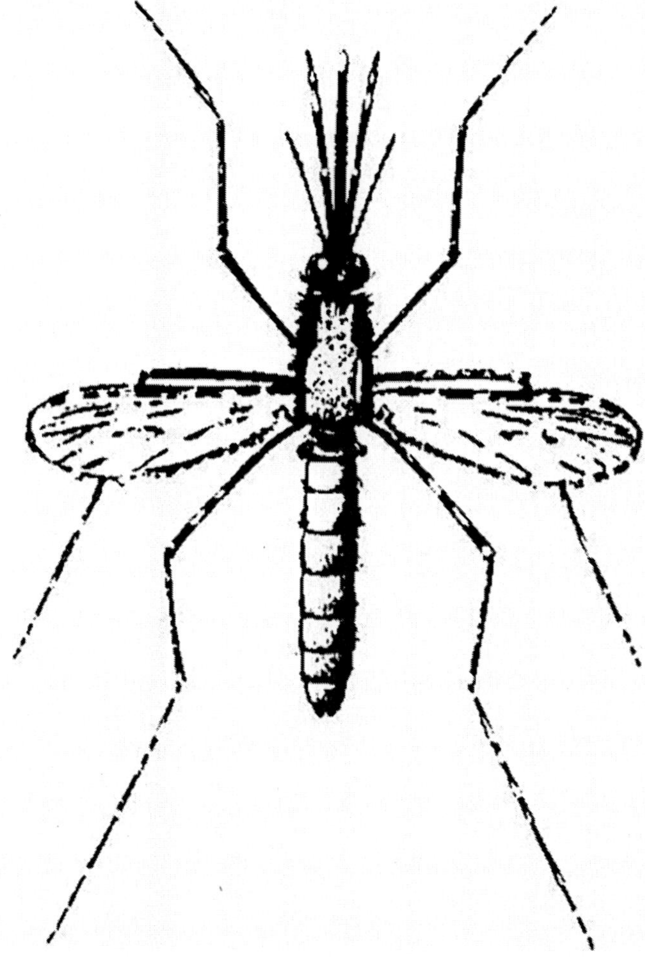

Fig. 2.2. General morphology of a mosquito. (*Source:* Free Wikipedia)

Mosquitoes have the capability to bite by day or by night, depending on species. They breed in all kinds of water fresh and salt, flowing or stagnant, foul or potable, water in tin cans, tree holes (phytotelmata) and salt marshes. Large coastal areas are occasionally inhabitable because of salt marsh mosquitoes. Mosquitoes may render recreation centres unusable and make normal living in and around homes extremely unbearable and vexing. Ironically, real estate values may suffer where mosquitoes are present, and losses caused by reduced industrial efficiency can be considerable because of the mosquito menace. As a principle, this is only the female of the vector mosquito which feeds on the blood of human or animal host, although females of some species are known to supplement their meals with plant sugars, too. Female mosquitoes may feed on the blood of a wide variety of mammals, birds, reptiles and amphibians as well as fishes. Recently adult mosquito feeding on invertebrate groups of annelids and insects (precisely a dragonfly) were also reported (Reeves *et al.*, 2018; Tyagi & Nazni, *pers. comm.*). Mosquitoes are generally host-specific for their blood-feeding requirements but they may quickly shift in case of any exigency.

Mosquitoes have their body clearly divided in three parts: head, thorax and abdomen. The head bears a pair of large compound eyes, a pair of antennae and mouthparts which are piercing and sucking type in females suited primarily for blood feeding, while in males they are developed to suck sap from the plants. Antennae too differ being shorter and sparsely haired in females while in males they are thickly haired and bushy. There is a pair of fully developed forewings, while the hind wings are reduced to rudimentary halteres or balancers. The males are equipped with external genitalia, called hypopygium. Following mating, males pass spermatozoa to the females where they are stored in spermatheca. Eggs are fertilized as they pass out of the ovary of the female. Females generally lay eggs in mildly moving or stagnant fresh (*e.g.*, *Anopheles*) or polluted water bodies (*e.g.*, *Culex*). Eggs are laid singly directly on water surface (Anopheles), in batches/groups approximated in a boat or raft-like shape (*Culex, Culiseta, Mansonia* and *Uranotaenia*) or along the moistened surfaces, like soil etc., close to water or on damp soil that will be flooded by water (*Aedes* and *Ochlerotatus*). Eggs of the genus *Toxorhynchites* are also laid singly on the water surface. Most eggs hatch into larvae within 48 hours, others might withstand subzero winters before hatching. Some mosquitoes (*e.g.*, *Aedes, Toxorhynchites*) are specialized to lay eggs in plant materials or tree-holes (phytotelmata). Some species may survive in the egg stage for 1-2 years before flooding occurs. *Aedes, Orthopodomyia* and *Psorophora* eggs are commonly laid in this manner. Water is a necessary part of their habitat. In general, development of mosquitoes is essentially the same.

There are four stages in the life cycle of mosquitoes; egg, larva, pupa and adult. The development is holometabolous. The immature stages are aquatic in nature. The eggs are usually white when first deposited, becoming dark within an hour or two. Anopheline eggs are oval-shaped and float horizontally. They are usually pointed on one end and have a pair of lateral floats. The larvae of all mosquitoes get their food from the water where they live, but to respire they have to come to the surface. Most larvae have siphon tubes for breathing and hang upside down from the water surface. Anopheles larvae do not have a siphon and lie parallel to the water surface to get a supply of oxygen through a breathing opening or spiracle. Larvae of the genus *Mansonia* and *Coquillettidia* are adapted to take oxygen under water from the air-conducting parts of plants. The larval developmental period includes four instars (*i.e.,* during growth the larva moults, or sheds its skin, four times and the stages between moults are called instars) which usually take 5-15 days to complete. At the end of the larval development pupa is formed by discarding the skin of the fourth instar larva. Only larvae feed while pupae do not feed. The larvae feed on microorganisms and organic matter in the water. Larva of *Toxorhynchites* species feeds on larvae of other mosquitoes sharing the habitat. *Toxorhynchites* larvae are much larger than other mosquito larvae. The larval body is characteristically divisible in three parts, but the pupa has only two parts; the fused cephalothorax and the curled and segmented abdomen. Pupae, commonly called tumblers, too, occur in water and are actively motile. When disturbed, it dives in a jerking, tumbling motion toward protection and then floats back to the surface. For respiring all pupae, except those of *Mansonia* and *Coquillettidia,* move (tumble) with a flip of their tails (caudal fins) to the surface where with the help of a pair of trumpets atop head region the oxygen is taken in. The pupal stage lasts for one day to several days. The metamorphosis of the mosquito into an adult is completed within the pupal case. When the pupal growth is complete, the pupal skin or exuvium breaks, the adult works its way out, crawls on to the surface of the water for a short time to allow itself to dry and all its body parts to harden. The wings have to spread out and dry properly before the newly emerged adult can fly.

Though not a general rule, blood feeding and mating does not occur for a couple of days after the adult emerges. The adult mosquito in all genera with vector species is a small fragile insect with a slender abdomen, one pair of narrow and delicate wings, covered or thinly so with scales, and three pairs of long slender legs. As the case may be, species may greatly vary in their size varying from slightly over 4 mm to about 12 mm. The three body parts (*i.e.,* head, thorax and abdomen] are clearly visible (Fig. 2.3).

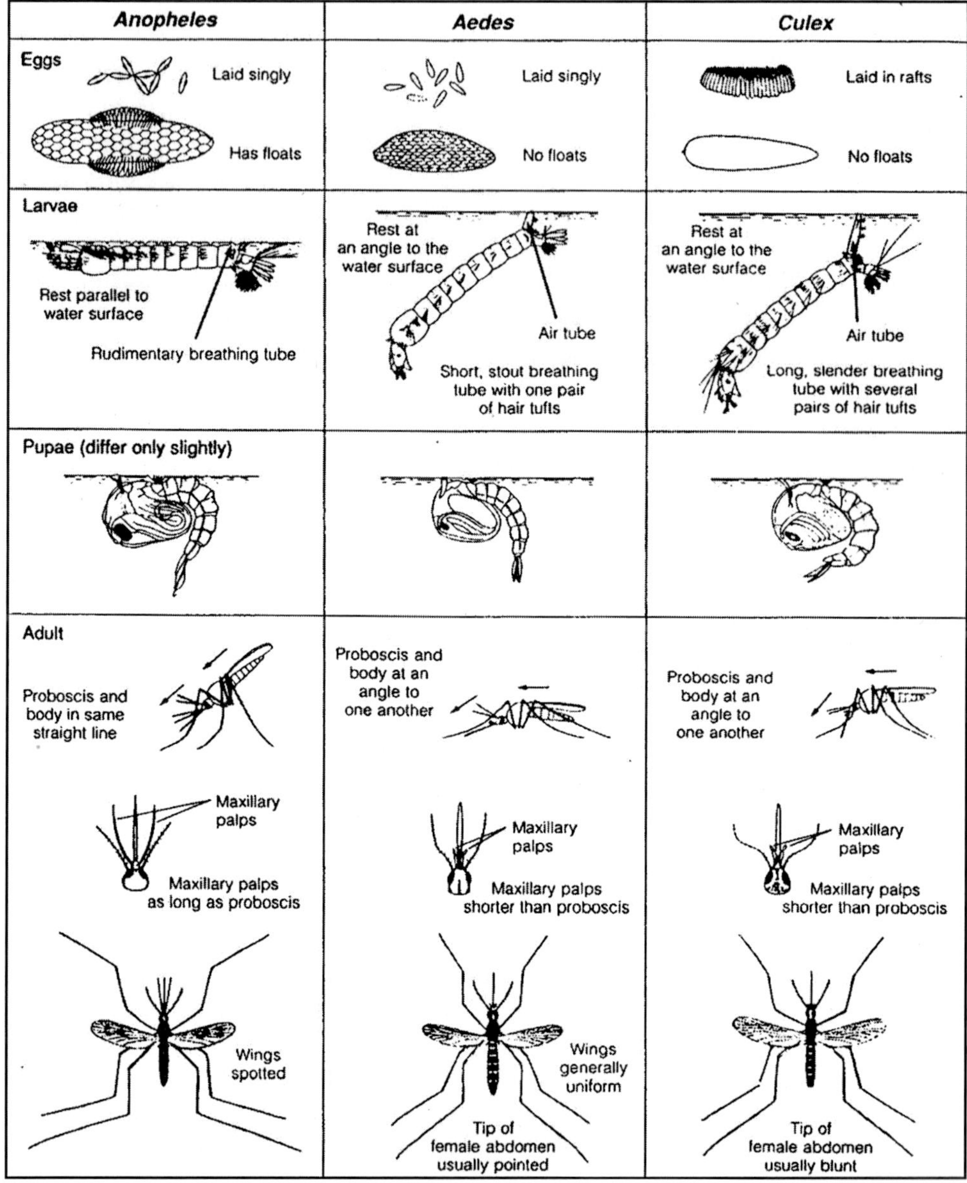

Fig. 2.3. Some of the main characteristics for differentiating *Anopheles, Aedes* and *Culex* mosquitoes. (*Source:* Tyagi, 2004)

The egg, larva and pupa stages depend on temperature and species characteristics to determine how long they take for development. For instance, *Culex tarsalis,* a common mosquito in USA might go through its life cycle in 14 days at 70 °F and take 10 days at 80 °F. On the other hand, some species have naturally adapted to go through their entire life cycle in as little as four days or as long as one month (Fig. 2.4).

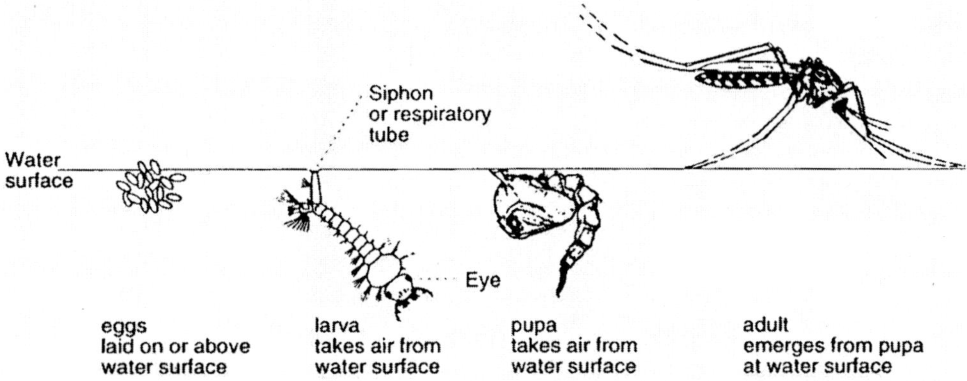

Fig. 2.4. The life cycle of a mosquito. (*Source:* Free Wikipedia)

Only female mosquitoes require a blood meal (protein) and bite animals warm or cold blooded. Stimuli that influence biting (blood feeding) include a combination of carbon dioxide, temperature, moisture, smell, colour and movement. Most mosquitoes are nocturnal in their feeding behaviour (*e.g.*, Anopheles), but some are diurnal or crepuscular (*e.g.*, Aedes). When the female finds a suitable host, such as the man, it introduces the piercing mouthparts through skin deep into body and penetrates a capillary, to suck the blood directly from the capillary or from the pool of the blood oozed outside the punctured capillary. Females of certain mosquitoes are single-time feeder while many species tend to feed in shift or multiple occasions on a given night or daytime. The number of the egg formation and development in ovary of the female is directly dependent on the quantum and nature of supply of blood meal received by the female mosquito. Depending on a particular species, an engorged female may like to rest inside house or prefer outside environs. This characteristic of the vector species has for long been exploited for their control through indoor spraying of residual insecticides.

Mosquitoes are the sole vectors of malaria, yellow fever and dengue, and they participate significantly in transmitting lymphatic filariasis and the encephalitides of various different kinds. Besides, their bites may itch for days, causing much loss of sleep and tranquillity, and serious nervous irritation and restlessness. Their saliva contains proteins alien to the human body and repeated bites result in sensitivity to it. Mosquitoes often cause serious economic losses limiting outdoor movements. Male mosquitoes do not bite, but feed on the nectar of flowers or other suitable sugar source. Acquiring a blood meal, *i.e.,* protein, is essential for egg production in females, but mostly both male and female mosquitoes are nectar feeders. Of those female mosquitoes capable of blood feeding, human blood meals are first or second choices. Horses, cattle, smaller mammals and/ or birds are preferred for blood feeding. Female *Toxorhynchites* mosquitoes cannot actually obtain a blood meal and are restricted to a nectar diet.

2.5. Mosquitoes' blood-feeding behaviour

Because some mosquitoes are transmitter of deadly human and animal diseases, blood-feeding is the most important characteristic of their life cycle which offers a unique amalgamation of various sophisticated processes. Adult female culicids feed according to their state in the gonotrophic cycle. Following the oviposition, the mosquito will seek a blood meal. Blood ingestion begins and continues until stretch receptors in the abdomen signal that the female mosquito cannot hold anymore. This is referred to as distention-induced host-seeking inhibition. Blood is then digested and egg maturation begins. Host-seeking is then further inhibited in oocyte-induced host-seeking inhibition. Following the oviposition of this batch of eggs, the cycle begins again. Variations in the cycle occur in mosquito species which blood-feed more than once during egg maturation. Vertebrate blood contains essential proteins necessary for eggs to mature. Egg production is dependent upon amino acids and does not require the presence of other constituents in the blood. It has been found that female *Aedes aegypti* fed on an artificial diet high in protein developed equal amounts of eggs to those fed on bovine blood (Kogan, 1990). Feeding is also dependent upon the circadian rhythms of the specific mosquito species. Circadian rhythms determine the time of day or night that a mosquito will feed. This is of great importance in determining a means of mosquito control; for example, bed netting can be used to prevent blood feeding while prospective hosts sleep. Mosquito behaviour is also dictated by a species' tendency to blood-feed on humans (anthropophagic) or animals (zoophilic) and whether the mosquito prefers to feed indoors (endophagic) or outdoors (exophagic). *Anopheles gambiae*, a vector of malaria in Africa and *An. culicifacies*, a vector of malaria in India, feed on sleeping hosts inside their homes. Host-seeking behaviour is dependent upon chemical and visual cues specific to a desirable host, as well as their internal species-specific cues.

The entire process of mosquito's blood feeding is a highly intricate activity which is accomplished by a melange of coordinated functions by various different organelles as well as chemical cues. This complexity of their mechanical and chemical systems for finding human host, finding a commensurate site to feed once they land, and then manipulating the body chemistry while they draw their nourishment are all an example of remarkable feat of evolution. To understand the whole *modus operandi* it is inevitable to first understand the feeding apparatus of a mosquito (Fig. 2.5).

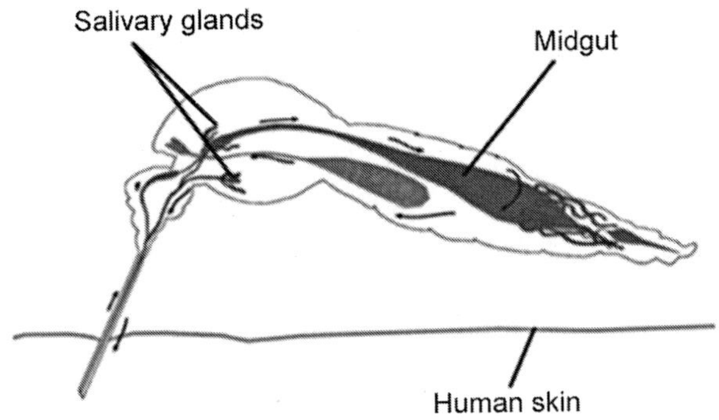

Fig. 2.5. Diagrammatic representation of an unfed mosquito initiating blood feeding, to exhibit pathway of parasite ingestion along with the blood-meal, and subsequent crossover through gut wall into hemocoel and entry into the salivary glands, and finally exiting into the human host's body through saliva-spit. (*Source:* Free Wikipedia)

The feeding mechanisms in mosquitoes revolve around two main structures, (i) the food canal, and (ii) the salivary duct. Both the food canal and salivary duct are separate, independent and parallelly running structure, never meeting on course. The mosquito squirts saliva before initiating feeding through one passage (and never the remnants of its last blood meal, as one might wrongly infer!) and draws blood up the another passageway. As a consequence, the food canal is not flushed out completely like a drug-injecting syringe does and the blood flow is always unidirectional (Fig. 2.6).

Mosquito's feeding apparatus then begins to work. It is a complex seven-piece set of devices for doing a variety of actions. One of those parts is a tube that allows the mosquito to secrete its saliva into the body being bitten. That saliva, remarkable in its action, contains a set of chemicals that act like an anaesthetic as well as anticoagulant. On one hand, the mosquito causes numb the biting spot on the skin so that hosts do not sense presence of the mosquito probing around on skin, and, on the other, make the blood flowing intensely so as to enable her continue to suck in a gluttonous manner! Meanwhile, the other parts begin an intricate pattern of coordinated movements that let her hack into man's flesh. The mandibles, on the edges, are serrated on the tip, like tiny steak knives, to carve their way in. Just below the surface, the whole package of savagely armoured parts bends and takes a 90° turn, to begin the attack (Fig. 2.7).

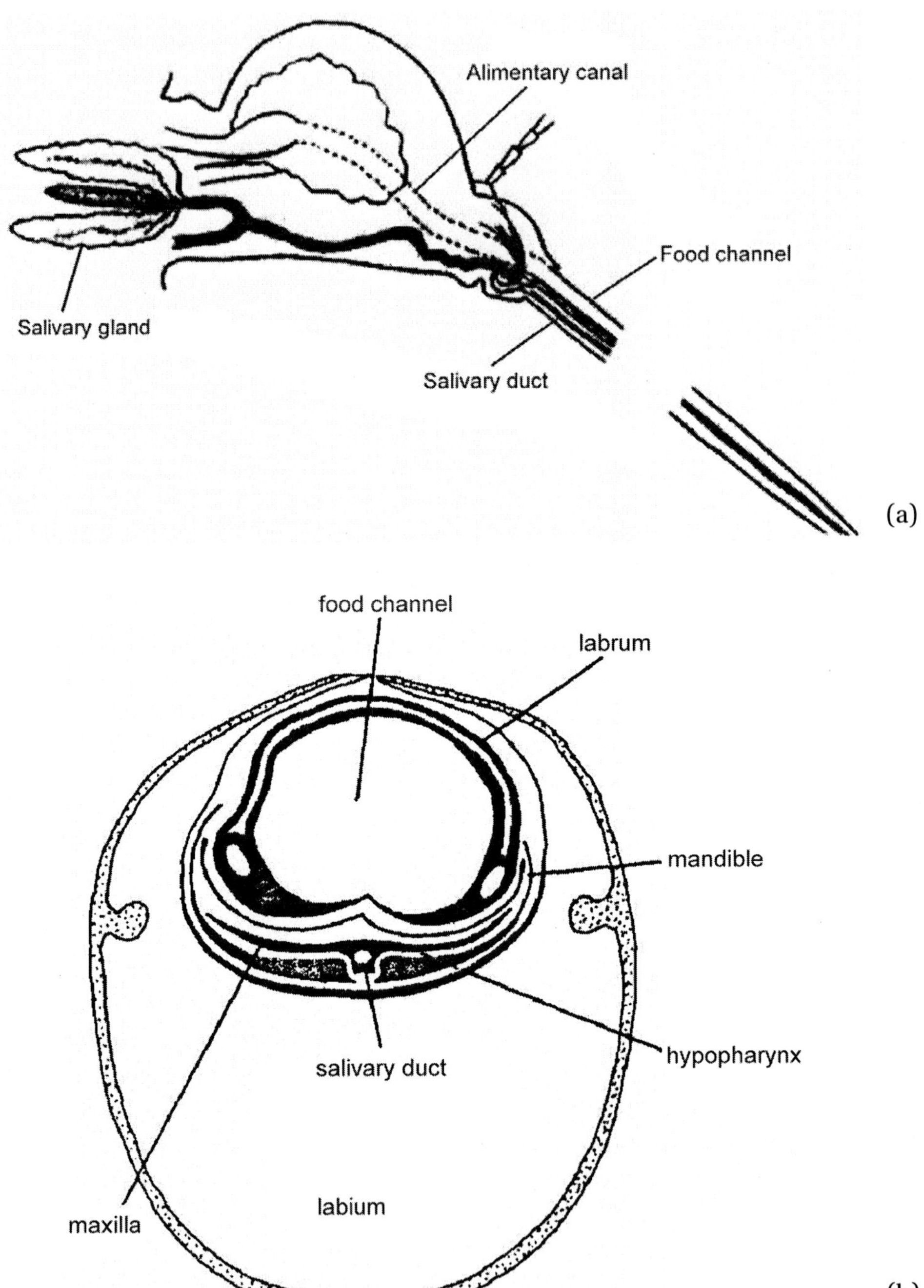

Fig. 2.6. (a) Anatomy of a mosquito showing separate passageways of each, the food channel and the salivary duct; and (b) structures, including the food channel and the salivary duct. (*Source:* Tyagi, 2004)

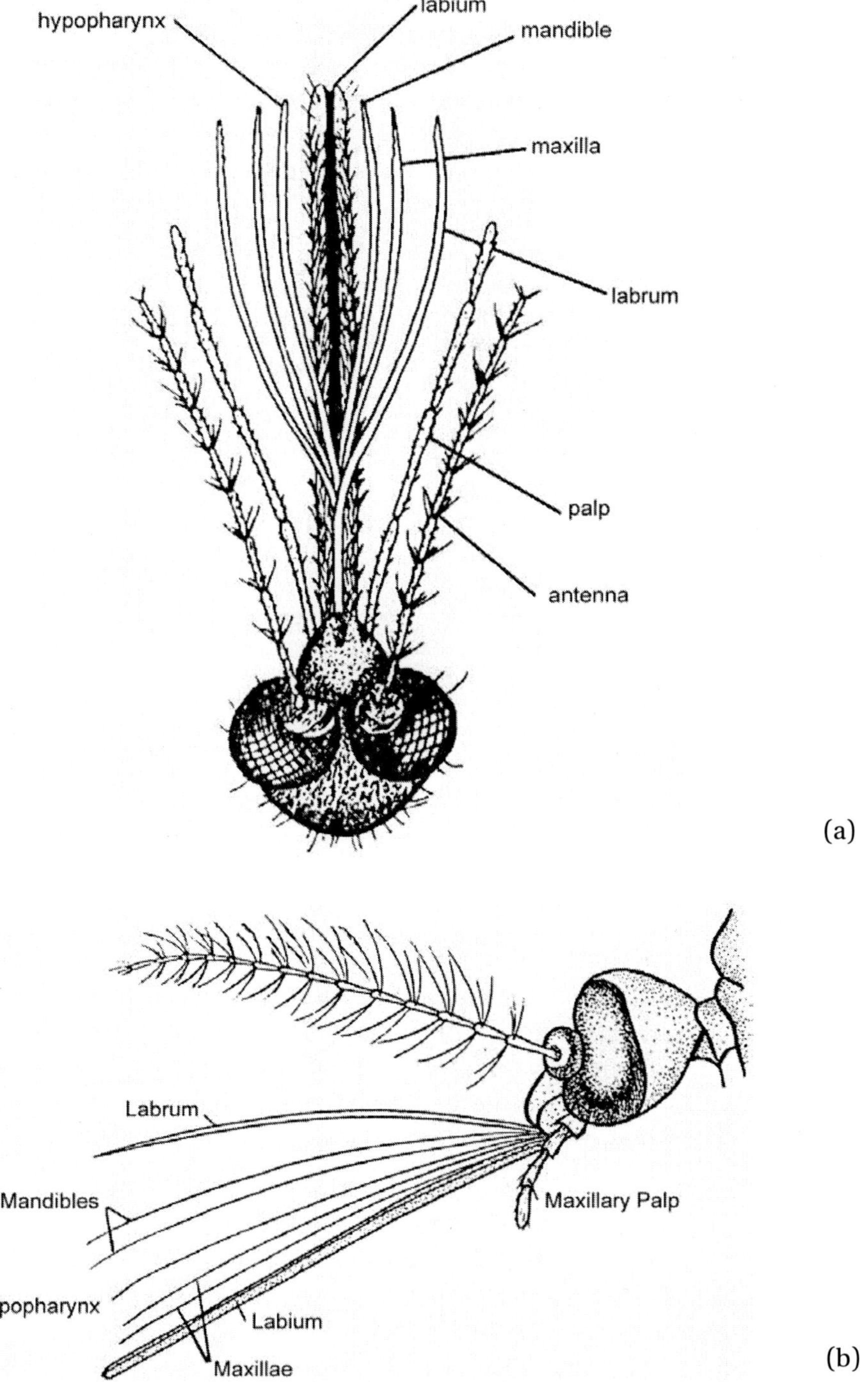

Fig. 2.7 (a, b). Piercing-sucking mouthparts of a mosquito: (a) dorsal and (b) lateral views of stylets removed from labial sheath. (*Source:* Tyagi, 2004)

That set of needle-like parts begins to randomly poke away, anywhere between 12 and 18 times in just several seconds. It is not actually looking for a venule or arteriole just yet. It's indiscriminately poking around to try and poke into anything in the area that may bleed. That tube in the middle of their mouth assembly that does the spitting is actually two straws in one, one inside the other. As they continue to spit out their anaesthetic, the other wider tube in the straw is sucking in samples of fluid to see if it has hit pay dirt (Fig. 2.8).

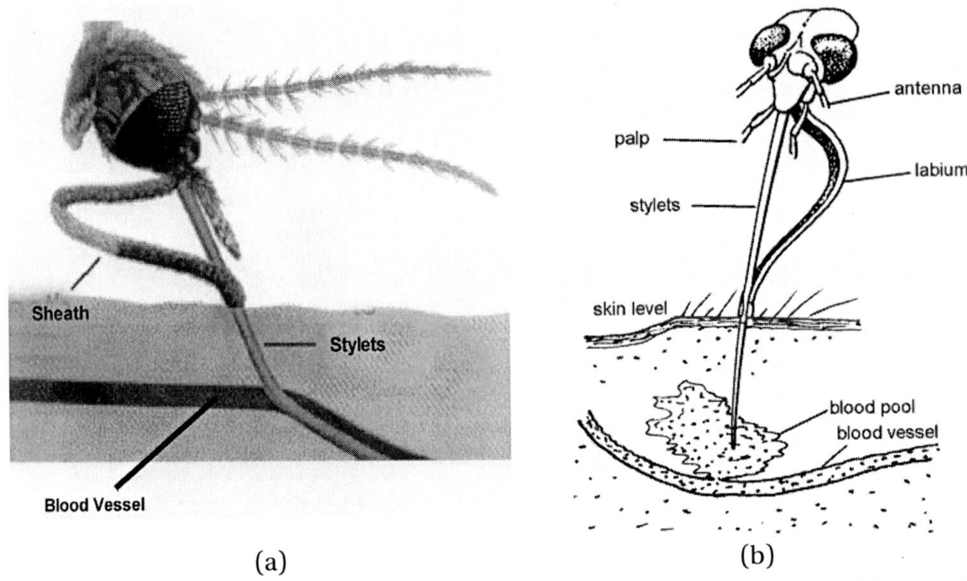

Fig. 2.8. Head of female mosquito at commencement of blood feeding (a) directly from a venule and (b) from a pool of blood formed as a result of strategic puncturing of the blood vessel. (*Source:* Tyagi, 2004)

Mosquitoes are the only known creatures that can spit and suck at the same time. If they don't find blood, they draw their probing mouth parts back a little, not all the way out, and dig into another spot. This process can last up to two minutes, if somehow the human body remains unaware that the mosquitoes are there.

Selecting a biting site is made on 'trial and error' basis not without a meticulous planning behind their host-seeking strategy. For zoophilic mosquitoes, host-accessibility is dependent upon physical barriers to blood feeding. Many mosquito species prefer to feed on relatively hairless, or featherless areas of a host. These areas are conveniently more vascularized to keep the host warm. Chipmunks and squirrels are more likely to be bitten on the ear (versus the back or feet). As for relatively hairless beings, it seems unlikely that mosquitoes would prefer any one biting site over another. This is not a universal truth since it is proved otherwise. Jong and Knols (1996) used a test subject inside a bed net, wearing only close-fitting underwear, to determine body regions most likely to be bitten by specific mosquito species. Mosquitoes were released into the bed net, and were recorded if they bit within three minutes. *Aedes aegypti* was found to prefer the shoulder and head region, while *Anopheles gambiae* preferred the feet. *Anopheles gambiae* has also been proven to be more attracted to Limburger cheese than human breath in wind tunnel experiments (Jong & Knols, 1995). This suggests that bacterial stimuli may also stimulate host-seeking behaviour, and is an interesting possible clue to better understanding host-seeking behaviour in mosquitoes. Selection of a biting site can also be further dictated by the presence of a parasite. Host-seeking behaviour may also be enhanced by the presence of a parasite in the vector's body. To increase their dispersal, pathogens within a mosquito vector may alter the host's behaviour. It is clear that by altering vector behaviour to increase hematophagy, a pathogen will gain increased dispersal. Pathogens can affect host attractiveness, increase probing behaviour, and reduce engorgement time (Edman *et al.*, 1984), as well as alter the vector's circadian rhythms. Filarial worms, for example, develop according to the activity of the vector and are successfully dispersed. Thus, in the mainland India, where the night-biting *Culex quiquefasciatus* is the vector, the microfilarae of *Wuchereria bancrofti* are freely flowing in the man's peripheral blood vascular system in the late evening/early night hours (nocturnal) and in the Andaman & Nicobar ils., where the day-biting *Ochlerotatus (Aedes) niveus* is the vector, the same appears in peripheral blood during the day time (diurnal; subperiodic).

Many factors work in unison to inveigle a haematophagous mosquito toward its host. After orienting toward a host according to chemical cues and using upwind optomotor anemotaxis, a mosquito has to have cues to land. Arresting stimuli, such as sweat, humidity, heat, colour, and movement are possible factors which lead to the termination of host-seeking, and the beginning of probing behaviour. However, as mentioned above, host-seeking behaviour may be prevented by certain stages of the gonotrophic cycle. Some researches have indicated that the host-seeking inhibitory responses of *Ae. aegypti* result from the neuropeptide known as Head Peptide. Increased amounts of this neuropeptide then act to desensitize sensory receptors to inhibit host-seeking behaviour. It has been further determined that nutritional and mating states, and other hormones released during egg maturation control the inhibition of host-seeking behaviour. There is a great deal of research yet to be done on hormonal control of host-seeking behaviour as understanding of these mechanisms can lead to genetic manipulation for vector control. If however a mosquito does taste blood, it turns on more powerful pumps in its mouth, increasing the suction on its "drinking" straw. Then she draws on the tiny cloud of blood that has formed under the skin. The nearest venule or arteriole, even if it's not the one already wounded, is pulled over to the needle-like tip of the whole apparatus, where it's impaled. That opens up the hole through which the mosquito starts to drink.

Generally, there are two types of blood feeders: vessel feeders and pool feeders. Vessel feeders take blood directly from venules or small veins, while pool feeders take blood from a blood pool resulting from the laceration of blood vessels (James and Harwood, 1969) (see Fig. 2.8).

The process of a mosquito questing for a blood meal involves a complex, interconnected cascade of behaviours, each probably having its own cadence of cues, be they visual, thermal, or olfactory. The complexity of these questing behaviours may account for the bewildering variations in trapping efficiency noted for certain species of mosquitoes at different times. The finely toothed maxillae of the fascicle begin sawing into the tissue of the skin with fine back-and-forth movements. A fascicle is "a bundle of muscle or tendon fibres". The fascicle is guided into the skin between the labella. As it goes into the skin the labium folds back like a hairpin and the mosquito shifts its legs closer to its body. When about half of the length of the fascicle has been inserted into the skin, the mosquito begins to draw blood. After the mosquito's abdomen is filled, she straightens her front legs to quickly withdraw the fascicle. The fascicle springs upward and forward out of the wound and is fitted back into the deep groove in the labium. Figure 2.6 above shows the stylet bundles of mosquitoes piercing, while their sheaths [labium] bend and the stylets come out of the grooves. A stylet is "a long thin pointed instrument".

To better understand the complexity of what is involved in the act of blood feeding, we need to know something about the anatomy of the mosquito's proboscis and alimentary canal. The proboscis projects directly in front of the insect. It consists of a large, scaly outer lower lip called the labium, which terminates at the tip in two hairy lobes called the labella. The labium forms a deep trough in which is concealed a smaller bundle of long tapering pale yellow feeding stylets that are collectively called the fascicle. The largest stylet is the sharply pointed labrum, which forms an inverted gutter up which the blood is drawn. At the side of the labrum are two long, thin mandibles. Under them are two much larger needle-like maxillae with fine saw-toothed tips. Under them in turn is a flattened stylet, the hypopharynx, down the center of which runs a single salivary canal.

The fascicle acts as both a piercing mechanism and a food canal. Then the mosquito is poised for the moment of piercing and it has all six legs on the surface of the victim's skin, its hairy labella touching the skin and its maxillary palps raised. It is believed that after the mosquito spreads its labella the finely toothed maxillae of the fascicle literally saw their way into the tissue of the skin with extremely rapid back-and-forth movements. Next, the large labium is folded back like a hairpin. The skin is actually punctured without any obvious thrusting of the mosquito's body. About half the length of the fascicle is inserted into the skin. As it is inserted, the mosquito gradually shifts its legs closer to its body. It takes an average female mosquito about fifty seconds to insert her fascicle into the human skin, and unless she is interrupted (perhaps by being swatted by the victim) she usually sucks blood for about two and a half minutes. When the female mosquito has drunk her fill, she withdraws the fascicle rather quickly (in about five seconds). The procedure is to straighten her front legs, leaning slightly backward, and moving the labium from side to side. Once the fascicle has been removed from the skin, it springs upward and forward and then returns to the deep groove of the labium (Fig. 2.9).

The feeding behaviour and habits of mosquitoes vary widely from one species to the next. *Aedes vexans,* the most common plague of Americans, both inside country and city, is different from *Aedes excrucians* of northern wooded areas, as well as from *Aedes tormentor* of the southern woods; *Culex pipiens,* the common house mosquito of the tropics, *Coquillettidia perturbans,* an inland swamp mosquito; *Toxorhynchitis rutilus* mosquitoes, which eat the larvae of *Aedes aegypti,* the yellow fever mosquito, the *Anopheles,* and so on down through more than 3500 different species. To complicate matters even more, the species of mosquitoes that predominate vary widely from area to area, so that the type that hovers around a city bedroom or other parts of the home is likely to be quite different from the one that follows people around on the golf course or in the park.

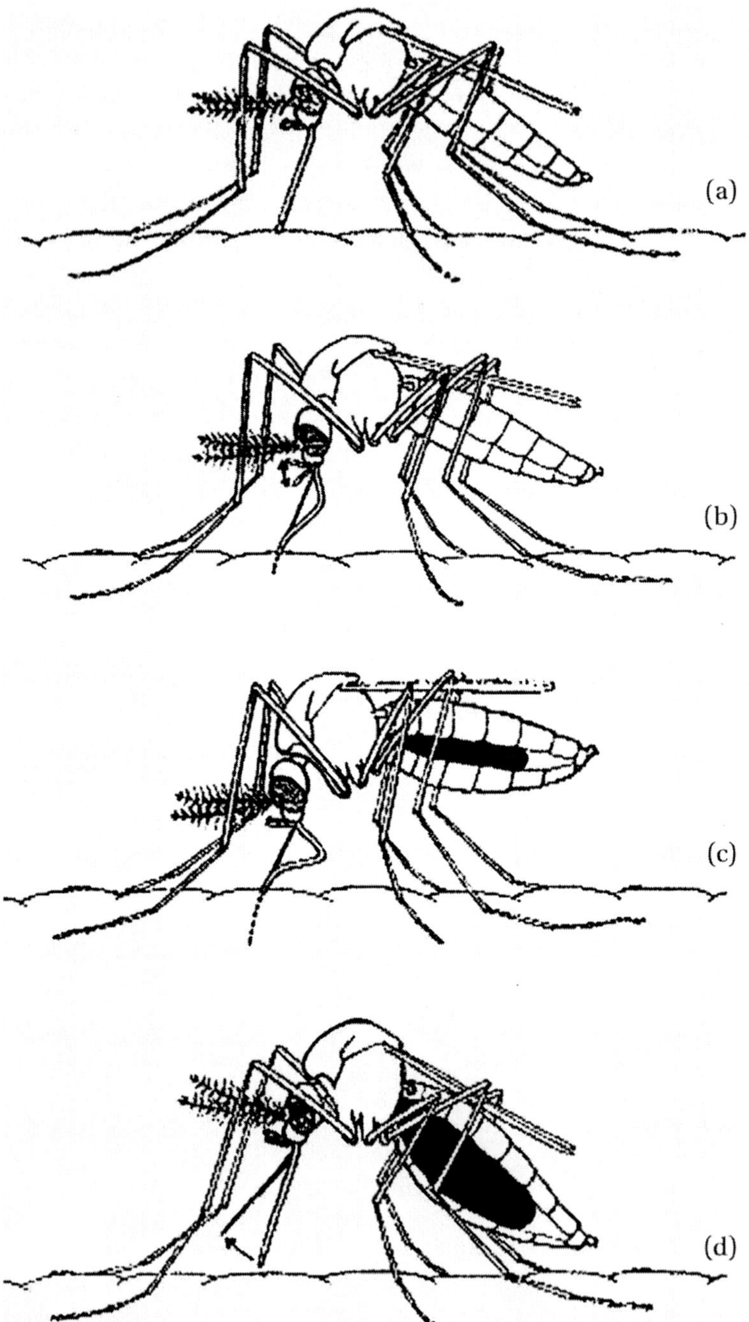

Fig. 2.9. Feeding mechanism of a mosquito: (a) a female mosquito bends and settles to probe, (b) it forces its mouthparts through the skin, (c) the mouthparts are guided deeper to suck blood with the support of labial sheath sticking outside host's body, and (d) after engorgement. (*Source:* Free Wikipedia)

There are mosquitoes that rest during the day and emerge in the dim light of dawn and evening. Some feed at night, other during bright daylight; still others, like the vector of La Crosse Encephalitis, are most active in late afternoon. After feeding on blood mosquitoes lay eggs in thousands of carefully selected places: tree holes, discarded tires, decorative fountains, flowerpot trays, elephant footprints, pitcher plants, empty cans, forest ponds, salt marshes, low spots in the yard; anywhere that water is likely to accumulate in the near future. It is said that some species of *Aedes* lay their eggs on dry ground that will be subject to future flooding, and the eggs may survive for as long as two years without water. The whole process from egg to blood-thirsty adult is accomplished in about ten days, and four to five days after her last blood meal, the female is ready to feed and lay eggs again (about 100 at a time).

At the juncture of inserting mouthparts in a host's body a more highly evolved chemistry comes into play. Mosquitoes inject chemicals that act as vasodilators, expanding the blood vessel to increase the flow. They also inject anti-clotting chemicals, to keep the tiny wound from sealing up before they have had its fill. Mixed in with these chemicals are parasites that the mosquitoes had earlier picked up while feeding on the host. The parasites are lodged in the mosquitoes' salivary glands, waiting for the mosquitoes to bite on their next victim. When the mosquitoes squirt their chemical payload into human body, out come the parasites. This is how mosquitoes spread diseases.

Mosquitoes feed from a wide range of different host animals including mammals, birds, reptiles, amphibians, fish, and even insects, arachnids and annelids. Blood feeding is a risky venture for insects because vertebrates are very large, attempting to blood feed could be lethal. However, the nutritional materials derived from blood can support present and future life for the mosquito, and all this fascinating and yet intricate activity is accomplished by the saliva present in the salivary glands of mosquitoes (Ribeiro, 1995). The saliva of mosquitoes contains a variety of pharmacologically active substances that counteract the normal hemostatic response to injury in vertebrate hosts. Bioactive salivary compounds exist in the salivary glands of mosquito vectors. These compounds have anticoagulant, antihemostatic and immunomodulatory properties that facilitate blood feeding. They can suppress host's innate and specific acquired immune defences. The fact that mosquito saliva can enhance pathogen transmission leads to the possibility that vaccinating the host against the components of vector saliva will inhibit pathogen transmission. Mosquitoes produce bioactive molecules in their salivary glands that are inoculated into host skin via saliva during feeding. The salivary bioactive molecules have antihemostatic and immunomodulatory functions. Such functions allow mosquitoes to live with and contend with the host immune system. Vasodilators in the saliva of *Aedes aegypti* are in fact tachykinins, which are peptides that have diverse pharmacological actions in the central nervous system, cardiovascular, and glandular tissues (Ribeiro, 1992). This vector secretes an active enzyme called apyrase that inhibits ADP-dependent platelet aggregation. The amino acid sequences obtained match the conceptual translation product of a cDNA clone isolated from an adult female salivary gland library. Sequence comparisons indicate similarities with a ubiquitous family of 5'-nucleotidases. The mosquito protein differs from other members of the family by lacking a carboxyl-terminal hydrophobic domain. The apparent conversion of a gene encoding an enzyme involved in a common metabolic event at the cellular level to a gene involved in the antihemostatic response of mosquitoes illustrates one way this particular insect has adapted to the challenges of blood-feeding. Mosquito anticoagulants seem to possess different potencies, too. For instance, the filariasis vector, *Culex quinquefasciatus*, exhibited an anticoagulant activity twice as high as that of *Aedes aegypti*. New World *Culex* mosquitoes are commonly thought of as bird feeders that have expanded their host range to include humans and other vertebrates. The observed differences between the two mosquito species may reflect the fact that bird hemostasis relies more heavily on coagulation than does mammalian hemostasis.

Bites of some mosquitoes itch sorely, *e.g.*, *Aedes, Armigeres* and *Mansonia* species etc. A mosquito's bite often leaves behind an imprint, evanescent though, of a slowly developing itching wheal. However, if it is a first mosquito bite ever, the roundish wheal fails to appear. The immune system in human body begins thereafter, and develops antibodies to recognize and flag the foreign chemicals. But after just one dose of mosquito spit, the familiar itchy bump left behind by subsequent bites is caused by an allergic reaction to all those chemicals the mosquitoes are squirting into the body. The immune system releases histamines, chemicals that make the cells in the lining of human's blood vessels spread apart. That lets fluid leak out into the skin. That fluid is carrying components of man's immune system that attacks and destroys the foreign chemicals from the mosquito. The extra fluid from the leaking blood vessels causes a bump, which irritates other receptors in the skin, leading to the itching. Children get bigger bumps from mosquito bites than adults because their immature immune systems respond with larger amounts of histamines. More leaking blood vessels equal bigger bumps, which mean more itching and scratching. However, following repeated mosquito biting, the human immune system also become more stronger and exhibits a different reaction to mosquito saliva instead of triggering histamines, another part of the immune system grabs onto the foreign chemicals and drags them away to macrophages, which destroy the chemicals in the mosquito's saliva.

Among mosquitoes, it has been established that only the females desire and extract blood. Male mosquitoes drink only sugary fluids such as flower nectar. Both in the wild and in the laboratory, mosquitoes will visit certain flowers and will feed on fruit placed in their cage. Since they (*Toxorhynchites*) vigorously probe the flowers of some plants, and can distinguish between different types of sugars, they play a certain role in the pollination activity of certain plants. The haematophagous females will also drink sugary fluids, but when hungry, if females are given a choice between sugar water and blood, they will always choose blood. If males are offered the same choice, they will always drink the sugar water as their mouthparts are not suited (or designed) for blood-sucking. Since male mosquitoes do not suck blood, they also do not transmit diseases. Like the males of many other insect families, they are important for just one reason - the mating, and then they become superfluous. The female usually needs to mate just once in her life, although she invariably blood feeds following each oviposition which may, under favourable conditions, go up to as many as six or more in her life time. She stores sperm in her body and fertilizes her eggs at the moment when she lays them.

Haematophagous mosquitoes do not really sting, in the sense of a hypodermic needle piercing the skin. Instead, they saw their way through, using four cutting stylets kept in a protective sheath along with a duct that carries anticoagulant into the wound and a tube that carries blood out. Mosquitoes fly into the wind, picking up scents which lead them to their victims. Just about everything attracts them, including the carbon dioxide we exhale. Mosquitoes will even suck the blood of birds, frogs, turtles, snakes and just about every warm-blooded animal. In fact, some mosquitoes prefer their sources of blood to be other than human. There are many species of mosquitoes that prefer the blood of birds or other animals (*e.g.*, many culicines). In fact, some species have been observed to feed on numerous mammalian groups, on a variety of reptilian species, and even on other kinds of insects. Anthropophilic mosquitoes are attracted to certain ranges of temperature and humidity, and to carbon dioxide from the exhalation of humans. They seem to prefer dark-coloured objects to light. Certain chemical smells may also come into play; for example, components of blood and sweat such as haemoglobin and amino acids. The best explanation of what brings the mosquito to its victim is that she simply flies upwind until she comes upon a potential victim exuding a "host beam" of warm, moist air laden with carbon dioxide. However, this piece of information is far from complete and a lot more research is needed to explore the facts. The mosquito proboscis (long tubular mouth parts used for feeding) consists of six different shafts. Four are cutting and piercing tools; a fifth transports blood from the host; and the sixth transports saliva, thought to act as an anticoagulant for the blood going in the other direction. The saliva also facilitates transportation of the organisms of malaria, yellow fever, dengue, and most of the other diseases for which mosquitoes are notoriously remembered. When a mosquito punctures ("bites") into the flesh, one usually feels an allergic reaction to the saliva, which causes the swelling and the itch. The fact that this reaction is allergic helps explain why some people suffer more than others when their skins are invaded. Some mosquito species are nocturnal, diurnal, or crepuscular (active at dawn and twilight). They also differ in their preferences for altitudes.

2.6. Reproduction

Mosquitoes exhibit holometabolous metamorphosis and therefore have a life cycle consisting of egg, larvae, pupa, and adult. Ovary ecdysteroidogenic hormone activates egg maturation in female mosquitoes (Fig. 2.10).

Adult female culicids feed according to their state in the gonotrophic cycle, and lay in a variety of habitats depending on the species. Following the oviposition, the mosquito will seek a blood meal. Blood ingestion begins and continues until stretch receptors in the abdomen signal that the female mosquito cannot hold anymore. This is referred to as 'distention-induced host-seeking inhibition'. Blood is then digested and egg maturation begins. Host-seeking is then further inhibited in 'oocyte-induced host-seeking inhibition'. Following the oviposition of this batch of eggs, the cycle begins again. Variations in the cycle occur in mosquito species which blood-feed more than once during egg maturation. Vertebrate blood contains essential proteins necessary for eggs to mature. Egg production is dependent upon amino acids and does not require the presence of other constituents in the blood. It has been found that female *Aedes aegypti* fed on an artificial diet high in protein developed equal amounts of eggs to those fed on bovine blood (Kogan, 1990). Female mosquitoes require the amino acids in blood proteins to mature their eggs. It is during the process of taking blood that the mosquito can transmit diseases such as malaria and dengue fever, which are re-emerging as very significant health problems. It is interesting to know that the female mosquito is not able to digest the proteins at the moment the blood meal is ingested because she has no active trypsin in her gut. Trypsin-synthesis requires a series of complex mechanisms following blood feeding. In the first four-six hours following a blood meal, the female gut produces a insulin-like special type of trypsin called 'early trypsin'. Although only small amounts of 'early trypsin' are made, the activity of 'early trypsin' plays a vital role in the regulation of digestive enzyme synthesis by the gut. Somehow 'early trypsin', acting on the meal proteins, produces a signal that induces translation of the several proteolytic enzymes required for blood meal protein digestion. The nature of this signal, which is not free amino acids, remains a secret that the mosquito has refused to share despite many attempts to identify it. The synthesis of 'early trypsin' itself is regulated in a unique manner.

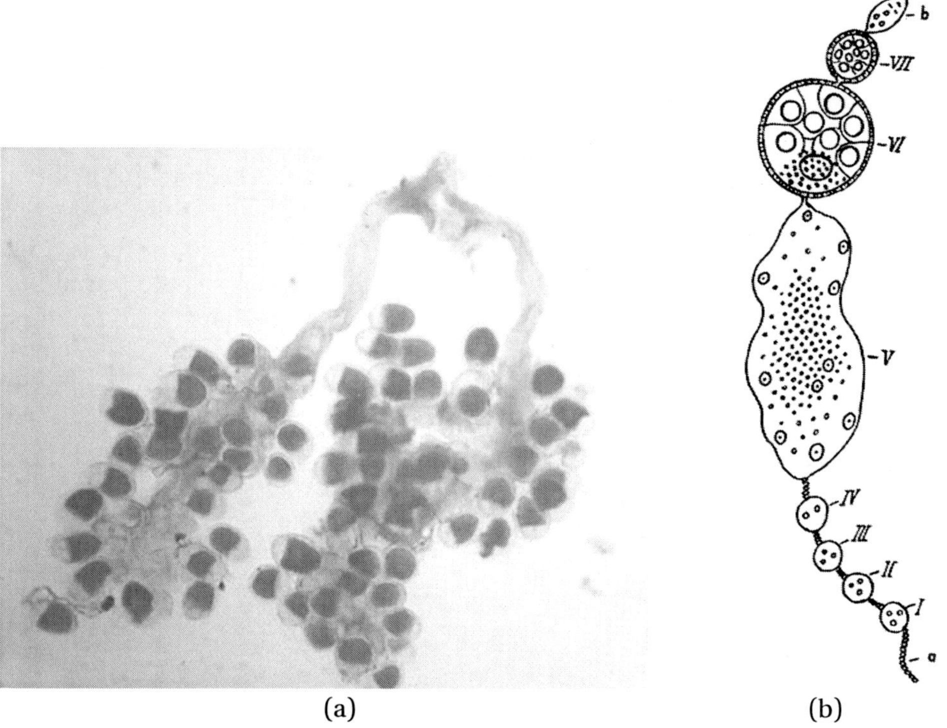

Fig. 2.10. (a) Ovaries with developing ova in *Anopheles sundaicus*, a major malaria vector in Andaman & Nicobar islands (India) (*Source:* Vidhya Thiruvoth), and (b) Follicular tube with follicles at fifth gonotrophic cycle of anopheline mosquito (I-IV - pigmented dilatations left by previous four follicles; V - degenerated fifth follicle; VI - sixth follicle about to develop; VII - undifferentiated seventh follicle; a – terminal stalk; b – germinal zone). (Adapted from Beklemishev *et al.*, 1959).

Transcription of the 'early trypsin' gene starts a few hours after the adult emerges from the pupal state. However, the 'early trypsin' mRNA is stored in the midgut epithelium and remains untranslated until a blood meal is taken. It seems that free amino acids in the blood are the signal for translation, apparently by providing amino acids to charge tRNAs in the gut, which in the absence of a blood meal are uncharged, thereby limiting translation. Thus, 'early trypsin' is part of a fascinating signalling system that conveys the presence of protein in the gut lumen to the epithelial cell. Once the signal is received, the epithelial cells can produce the enzymes required for protein digestion; in the absence of the signal the cell is quiescent. These regulatory mechanisms ensure that proteases are made only when needed. Given the importance of amino acids for producing the proteins of the egg, one might suspect that the mosquito would be quite efficient in using the amino acids. In fact, the contrary is true. A female mosquito uses only 10-15% of the amino acids to make egg proteins. Some of the rest are used to make egg lipids, and some are used to build energy reserves of the female. But more than 80% of the amino acid carbon is oxidized to CO_2 to provide the energy needed for egg production. Matters related to the partitioning of the meal proteins into various products and how the nutritional state of the female affects these processes are a major subject of research now. It is probable that females who had poor larval nutrition will need more than one blood meal to mature a batch of eggs, because the first or even the second meal is needed to build up the energy reserves effecting egg production. Under such circumstances the possibility of disease transmission increases, so it becomes important to understand how the female manages the blood meal. Because the female oxidizes 80% of the amino acids, she has a very large amount of ammonia to deal with. The female mosquito seems to place the ammonia in proline, which she stores in her hemolymph (blood), until she can convert the ammonia to urea or uric acid for secretion. This is a unique process, still not understood adequately, in mosquitoes and the hemolymph proline concentration can be as high as 80 mM.

♦ ♦ ♦ ♦ ♦ ♦

Chapter 3

BIOMEDICALLY IMPORTANT MOSQUITOES: *VECTOR DIVERSITY*

3.1. Introduction

No creature has touched directly the lives of more human beings than the mosquito (Spielman & D'Antonio, 2001). Mosquitoes are cosmopolitan in distribution and are found in all climatic zones and zoogeographical regions. There are hundreds of different species of mosquitoes - so many that even scientists don't agree on how they should be classified. Researchers have spent thousands of hours exploring, catching and staring at mosquitoes under magnifying glasses and microscopes, grouping them into genera (the plural of genus) and species. Scientists who dedicate their lives to learning about these little insects are actually doing important work, because each kind of mosquito has a slightly different secret for surviving cold, drought, predators and other dangers, and different types of mosquitoes carry different diseases. By classifying them into species and studying their habits, scientists can figure out the best way to get rid of the ones that carry the worst diseases. Though tentatively figured out, yet there seems to be a little over 3,500 species of mosquitoes worldwide, under 112 genera. Listed below are the major genera of mosquitoes found in India, with some of the important vector species within each genus indicated.

1. Anopheles
   i. *An. culicifacies* ii. *An. stephensi*
   iii. *An. fluviatilis* iv. *An. minimus*
   v. *An. baimaii* vi. *An. sundaicus*
   vii. *An. philippinensis*
2. Aedes
   i. *Ae. aegypti* ii. *Ae. albopictus*
3. Culex
   i. *Cx. quinquefasciatus* ii. *Cx. tritaeniorhynchus*

iii. *Cx. vishnui*　　iv. *Cx. pseudovishnui*

v. *Cx. gelidus*

4. Mansonia

i. *Ma. uniformis*　　ii. *Ma. annularis*

iii. *Ma. indiana*

5. Armigeres

i. *Ar. subalbatus*

Illustrations of the main differences between the eggs, larvae, pupae and adults of three common genera, *Anopheles, Aedes* and *Culex* are given in following pages on respective places of description.

3.2. Biology of important mosquitoes

Mosquitoes are classified under the family Culicidae of the Order Diptera, and are by far the most thoroughly researched haematophagous insects. They concern humans more than any other blood-sucking arthropod mainly because of their involvement in both transmitting numerous deadly and/or debilitating diseases (malaria, filariasis, dengue, Japanese encephalitis, Rift valley fever, Chikungunya, Zika virus, West Nile virus and O'Nyong-nyong etc. etc.) and creating biting nuisance of great public health significance. Consequently, out of the many species of blood-sucking insects, mosquitoes particularly belonging to genera *Anopheles, Culex, Mansonia* and *Aedes* as well as, to some extent, *Armegeres,* are amongst the biomedically most important species as they are responsible for nearly 10% of all the sickness of humans. Although, there are about 3500 species all over the world, however only fewer than 100 species under 8 genera are vectors of diseases (Service, 2008). These vectors are organized in two of the three subfamilies, Anophelinae and Culicinae, while the third subfamily Toxorhynchitinae has members with mouthparts suited to feed on plant sap or juice only. Some of important genera of mosquitoes are described below:

3.2.1. Anopheles mosquitoes

From medical standpoint this genus is the most important of all genera of mosquitoes. While many species are serious vectors of the deadly infection, malaria, to humans, several others are painful biters as well as nuisance mosquitoes. Only Anopheles species can transmit human malaria parasites, i.e., *Plasmodium vivax, P. falciparum, P. ovale, P. malariae* and *P. knowlesi.* There are over 400 species of Anopheles, of which nearly 65 species occur in India (Nagpal & Sharma, 1995; Tyagi *et al.*, 2015). Species of this genus are found in all geographical regions of the world, and the Oriental Region is the richest (Rao, 1984).

Important taxonomic and biological features of each larval stage, pupa and adult are described below:

(i) *The Egg*: Eggs of Anopheles mosquitoes are boat-shaped with an upper deck having a micropile at one end and a fringe all around, and usually with lateral floats with transverse striations called ridges. Eggs are laid singly. In optimal temperature it takes about 36-48 hours for eggs to hatch (Fig. 3.2).

(ii) *The Larva*: This is very agile from the time it hatches out. There are four larval instars before it converts into pupal stage. It rests characteristically parallel to the surface of water. The head of a larva is capable to rotate at 180° to enable it feed more intensely and vigorously. Characteristic palmate hairs, present in many species, occur on both thorax and abdomen. Two (one on each side) reversible organs, called Nuttal's organs, occur on the first segment of the thorax. These are best seen in the live specimens as they are fleshy and protrusible. Larvae have a dorsal spiracular plate supporting two spiracles meant for bringing about respiration. The total duration taken by larvae under normal conditions of temperature ranges from 5-7 days.

(iii) *The Pupa*: This has a peculiar 'comma' form with highly active up and down vertical movements. The paddle is oval in outline. It may be either undivided or only slightly invaginated, with the tip having a pair of setae one of which is ventral and sub-apical. On the abdominal segments 3rd to 7th there are located spine-like seta No. 9 (Belkin's system) at the postero-lateral corners usually with branches on 8th abdominal segment. The trumpet, located on the top of the cephalothorax, often bears at least one cleft. Pupal development completes in about 48 hours.

(iv) *The Adult*. Soon after the adult emerges from pupae it flies away to take shelter in some suitable place, where it rests for some time before taking its first blood meal. Males generally emerge earlier than females, and have shorter lifespan than their counterparts (Fig. 3.1).

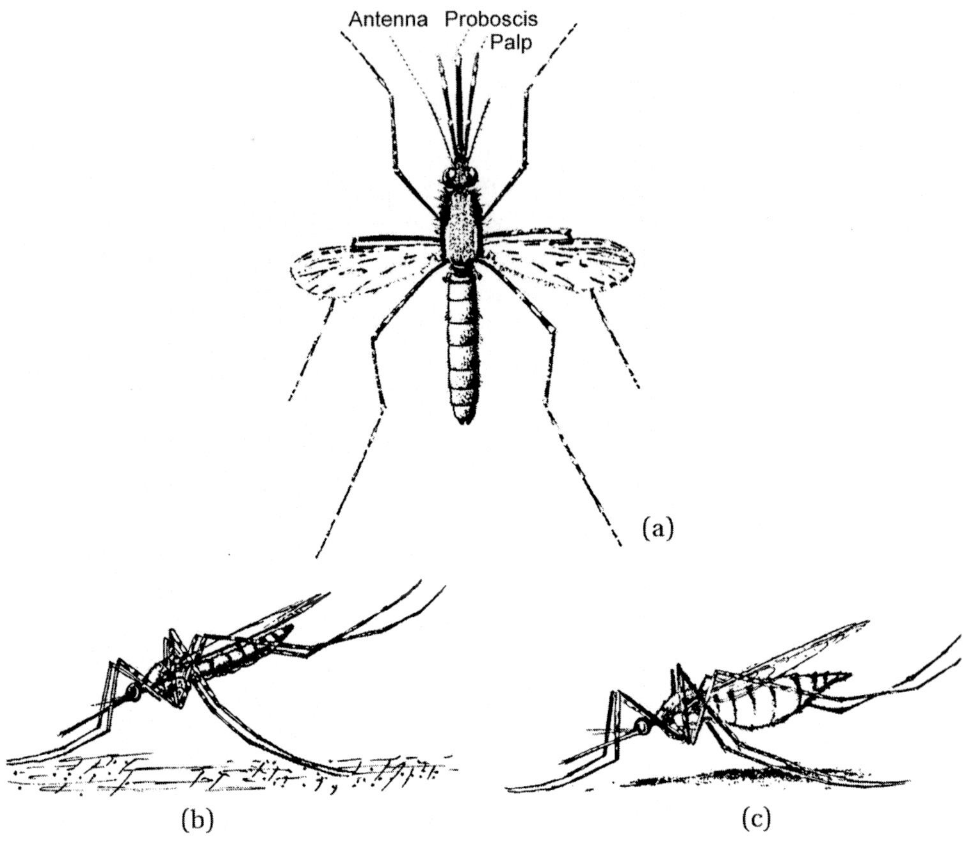

Fig. 3.1. Anopheles mosquito: (a| general format, (b) at rest, (c) after engorgement. (*Source*: Free Wikipedia)

Anopheline adults are identifiable from the rest of the mosquitoes due to their very characteristic colour patterns mostly on the palpi, wings and legs, beside other prominent traits as seen in their posture at rest etc. The palpi of both sexes are of similar length, almost equal to that of the proboscis. In males, however, the two distal segments in palpi are enlarged giving them a club-like appearance. The posterior edge of scutellum is only slightly curved and is continuous without any marked division into lobes. Hairs on these lobes are in continuous series.

The axes of head, thorax and abdomen are nearly in a straight line. Anopheline (Spielman & D'Antonio, 2001) mosquitoes rest on an object in a characteristic posture. The angle of resting, varying from species to species, is close to 60° with the tip of the abdomen directed away from the surface (*e.g., An. hyrcanus, An. barbirostris*). In a few species like *An. culicifacies, An. culiciformis, An. sintoni* etc. the body is almost parallel to the surface, giving adult a semblance of a Culex-like posture.

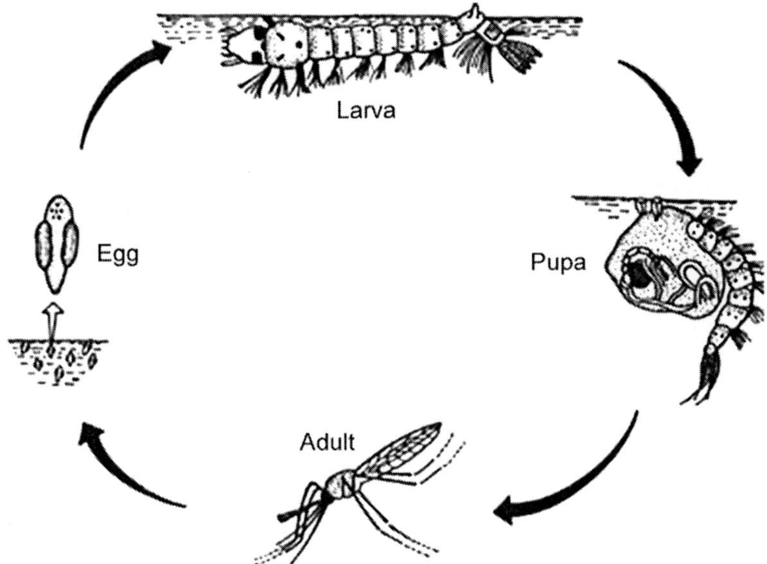

Fig. 3.2. Different developmental stages in the life cycle of an Anopheles mosquito: eggs, larva, pupa, and adult. (*Source*: Free Wikipedia)

Of the three acini of the salivary glands of each side, the middle is sac-like instead of being tubular as is noticed in culicines. Wings are spotted or unspotted; the former condition being very common, though. On the other hand, members of subgenus Anopheles, like *An. aitkenii*, have completely dark wings. Males have only one claw in the fore tarsi. Females have only one spermatheca.

Most vector species are endophilic and endophagic in behaviour (*An. culicifacies, An. subpictus, An stephensi*), although a few are sylvatic in nature, too (*An dirus, An. pulcherrimus*). Although, there is no direct proof of hibernation exhibited by any Indian species, there are species in colder temperate regions elsewhere in the world which manifest overwintering. In India, yet, *An subpictus* is suspected to possess the potential to overwinter.

Some species exhibit 'sibling species complexes", a phenomenon which carry tremendous medical significance in the sense that while some of them may be vectors, others are either less vectorial or are not able to transmit the pathogen at all. The most widely examined species complexes belong to *An, gambiae* in Africa and *An. culicfacies* in India (Subbarao *et al.*, 1999).

In India as many as eight Anopheles species complexes exist (Table 3.1). Strangely, An. stephensi has two races, forms or subspecies but not sibling species. Tyagi *et al.* (1991) has advocated presence of two 'forms', bigger and smaller, in the Thar Desert ecosystem - the original home in India for the species.

Table 3.1 Sibling species complexes among Indian anophelines (*Source:* Subbarao *et al.*, 1999).

Species Complex	No. of sibling species identified	Sibling species found in India
An. annularis	2	A, B
An. culicifacies	5	A, B, C, D, E
An. dirus	7	D, E (D in north-eastern States and E in Karnataka)
An, fluviatilis	4	S, T, U, V
An. minimus	3	A
An. sundaicus	4	A new cytotype D
An. philippinensis-nivipes	3	nivipes A, B
An. subpictus	4	A, B, C, D

Knowledge of species complexes is highly vital particularly to malaria control strategy, especially when target-specific selective and sustainable vector control is urgently needed. It is important that vector control, as a part of the global as well as the regional malaria control strategy, should succeed. Its success would depend on our total knowledge of biology, ecology and behaviour, as also governed by genetics, of the vector species involved and of the vector control options and their selective use. Most of anopheles known to transmit malaria in India have been determined as species complexes. Members of a species complex, commonly known as sibling species, are reproductively isolated evolutionary units with distinct gene pools and, therefore, differ in the biological characteristics. Differences in biological characteristics of members of the complexes have

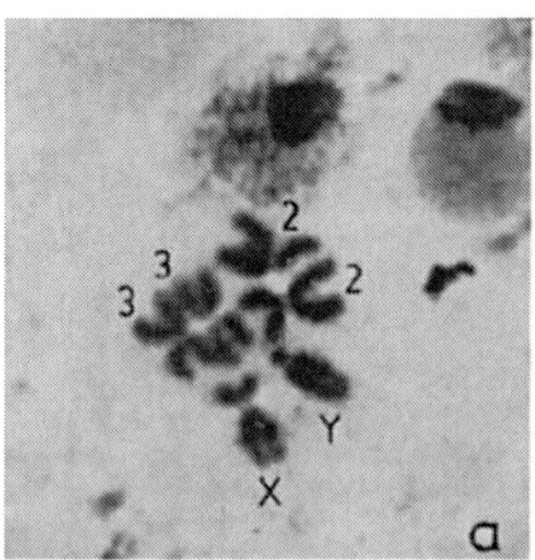

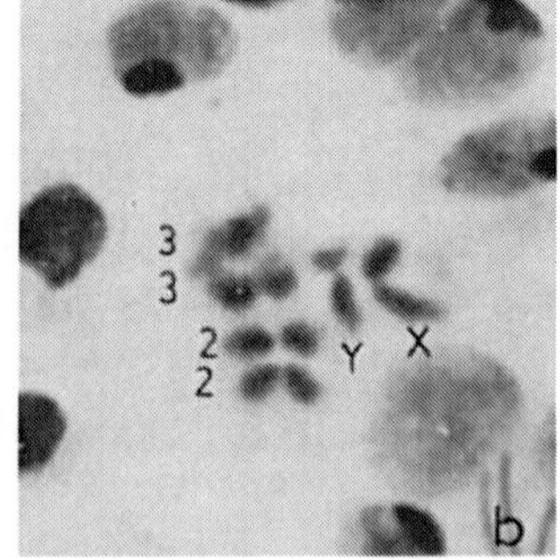

Fig. 3.3. Male mitotic karyotypes of *An. culicifacies* from Rameswaram island (a) with acrocentric Y-chromosome (species B); (b) with sub-centric Y-chromosome (Species E). (*Source:* WHOSEARO, 2007).

an important bearing on malaria transmission dynamics. It is therefore imperative to determine sibling species composition and their bionomics as well as their roles in the transmission of malaria. At cellular level, sibling species among mosquitoes were first determined through cyotaxonomic techniques where as a rule either chromosomes under mitotic and/or meiotic divisions in ova of the developing ovaries in females or the testes in males became object (Fig. 3.3), but more characteristically, these were the polytene chromosomes from the germinal layer of the developing ovaries or the salivary glands which exhibited more clearly all the necessary features of identification (Fig. 3.4). Later molecular techniques have replaced the cytotaxonomic techniques because they are quicker and more accurate (Fig. 3.5). Time taken to arrive a decision on the correct determination of a sibling species by any of these methods, however, is still a crucial factor.

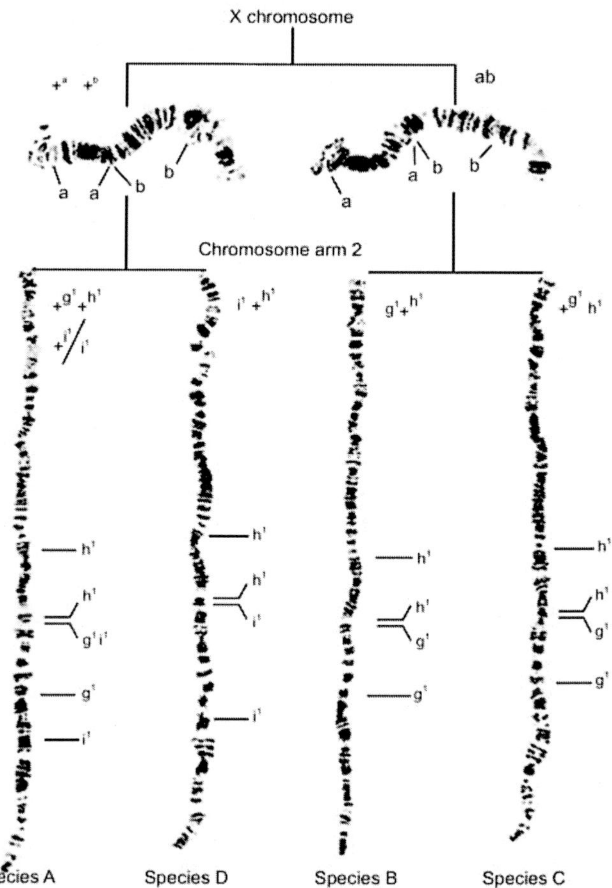

Fig. 3.4. Schematic representation of polytene chromosomes of *Anopheles culicifacies* sibling species (*Source:* WHOSEARO, 2007).

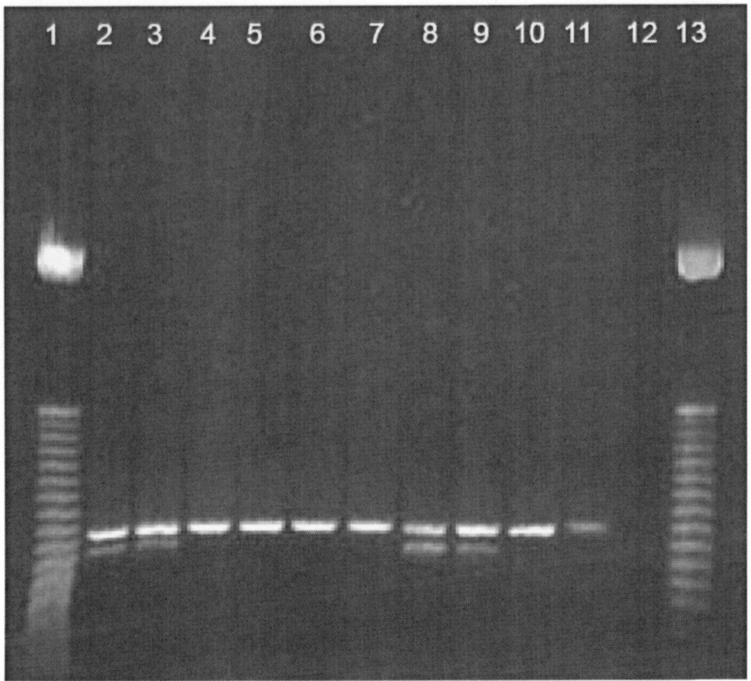

Fig. 3.5. PCR assay using primers designed from D2 region of 28S rDNA which differentiates species A and D from species B,C and E of *An. culicifacies.* Lanes 1 & 13: DNA ladder; Lanes 2 & 3: species A; Lanes 4 & 5: species B, Lanes 6 & 7: species C; Lanes 8 & 9: species D; Lanes 10 & 11: species E, Lane 12: negative control. (*Source:* WHOSEARO, 2007).

3.2.2. Aedes mosquitoes

Several species of the genus Aedes are of tremendous medical and public health importance, *e.g., Ae. aegypti, Ae. albopictus, Ae. sollicitans, Ae. nigramaculis, Ae. vexans* etc. In as far as India is concerned, the most important vector species are *Ae. aegypti* and *Ae. albopictus,* due to their involvement in transmission of dengue, chikungunya and Zika fever. In Africa and South America where deadly Yellow Fever prevails the vector is *Ae. aegypti.*

Biology and life cycle of *Aedes aegypti* are described below:

Main biological and taxonomical features are described for the two important vector species (i.e., *Ae. aegypti* and *Ae. albopictus*). In general, the biology of *Ae. aegypti* follows the pattern of other culicid mosquitoes. The life cycle is completed through four distinct stages; egg, larva, pupa and adult. The larval development passes through four instars, each undergoing progressive increase in size till the pupal stage. Pupae and adults do not show any growth. Only female adults of Ae. aegypli feed on blood, sugar and water for maturation of eggs, energy requirements and maintenance of the osmotic balance of the body (Fig. 3.6).

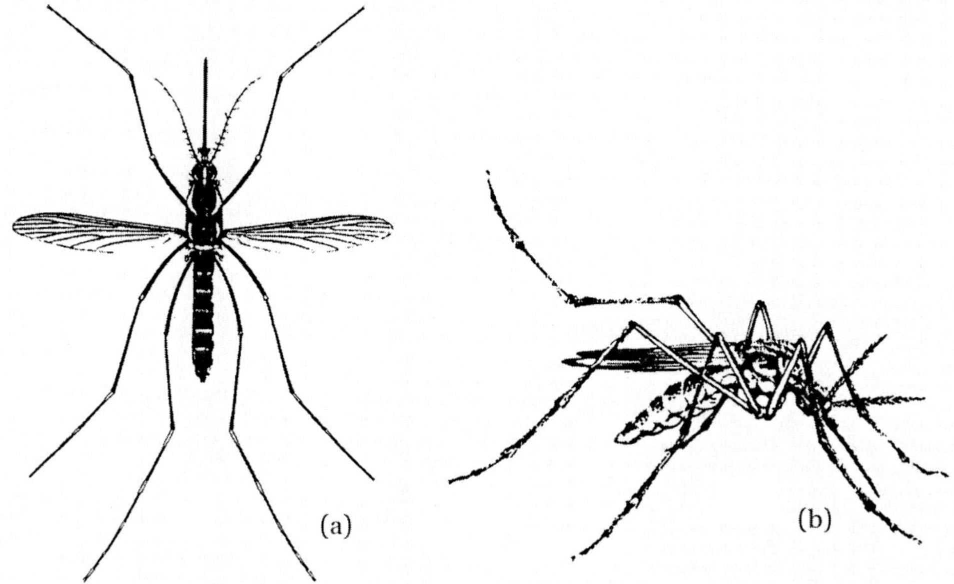

Fig. 3.6. Female *Aedes aegypti* mosquito: (a) dorsal, and (b) lateral views. (*Source*: Free Wikipedia).

(i) *The Egg:* Eggs are small, about 1 mm in length, torpedo shaped and deeply black in, although when freshly laid they are translucent and pure white. The eggs are capable to remain viable for weeks or even months and can withstand prolonged desiccation. Like eggs of Ochlerotatus mosquitoes, Aedes mosquito eggs, too, are more resistant to drying out (in fact, some complete drying out before the eggs will hatch) and hatch only when flooded with water (salt water high tides, irrigated pastures, tree holes flooded by rains, flooded stream bottoms). Eggs house fully developed larvae ready to hatch. The head of the larva occupies nearly one-third of the egg-volume, while the siphon and other terminal parts fill most of the remaining posterior pole. The egg has very characteristic polygonal markings, lined with a network of milky white membranous channels filled with air at the base of the exochorion. Beneath exochorion is a dense, tough and hard endochorion membrane which gives the egg its shape and rigidity.

(ii) *The Larva*: As in all other mosquitoes, there are four larval instars, each instar ending in a moult or ecdysis. The first instar larva hatches out from the egg by first widening the transverse crack at the dorsal side anteriorly. Finally the larva breaks the eggs by means of a thorn-shaped egg breaker. Initially the freshly hatched larva is unable to move but shortly the young larva wriggles its way to the water surface to make the contact with the free air. lt takes about 2 -3 days at 27-29°C for the laiya to hatch out from the egg. However, the larva may take few minutes to several hours to emerge at the surface. The first instar larva is identifiable on the basis of the presence of the egg breaker on the head. The larva carries a characteristic siphon. The pectin spine arising on the soft membrane number 4-5. Siphon carries the trachea. The comb teeth, which are delicate, number only five.

Table 3.2. Some identification characteristics of various larval instars as well as their time of development. (Tyagi, 2004)

Larval instar	Transverse diameter of head (mm)	No. of pectin teeth	Period for completion of four instars (in hrs)
1st	0.3	6	24
2nd	0.45	7-10	24
3rd	0.65	12-15	24
4th	0.95	12-20	96-240

The second and third instars can be identified by their size and general appearance. Little change occurs until fourth instar when the development of the future pupal structure begins. The compound eyes or the optic lobes begin to form in the late third instar. The imaginal buds at the bases of the pleural hairs become apparent in the second instar, and are quite conspicuous in the third, though their development is a feature of the fourth instar. The gonads are already visible in the 6th abdominal segment in the third instar, when also the buds of the developing gonoxites become apparent. During the fourth instar period a condition is reached in which the larva is little more than the roughly sculptured body of the future adult disguised in the larval skin. Such a condition has been termed a prepupa. The duration of the larval stages is influenced by the water temperature. The most common period lies between 6 and 8 days (Fig. 3.7).

(iii) *The Pupa*: The pupa of *Ae. aegypti* resembles most of other culicines. However, it differs from the anopheline pupa in having more rounded body and cylindrical breathing trumpets. The main characters used to identify the pupa of *Ae. aegypti* from other culicine pupae are: (a) form of tlre respiratory trumpets, (b) arrangement and characters of the hairs, particularly those in the abdominal region, and (c) characters of the tail fins or paddles. The pupal body consists of a large globular anterior region, the cephalothorax, and a narrower abdomen which is normally kept flexed under the former to propel the insect when swimming.

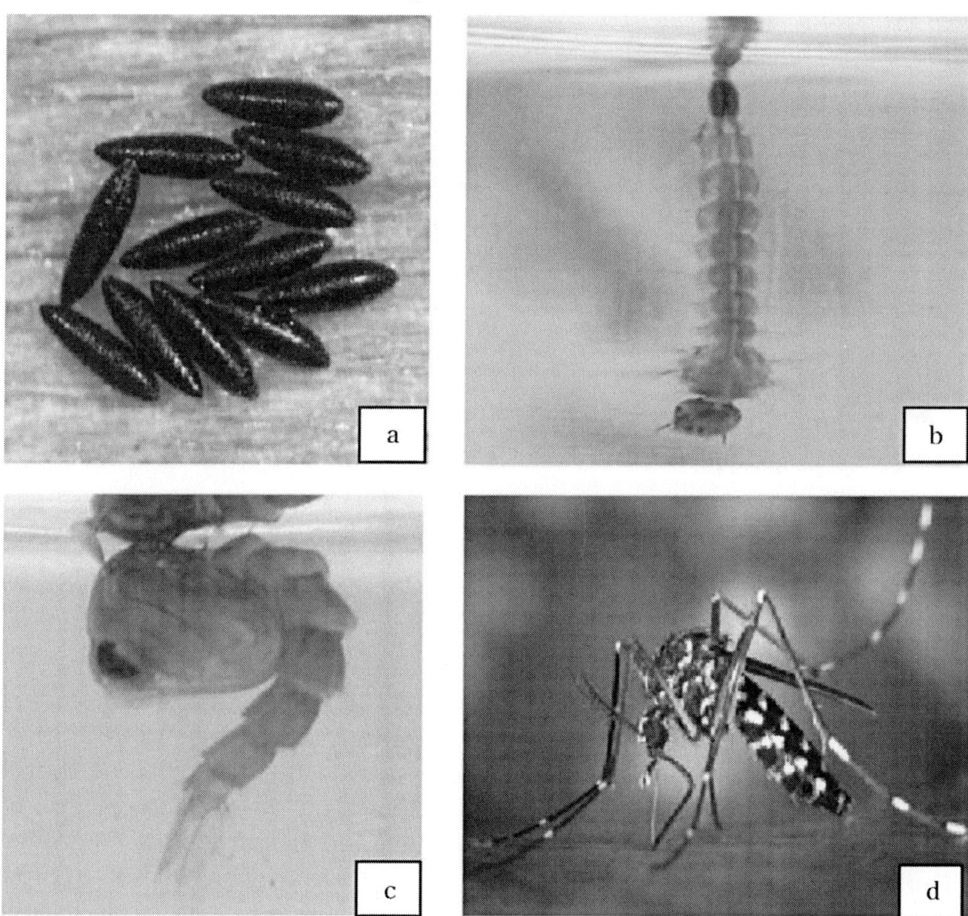

Fig. 3.7. Different developmental stages of an aedine mosquito, *Aedes albopiclus*: [a] Eggs, (b) Larva, (c) Pupa, and (d) Adult female. (*Source*: Free Wikipedia)

(iv) *The Adult*: It is a small dark structure easily identifiable by the lyre-shaped, silvery white lines on the thorax and the white bands on the tarsal segments. The adult is both vector as well as pest when present in large numbers and therefore deserve special attention in medical entomology. This is a thoroughly domesticated mosquito and breeds almost exclusively in artificial containers in and around human habitations. The females lay their eggs singly on the water just at the margin, or on the sides of the container above the water line. Given the opportunity, the adult prefers human blood to that of other animals. Adults readily enter houses to find suitable hosts. The adult quietly hails to settle for feeding, preferring to bite around the ankles, under shirtsleeves, or in the back of the neck. Males can be easily distinguished from females on the basis of antennae and maxillary palps (Fig 3.8).

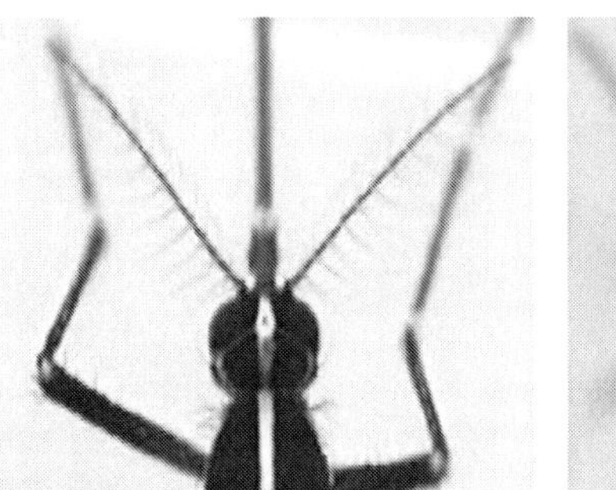
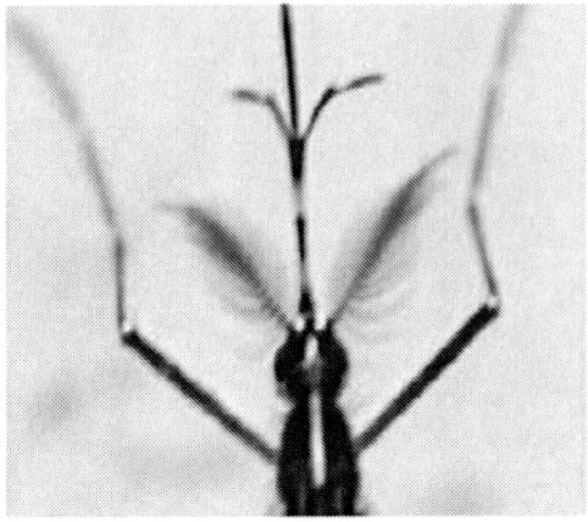

Fig 3.8. Mouthparts and antennal differences between female and male mosquitoes of *Ae. albopictus*. (*Source*: Free Wikipedia)

Like *Ochlerotatus*, the *Aedes* mosquitoes are also painful and persistent biters. They search for a blood meal early in the morning, at dusk (crepuscular feeders) and into the evening. Some are diurnal (day-time biters) especially on cloudy days and in shaded areas. They are strong flyers and are known to fly many miles from their breeding sources.

The adult of *Ae. aegypti* resembles markedly to that of *Ae. albopictus*, commonly called as the Asian Tiger mosquito, another mosquito of great medical significance, except that the lyre-shaped markings on the thorax are replaced by a stripe of silvery-white scales on the dorsum of thorax. *Aedes albopiclus* is a forest dwelling species and has adequately adapted itself to urban environments as well. A native of Asia, it has spread to most tropical parts of the globe. ln Kerala State, southern India, it breeds abundantly in the latex collecting cups associated with rubber plantations, besides cocoa and coconut shells. It has the nasty habit of displacing other species, including *Ae. aegypti*, from their breeding strongholds and has fast adapted to breed in all types of discarded containers, including tyres and littered plastic tea-cups (Hiriyan *et al.*, 2004).

3.2.3. Culex mosquitoes

The genus Culex comprises several medically important species, such as *Cx. pipiens, Cx. quinquefasciatus, Cx. tarsalis, Cx. nigripalpus, Cx.*

tritaeniorhynchus, Cx. bitaeniorhynchus, Cx. pseudovishnui, Cx. restans and *Cx. salinarius*. In Indian context, of greatest importance are *Cx. quinquefasciatus* and *Cx. tritaeniorhynchus* which, besides being serious pests, are in the first place responsible for transmitting bancroftian filariasis and Japanese encephalitis, respectively. These are brown coloured mosquitoes of medium size and like the rest of their group these have four stages of development in their life cycle (Figs. 3.9, 3.10).

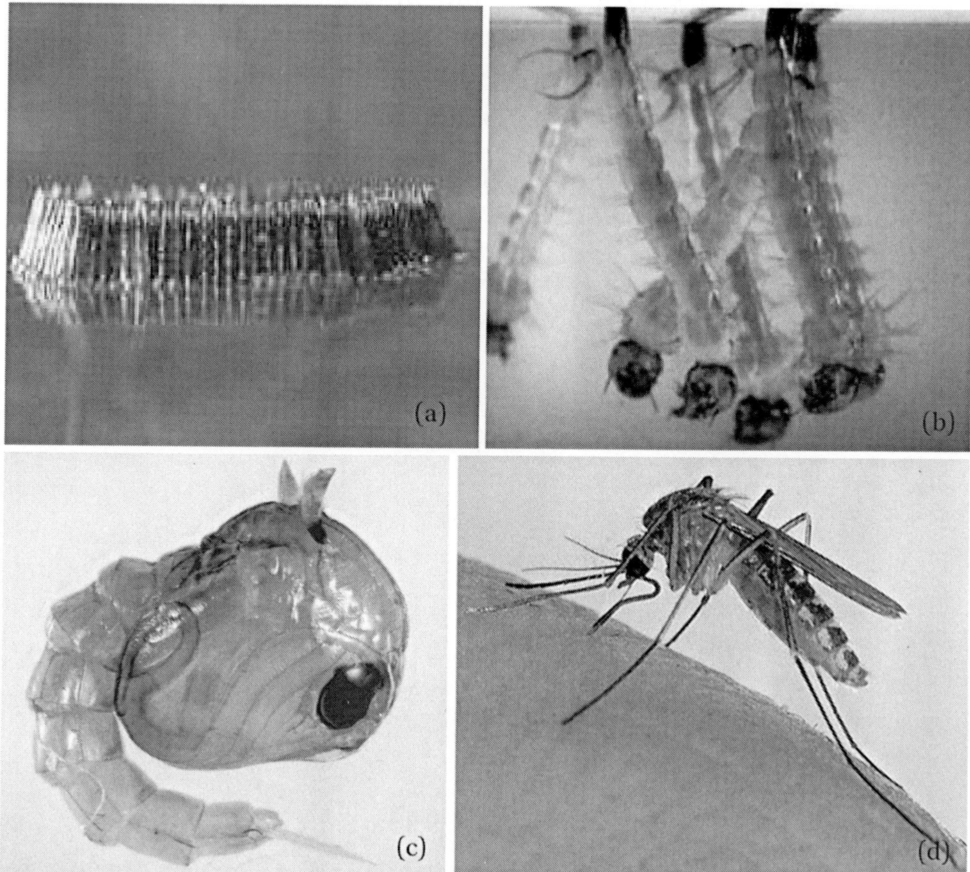

Fig. 3.9. The four stages of life cycle of a Culex mosquito; (a) Eggs (arranged in a egg-raft), (b) larva, (c) pupa, and (d) adult (in feeding process). (*Source*: Free Wikipedia)

(i) *The Egg*: Many mosquitoes, such as *Culex quinquefasciatus*, lay their eggs on the surface of fresh or stagnant water. Culex mosquitoes usually lay their eggs at night over a period of time, sticking them together to form a raft of 100 to 300 eggs. A raft of eggs look like a speck of soot floating on the water and is about 1/4 inch long and 1/8 inch wide. A female mosquito may lay a raft of eggs every third night during its life span. Like the eggs of Anopheles and Mansonia mosquitoes, Culex eggs too are more susceptible to long periods of drying out.

(ii) *The Larva*: Tiny mosquito larvae emerge from the eggs within 24-48 hours almost in unison. Mosquito larvae, commonly called "wrigglers", live in water from four to 14 days depending on water temperature. They come to the surface at frequent intervals to obtain oxygen through a breathing tube called a siphon. They tend to feed constantly since maturation requires a huge amount of energy and food. They hang upside down at the water surface with the breathing tube up and the brushes by their mouth filtering anything small enough to be eaten toward their mouths to nourish the growing larva. They feed on algae, plankton, fungi and bacteria and other microorganisms.

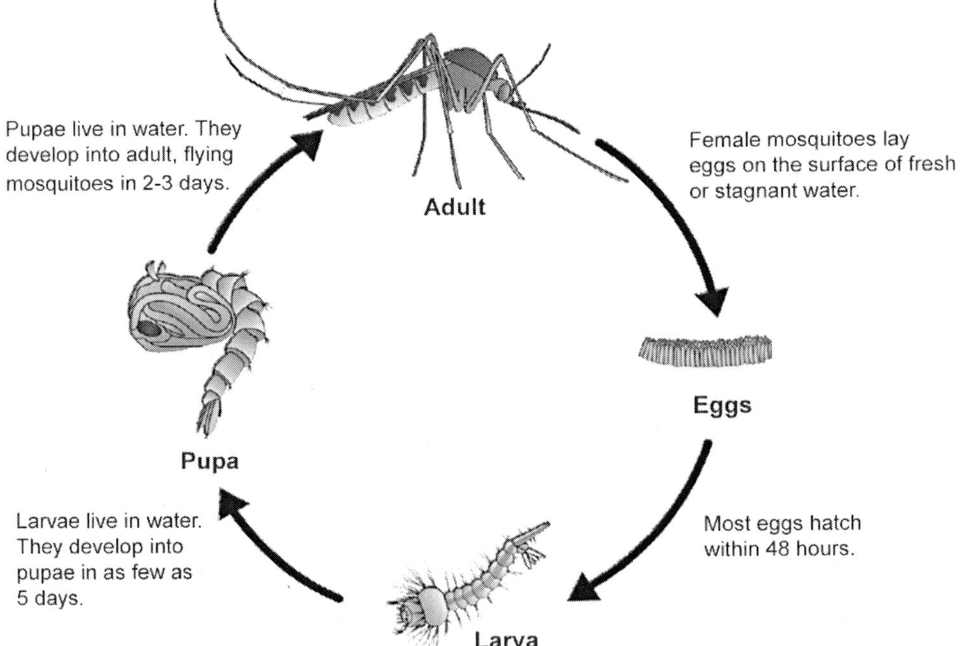

Fig. 3.10. Life cycle of a Culex mosquito. (*Source*: Free Wikipedia)

(iii) *The Pupa*: A typical culicine pupa (cf. *Aedes* mosquito pupa).

(iv) *The Adult*: *Culex quinquefasciatus* is fondly referred to as house mosquito due to its abundant occurrence in houses. It breeds in cesspools, storm-sewer catch basins, street gutters, open septic tanks and effluent from sewage disposal plants. Adults are active only at night. When disturbed, they can be easily spotted flying out of their hideouts in and around houses, outbuildings and other shelters. *Culex tritaeniorhynchus* breeds mainly in rice fields and it is both an exophilic and exophagic mosquito.

Culex mosquitoes are painful and persistent biters also, but prefer to attack at dusk and after dark. They readily enter dwellings for blood meals. Domestic and wild birds usually are preferred over man, cows and horses. *Culex nigripalpus* is known to transmit St. Louis encephalitis to man in the USA.

Culex mosquitoes are generally weak flyers and do not move far from home, although they have been known to fly up to 4 km. They live only a few weeks during the warm summer months. Those females that emerge in late summer search for sheltered areas where they "hibernate" until spring. Warm weather brings them out again in search of water, on which to lay their eggs. *Culex pipiens* and *Cx. quinquefasciatus* are found in most of the tropical countries. All culicines generally rest with their entire body more or less close to the surface. The scales are narrow and not metallic coloured. There are no distinct patterns or markings on the body. The abdomen ends bluntly. Prespiracular and postspiracular areas are without bristles. Forelegs in males bear toothed claws. In females, the claws are there but simple. The tarsi in females have large pulvilli.

3.2.4. Mansonia and Coquillettidia mosquitoes

Large sized, yellowish brown to dark brown mosquitoes with very painful bites. Mansonia occurs in all zoogeographical regions, while *Coquillettidia* occurs mainly in Afro-tropical region but also in Nearctic, Neotropical, Oriental and Australasian regions (Fig. 3.11).

Fig. 3.11. Mansonia and Coquillettidia mosquitoes. (*Source*: Free Wikipedia)

Mansonia mosquitoes are characterized by possessing postspiracular setae, but no prespiracular setae are found. They are in the habit of biting and resting both inside and outside the houses. They breed in the heavily vegetated pond water growing aquatic plants such as Pistia spp. etc. Larvae respire on air-conducting parts of the aquatic plants. *Coquillettidia* mosquitoes have neither pre- or postspiracular setae, but have an overall yellow colour. The adult body has a covering of broad scales, which may form fine speckled patterns of colour on the wings. The abdomen is blunt ended. The scales on the wings are very broad. The claws are simple and pulvilli are small or even absent. Some species like *Coquillettidia perturbans* breeds in marshes, ponds and lakes having a thick growth of aquatic vegetation. Larval development is generally slow, lasting several months. Just like that in *Mansonia* species, larvae in this species too breathe by penetrating air conducting plant tissues with their breathing tubes. They readily enter the house and bite viciously. Some of them, such as *Ma. annullifera* and *Ma. uniformis* are vectors of human lymphatic filariasis caused by *Brugia malayi* ln India, this type of lymphatic filariasis is mostly restricted in Kerala State.

3.2.5. Haemagogus mosquitoes

These mosquitoes occur in Neotropical region only where they are of much importance in public heath. The females of Haemagogus mosquitoes are day-flying and most probably suck human blood. These mosquitoes are identifiable due to certain specific characteristics. Adults are densely scaly mosquitoes, whose general colouration is black with brilliant silvery markings. Scales on their wings are narrow, but they are fairly broad on the body. The anterior pronotal lobes are large, collar like, closely approximated behind the head and nearly meet at the top. The abdomen of these mosquitoes is blunt ended. Large part of dorsum of thorax without bristles; none present even immediately in front of scutellum. Forelegs of the sexes possess toothed claws. There are no prespiracular or postspiracular areas. Proboscis is slender, distinctly longer than fore femur. The palpi not more than $1/6^{th}$ of length of proboscis, in males these likewise are shorter than the proboscis.

3.2.6. Sabethes mosquitoes

Members of this genus occur only in Neotropical region. They are identifiable due to scutum and abdomen being covered with scales having a metallic colour. The scales of wings are narrow, but on the body they are fairly broad. The abdomen is blunt ended. The anterior pronotal lobes are very large. The mid or hindlegs have a conspicuous area of long scales in the shape of a paddle. Prespiracular bristles are present but there are no postspiracular bristles. There is a group of bristles on the postnotum. The palps of males are shorter than the proboscis.

3.2.7. Psorophora mosquitoes

This genus has several species which are of immense public health importance largely due to their nuisance value particularly when in large numbers. Many species like *P. ciliata* and *P. columbaie* are large sized with shaggy legs. The scales on body and wings are narrow and without metallic colour. Prespiracular and postspiracular bristles are present. The abdomen is pointed at the tip. Both sexes have forelegs with toothed claws. The male palps are longer than the proboscis. They are vicious biter and because of their gigantic appearance they often offer a frightening look. Some species do not lay eggs directly in water but on the ground that is subject to flooding from rainfall, overflow or irrigation. They breed in temporary pools, irrigation ditches and seepage pools, and larvae survive predating on other aquatic insects including mosquito larvae. Members of this genus occur only in the Neartic and Neotropical regions. Like Coquillettidia and Mansonia mosquitoes, the Psorophora mosquitoes too are becoming more pestiferous as an ever-expanding human population invades their natural habitats.

3.2.8. Culiseta mosquitoes

Found in all zoogeographical regions in the world, they are identifiable due to presence of narrow scales on the body and wings which are non-metallic in colour; there are, however, coloured patterns on the body. The abdomen is blunt ended. Prespiracular bristles alone are present. Forelegs of males have toothed claws; in female's forelegs they are simple. The stem vein (base of subcostal vein) of the wing usually has bristles on the ventral surface. Most species are moderate to highly aggressive biters, attacking in the evening hours or in the shade during the day.

3.2.9. Armigeres mosquitoes

This is a very painful biter, both in day and at dusk, and occurs mainly in the Oriental region, but a few species are found in Australian region and one as far as Japan. A medium sized to rather large-sized mosquito it tends to seek blood by attacking and obstinate feeding. Mosquitoes of this genus are not overtly ornamental and resemble in general appearance and structure those of Aedes (Subgenus Stegomyia). Scales of head broad and flat, a few upright scales on nape, at least in some species, scutellar scales are all broad and flat. Proboscis rather stout, slightly curved downwards towards tip, the curved part somewhat laterally flattened. Palpi of male as long as, or longer than, proboscis, of 3-segments, the last two of which are long, slender, turned upwards, and without hair-tufts. Palpi of male of variable length, form about 1/2 to 3/4 of length of proboscis. Antennae of male plumose, the plume-hairs projecting fairly evenly round shaft. Abdomen of female somewhat tapering, segment 8 partially retractile, cerci short. Coxite of male hypopygium with lobe on inner surface bearing spines. Style of variable form, but usually with spines along apical part or at tip. Phallosome divided into small lateral plates with teeth. Paraprocts without teeth or hairs at crown. The only character distinguishing the genus, as a whole, from Aedes appears to be the form of the proboscis.

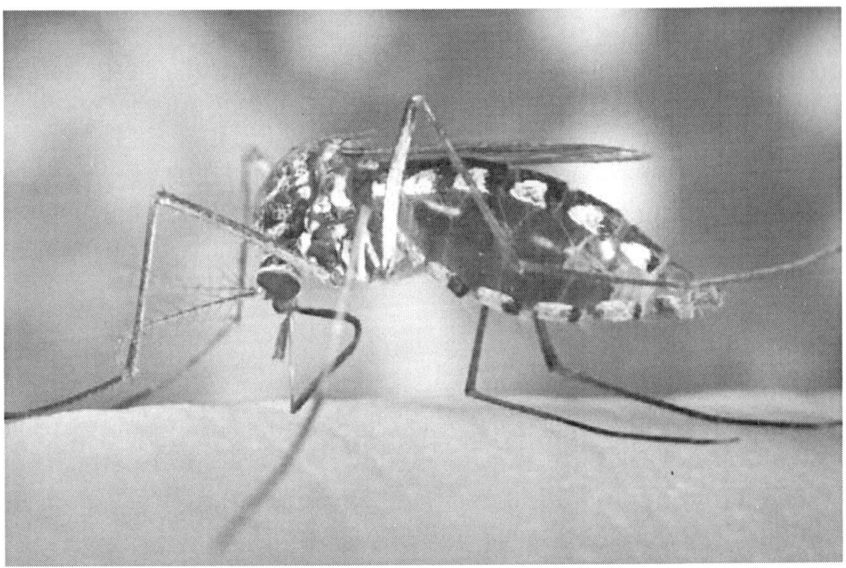

Fig. 3.12. *Armigeres subalbatus* mosquito. (*Source*: Free Wikipedia)

Of the nearly 30 species known under the genus, 13 have so far been recorded in India, of which *Ar. subalbatus* is very common in the Western Ghat ranges in Kerala State (Fig. 3.12). All species chiefly occur in forested areas with high rainfall, and the larvae live mainly in bamboos, coconut shells etc. In Kerala's Western Ghat region where latex producing trees are afforested, *Ar. subalbatus* has often been collected in the latex collecting cup fastened along with the trunk of the tree. Two species, *Ar. subalbatus* and *Ar. flavus*, have been occasionally incriminated with the Japanese encephalitis virus.

Chapter 4

MOSQUITO AND DISEASES - *DEADLY AND DEBILITATING*

4.1. Introduction

Vector-borne diseases, led by mosquitoes, account for 17% of all the infectious diseases, with 700,000 deaths per year (Fig. 4.1). Mosquitoes can be an annoying and serious problem in man's domain. In fact, the animal responsible for the most human deaths world-wide is the mosquito. They interfere with work and spoil hours of leisure time. Their attacks on farm animals can cause loss of weight and decreased milk production. Some mosquitoes are capable of transmitting diseases, such as malaria, yellow fever, dengue, chikungunya, Zika, filariasis and various different types of encephalitides [*e.g.*, St. Louis encephalitis (SLE), Western Equine encephalitis (WEE), Eastern Equine encephalitis (EEE), LaCrosse encephalitis (LAC), Japanese encephalitis (JE), West Nile virus (WNV)] to humans. Malaria, the severest of all, is also humanity's greatest bane triggered by mosquito. To animals mosquitoes of course do transmit an array of inflictions.

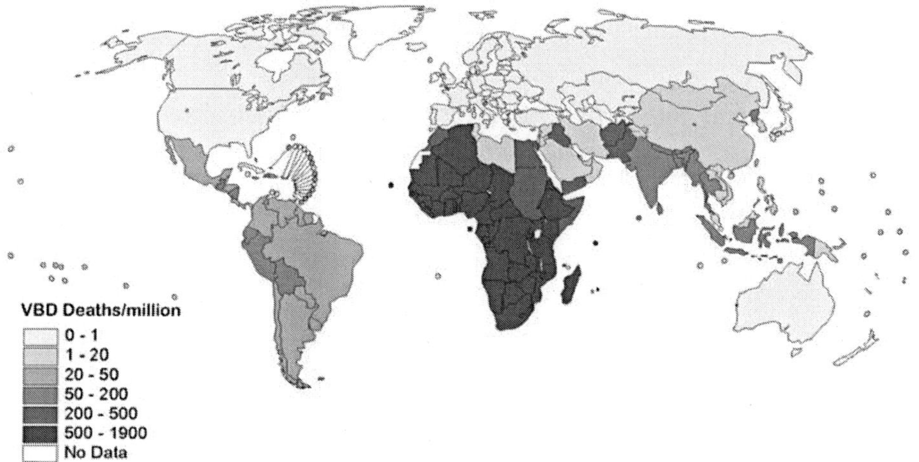

Fig. 4.1. Map showing the global incidence of deaths caused by vector borne diseases. (*Source:* WHO, 2004).

Once seen as little more than a small sacrifice for the lush vistas of tropical climes, mosquitoes, and the diseases they transmit, have international agencies and the medical community up in arms. Mosquito-borne diseases exert unfathomable economic losses to humans. These losses are not merely monetary but relate to health as well. Diseases like malaria, filariasis, dengue and Japanese encephalitis incur annual losses worth several trillions of US dollars all over the world, if only expressed in terms of Disability Adjusted Life Years (DALYs) (Table 4.1). In Sub-Saharan Africa, a major malaria infested region, only malaria-related annual economic burden in 1987 was to the tune of US$ 0.8 billion (0.6% of gross domestic product) which further hiked to US$ 1.7 billion (1.0% of GDP) by 1995. Economic losses have a tendency to rise along with the national income and improved socioeconomic conditions. Also, there are a number of direct and indirect factors which are considerable when to calculate economics of malaria. Often, the related documentations are faulty and grossly incomplete. In spite of this, even if only the recorded facts are taken into consideration, the economic loss due to malaria in India comes somewhere between Rs. 15,777.85 and Rs. 29,934.85 million at the 1991 level. This estimate obviously did not consider major areas like enormous losses in agriculture, industry, mining, railways and other important productive sectors in the country susceptible to malaria-related perils, in addition to indiscernible yet apparent losses to the physical and intellectual growth of the population living in endemic areas. The above information is a clear indication about the enormity of seriousness malaria disease needs to be dealt with in India. It needs be understood emphatically that the management of mosquito-borne diseases in India takes away more than half of the financial layout earmarked for the entire health sector!

Table 4.1. Global public health importance of mosquito-borne diseases (prevalent in India) with their vectors and distribution worldwide [1](Hay *et al.*, 2017).

Disease	Vector	Disease burden in DALYs	Distribution
Malaria	*Anopheles* spp.	56,201,200	Tropics, main burden Sub-Saharan Africa
Filariasis			
Lymphatic filariasis	*Culex* spp.	1,189,000	Tropical urban areas
Arboviral diseases			
Dengue	*Aedes* spp.	29,56,900	Urban tropics

Disease	Vector	Disease burden in DALYs	Distribution
Chikungunya	*Aedes* spp.	25,588*	Urban tropics
Yellow Fever	*Aedes aegypti*	374,000	Africa, Central & Southern Americas
Zika Virus	*Aedes* spp.; also of lesser importance are certain other genera	Each microcephaly and GBS case incurs the loss of 29.95 DALYs and 1.25 DALYs per case (Alfaro-Murillo *et al.*, 2016), in Latin America.	Recently sporadic cases occurred in Asia South America
Japanese encephalitis	*Culex* spp.	426,000	Spread sporadically in many countries (also India) across the world

[1] Global estimate of Disability Adjusted Life Years (DALYs) in 2017.

* Chikungunya DALYs are for India; Source: Krishnamoorthy et al., 2009

4.2. Mosquito-borne Diseases in India

4.2.1. Malaria

Throughout the world more than 40% of population is exposed to varying levels of risk for malaria in 91 countries. Of a total world population of approximately seven billion, 1.7 billion (35%) live in areas where endemic malaria has been considerably reduced or even eradicated, but where transmission has begun again and in which the situation is now unstable or deteriorating. In some areas, severe malaria problems developed following major ecological or social changes, particularly the agricultural or the exploitation of forests or socio-political interests. Each year 300-500 million cases of malaria, 90% of them in Africa, and between 1.5 and 2.7 million deaths are reported worldwide (WHO, 2013). Severe malaria morbidity and mortality are linked with areas endemic for *Plasmodium falciparum*, the predominant species in Africa. According to a report malaria kills a child somewhere in the world every 30 seconds (Alnwick, 2000), whereas it has been estimated that 800,000 children die each year of malaria in Africa alone. Mortality is lower but nevertheless important in the Americas and Southeast Asia, especially where *P. falciparum* drug resistance is widespread. The Southeast Asia region embodying India is of particular interest to us as there occur approximately 3.54 million cases every year.

Malaria offers perhaps the best example for determining the economic burden levied by any arthropod-borne disease in a country, since (i) among all the vector-borne diseases malaria continues to be one of the biggest contributors to disease burdens in terms of death and suffering, and from the public health point of view it is undoubtedly the most devastating parasitic disease in the tropics, and (ii) it is definitely the most extensively studied subject. By undermining the health and capacity to work of hundreds of millions of people, it is closely linked to poverty and contributes significantly to stunting social and economic development. Nearly four decades ago there were hopes that malaria could be eradicated in a fairly short time, mainly through the intensive use of insecticides. However, malaria still remains a serious threat in endemic countries. Those at greater risk of dying from the disease are children under age 5 in malaria-endemic areas, pregnant women, people moving from non-malarious to endemic zones for reasons of work, migration, refuge, war or tourism, and travellers who visit endemic countries and return home with the disease. ln many parts of the country, malaria is becoming even a greater problem than it has been before. Epidemics are recurring in areas where transmission has been interrupted, and are generally associated with deteriorating social and economic conditions.

Malaria control in India is presently hampered mainly by two reasons: (i) drug resistance development in malarial parasites, and (ii) insecticide resistance development in malaria vectors. *Plasmodium vivax* and *P. falciparum* are dominant species in India, with the former responsible for 60-65% cases and the latter for 35-40% cases (Sharma, 1999). Among the six major vector species prevalent in India, *An. culicifacies*, the main transmitter of rural and peri-urban malaria in the peninsular India, is alone responsible for 65% malaria and 55% *P. falciparum* annually. This mosquito is resistant to DDT and HCH all over the country, and to malathion in Maharashtra, Gujarat and certain other States. *Anopheles stephensi* transmits nearly 12% malaria cases in the country, mostly in the urban and industrial areas. This species, too, has developed multiple resistance against various insecticides, many of which are however used as larvicides. Yet, another vector species, *An. fluviatilis*, common in mountainous and sylvatic areas, transmits 15% of total malaria cases and 30% *P. falciparum* in the country.

India became the nucleus of major malariological investigations following the path-breaking discovery of 'malaria-mosquito connection' by

Dr Ronald Ross in 1897 for which he was awarded the Nobel Prize (1902). In those days mysteries surrounding malaria, rampant and devastating in nature, were being attempted to be solved in many countries, both in the tropic and temperate countries, much before the entry of Dr Ronald Ross in the arena. Many European nations especially Britain, France, Italy and American nations particularly United States had under their rule several countries in the fragmented Asia in the West and the Africa, but the mosquito-borne diseases like malaria and filariasis always dreaded them to make their effective governance and living. While Dr Patrick Manson had already determined the role of mosquito in transferring filarial worm from man to man in China in early 1880s and a team of great scientists like Prof. Grassi, Dr Golgi and Dr Bignami were working on malaria (and there the possible role of mosquitoes) for several years in Italy around the same period – the seed for which were sown about one hundred years ago by Giovanni Lancisi, and even the most plausible interpretation on possible mosquito-association with malaria dissemination was already sounded by Dr Daniel Beauperthuy in Venezuela and Dr Josiah Nott in Alabama more than two decades ago, still the 'mosquito malaria theory' was destined to be originated in India first, by Ronald Ross. Indian scientific community take upon themselves this moment of great discovery, as well as his grand research legacy as a matter of pride and dignity. No wonder, therefore, that when the Indian scientists led by Dr V.P. Sharma, himself an internationally acclaimed malariologist, took the initiative to organize an international symposium on malaria in 1997 at Hyderabad, and commemorated one hundred years of Ross's discovery of malaria-mosquito connection (20th August, 1897). Not only this, very recently a malaria seminar was organized by the Godrej Sara Lee Pvt. Ltd., a great commercial name in manufacturing mosquito-repelling products, on 20th August, 2003 in Mumbai (old-named Bombay), to coincide with the "Mosquito Day" coined by Ross himself.

Ever since Ross's famous 1897 discovery, India has been in forefront of malaria related researches ably contributed by internationally famed malriologists or culicidologists like Drs R.S. Christophers, P.J. Barraud, J.A. Sinton, G. Covell, G. MacDonald, D.K. Viswanathan, I.M. Puri, Dr T. Ramachandra Rao, Dr P.K. Rajagopalan, Dr V.P. Sharma etc., just to name a few for example (for details see Chapter 8). Thus, the mosquito continues to be incredibly closer to Indian psyche, by virtue of its place both in their life and the research programmes aiming at her control. Interestingly, in Indian life, one is aware of quite a few adages referable to mosquitoes for their impact on man, for example, when a man is extremely emaciated in physical appearance, he is referred to as to resemble a 'mosquito'; when somebody is annoyingly entangled with an opponent, the former warns the latter for making him crushed like a 'mosquito' is done, when a person is uncontrollably, but secretively, talking into the ears of another person, the former is reminded to not whine like a 'mosquito'; and, when an already big human population continues to grow unchecked they are often referred to increasing like flies and 'mosquitoes' do.

In spite of great early discoveries and encouraging, though fragmented, anti-malaria campaigns in hyperendemic regions in the country, malaria continued to surge unabated throughout the country. Unfortunately, no systematic country-wide studies were performed prior to 1961, and details are available through scattered surveys in different parts of India. Malaria epidemics occurred throughout India with cadence (Covell, 1927). Sir Gordon Covell, the one-time head of the Malaria Institute of India, recorded epidemics of "great intensity" in Murshidabad (Bengal) in 1821. In 1852, one malaria epidemic wiped out the entire village of Ula and then spread across the Bhagirathi River to Hooghly and continued to devastate populations for many years in Burdwan. The development of the Indian railways under the British administration contributed a great deal to the spread of malaria. The rather haphazard construction of railway embankments provided a number of breeding sites for the malaria vectors. It is also likely that the labourers that worked on the construction of railways introduced different strains of the parasite to the areas in which they worked, possibly increasing the incidence of the disease. The year 1908 saw a countrywide pandemic during which huge mortality and morbidity occurred. In 1921, Bengal suffered a severe malaria epidemic which resulted in over 730,000 deaths alone. Thereafter, the number of deaths from malaria slowly decreased to within 300,000 to 400,000 per annum. During the Second World War however malaria deaths rose again, when Bengal recorded over 680,000 deaths in 1943, and 763,220 deaths in 1944. In Bombay Presidency, there were significant outbreaks of malaria during the construction of the Colaba causeway between 1821 and 1841 during the construction of Alexander Dock and Hughes Dry Dock. The number of malaria cases in Bombay increased relatively sharply after 1901, during which time the construction of the railway was completed, ensuring an increase in traffic and a greater movement of people. Among British troops in 1921, over 30% of all hospital admissions were for malaria and in 1922, 10,552 cases of the disease were reported, with 26 deaths out of a total military force of 60,166. This represents an incidence rate of 175 per 1000 which increased slightly to 180 per 1000 in 1923. The number of cases among the Indian troops is broadly similar to that of the British troops, with 173 cases per 1000 in 1922. However, the mortality rate among Indian soldiers was around 3 times higher than for British soldiers; the Indian soldiers possibly did not have access to the same level of medical care and treatment offered to the British soldiers.

Early malaria control efforts tended to concentrate on the removal of breeding sites and later used chemicals such as the Paris green and kerosene for larviciding mosquito immatures. One of the first formal operations to control the disease took place at Mian Mir, near the city of Lahore. Mian Mir later became a British military post and the focus of a concerted effort to eradicate malaria. Mian Mir had an intricate system of irrigation canals which provided excellent breeding grounds for the malaria vectors. The incidence of malaria was astonishing, with an average incidence of slightly under 100%, rising to over 300% during epidemic years. The malariologists Dr J.W.W. Stephens and Dr S.R. Christophers, arrived at Mian Mir in 1901 with ambitious plans to remove all the breeding sites, evacuate the infected people and administer quinine as both a curative and preventative measure. Their programme began in 1901 and eventually developed into a massive effort, with between four and five hundred soldiers set to work full time at filling in the irrigation canals. The programme of constantly filling in ditches and removing puddles and any other potential breeding site continued until 1909, when to their dismay a serious malaria epidemic broke out, forcing to abandon the programme mainly on accounts of high costs and low output. The lesson learnt from the Mian Mir debacle was that mosquitoes, well entrenched, tenacious, adaptable, are difficult to control. The enemy (mosquito) in Mian Mir proved more formidable than had previously been admitted, especially in its adaptability to attack, its capacity to cover considerable distances and make use of secondary breeding places, when those of first choice were closed. It proved above all that in malaria control every local situation needed to be thoroughly studied and a specific method employed (Harrison, 1978).

Similar failures were conceivable in Bombay, Jhansi, Poona, Meerut, Secunderabad and at all other military posts where similar activities concurrent to Mian Mir were initiated. In 1917, the Bengal Nagpur Railway (BNR) and the East India Railways formed a separate malaria control organization, specifically to control the disease in and around stations. There every person suffered 3 bouts of malaria during a single year. Similar larviciding and breeding pool removal programmes were undertaken during the 1920s in the tea plantations of Assam and in Mysore by the Rockefeller Foundation. The Rockefeller Foundation began using pyrethrum sprays experimentally in India to great success and this method of malaria control was recognised as enormously valuable. Pyrethrum was first used in malaria control in Delhi in 1938, following its phenomenal successful usage by William Gorgas in Cuba about two decades ago. The use of pyrethrum was then expanded to Assam in 1942. All of the above interventions would have had some impact of the incidence of malaria, yet they were unable to sustain the control of the disease. With availability of DDT, the wonder insecticide, in late 1940s, however, changed for a moment the whole concept of malaria control through vector control.

Because of a constantly worsening state of malaria disease in the country the government of India, soon after independence, initiated the first ever countrywide control programme. Thus came into being the National Malaria Control Programme in 1953 which, inspired by huge success in controlling malaria incidence, was transformed into a long-term National Malaria Eradication Programme (NMEP) with a clear vision to accomplish the task of overpowering the disease in four phases, i.e., preparatory phase, attack phase, consolidation phase and maintenance phase. The strategy has been to cause interruption in the disease transmission by spraying residual insecticides in the rural areas (351.8 million population) and by source reduction and larviciding in urban areas (62.1 million population). Because most of the primary malaria vectors in India are endophilic and/ or endophagic in behaviour, indoor residual spraying of insecticides, DDT, HCH and malathion, in the rural areas and anti-larval operations in the urban set-ups have been the major means to control malaria in the country. Systematic hard work of NMEP showed up spectacular results, reducing the malaria cases in 1963 to about 0.1 million without any death (Fig. 4.2).

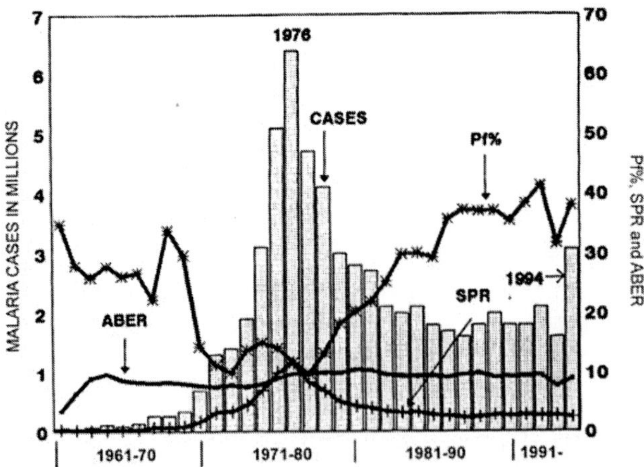

Fig. 4.2. Malaria incidence between 1960s and 1990s. (*Source:* Adapted from Sharma, 1999)

However, complacency that followed resulted malaria to spring back with a vengeance shooting up malaria cases much above the 10 million mark and several hundreds of deaths. This led Government of India to reorganize the functioning of NMEP, at different time periods, to re-achieve the success which however eluded for several years in the same manner and intensity as happened earlier.

Major anti-malaria activities launched in India since the independence can be summed up briefly as follows:

(i) *National Malaria Control Programme* (NMCP): It was launched in 1953 with comprehensive indoor residual spray with DDT. By 1957, 165.7 million population was effectively protected.

(ii) *National Malaria Eradication Programme* (NMEP): The success achieved under the NMCP was so spectacular and encouraging that the Government of India launched National Malaria Eradication Programme, in 1958, with the objective of eradicating the disease from the country. This brought down malaria cases to 0.1 million and no deaths due to malaria were reported in the country in 1964. The euphoria of victory however proved to be ephemeral as malaria incidence began to escalate in the late sixties owing to multiple factors comprising financial, logistic, administrative and technical constraints, on one hand, and a feeling of complacency due to a near victory over the disease in the country, on the other. Consequently, malaria resurged with a vengeance and reached its peak in 1976, when about 6.47 million cases of malaria were recorded in the country. The NMEP originally began as a completely centrally sponsored health programme and, till 1979, the entire expenditure on malaria control throughout the country was borne by the Centre. Thereafter, the programme was implemented on a 50:50 cost sharing basis with the States, with the exception of the seven north-eastern States which, from December 1994 onwards, were brought under 100% Central assistance as a special case. All other States in the country, while meeting out the entire operational costs for malaria control as their part of the cost sharing, continued to receive 50% of the costs from the Centre in the form of required equipment and material. The Indian Government spend nearly Rs. 350 crores on malaria control annually. Additionally, the World Bank also recently provided a sum of Rs. 1075 crores for five years for the Enhanced Malaria Control Programme.

(iii) *Modified Plan of Operation* (MPO): The successes and failures under the NMEP were reviewed in depth and a Modified Plan of Operation was introduced in the year 1977 with three main objectives (i) prevention of deaths due to malaria, (ii) reduction of morbidity due to malaria, and (iii) maintenance of industrial and green revolution due to freedom from malaria as well as retention of achievement gained so far. Under the MPO a new strategy was adopted from eradication of malaria to the containment of the disease in the country, i.e., prevention of death and reduction in morbidity due to malaria. As a strategy under this Plan areas with 2 annual parasite incidence (API) or more were earmarked for regular rounds of spray. Surveillance activities in the whole country were decentralized to Primary Health Centres (PHC) along with laboratory services. People's participation was solicited by involving volunteers in the distribution of antimalarials through Drug Distribution Centres (DDC) and Fever Treatment Depots (FTD). Malaria control in urban areas (<40,000 population) is carried out under the Urban Malaria Scheme (UMS) of the NMEP, launched in 1971-72. This led to significant reduction in malaria incidence in the country to a level of around 2 million cases by 1987 for one decade. Since 1994, however, focal outbreaks of malaria have been continuously reported from different parts of country, particularly the north-eastern States and the Thar Desert region in Rajasthan State in the north-west India. This unwarranted situation resulted in resurgence of malaria, generally pronounced by *P falciparum* dominance, as well as deaths due to malaria in the country.

(iv) *Plasmodium falciparum Containment* Programme (PfCP): The *P. falciparum* problem attained new dimension when a focus of chloroquine-resistant strain was discovered in Assam in 1973. Subsequently, new foci of chloroquine resistance were traced in different north-eastern States, Orissa, Madhya Pradesh, Maharashtra, Rajasthan etc. This situation led to the creation of PfCP in 1977 to tackle the rising threat of *P. falciparum*.

(v) *Malaria Action Programme Based on Expert Committee Report - 1995*: An expert committee that took cognizance of the fact of availability of appropriate technologies for control of malaria in different epidemiological situations or paradigms in the country attributed the various setbacks in malaria eradication programme, particularly recurrence of periodic epidemics and high mortality due to various different imminent factors like the administrative indifference, the organizational weakness, the low prioritization to malaria under the health services and apathy of middle level and peripheral workers in the States. Accordingly, the expert Committee recommended intensification of malaria control activities throughout the country with focus on high risk areas through an integrated malaria control strategy, with following mentioned components: (i) Early case detection and prompt treatment, (ii) Selective vector control, (iii) Promotion of personal protection methods, (iv) Early detection and containment of epidemics, (v) Information, education and communication towards personal prevention and community participation, and (vi) Institutional and management capacity building, trained manpower development and efficient management information system.

(vi) *Enhanced Malaria Control Project* (EMCP): It was launched in September 1997 with the help of World Bank assistance and covered, in addition to 19 towns in 10 States, 62.2 million high risk tribal population living in 100 districts and 1045 PHCs of the seven peninsular States, *viz.*, Andhra Pradesh, Bihar, Gujarat, Madhya Pradesh, Maharashtra, Orissa and Rajasthan. These PHCs, which a tribal population of more than 25%, had contributed 65.5% malaria cases and 78.60% *P. falciparum* cases in 1997. The benefits of the programme are conceived to reach other malaria endemic areas as well since the strengthening of the components of the Information, Education and Communication (IEC), trained manpower development and Management Information System (MIS) covers the whole country.

(vii) *Roll Back Malaria (RBM):* In January 1998 the World Health Organization proposed Roll Back Malaria initiative taking into consideration large scale morbidity and mortality due to malaria in developing countries and also observing the changing epidemiological scenario, and the experience of the past control efforts. The basic concepts of the RBM initiative are not different from the existing control strategy in India, but it lays more emphasis on social mobilization, inter-sectoral collaboration, effective partnerships, strengthening of Primary Health Care, investment in development of effective methods of control, shaping of existing tools by operational research and it is expected that this will result in sustained reduction in disease burden soon (Alnwick, 2000).

During 1980s malaria was reduced to less than 7 million which dramatically plummeted in 1990s and 2000s to less than 3 million, although, ironically the WHO-SEARO claimed the cases could be nearly 8-10-times more whereas Vas Dev & Sharma (1913) have suspected deaths and cases to be as high as 10 to 20 times more (Fig. 4.3). Currently malaria cases in India are reported to be less than one lakh.

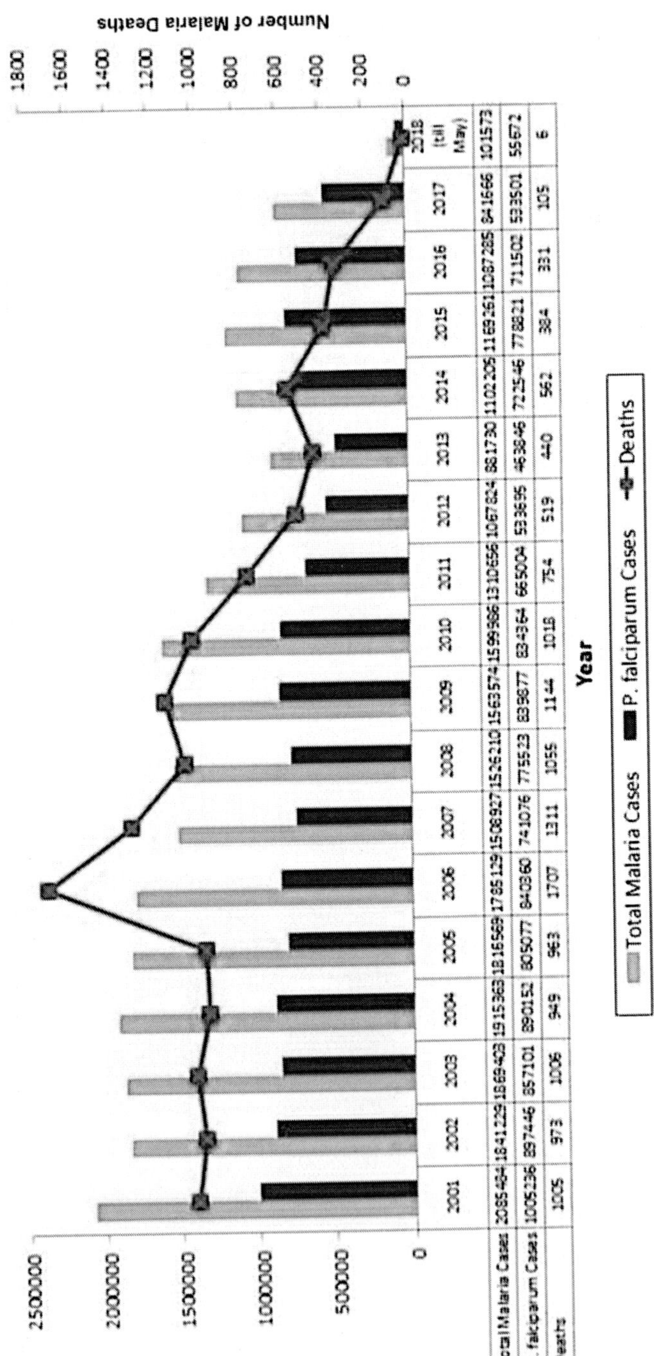

	2001	2002	2003	2004	2005	2006	2007	2008	2009	2010	2011	2012	2013	2014	2015	2016	2017	2018 (till May)
Total Malaria Cases	2085484	1841229	1869403	1915363	1816569	1785129	1508927	1526210	1563574	1599986	1310656	1067824	881730	1102205	1169261	1087285	841666	101573
P. falciparum Cases	1005236	897446	857101	890152	805077	840360	741076	775523	839877	834364	665004	533695	463846	722546	778821	711502	533501	55672
Deaths	1005	973	1006	949	963	1707	1311	1055	1144	1018	754	519	440	562	384	331	105	6

Fig. 4.3. Trend of malaria cases and deaths, 2001-2018 (till May). (*Source:* Free Wikepedia)

Led by DDT's widespread usage in controlling vector population, now largely jeopardized owing to development of resistance in vector mosquitoes, other insecticides like HCH (later banned in 1997), malathion, and certain synthetic pyrethroids such as cyfluthrin, deltamethrin and lambdacyhalothrin etc. have also been selectively employed in the control of malaria, although spray targets were rarely achieved due to logistic and operational constraints. This forced a new technology of Long Lasting Insecticide Impregnated Bed-nets (LLIN) introduced in tackling the adult vector population which in hyperendemic north-eastern States and Odisha (Orissa) catapulted the malaria scenario and mitigated a great deal the impact of malaria there, despite a serious concern of steady rise in *Plasmodium falciparum* cases (Fig. 4.3).

In recent times the progress of National Vector-Borne Diseases (NVBDCP) has been independently evaluated twice, in 2008 and 2013, and clearly the reduction in malaria incidence in the country has been vouched, though far from satisfactory. India has set the target for elimination for malaria by 2030, with less than one million cases reported in 2018.

4.2.2. Lymphatic Filariasis

Lymphatic Filariasis is commonly known as elephantiasis. About 90% of infections worldwide are caused by *Wuchereria bancrofti*, 9.9% by *Brugia malayi* and remaining (0.1%) by *Brugia timori. Culex quinquefasciatus* is the vector of *W. bancrofti.* WHO estimated that about 1.43 billion people are at risk globally where 65% of people at risk reside in South-East Asia, 30% in the African Region and the remainder in other parts of the tropical world. More than 120 million infections worldwide, with over 40 million patients were affected severely, in which SE Asia contributed 78 million cases (40% in India). Around 600 million populations residing in 250 districts from 20 States/UTs in India are endemic and at risk of LF infection. In India, 99.4% of infections are caused by *Wuchereria bancrofti* species whereas *Brugia malayi* is responsible for remaining 0.6% of infection. LF infection is prevalent in all regions except the Northern states in India where southern region was reported with high number of states affected than other regions. National Filaria Control Programme (NFCP) was launched in India in 1955 and the National Health Policy 2002 aims at elimination of LF by 2015. A gradual decreasing trend of microfilaria rates was observed over a period of years where it was reported 1.24 % in 2004 and 0.34% in 2011. The main control measures were mass Diethyl Carbamazine Citrate (DEC) administration, anti-larval measures in urban areas and indoor residual spray in rural areas undertaken in India.

Lymphatic Filariasis is presently endemic in 255 districts in 16 states and 5 UTs. To achieve the goal of elimination, GoI in year 2004, launched the strategy of Annual Mass Drug Administration (MDA) with single dose of Diethyl Carbamazine Citrate (DEC) to population living at the risk of filariasis except children below 2 years, pregnant women and seriously ill persons. The co-administration of DEC + Albendazole was introduced for MDA since 2007. The current strategy of elimination of Lymphatic Filariasis with annual Mass Drug Administration (MDA) was to reduce microfilaria rate below 1% so that transmission is interrupted and the new generation will not contract the disease. The affected people with this disease mainly known as elephantiasis will be provided home based lymphoedema management procedure by simple washing and drying. The people affected with hydrocele will be motivated for surgical intervention. The programme has already listed about 8 lakh lymphoedema and 4 lakh hydrocele cases. About 1.3 lakh hydrocele cases have been operated.

In course of time the programme scaled up to reach about 600 million population achieving 100% geographical coverage in all 255 endemic districts. Coverage rates also improved from 73% in 2004 to 88% in 2015. The efforts to scale up MDA and tackle LF have shown success. Out of 255 endemic districts 222 have microfilaria prevalence below 1% qualifying them for validation through Transmission Assessment Survey (TAS). Seventy two districts have already passed the first round of TAS. The first round of TAS was slated to take up in all 255 districts by end of 2017. However, as per WHO guidelines, three round of TAS at interval of 2-3 years have to be conducted to ensure elimination. Therefore, in spite of the achievement of microfilaria rate below 1% in all the 255 districts, the elimination process will start after 3-4 years. Yet, there are last mile challenges like survey of non-endemic districts for current status of LF endemicity, liquidating foci in non-endemic districts with treatment and vector control under Integrated Vector Management as the whole districts may not require MDA, sustaining the achievements gained so far and preventing failure in TAS etc.

4.2.3. Mosquito-borne arboviral diseases

Worldwide more than 130 arboviruses of public health importance are known to cause human diseases which belong to one of three virus genera: *Flavivirus, Alphavirus* (Togaviridae) and *Bunyavirus*. Major mosquito-borne arboviral diseases are described here.

4.2.3.1. FLAVIVIRUSES

Flavivirus is a genus of viruses belongs to the family Flaviviridae which contains 70 recognised viruses includes human and an animal pathogenic virus of global importance, of which nearly 50% produces clinical disease in humans. Human flaviviruses are West-Nile virus (WNV), dengue virus (DENV), Japanese encephalitis virus (JEV), tick-borne encephalitis virus (TBEV) and yellow fever virus (JEV) as well as hepacivirus hepatitis C virus (HCV). Many flaviviruses of human importance, nearly more than 80% are transmitted to vertebrates by the bite from an infected arthropods (mosquitoes or ticks). The flaviviruses can be divided into 2 clades: one with the vector borne viruses and the other with no known vector. The vector clade can be subdivided into a mosquito borne clade and a tick borne clade. The mosquito borne clade has two branches- one branch contains neurotropic viruses, which are spread by the Culex species and associated predominantly with encephalitic disease. The second branch is the non-neurotropic viruses, often associated with haemorrhagic disease and shock syndrome in humans where *Aedes* species act as vectors. Here under dengue, Zika, Japanese encephalitis and West Nile Virus are described being the major mosquito-borne infections in India.

(i) Dengue

It is the most important vector-borne disease of humans and likely to be more important than malaria globally in terms of morbidity with almost half of the world's population (>70% live in Asia Pacific countries) at risk of infection and is endemic in more than 125 countries. Although the full burden of disease is underrepresenting and difficult to estimate, it affects almost all the regions of WHO (Fig. 4.4).

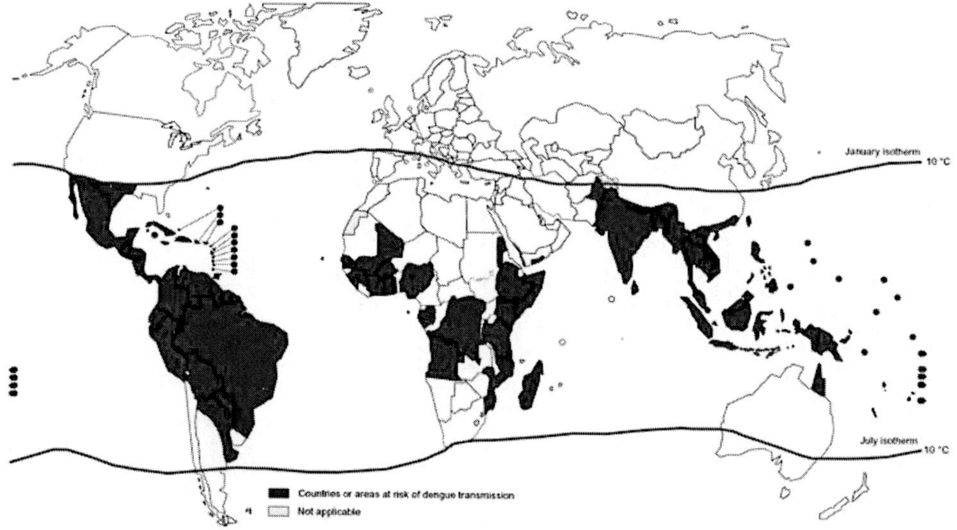

Fig. 4.4. Global distributions of countries at risk of dengue transmission, 2011. (*Source*: WHO report on Neglected Tropical Diseases, 2013)

DENVs-1–4 are transmitted from one human host to another by mosquitoes of the Aedes genus, principally *Aedes aegypti* and *Ae. albopictus*. A study estimates that there are 390 million dengue infection including 96 million apparent dengue infections, and 294 million inapparent infections per year globally which is threefold higher than the estimate given in the official WHO figure. In the global burden of apparent dengue infection, Asia accounts for 70% (67 million infections) and in which, India alone contributed 49% and 34% (33 million infections) of the respective Asia and global total disease burden. India is significantly contributing high dengue burden compared to other countries.

In India, the first epidemic of clinical dengue-like illness was recorded in Madras (now Chennai) in 1780 and the first virologically proved epidemic of dengue fever (DF) occurred in Calcutta (now Kolkata) and Eastern Coast of India in 1963-1964. The first major wide spread epidemics of DHF/DSS occurred in India in 1996 followed sooner in 2004 where after the infection has been on the rise involving at present nearly all (35/36) but one State/Union Territory by the year 2012 and 2013; the possible exception is Leh, without a case then, created as a UT in 2019. The virus, DEN3, was for the first time in India isolated in *Aedes aegypti* from Vellore. Later the virus was detected in *Ae. albopictus* as well (Reuben *et al.*, 1988). During 2003 epidemic Kerala was among the most affected States whose history with dengue started since mid-1990s only (Table 4.2).

Table 4.2. Dengue incidence and deaths in Kerala State since 1997 up to 2003.

Year	Cases	Proportionate increase (talking 1997 as base year)	Deaths
1997	14	–	4
1998	67	1 : 4.8	13
1999	1	–	0
2000	0	–	0
2001	70	1 : 5	1
2002	219	1 : 15.6	2
2003	3546	1 : 253.3	68

India reported a total of 1,97,440 dengue cases and 2049 deaths occurred during 1996-2012. An average of 11,614 dengue cases with case fatality rate (CFR) 1.04% was estimated per year for the country. The CFR was 3.3% in 1996, but thereafter it declined to about 1.0% till 2007. Maximum cases were contributed by Northern States (35.11%), followed by those in South region (34.23%). A maximum dengue mortality was reported in the North (46.66%), followed by South (22.94%), West (19.96%), East (5.95) and Central region (4.49%) in that order. During the last two decades (1991 to 2012), more States have consistently reported dengue cases and number of such States have increased from 4 to 8 by 2001 and to 34 by 2012.

Table 4.3. Dengue Cases and Deaths in the Country since 2015

S. No.	Affected States/UTs	2015		2016		2017		2018 (Prov.)		2019 (till Nov.)	
		C	D	C	D	C	D	C	D	C	D
1	Andhra Pradesh	3159	2	3417	2	4925	0	4011	0	4647	0
2	Arunachal Pradesh	1933	1	13	0	18	0	1	0	123	0
3	Assam	1076	1	6157	4	5024	1	166	0	167	0
4	Bihar	1771	0	1912	0	1854	0	2142	0	6193	0
5	Chattisgarh	384	1	356	0	444	0	2674	10	681	0
6	Goa	293	0	150	0	235	0	335	1	874	0
7	Gujarat	5590	9	8028	14	4753	6	7579	5	14835	16
8	Haryana	9921	13	2493	0	4550	0	1898	0	937	0
9	Himachal Pradesh	19	1	322	0	452	0	4672	7	320	2
10	J & K	153	0	79	1	488	0	214	0	435	0
11	Jharkhand	102	0	414	1	710	5	463	1	803	0
12	Karnataka	5077	9	6083	8	17844	10	4427	4	15232	13
13	Kerala	4075	25	7439	13	19994	37	4083	32	3940	16
14	Madhya Pradesh	2108	8	3150	12	2666	6	4506	5	3645	2
15	Meghalaya	13	0	172	0	52	0	44	0	61	0
16	Maharashtra	4936	23	6792	33	7829	65	11011	55	12374	25
17	Manipur	52	0	51	1	193	1	14	0	334	0
18	Mizoram	43	0	580	0	136	0	68	0	42	0
19	Nagaland	21	1	142	0	357	0	369	0	8	0
20	Odisha	2450	2	8380	11	4158	6	5198	5	3251	0
21	Punjab	14128	18	10439	15	15398	18	14980	9	8949	0
22	Rajasthan	4043	7	5292	16	8427	14	9587	10	12664	14
23	Sikkim	21	0	82	0	312	0	320	0	243	0
24	Tamil Nadu	4535	12	2531	5	23294	65	4486	13	6577	5
25	Tripura	40	0	102	0	127	0	100	0	100	0
26	Telangana	1831	2	4037	4	5369	0	4592	2	12072	6
27	Uttar Pradesh	2892	9	15033	42	3092	28	3829	4	9280	20
28	Uttrakhand	1655	1	2146	4	849	0	689	3	10500	8
29	West Bengal	8516	14	22865	45	37746	46			-	-

S. No.	Affected States/UTs	2015		2016		2017		2018 (Prov.)		2019 (till Nov.)	
		C	D	C	D	C	D	C	D	C	D
30	A & N Island	153	0	92	0	18	0	49	0	168	0
31	Chandigarh	966	1	1246	0	1125	0	301	0	235	0
32	Delhi	15867	60	4431	10	9271	10	7136	4	4155	0
33	D&N Haveli	1154	0	4161	2	2064	0	493	0	954	2
34	Daman & Diu	165	0	89	0	59	0	163	0	128	2
35	Puduchery	771	0	490	2	4568	7	592	2	1495	1
	Total	**99913**	**220**	**129166**	**245**	**188401**	**325**	**101192**	**172**	**136422**	**132**

C= Cases | D= Deaths | NR= Not Reported

The high prevalence, lack of an effective vaccine, and absence of specific treatment conspire to make dengue fever a global public health threat. After 2015 dengue cases have made a steep rise breaking all previous records (Table 4.3).

The goal of global strategy for dengue prevention and control 2012-2020 was to reduce at least 25% of morbidity and 50% of mortality by 2020 and to estimate the real burden by 2015. Many countries have national dengue prevention and control programme with an aim of reduction of vector population whereas in India, there is no such specific national programme especially for dengue but biological and chemical vector control strategies as one of the component under Integrated Disease Surveillance Project (IDSP) are implemented for vector management. In addition, inter-country collaboration was formed recently between the Serum Institute of India with the University of Mahidol in Thailand to develop a dengue vaccine which shows the significant changes in the policy formulation towards control and elimination of the disease.

(ii) Zika Virus

Zika virus (ZIKV), a previously obscure flavivirus closely related to dengue, West Nile, Japanese encephalitis and yellow fever viruses, has emerged explosively since 2007 to cause a series of epidemics in Micronesia, the South Pacific, and most recently the Americas. After its putative evolution in sub-Saharan Africa, ZIKV spread in the distant past to Asia and has probably emerged on multiple occasions into urban transmission cycles involving Aedes (Stegomyia) spp. mosquitoes and human amplification hosts, accompanied by a relatively mild dengue-like illness. The Zika virus (ZIKV) global epidemic prompted the World Health Organization to declare it a 2016 Public Health Emergency of International Concern. The unprecedented numbers of people infected during recent outbreaks in the South Pacific and the Americas may have resulted in enough ZIKV infections to notice relatively rare congenital microcephaly and Guillain–Barré syndromes. Another hypothesis is that phenotypic changes in Asian lineage ZIKV strains led to these disease outcomes. ZIKV and the associated neurological complications represent a long-term world-wide challenge to public health. Re-emerging viruses like ZIKV are an ever-present threat in the 21st century where fast transcontinental travel lends itself to viral epidemics such as that in India during late 1910s (Singh *et al.*, 2019).

Zika is an RNA virus containing 10,794 nucleotides encoding 3419 amino acids. It is closely related to Spondweni virus. The two viruses are the only members of their clade within the mosquito-borne cluster of flaviviruses. Studies in the Zika Forest suggested that ZIKV blunted the viremia caused by Yellow Fever Virus in monkeys but did not block transmission of YFV.

Compared to other allied viruses, Zika can be paralleled with following specific characteristics: (i) Zika is spread mostly by the bite of an infected Aedes species mosquito (*Ae. aegypti* and *Ae. albopictus*). These mosquitoes bite during the day and night, (ii) Zika can be passed from a pregnant woman to her fetus. Infection during pregnancy can cause certain birth defects, (iii) Among many other illnesses or syndromes, microcephaly in newborns is generally reported as a result of Zika infection, and (iv) There is no vaccine or medicine for Zika.

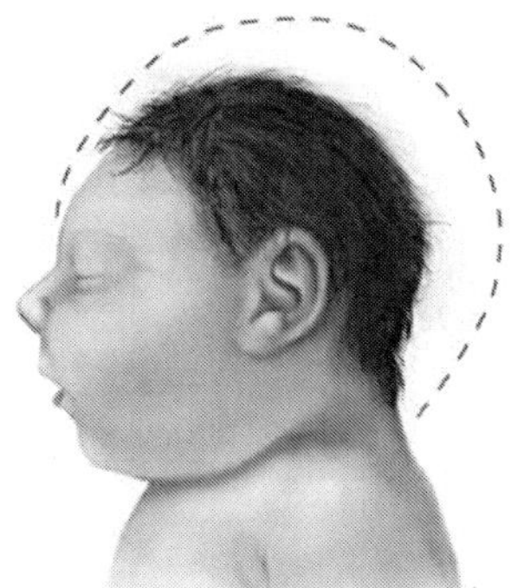

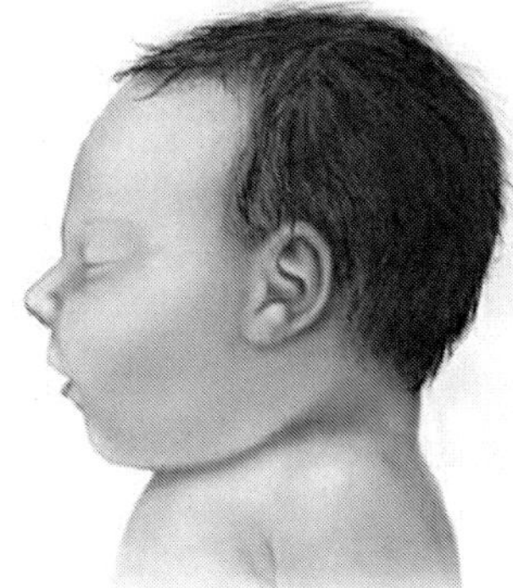

Fig. 4.5. Baby (a) with microcephaly, compared to (b) normal baby head. (*Source:* Free Wikipedia)

Microcephaly is characterized by (i) Symptoms include below-average head size (ii) Its often caused by failure of brain to grow at normal rate, (iii) Head circumference: Less than 42 cm at full growth, (iv) Affects 25, 000 children in US each year (Fig. 4.5).

Since 1950s the Zika Disease has been known to occur within a narrow equatorial belt from Africa to Asia. However, in 2014 the virus (ZIKAV) spread eastward across the Pacific Ocean to French Polynesia, then to Easter Island and in 2015 to Mexico, Central America, the Caribbean, and South America, where the Zika outbreak has reached pandemic levels. India has reported a few cases of Zika virus (Singh *et al.*, 2019) Fig. 4.6.

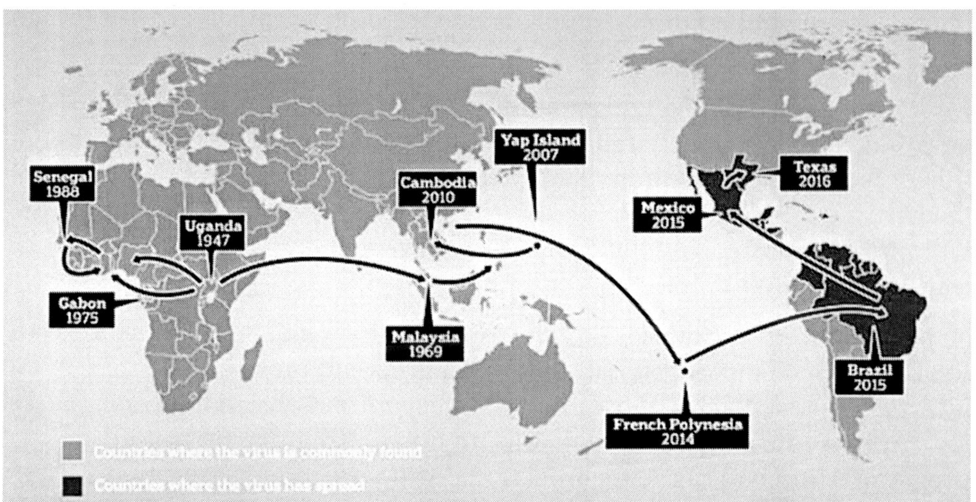

Fig. 4.6. Pathway of spread of Zika virus from its cradle in Africa all over the world. (*Source*: Wikipedia)

Zika virus was first discovered from the Rhesus monkey in the Zika forest of Uganda in 1947. The first human case was reported in Nigeria seven years later. It has now become a global public health concern. Currently, 58 countries and territories are affected by the Zika virus. ZIKV is a flavivirus and is maintained in a sylvatic cycle which involves nonhuman primates and the Aedes mosquitoes as vectors (*Aedes africanus, A. aegypti*). ZIKV was only known to cause infection in Africa and Southeast Asia. ZIKAV could be transmitted through blood, but this is an infrequent mechanism. The virus has also been isolated in semen, but person-to-person sexual transmission is unconfirmed (Fig. 4.7).

Mosquito-transmitted Zika virus (ZIKV) is a member of the Flavivirus genus in the Flaviviridae family. Similar to other flaviviruses, ZIKV is transmitted by several mosquito species including *Aedes africanus, Ae. aegypti, Ae. albopictus,* and *Ae. hensilli.*

ZIKV was first isolated in a sentinel rhesus monkey in the Zika forest of Uganda in 1947. Prior to its rapid spread across the Pacific Islands in the

21st century, ZIKV was also documented in a handful of individuals in sub-Saharan Africa and then in Southeast Asia by the mid-20th century. These cases were rare, geographically isolated events and many others may have gone undocumented. After its original identification, the first major ZIKV outbreak outside of Africa took place in 2007 on Yap Island with approximately three-fourths of the population over the age of three suspected to be infected. *Ae. hensilli* was the predominant mosquito species identified in the transmission of ZIKV in this event and infection was characterized by rash, arthralgia, and conjunctivitis. In October 2013, a larger ZIKV outbreak (29,000 cases) occurred in French Polynesia. While most patients exhibited mild symptoms, one patient developed Guillain-Barre Syndrome (GBS) one week post-infection. GBS is an auto-immune disease that attacks the peripheral nervous system leading to weakness in the extremities and, in severe cases, paralysis and death. Alarmingly enough, the total number of the GBS cases that followed this outbreak was roughly 20-fold higher than baseline levels, suggesting a direct association between ZIKV and GBS. Following this event, outbreaks were reported in New Caledonia, the Cook Islands, and the Easter Island. Phylogenetic analyses of ZIKV variants isolated from French Polynesia (2013), Cambodia (2010), and Yap Island (2007) confirmed that these outbreaks were linked to the expansion of the Asian lineage of ZIKV.

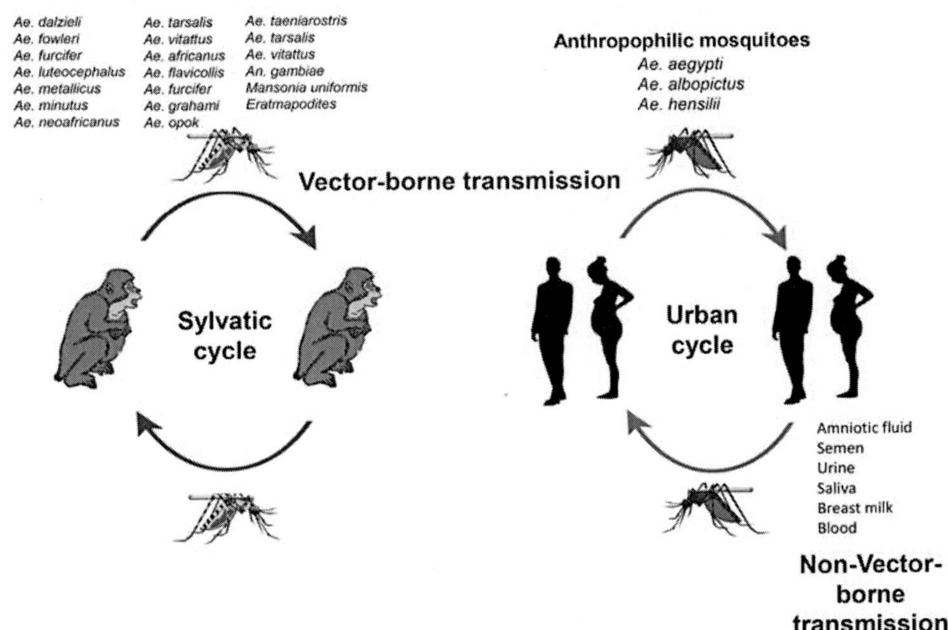

Fig. 4.7. Zika virus transmission cycle. (*Source:* Free Wikipedia)

The beginning of a string of severe outbreaks in the Americas ensued in March 2015 when ZIKV was detected in Bahia, Brazil. Between January and July 2015, the state of Rio de Janeiro also experienced its first ZIKV outbreak with patients who presented with headache, arthralgia, myalgia, non-purulent conjunctivitis, and lower back pain. After reaching Brazil, the virus then rapidly spread throughout the Americas. Puerto Rico experienced its first case on December 31, 2015, and since then has experienced a total of 37,000 locally transmitted symptomatic or probable cases. By February 2016, the CDC received nine confirmed ZIKV cases in pregnant women who had traveled to South America. By April of 2016, ZIKV transmission had been registered in 27 countries in the Americas. The first locally transmitted ZIKV case in the United States mainland was reported in July of 2016. By July 22, 2016, the Florida Department of Health had identified 321 ZIKV disease cases in Florida residents and visitors. Currently, the CDC reports there have been 5,716 cases reported on the United States mainland, with 231 of those cases transmitted locally through mosquitos (CDC Cumulative Zika Disease Case Counts in the United States, 2015–2018). There are currently 86 countries, territories, or subnational areas with evidence of ZIKV transmission. The actual number of infections is certainly higher as approximately 80% of infections are asymptomatic and the community lacked suitable diagnostic methods during the early outbreaks. The recurrence of ZIKV around the world may have several causes including the generation of more virulent strains, new routes of transmission, and novel modifiers of the disease.

Zika virus infection frequently goes unnoticed or is asymptomatic in approximately 80% of cases and most patients present with only mild infection. ZIKV is most commonly transmitted from the bite of a mosquito, where initial infection most likely occurs in human skin cells directly affecting permissive human dermal fibroblasts, epidermal keratinocytes, and immature dendritic cells. Experimental evidence suggests that in mice, ZIKV infects placental cells via a trans-placental route that results in restricted intrauterine. This route has been suggested to involve viral spread to the chorionic villi, amniochorionic membranes, and from the basal to the parietal decidua. In 2016 it was estimated that 29% of babies born to infected mothers exhibit developmental abnormalities, suggesting that the trans-placental route of transmission, also termed vertical transmission in mice, is relevant to ZIKV infection in humans. A very recent CDC evaluation of past cases in newborns states that one in seven is affected by vertical transmission (CDC Vital Signs report, August 8, 2018). Placental ZIKV infection induces proliferation and prominent hyperplasia of Hofbauer cells in the chorionic villi, but does not elicit villous necrosis or a lymphoplasmacellular or acute inflammatory reaction in maternal or fetal cells. These developmental abnormalities tend to occur more frequently when infection takes place during the first trimester. Within the past 2 years, however, two cases were described in women infected during the third trimester, with fetuses displaying the first signs of congenital brain malformations at the 36th week of pregnancy. While one study found that primary human trophoblasts from full-term placenta do not host ZIKV because of their interferon gamma 1 response, a different study revealed that ZIKV infection increased the risk of latent microcephaly because of the virus' ability to replicate in human placental trophoblasts without causing any observable host cell destruction. Therefore, the congenital transfer of the virus by transplacental infection may cause this spectra of abnormalities. One clinical study found that 14% of ZIKV-infected symptomatic semen samples were positive for ZIKV RNA and persisting for more than six months. Two other non-vector routes of ZIKV transmission include perinatal and sexual transfer. A mouse model revealed that the infectious virus was present in the semen at days 7 to 21 post-inoculation. Another report found human semen with viral up to 38 days after initial detection in humans. Evidently, the vaginal tract in mice is also highly susceptible to ZIKV replication. There is high probability that this is also true for humans. Blood transfusion represents another worrisome mode of transmission, as two cases of probable transmission have already been reported in Brazil.

Accentuating the problem, presentations of ZIKV are appearing in cells and tissues that have not been previously seen in humans. In Brazil, a study detected the infectious ZIKV particles in the urine and the saliva of patients during the acute phase of infection. In a couple returning from Martinique, French West Indies, ZIKV remained detectable in the plasma for roughly two weeks; their urine samples tested positive at 39 days. Furthermore, it has been reported that two patients with ZIKV infection also had cases of severe thrombocytopenia. The first patient exhibited only 1,000 platelets/mm3 and died following multiple hemorrhages. The second patient had 2,000 platelets/mm3, melena, and ecchymosis, but, luckily, recovered after receiving intravenous immunoglobulin. Another case reported severe immune thrombocytopenic exacerbation and antinuclear antibody positivity induced by ZIKV infection. In addition to thrombocytopenia, ZIKV is likely associated with ocular abnormalities. A case study in Salvador, Bahia, Brazil, revealed that in 29 examined infants born to ZIKV-infected pregnancies, ocular abnormalities were present in 17 eyes of 10 children. ZIKV was also detected in the conjunctival fluid suggesting that ZIKV affects the permeability of the retinal pigment epithelium. A recent clinical case also associated the virus with hearing loss. A newborn from Brazil, whose mother was infected with ZIKV during pregnancy, was confirmed to have bilateral profound hearing loss.

Normally, ZIKV infection is not fatal. However, the first fatal case of ZIKV-associated encephalitis was reported in 2016 in a 47 year old non-pregnant woman soon followed by the report of three additional ZIKV-related fatalities with one of the patients being severely immunocompromised. An additional risk of ZIKV in adults is damage to the testis. A study in mice reported the persistence of ZIKV in the testis and epididymis leading to extensive tissue damage. Male mice were reported to exhibit oligospermia, diminished testosterone, and inhibin B levels. A more recent study reinforced these observations and revealed that peritubular spermatogonium cells are vulnerable to ZIKV infection. Furthermore, even an acute, uncomplicated, symptomatic ZIKV infection may result in microhematospermia in the absence of hematuria. In women, primary human endometrial stromal cells are also infected by ZIKV. In addition to widely studied congenital infection with ZIKV, recently studies have documented ZIKV infection in immunocompetent neonates. Curiously, BALB/c neonatal mice were resistant to ZIKV infection, whereas Kunming, ICR and C57BL/6 neonatal mice were fatally susceptible to ZIKV infection. Similar to congenital infection, different areas of CNS including gray matter, hippocampus, cerebral cortex, and spinal cord (but not olfactory bulb) were severely affected. ZIKV was replicated and caused pathogenesis in liver and testis, implying that ZIKV infection may engender a high health risk in neonates by postnatal infection.

Evidence for a ZIKV-induced dysregulation of neural development comes from the studies in which ZIKV was injected directly into the lateral ventricle of the fetal mouse brain. The virus readily homed to the dorsal telencephalon while other brain regions were spared. There was also inhibition of cell division in the ventricular, sub-ventricular and intermediate zones; the sub-ventricular zone is a home to many radial glial progenitor cells that differentiate into cortical neurons. The infected cortex became shorter and the size of the lateral ventricle cavities was noticeably reduced as compared with uninfected mice. In addition, after entering the CNS, ZIKV infects astrocytes, visual system nuclei, and the retina. Undifferentiated neurons, which are plentiful in the fetal brain, are highly susceptible to ZIKV infection, especially during the early stages of neurogenesis. Intriguingly, only 40% of human fetal neural progenitors become infected by ZIKV, thus limiting the anti-ZIKV immune and inflammatory response. However, ZIKV infection readily leads to the death of these infected cells. In contrast, differentiated neurons in the adult brain are relatively resistant to the virus. These observations explain the infrequent neurological pathologies in the infected adults. The inconsistent results between different research groups indicates this aspect of ZIKV infectivity is not completely understood and warrants further study, particularly for the blood-brain barrier permeability to viruses. Differences in blood-brain barrier permeability in different stages of fetal development may help to explain the greater susceptibility of fetuses to ZIKV in the earlier trimesters.

There is overwhelming experimental support for a direct link between ZIKV infection and microcephaly in both rodent models and humans. Thus, Asian ZIKV (Asian strain SZ01) infects neural progenitor cells (NPCs) in vivo and affects brain development resulting in cell-cycle arrest, cell apoptosis, differentiation inhibition, cortical thinning, and, ultimately, microcephaly. Further evidence comes from the genome-wide profiling analyses of the infected brain suggesting the up-regulation of the viral entry receptor genes and aberrations in the immune response, apoptosis, and microcephaly pathways. Naturally, ZIKV shares some common biology with other flaviviruses such as West Nile (WNV), Dengue (DENV), tick-borne encephalitis, and Yellow fever, which may also cause neurological damage and encephalitis. However, the striking and unusually dangerous epidemiology of ZIKV and its unique interactions with humans suggest the presence of a modified underlying molecular makeup of this formidable virus. It has now been firmly established that there is a causal relationship between prenatal ZIKV infection and microcephaly, along with other extensive neurological and brain anomalies. It is likely that the Toll-like receptor 3 (TLR3) is linked to ZIKV neuropathology as TLR3 was upregulated in human ZIKV-infected organoids and mouse neurospheres. Conversely, TLR3 inhibition reduced the phenotypic effects of ZIKV infection. The pathway analysis suggested that ZIKV-related TLR3 activation affected 41 neurodevelopmental genes. Based on this and similar studies, there is now a consensus among researchers that ZIKV predominantly targets neural progenitors and affects neurogenesis during the first trimester and, as a result, abrogates normal brain development. In agreement, immunocytochemistry and electron microscopy demonstrated that ZIKV infects neurospheres and brain organoids, reducing their viability and growth. In order to mimic ZIKV infection during the course of human cortical development, researchers employed organotypic cultures and demonstrated that ZIKV preferentially infected neural progenitors in early stage cortical organoids. Using cost-effective miniature spinning bioreactors, the authors demonstrated that ZIKV infection resulted in suppressed proliferation and increased cell death in the early stage cortical organoids. Not surprisingly, these organoids exhibited certain macroscopic features resembling microcephaly. Overall, there is a well-justified conclusion that ZIKV selectively affects the fetal brain development.

Intriguingly, ZIKV has evolved to contain a nucleotide composition and RNA modifications, such as methylation and the formation of G-quadruplexes that allow effective replication in mosquito and primates. The ZIKV genome produces non-coding subgenomic flavivirus RNAs (sfRNAs) due to stalling of host 5′-3′ ribonucleases on viral RNA structures in the 3′ untranslated region (UTR). These sfRNAs exert important proviral functions such as antagonizing the innate interferon response and RNA interference; see excellent recent review for further details. A global proteome- and gene-expression wide view of ZIKV infection may provide the field with potential biomarkers and indicators of pathogenesis unseen with existing studies. Further, applying this approach to vertical transmission may provide insights into the critical proteins in the brain involved in the development of microcephaly. To this end, our groups have embarked on a mission to characterize the changes to the proteome in ZIKV-infected mothers and their pups. Pathway analysis of ZIKV-infected pups in utero can uncover the alterations in key subsystems. Proteomics can also reveal essential molecules like kinases, receptors, and transcription factors to target with ZIKV therapy. Indeed, the major goal of this work will be to identify biomarkers for ZIKV-induced microcephaly.

(iii) Japanese encephalitis (JE)

Japanese encephalitis, a zoonotic disease (mainly ardeid birds and swine being the amplifier), is widespread over South East Asia and Pacific regions where 3 billion people are at risk of infection (Fig. 4.8). JE virus is the leading cause of viral encephalitis in the Asia which accounts for 68000 clinical cases and 25-30% of CFR annually. It was estimated in 2011 that approximately 67 900 JE cases typically occur each year in the 24 JE-endemic countries, for an incidence of 1.8 per 100 000 overall. It is primarily a disease of children and approximately 51000 (75%) cases occur among children (0-14 years). CFR in humans ranges from 25-30%, and as many as 22% of JE patients having objective neurological deficits and 28% having subnormal intelligence quotients.

JE disease was first reported in Japan in 1924, was subsequently reported in other Asian countries, whereas in India, a first case was reported in 1955. It is transmitted by infective bites of female mosquitoes mainly belonging to *Culex tritaeniorhynchus, Culex vishnui* and *Culex pseudovishnui* group. In India, *Culex vishnui group (Culex tritaeniorhynchus, Culex vishnui and Culex pseudovishnui)* are the chief vectors of JE. JE is not considered as neglected tropical disease (NTD) globally by WHO but still an important neglected disease in India. JE viral activity has been widespread in India but most commonly found in rural rice-growing areas where flooded fields and irrigation systems favours ideal larval habitats for Culex vector mosquitoes. Till 2012, about 17 states/UTs in India have reported incidence of JE. Since 2008, a maximum no. of JE cases were reported in the years 2011 (1214) and 2012 (745) compared to previous years. It was estimated that an average incidence of JE cases and deaths were about 719 and 121 respectively can occur annually. Eastern and central regions of India are highly affected with maximum no. of cases and deaths compared to other regions. One of the historically most significant area or region for periodic episodes of Japanese encephalitis in India has been the Gorakhpur Division in eastern Uttar Pradesh, experiencing country's worst epidemics on continuous basis. There are some effective vaccinations (Mouse Brain-Derived JE Vaccine) available in the Market against JE in India. Government of India also launched JE vaccination campaign now and then in the risk prone JE endemic districts.

Fig. 4.8. Global distributions of countries at risk of Japanese encephalitis (*Source:* WHO, Fact sheet on JE (No 386), 2014)

The JEV manifests in five strains all over the affected regions, i.e. JEV1-5. In India until mid-2010 JEV3 was the only strain prevailing, but later JEV1 also was added.

In India, the JEV has been isolated from 17 species of mosquitoes; 10 species of Culex, 03 species each of Anopheles and Mansonioides, and 01 species of Armigeres (Table 4.9). The Culex vishnui subgroup of mosquitoes consisting of *Culex tritaeniorhynchus* Giles, *Cx. vishnui* Theobald and *Cx. pseudovishnui* Colless have been implicated as major vectors of JE virus (JEV)1.

Table 4.9. Mosquito species so far incriminated with JEV in India. (B.K. Tyagi, pers. archive)

S. No.	Mosquito Species	No. of Isolations	State
1	*Anopheles peditaeniatus*	1	Karnataka
2	*An. barbirostris*	1	West Bengal
3	*An. subpictus*	9	Karnataka, Kerala, Tamilnadu
4	*Culex tritaeniorhynchus*	79	Karnataka, Kerala, Tamilnadu
5	*Cx. vishnui*	30	Karnataka, Kerala, West Bengal
6	*Cx. pseudovishnui*	8	Karnataka, Goa
7	*Cx. epidesmus*	1	West Bengal
8	*Cx. bitaeniorhynchus*	3	Karnataka, West Bengal
9	*Cx. infula*	6	Tamilnadu
10	*Cx. gelidus*	8	Tamilnadu, Karnataka
11	*Cx. whitmorei*	4	Tamilnadu, Karnataka, Andra Pradesh, West Bengal
12	*Cx. fuscocephala*	7	Tamilnadu, Karanataka
13	*Cx. quinquefasciatns*	1	Karnataka
14	*Ma. annulifera*	2	Kerala, Assam
15	*Ma. indiana*	3	Kerala
16	*Ma. uniformis*	4	Kerala, Karnataka
17	*Armigeres subalbatus*	1 (male)	Uttar Pradesh (Gorakhpur)

One of earliest and foremost identification guide to the JE vectors in field was offered by Reuben *et al.* (1994) which still remain a reference work on the desk of a serious researcher.

(iv) West Nile Virus (WNV)

West Nile virus infections have been commonly found in Africa, Europe, Asia, and the United States. It usually causes sub-clinical infection or mild infection called West Nile fever (WNF) in human and horses but in India WNV in horses has not been documented. Various Culex species such as *Culex tritaeniorhynchus, Cx. vishnui, Cx. bitaeniorhynchus* and *Cx. univittatus, Culex pipiens fatigans* and *Aedes albopictus* have acted as potential vectors of WNV. Apart from Mosquitoes, certain types of migratory birds also playing a possible role in the maintenance of WNV has been described in India. Unlike Dengue and JE, there was no serious epidemic of WEN reported in India. In India, antibodies against WNV were first detected in humans in Bombay (Mumbai) in 1952. WNV neutralizing antibodies (about 20-30%) have been detected in human sera collected from Tamil Nadu, Karnataka, Andhra Pradesh, Gujarat, Maharashtra, Madhya Pradesh, Orissa and Rajasthan and mostly virus activity has been reported from southern, central and western India. Prevalence and Epidemiological data on WNV in India as a whole has not been documented well to assess the overall disease burden of WN infection but some district and state based serological study reports the evidence of prevalence of WN virus activity in India. As of now the vaccination is under laboratory trial and not available for the control of WNV infection.

4.2.3.2. ALPHAVIRUSES

Alphavirus is one of two genera in the family Togaviridae; the other togavirus genus is Rubivirus (rubella virus). The 27 alphaviruses are classified on the basis of antigenic properties. All alphaviruses share antigenic sites on the capsid and at least one envelope glycoprotein, but viruses can be differentiated by several serological tests, particularly neutralization assays. Disease's clinical manifestation occurs in either of two general forms, depending upon the virus: one is typified by fever, malaise, headache, and/or symptoms of encephalitis (*e.g.*, eastern, western, or Venezuelan equine encephalitis viruses) and the other by fever, rash, and arthralgia (*e.g.*, chikungunya, Ross River, Mayaro, and Sindbis viruses). Here under Chikungunya is described being a major mosquito-borne infection in India.

(i) Chikungunya

Chikungunya fever (CHIK fever) is caused by chikungunya virus (CHIKV) which is geographically distributed in Africa, Southeast Asia and India. *Aedes albopictus* , along with *Aedes aegypti,* is the main vector for the transmission of CHIV infection. Chikungunya causes febrile illness like similar features of Dengue virus infection but it primarily affects the peripheral small joints and leads to prolonged arthralgic syndrome. CHIKV was first isolated in Tanzania (formerly Tanganyika) in 1953 during an outbreak of dengue like illness. Then the CHIKV was repeatedly detected from various regions in Africa and Asian countries. In South East Asia, frequent outbreaks were reported from the 1960s through to 2003 in India, Malaysia, Indonesia, Cambodia, Vietnam, Myanmar, Pakistan and Thailand. The entry point of CHIKV in India was unknown but had a history of outbreak in the year 1963 in Kolkatta and Tamil Nadu (Chennai) in 1965 when more than 3 lakh population were affected with adverse complication of haemorrhagic manifestation. Since 1973,there was no CHIK cases were reported till the end of 2005 and perceived it as almost disappeared from India. Again it was re-emerged in the year 2006 in Karnataka and Andhra Pradesh, Southern India when WHO also declared that the CHIK outbreak in India. About 1.11 million people were affected in 2006 with an attack rate of 4-45%. Since 2006, CHIK virus activity was prevalent and continuously affects a proportion of people every year with an average of 52,078 cases annually. Between 2007 and 2012, a highest no. of cases was documented in the year 2008, followed by 2009. Table 4.10 shows the prevalence of Chikungunya in India since 2015.

Table 4.10. Epidemiological Profile of Chikungunya Fever in the Country since 2015.

S. No.	Affected States/UTs	2015		2016		2017		2018		2019 (Prov.)	
		No. Susp. Chikun-gunya cases	No. of conf. cases	No. Susp. Chikun-gunya cases	No. of conf. cases	No. Susp. Chikun-gunya cases	No. of conf. cases	No. Susp. Chikun-gunya cases	No. of confir-med cases	No. of conf. cases	No. of conf. cases
1	Andhra Pradesh	817	83	960	147	1162	108	622	79	700	67
2	Arunachal Pradesh	35	6	239	8	133	0	507	1	332	55
3	Assam	0	0	40	40	41	41	3	3	0	0
4	Bihar	3	1	566	566	1251	1251	156	156	532	532
5	Goa	561	32	337	49	509	48	455	77	796	319
6	Gujarat	406	42	3285	847	7953	1363	10601	1290	6491	496
7	Haryana	1	1	5394	1970	220	6	62	3	0	0
8	J&K	0	0	1	1	0	0	1	1	0	0
9	Jharkhand	21	0	47	14	269	17	3405	851	1679	166
10	Karnataka	20763	2099	15666	1528	32831	3511	20411	2546	35366	2970
11	Kerala	175	152	129	129	78	74	77	77	52	52
12	Madhya Pd.	67	11	2280	862	2477	858	3211	1609	2248	620
13	Meghalaya	78	15	360	68	236	45	44	2	9	1
14	Mizroram	0	0	0	0	0	0	93	10	0	0
15	Maharashtra	391	207	7570	2949	8110	1438	9884	1009	4382	1378
16	Manipur	0	0	0	0	0	0	2	0	40	3
17	Odisha	81	46	51	15	0	0	0	0	0	0
18	Punjab	180	18	4407	2054	3251	201	736	25	243	2
19	Rajasthan	7	7	2506	2215	1612	1612	254	254	169	169
20	Sikkim	0	0	30	5	130	8	384	28	641	62

S. No.	Affected States/UTs	2015		2016		2017		2018		2019 (Prov.)	
		No. Susp. Chikun-gunya cases	No. of conf. cases	No. Susp. Chikun -gunya cases	No. of conf. cases	No. Susp. Chikun-gunya cases	No. of conf. cases	No. Susp. Chikun-gunya cases	No. of confir-med cases	No. of conf. cases	No. of conf. cases
21	Tamil Nadu	329	329	86	86	131	131	284	284	345	345
22	Telangana	2067	149	611	71	1277	58	1954	489	4816	1078
23	Tripura	180	7	311	70	574	64	683	75	953	106
24	Uttar Pradesh	0	0	2458	2458	103	103	58	58	5	5
25	Uttarakhand	0	0	35	10	0	0	29	7	1	1
26	West Bengal	1013	61	1071	117	2103	577	52	23	-	-
27	A&N Island	68	3	18	0	93	17	205	27	702	53
28	Chandigarh	1	1	2857	272	1810	54	357	4	82	0
29	D&N Haveli	0	0	0	0	0	0	0	0	0	0
30	Delhi	64	64	12279	9793	940	940	407	407	442	442
31	Lakshadweep	0	0	0	0	0	0	-	-	-	-
32	Puduchery	245	8	463	20	475	23	2876	361	4191	555
	Total	**27553**	**3342**	**64057**	**26364**	**67769**	**12548**	**57813**	**9756**	**65217**	**9477**

CHIKV infection became, like Dengue, a leading vector borne disease of public health importance in terms of high morbidity. In Kerala both *Ae. aegypti* and *Ae. albopictus* were incriminated with the CHIKV. Since there is no vaccination available, the only way of preventing the disease is use of physical means of using protective clothing. Paramasivan *et al.* (2009) have discovered chikunhunya in *Lakshadweep* ils. for the first time. *Aedes albopictus* is the major vector in Kerala, just like in most other rural and sylvatic parts of the country, but also in many just like in other regions. Chikungunya has presently assumed a serious proportion as being reported with lots of human disabilities and even deaths in Europe, Americas, Indian and Pacific Ocean island nations or otherwise places, *e.g.*, Le Reunion etc.

Chapter 5

MOSQUITOES OF INDIA - *VECTORS AND PESTS OF PUBLIC AND VETERINARY IMPORTANCE*

5.1. Introduction

Currently, 3,543 species of Culicidae are formally recognized and distributed over a diverse range of habitats around the world (Harbach 2007, 2014). India, one of the world's megadiverse countries, has a highly diverse mosquito fauna and a high prevalence of mosquito-borne diseases (Tyagi, 2004). A tremendous variety of ecosystems in India provides environmental conditions that favour the immigration, adaptation, development and persistence of a great diversity of mosquito species including vectors (Fig. 5.1 and 5.2) (Das *et al.*, 1990; Tyagi, 2002, 2003; Rajavel *et al.*, 2005a-f; Dutta *et al.*, 2003; Bhattacharyya *et al.*, 2004a,b, 2009). Since the discovery of 'malaria-mosquito connection' by Dr Ronald Ross in Secundrabad, India in 1897, mosquitoes had been surveyed in almost all parts of the country during past more than 130 years, mostly by the scientists of the erstwhile National Institute of Communicable Diseases (now NCDC), Delhi, Malaria Research Centre (now NIMR), Delhi, Vector Control Research Centre, Puducherry, Centre for Research in Medical Entomology, Madurai, National Institute of Virology, Pune, Vector Control & Zoonotic Diseases, Hosur, Regional Medical Research Centre, Dibrugarh, and Desert Medicine Research Centre, Jodhpur. During early years following the pathbreaking discovery by Ross almost entirely these surveys were directed towards sampling the malaria vectors (*cf.* Christophers, 1933 and Barraud, 1934). Sampling of mosquitoes was made from all available habitats in high altitudes, forests, deserts, plateaus, marshes, estuarine coastal areas etc. Most of these collections are preserved in the museums of above said institutes, although in some of these places much of the specimens are already dilapidated or have not been systematically accessioned for future reference work. Also, ironically, much of these specimens come from the adult mosquito collections, there too the females, and there is but abysmally referenced collections of aquatic stages of eggs, larvae and pupae. While, on one hand, pupae, whose characters of taxonomic importance are viewed with a great degree of certitude (Harrison & Scanlon, 1975; Nagpal & Sharma, 1995) are clearly least sampled and preserved, only fourth instar larvae have been given prominence for taxonomic purpose, on the other (Christophers, 1933; Barraud, 1934; Puri, 1960). Recently Tyagi *et al.* (2012, 2014) have provided keys for identification of all the major vector species for malaria, dengue/chikungunya/Zika, lymphatic filariasis and Japanese encephalitis, based on both larval and adult characteristics.

Fig. 5.1. Some of the various different kinds breeding habitats of mosquitoes in the Western Ghat region (*Source*: Tyagi *et al.*, 2015).

Fi g. 5.2. Some typical breeding habitats of mosquitoes in the Thar Desert region's sandy, canalized (marshy) and rocky (mine) areas (*Source*: Dr B.K. Tyagi *pers. archive*).

Molecular techniques have in recent decades contributed significantly to our understanding of the phylogeny and decipher the enigma associated with various different species complexes in mosquitoes and virtually revolutionized the whole taxonomic system. Although, in as far as Indian fauna is concerned, there has not been a single treatise where molecular barcode was incorporated as a characteristic to compliment the conventional taxonomic methodology for distinguishing either a distinct species, or closely allied sibling species in a complex or even confusing taxa in the family Culicidae. Only some scattered efforts have been made to demonstrate how useful are they in identifying even the morphologically most close taxa (Fig. 5.3).

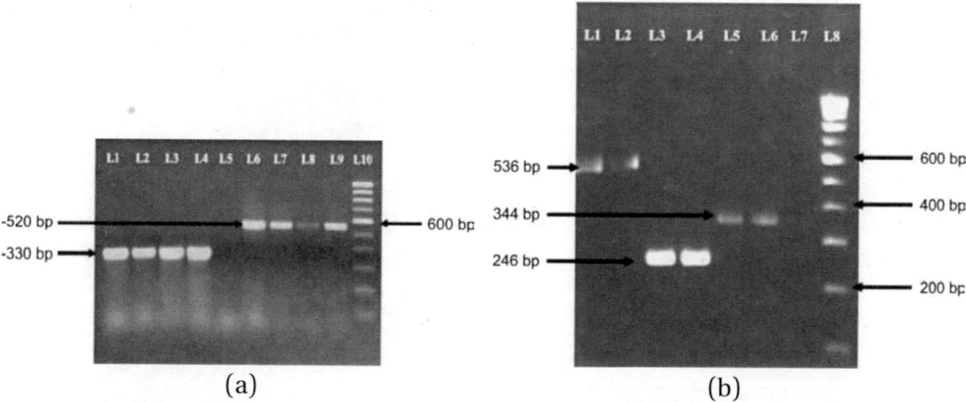

Fig. 5.3. (a) Gel picture showing ITS2 PCR differentiating *Aedes aegypti* (330 bp) and *Ae. albopictus* (520 bp), and (b) Gel picture showing differential PCR results identifying *Culex triteaniorhynchus, Cx. vishnui* and *Cx. pseudovishnui* with 536 bp, 246 bp & 344 bp, respectively. (*Source:* Dhananjeyan *et al.*, 2014).

In the field of medical entomology, it is very important that the vector species be identified precisely, so that the vector control strategies can be designed properly and can curtail an epidemic. Precise identification of insect species is essential in many fields such as agriculture (pest management); Forensic Sciences (understanding the time of crime based on the occurrence of the species and its life stage at the time of investigation); Conservation of biodiversity (identifying species those are endangered) etc. Therefore, taxonomy forms the basis for every biological study. Many a insect species, particularly sibling species complexes and those sylvatic taxa whose adults are difficult to catch with information on breeding habitats, pose challenge to their correct identification. Many cytogenetical techniques such as those based on the polytene chromosomes have earlier been attempted on mosquito species, largely at best to distinguish cryptic sibling species among species complexes (Subbarao *et al.*, 1983), but without any success in complimenting existing dichotomous keys. Application of modern molecular tools have not only overcome this pitfall but also helped draw a better evolutionary trend of various taxa. DNA based molecular taxonomy could be used for specimens of any stage, sex and form. In circumstances where it is difficult to find the adult near the breeding site such as the case with sylvatic mosquitoes, Dhananjeyan et al. (2014) have found out a useful method to identify the mosquito on the basis of molecular analysis of the genomic DNA extracted from the exuvium's of the species and helped comprehend distinction among hard-to-identify species under genera Culex, Anopheles and Aedes. This methodologies has a visible advantage because of the fact that genomic DNA could be isolated from a wide variety of specimens such as immature, adult, fresh, ethanol preserved, frozen, physically damaged during collection or handling, museum specimens etc. (Fig. 5.3).

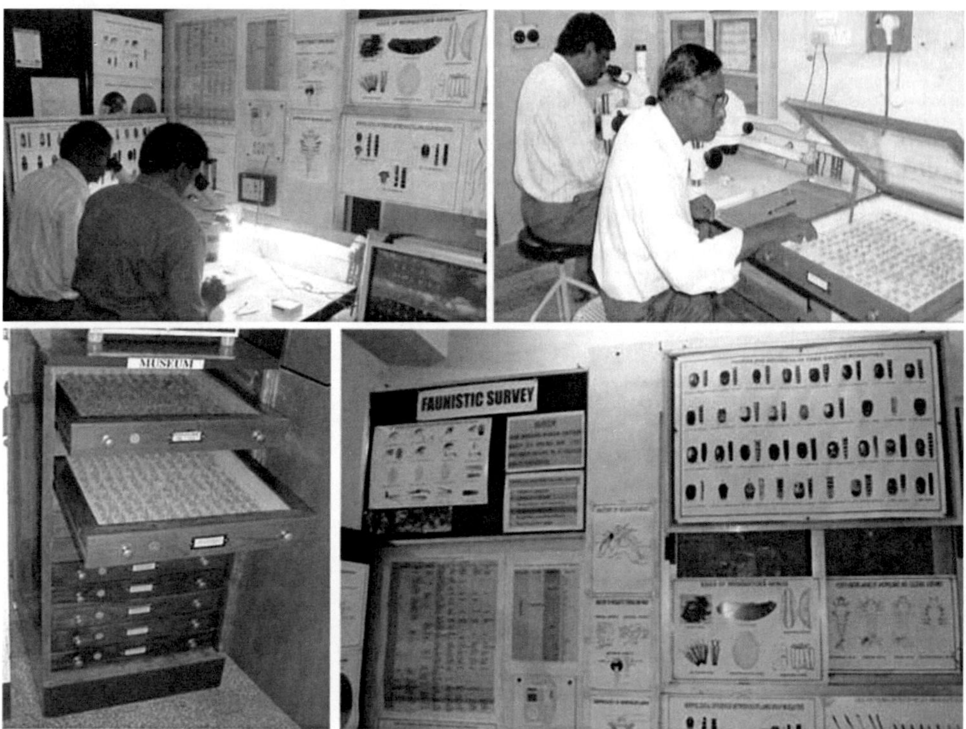

Fig. 5.4. Display of pinned specimens in cupboards and diagrammatic charts.

While the old collections of various 'Type' specimens are preserved in the museums of National Centre for Disease Control, Delhi and Indian Museum/Zoological Survey of India, Kolkata, the recent lot of collections from all over India is best preserved with a documented inventory in the mosquito museum of three Indian Council of Medical Research institutions, *viz.*, Vector Control Research Centre, Pondicherry, Regional

Medical Research Centre, Dibrugarh and Centre for Research in Medical Entomology, Madurai (Fig. 5.4) (Tewari *et al.*, 2007). The VCRC mosquito museum contains more than 14,800 adult mosquitoes, including 8,426 that are individually pinned. The collection also includes some 1,048 male and female genitalia preparations, and 815 larval and 444 pupal exuviae on microscope slides. Mosquitoes of 19 genera, 37 subgenera, and 181 species are housed in the museum, representing about 50% of the known species from India (Rajavel *et al.*, 2005e). The CRME mosquito museum houses 242 species, with a large number of types and paratypes etc. all authoritatively classified and accessioned, including the six new species described by the centre, *viz.*, (1) *Rhinoskusea portonovoensis* Tewari & Hiriyan, 1992, (2) *Tewarius agastyai*, Tewari & Hiriyan, 1992, (3) *Tewarius reubenae* Tewari & Hiriyan, 1992, (4) *Verrallina* (*Neomacleaya*) *assamensis,* Bhattacharya *et al.*, 2004, (5) *Anopheles* (*Cellia*) *pseudosundaicus* Tyagi *et al.*, 2009, and (6) *Toxorhynchites (Toxorhynchites) tyagii* Krishnamoorthy *et al.*, 2013. Reinert (2006) described a new genus, Tewarius (Type species: *Aedes agastyai* Tewari and Hiriyan, 1992), in tribe Aedini and the primary diagnostic characters of females, female genitalia, males, male genitalia, pupae, and fourth-instar larvae are provided. Four species are included in the genus: *Tewarius agastyai* (Tewari and Hiriyan) (type species), *Te. nummatus* (Edwards), *Te. pseudonummatus* (Reinert), and *Te. reubenae* (Tewari and Hiriyan). Two species groups are designated, the Agastyai and Nummatus Groups. The *gen. nov.* Tewarius, a masculine gender, is also the first ever generic name erected in the honour of an Indian worker, Satish Chandra Tewari (Centre for Research in Medical Entomology, Tamil Nadu, India) (Tyagi *et al.*, 2015).

The following two mosquito catalogues list species with a slight variance that occur in India: (1) Bhattachharyya *et al.* (2014) reported 393 species under 49 genera and 41 subgenera; the Subfamily Anophelinae containing 61 species under one genus while Culicinae contained 332 species under 11 tribes and 48 genera, and (2) Tyagi *et al.* (2015) reported 404 species and subspecies (which is >12% of the total number of taxa in the world, belonging to 50 genera and 2 subfamilies (12 tribes) (Table 5.1). The taxa are classified in a more modern and universally acceptable classificatory system proposed by Harbach (2004), with some consideration of the other popular systems earlier used by Christophers (1933) and Barraud (1934) in their respective magnum opuses 'Anophelini', and 'Culicini and Magarhirini' under Culicidae volumes of the internationally famous *The Fauna of British India* series (Table 1). The Catalogue has attempted to remove certain 'dark points' in the taxonomy and systematics of the important vector species prevalent in India.

Table 5.1. Mosquitoes species recorded in world and India (Source: Tyagi *et al.*, 2015)*

Subfamily	Tribes	World			India			CRME Museum		
		No. of Genera	No. Of Sub genera	No. of species	No. of Genera	No. Of Sub genera	No. of species	No. of Genera	No. Of Sub genera	No. of species
Anophelinae	Anophelini	3	10	481	1	2	63	1	2	45
	Aedeomyiini	1	2	7	1	1	2	1	1	1
	Aedini	81	49	1253	34	18	175	24	14	91
	Culicini	4	29	796	2	8	85	2	6	56
	Culisetini	1	7	37	1	2	3	0	0	0
	Ficalbiini	2	3	53	2	3	7	2	3	5
Culicinae	Hodgesiini	1	0	11	1	0	2	1	0	1
	Mansoniini	2	5	82	2	2	7	2	2	4
	Orthopodomyiini	1	0	36	1	0	5	1	0	4
	Sabethini	14	32	427	3	4	16	3	3	9
	Toxorhynchitini	1	4	89	1	1	9	1	1	7
	Uranotaeniini	1	2	267	1	2	29	1	2	17
2	**12**	**112**	**143**	**3539**	**50**	**43**	**403**	**39**	**34**	**240**

* At the time of writing this book even this catalogue needed an update (*vide infra*)

Historically, during pre-independence time, it was Theobald (1910a,b), a British entomologist, who first catalogued the Indian mosquito species by personally consulting at the Indian Museum, Kolkata. Thereafter, Christophers (1933) and Barraud (1934) systematically dealt with in two independent *magnum opuses* published under *The Fauna of British India* series on the taxonomy of Anophelini, and Culicini and Magarhirini (Toxorhynchitinae). Though largely outdated in modern context of classification, identification and description, both these classics still continue to be the most authentic reference works for a serious Indian researcher. Post-independence a crop of highly energetic entomologists, such as Drs I.M. Puri, B. Wattal, N.L. Kalra, T.R. Rao, V.P. Sharma, P.K. Rajagopalan, V. Dhanda, Bina Pani Das etc., to mention a few only for example, emerged and brought out significant check-lists of Indian mosquitoes. However, all these pieces of indispensable literature dealt exclusively with *Anopheles* mosquitoes possibly due to an unprecedented high significance tagged with malaria in the country! Attention to malaria control was so highly focussed that only one new species, *Anopheles pseudosundaicus* (Tyagi *et al.*, 2009), could be brought on record from India in more than 70 years; the last new species having been described was *An. pinjaurensis* in 1932.

Due to a fervent nationwide anti-malaria campaign between 1950 and 1980, hardly any attention was given to the non-anopheline mosquitoes, and the biomedical significance of culicine mosquitoes was almost completely overshadowed in terms of their taxonomic biodiversity, albeit many of the culicine mosquitoes played key roles in transmitting several deadly and/or debilitating diseases such as lymphatic filariasis, West Nile fever and Japanese encephalitis, apart from dengue and chikungunya diseases transmitted exclusively by *Aedes* mosquitoes in India. Yet, a few isolated attempts were made to comprehend distributional patterns of culicines on regional basis such as those in the Western Himalayas (Rao *et al.*, 1973), Southern India including Western Ghats and Eastern Ghats as well as Andaman & Necobar Islands (Reuben *et al.*, 1993,1994; Rajavel *et al.*,2005a-d), Northeastern states of seven sisters (Dutta *et al.*, 2003), the Thar Desert in western Rajasthan (Tyagi & Verma, 1991; Tyagi, 2002, 2004) and the tribal areas across the country (Sharma *et al.*, 2015). Now and then certain international workers also reported Indian culicines (Stone, 1961; Stone & Delfinado, 1973; Sirivanakarn, 1976; Knight & Stone, 1977; Rattanarithikul *et al.*, 2005), although the Indian fauna appeared more of a passing reference while describing *in extenso* the mosquitoes of the Oriental Region. The rather pathetic state of Indian mosquito taxonomy could be gauged from the fact that less than a dozen new species only had been described after the publication of the famous *FBI* volumes (Bhattacharyya *et al.*, 2004a,b,2009,2014; Tyagi *et al.*, 2009; Ramalingam, 1987; Tewari *et al.*, 1991,1992,1995, 2007; Natarajan *et al.*, 2009, 2016; Jagbir & Jagdish, 1999; Singh & Prakash, 2008; Krishnamoorthy *et al.*, 2013); Munirathinam *et al.*, 2014; Rajavel & Natarajan, 2006, 2011; Rajavel *et al.*, 1998, 2011.

CATALOGUE OF INDIAN MOSQUITOES

Order: Insecta
Family: Culicidae
Subfamily: Anophelinae
Tribe: Anophelini

I. Genus *ANOPHELES* Meigen, 1818

(i) Subgenus *Anopheles* Meigen, 1818

1. *Anopheles (Anopheles) ahomi* Chowdhury, 1929
2. *Anopheles (Anopheles) aitkenii* James, 1903
3. *Anopheles (Anopheles) annandalei* Prashad, 1918
4. *Anopheles (Anopheles) argyropus* (Swellengrebel, 1914)
5. *Anopheles (Anopheles) baileyi* Edwards,1929
6. *Anopheles (Anopheles) barbirostris* van der Wulp, 1884
7. *Anopheles (Anopheles) barbumbrosus* Strickland & Chowdhury, 1927
8. Anopheles (Anopheles) *barianensis* James, 1911
9. *Anopheles (Anopheles) bengalensis* Puri, 1930
10. *Anopheles (Anopheles) crawfordi* Reid, 1953
11. *Anopheles (Anopheles) culiciformis* Cogill, 1903
12. *Anopheles (Anopheles) gigas* Giles, 1901
13. *Anopheles (Anopheles) hodgkini* Reid, 1962
14. *Anopheles (Anopheles) insulaeflorum* (Swellengrebel & Swellengrebel de Graaf, 1920)
15. *Anopheles (Anopheles) interruptus* Puri, 1929
16. *Anopheles (Anopheles) lesteri paraliae* Sandosham 1959
17. *Anopheles (Anopheles) lindesayi* Giles, 1900
18. *Anopheles (Anopheles) nigerrimus* Giles, 1900
19. *Anopheles (Anopheles) nilgiricus* Christophers, 1924
20. *Anopheles (Anopheles) nitidus* Harrison, Scanlon & Reid, 1973
21. *Anopheles (Anopheles) peditaeniatus* (Leicester, 1908)
22. *Anopheles (Anopheles) pinjaurensis* Barraud, 1932
23. *Anopheles (Anopheles) roperi* Reid, 1950
24. *Anopheles (Anopheles) sinensis* Wiedemann, 1828
25. *Anopheles (Anopheles) sintoni* Puri, 1929
26. *Anopheles (Anopheles) umbrosus* (Theobald, 1903)

(ii) Subgenus *Cellia* Theobald, 1902

27. *Anopheles (Cellia) aconitus* Donitz, 1902
28. *Anopheles (Cellia) annularis* van der Wulp, 1884
29. *Anopheles (Cellia) baimaii* Sallum & Peyton, 2005
30. *Anopheles (Cellia) balabacensis* Baisas, 1936
31. *Anopheles (Cellia) culicifacies* Giles, 1901
32. *Anopheles (Cellia) dravidicus* Christophers, 1924
33. *Anopheles (Cellia) d'thali* Patton, 1905
34. *Anopheles (Cellia) elegans* (James, 1903)
35. *Anopheles (Cellia) fluviatilis* James, 1902
36. *Anopheles (Cellia) jamesii* Theobald, 1901
37. *Anopheles (Cellia) jeyporiensis* James, 1902
38. *Anopheles (Cellia) karwari* (James, 1903)
39. *Anopheles (Cellia) kochi* Donitz, 1901
40. *Anopheles (Cellia) maculatus* Theobald, 1901
41. *Anopheles (Cellia) majidi* Young & Majid, 1928
42. *Anopheles (Cellia) minimus* Theobald, 1901
43. *Anopheles (Cellia) mirans* Sallum & Peyton, 2005
44. *Anopheles (Cellia) moghulensis* Christophers, 1924
45. *Anopheles (Cellia) multicolor* Cambouliu, 1902
46. *Anopheles (Cellia) nivipes* (Theobald, 1903)
47. *Anopheles (Cellia) pallidus* Theobald, 1901
48. *Anopheles (Cellia) philippinensis* Ludlow, 1902
49. *Anopheles (Cellia) pseudojamesi* Strickland & Chowdhury, 1927
50. *Anopheles (Cellia) pseudosundaicus* Tyagi *et al.*, 2009
51. *Anopheles (Cellia) pseudowillmori* (Theobald, 1910)
52. *Anopheles (Cellia) pulcherrimus* Theobald, 1902
53. *Anopheles (Cellia) sergentii* (Theobald, 1907)
54. *Anopheles (Cellia) splendidus* Koidzumi, 1920
55. *Anopheles (Cellia) stephensi* Liston, 1901
56. *Anopheles (Cellia) subpictus* Grassi, 1899
57. *Anopheles (Cellia) sundaicus* (Rodenwaldt, 1925)
58. *Anopheles (Cellia) tessellatus* Theobald, 1901
59. *Anopheles (Cellia) theobaldi* Giles, 1901
60. *Anopheles (Cellia) turkhudi* Liston, 1901
61. *Anopheles (Cellia) vagus* Donitz, 1902

62. *Anopheles (Cellia) varuna* Iyengar, 1924
63. *Anopheles (Cellia) willmori* (James, 1903)

Subfamily: Culicinae

Tribe : Aedeomyiini Theobald, 1901

I. Genus *AEDEOMYIA* Theobald, 1901

(i) Subgenus Aedeomyia Theobald, 1901

64. *Aedeomyia (Aedeomyia) catasticta* Knab, 1909
65. *Aedeomyia (Aedeomyia) venustipes* (Skuse, 1889)

Tribe : Aedini Neveu-Lemaire, 1902

I. Genus *AEDIMORPHUS* Theobald, 1903

66. *Aedimorphus alboscutellatus* (Theobald, 1905)
67. *Aedimorphus caecus* (Theobald, 1901)
68. *Aedimorphus culicinus* (Edwards, 1922)
69. *Aedimorphus jamesi* (Edwards, 1914)
70. *Aedimorphus lowisii* (Theobald, 1910)
71. *Aedimorphus mediolineatus* (Theobald, 1901)
72. *Aedimorphus nigrostriatus* Barraud, 1927
73. *Aedimorphus pallidostriatus* (Theobald, 1907)
74. *Aedimorphus pampangensis* (Ludlow, 1905)
75. *Aedimorphus pipersalatus* (Giles, 1902)
76. *Aedimorphus punctifemoris* (Ludlow, 1921)
77. *Aedimorphus stenoetrus* (Theobald, 1907)
78. *Aedimorphus syntheticus* (Barraud, 1928)
79. *Aedimorphus taeniorhynchoides* (Christophers, 1911)
80. *Aedimorphus trimaculatus* (Theobald, 1905)
81. *Aedimorphus vexans* (Meigen, 1830)

II. Genus *ARMIGERES* Theobald, 1901

(i) Subgenus Armigeres Theobald,1901

82. *Armigeres (Armigeres) aureolineatus* (Leicester, 1908)
83. *Armigeres (Armigeres) durhami* (Edwards, 1917)
84. *Armigeres (Armigeres) joloensis* (Ludlow, 1904)
85. *Armigeres (Armigeres) kesseli* Ramalingam, 1987

86. *Armigeres (Armigeres) kuchingensis* Edwards, 1915
87. *Armigeres (Armigeres) mahantai* Bhattacharyya *et al.*, 2009
88. *Armigeres (Armigeres) pallithorax* Dong, Zhou & Dong, 2004
89. *Armigeres (Armigeres) subalbatus* (Coquillett, 1898)
90. *Armigeres (Armigeres) theobaldi* Barruad, 1934

(ii) Subgenus Leicesteria Theobald, 1904

91. **Armigeres (Leicesteria) annulipalpis** (Theobald, 1910)
92. **Armigeres (Leicesteria) annulitarsis** (Leicester, 1908)
93. **Armigeres (Leicesteria) dentatus** Barraud, 1927
94. **Armigeres (Leicesteria) digitatus** (Edwards, 1914)
95. **Armigeres (Leicesteria) dolichocephalus** (Leicester, 1908)
96. **Armigeres (Leicesteria) flavus** (Leicester, 1908)
97. **Armigeres (Leicesteria) inchoatus** Barraud, 1927
98. **Armigeres (Leicesteria) longipalpis** (Leicester, 1904)
99. **Armigeres (Leicesteria) magnus** (Theobald, 1908)
100. **Armigeres (Leicesteria) omissus** (Edwards, 1914)

III. Genus *AYURAKITIA* Thurman, 1954

101. *Ayurakitia peytoni* (Reinert, 1972)

IV. Genus *BRUCEHARRISONIUS* Reinert, 2003

102. *Bruceharrisonius aureostriatus* (Doleschall, 1857)
103. *Bruceharrisonius christophersi* (Edwards, 1922)
104. *Bruceharrisonius doonii* (Wattal, Bhatia & Kalra, 1958)
105. Bruceharrisonius greenii (Theobald, 1903)

V. Genus *CANCRAEDES* Edwards, 1929

106. *Cancraedes cancricomes* (Edwards, 1922)
107. *Cancraedes simplex* (Theobald, 1903)

VI. Genus *CHRISTOPHERSIOMYIA* Barraud, 1923

108. *Christophersiomyia annulirostris* (Theobald, 1905)
109. *Christophersiomyia gombakensis* (Mattingly, 1959)
110. *Christophersiomyia ibis* (Barraud, 1931)
111. *Christophersiomyia thomsoni* (Theobald, 1905)

VII. Genus *COLLESSIUS* Reinert, Harbach & Kitching, 2006

(i) Subgenus *Alloeomyia* Reinert, Harbach & Kitching, 2008

112. *Collessius (Alloeomyia) pseudotaeniatus* (Giles, 1901)

(ii) Subgenus *Collessius* Reinert, Harbach & Kitching, 2006

113. *Collessius (Collessius) elsiae* (Barraud, 1923)
114. *Collessius (Collessius) macdougalli* (Edwards, 1922)
115. *Collessius (Collessius) shortti* (Barraud, 1923)

VIII. Genus *DANIELSIA* Theobald, 1904

116. *Danielsia albotaeniata* Leicester, 1904
117. *Danielsia lepchana* (Barruad, 1923)

IX. Genus *DENDROSKUSEA* Edwards, 1929

118. *Dendroskusea kanarensis* (Edwards, 1934)
119. *Dendroskusea micropterus* (Giles, 1901)
120. *Dendroskusea periskelata* (Giles, 1902)
121. *Dendroskusea ramachandrai* (Reuben, 1967)
122. *Dendroskusea reginae* (Edwards, 1922)

X. Genus *DOWNSIOMYIA* Vargas, 1950

123. *Downsiomyia albolateralis* (Theobald, 1908)
124. *Downsiomyia albonivea* (Barruad, 1934)
125. *Downsiomyia mohani* (Knight, 1969)
126. *Downsiomyia nivea* (Ludlow, 1903)
127. *Downsiomyia niveoides* (Barruad, 1934)
128. *Downsiomyia novonivea* (Barraud, 1934)

XI. Genus *EDWARDSAEDES* Belkin,1962

129. *Edwardsaedes imprimens* (Walkar, 1860)

XII. Genus *FINLAYA* Theobald, 1903

130. *Finlaya flavipennis* Giles, 1904
131. *Finlaya poicilia* Theobald, 1903

XIII. Genus *FREDWARDSIUS* Reinert, 2000

132. *Fredwardsius vittatus* (Bigot, 1861)

XIV. Genus *GILESIUS* Reinert, Harbach & Kitching, 2006

133. *Gilesius pulchriventer* (Giles, 1901)

XV. Genus *HEIZMANNIA* Ludlow, 1905

(i) Subgenus *Heizmannia* Ludlow, 1905

134. *Heizmannia (Heizmannia) aureochaeta* (Leicester, 1908)
135. *Heizmannia (Heizmannia) chandi* Edwards, 1922
136. *Heizmannia (Heizmannia) chengi* Lien, 1968
137. *Heizmannia (Heizmannia) complex* (Theobald, 1910)
138. *Heizmannia (Heizmannia) covelli* Barraud, 1929
139. *Heizmannia (Heizmannia) funerea* (Leicester, 1908)
140. *Heizmannia (Heizmannia) greenii* (Theobald, 1905)
141. *Heizmannia (Heizmannia) himalayensis* Edwards, 1922
142. *Heizmannia (Heizmannia) indica* (Theobald, 1905)
143. *Heizmannia (Heizmannia) metallica* Leicester, 1908
144. *Heizmannia (Heizmannia) reidi* Mattingly, 1957
145. *Heizmannia (Heizmannia) viridis* Barraud, 1929

(ii) Subgenus *Mattinglyia* Lien, 1968

146. *Heizmannia (Mattinglyia) discrepans* (Edwards, 1922)
147. *Heizmannia (Mattinglyia) tripunctata* (Theobald, 1908)

XVI. Genus *HIMALAIUS* Reinert, Harbach & Kitching, 2006

148. *Himalaius gilli* (Barraud, 1924)
149. Himalaius simlensis (Edwards, 1922)

XVII. Genus *HOPKINSIUS* Reinert, Harbach & Kitching, 2008

(i) Subgenus *Yamada* Reinert, Harbach & Kitching, 2008

150. Hopkinsius (Yamada) albocinctus (Barraud, 1924)

XVIII. Genus *HULECOETEOMYIA* Theobald, 1904

151. *Hulecoeteomyia chrysolineata* (Theobald, 1907)
152. *Hulecoeteomyia formosensis* (Yamada, 1921)
153. *Hulecoeteomyia harveyi* (Barraud, 1923)
154. *Hulecoeteomyia pallirostris* (Edwards, 1922)
155. *Hulecoeteomyia saxicola* (Edwards, 1922)
156. *Hulecoeteomyia cherrapunjiensis* Natarajan *et al.*, 2016

XIX. Genus *INDUSIUS* Edwards, 1934

157. *Indusius pulverulentus* (Edwards, 1922)

XX. Genus *JIHLIENIUS* Reinert, Harbach & Kitching, 2006

158. *Jihlienius unicinctus* (Edwards, 1922)

XXI. Genus *KENKNIGHTIA* Reinert, 1990

159. *Kenknightia dissimilis* (Leicester, 1908)
160. *Kenknightia karwari* (Barraud, 1924)

XXII. Genus *LORRAINEA* Belkin,1962

161. *Lorrainea amesii* (Ludlow, 1903)
162. *Lorrainea fumida* (Edwards, 1928)

XXIII. Genus *MUCIDUS* Theobald, 1901

(i) Subgenus *Mucidus* Theobald, 1901

163. *Mucidus* (*Mucidus*) *laniger* (Wiedemann, 1820)
164. *Mucidus (Mucidus) quasiferinus* (Mattingly 1961)
165. *Mucidus (Mucidus) scatophagoides* Theobald, 1901

XXIV. Geunus *NEOMELANICONION* Newstead, 1907

166. *Neomelaniconion lineatopenne* (Ludlow, 1905)

XXV. Genus *OCHLEROTATUS* Lynch Arribalzaga, 1891 *Sensu*

(i) Subgenus *Finlaya* Theobald, 1903 *Sensu auctorum*

167. *Ochlerotatus (Finlaya) auronitens* (Edwards, 1922)
168. *Ochlerotatus (Finlaya) oreophilus* Edwards, 1916
169. *Ochlerotatus (Finlaya) sintoni* (Barraud, 1924)
170. *Ochlerotatus (Finlaya) suffusus* (Edwards, 1922)
171. *Ochlerotatus (Finlaya) versicolor* (Barraud,1924)

Subgenus uncertain

172. *Ochlerotatus pulcritarsis* (Rondani, 1872)
173. *Ochlerotatus pullatus* (Coquillett, 1904)

XXVI. Genus *PARAEDES* Edwards, 1934

174. *Paraedes barraudi* Edwards, 1934
175. *Paraedes chrysoscuta* (Theobald, 1910)
176. *Paraedes menoni* (Mattingly, 1958)
177. *Paraedes ostentatio* (Leicester, 1908)

XXVII. Genus PETERMATTINGLYIUS Reinert, Harbach & Kitching, 2009

(i) Subgenus *Petermattinglyius* Reinert, Harbach & Kitching, 2009

178. Petermattinglyius (Petermattinglyius) franciscoi (Mattingly, 1959)
179. Petermattinglyius (Petermattinglyius) iyengari (Edwards, 1923)

XXVIII. Genus *PHAGOMYIA* Theobald,1905

180. *Phagomyia assamensis* (Theobald 1908)
181. *Phagomyia cacharana* (Barraud, 1923)
182. *Phagomyia cogilli* (Edwards, 1922)
183. *Phagomyia deccana* (Barraud, 1923)
184. *Phagomyia feegradei* (Barraud, 1934)
185. *Phagomyia gubernatoris* (Giles, 1901)
186. *Phagomyia inquinata* (Edwards, 1922)
187. *Phagomyia khazani* (Edwards, 1922)
188. *Phagomyia lophoventralis* (Theobald, 1910)
189. *Phagomyia prominens* (Barraud, 1923)
190. *Phagomyia stevensoni* (Barraud, 1923)

XXIX. Genus *RHINOSKUSEA* Edwards, 1929

191. *Rhinoskusea longirostris* (Leicester, 1908)
192. *Rhinoskusea portonovoensis* (Tewari & Hiriyan, 1992)
193. *Rhinoskusea wardi* (Reinert, 1976)

XXX. Genus *SCUTOMYIA* Theobald, 1904

194. *Scutomyia albolineata* Theobald, 1904

XXXI. Genus *STEGOMYIA* Theobald, 1901

(i) Subgenus *Actinothrix* Reinert, Harbach & Kitching, 2009

195. *Stegomyia (Actinothrix) edwardsi* Barraud, 1923
196. *Stegomyia (Actinothrix) seampi* (Huang, 1974)

(ii) Subgenus *Heteraspidion* Reinert, Harbach & Kitching, 2009

197. *Stegomyia (Heteraspidion) annandalei* Theobald, 1910
198. *Stegomyia (Heteraspidion) craggi* Barraud, 1923

(iii) Subgenus *Huangmyia* Reinert, Harbach & Kitching, 2009

199. *Stegomyia (Huangmyia) mediopunctata* Theobald, 1905
200. *Stegomyia (Huangmyia) perplexa* Leicester, 1908

(iv) Subgenus *Stegomyia* Theobald, 1901

201. *Stegomyia (Stegomyia) aegypti* (Linnaeus, 1762)

(v) Subgenus *Xyele* Reinert, Harbach & Kitching, 2009

202. *Stegomyia (Xyele) desmotes* Giles, 1904

[subgenus unassigned yet for following]

203. *Stegomyia albopicta* (Skuse, 1895)
204. *Stegomyia flavopicta* (Yamada, 1921)
205. *Stegomyia gardnerii imitator* (Leicester, 1908)
206. *Stegomyia krombeini* (Huang, 1975)
207. *Stegomyia malayensis* (Colless, 1962)
208. *Stegomyia novalbopicta* (Barraud, 1931)
209. *Stegomyia patriciae* (Mattingly, 1954)
210. *Stegomyia pseudalbopicta* Borel, 1928
211. *Stegomyia scutellaris* (Walker, 1859)
212. *Stegomyia subalbopicta* (Barraud, 1931)
213. *Stegomyia unilineata* (Theobald, 1906)
214. *Stegomyia w-alba* Theobald, 1905

XXXII. Genus *TEWARIUS* Reinert, 2006

215. *Tewarius agastyai* (Tewari & Hiriyan, 1992)
216. *Tewarius nummatus* (Edwards, 1923)
217. *Tewarius reubenae* (Tewari & Hiriyan, 1992)

XXXIII. Genus *UDAYA* Thurman, 1954

218. *Udaya argyrurus* (Edwards, 1934)
219. *Udaya subsimilis* (Barraud, 1927)

XXXIV. Genus *VERRALLINA* Theobald, 1903

(i) Subgenus *Harbachius* Reinert, 1999

220. *Verrallina (Harbachius) abdita* (Barraud, 1931)
221. *Verrallina (Harbachius) consonensis* (Reinert, 1973)
222. *Verrallina (Harbachius) uniformis* (Theobald, 1910)
223. *Verrallina (Harbachius) yusafi* (Barraud, 1931)

(ii) Subgenus *Neomacleaya* Theobald,1907

224. *Verrallina (Neomacleaya) agrestis* (Barraud, 1931)
225. *Verrallina (Neomacleaya) andamanensis* (Edwards, 1922)
226. *Verrallina (Neomacleaya) assamensis* Bhattacharyya *et al.*, 2004

227. *Verrallina (Neomacleaya) atria* (Barraud, 1928)
228. *Verrallina (Neomacleaya) cauta* (Barraud, 1928)
229. *Verrallina (Neomacleaya) clavata* (Barraud, 1931)
230. *Verrallina (Neomacleaya) comata* (Barraud, 1931)
231. *Verrallina (Neomacleaya) hirsutipleura* (Barraud, 1928)
232. *Verrallina (Neomacleaya) indica* (Theobald, 1907)
233. *Verrallina (Neomacleaya) pseudodiurna* (Theobald, 1910)
234. *Verrallina (Neomacleaya) pseudomediofasciata* (Theobald, 1910)
235. *Verrallina (Neomacleaya) rami* (Barraud, 1928)
236. *Verrallina (Neomacleaya) seculata* (Menon, 1950)
237. *Verrallina (Neomacleaya) vallistris* (Barraud, 1928)
238. *Verrallina (Neomacleaya) unca* (Theobald,1901)

(iii) Subgenus *Verrallina* Theobald,1903

239. *Verrallina (Verrallina) butleri* (Theobald, 1901)
240. *Verrallina* (*Verrallina*) *dux* (Dyar & Shannon, 1925)
241. *Verrallina (Verrallina) lugubris* (Barraud, 1928)

Tribe: Culicini Meigen, 1818

I. Genus *CULEX* Linnaeus, 1758

(i) Subgenus *Barraudius* Edwards, 1921

242. *Culex* (*Barraudius*) *modestus* Ficalbi, 1890

(ii) Subgenus *Culex* Linnaeus, 1758

243. *Culex (Culex) alienus* Colless, 1957
244. *Culex (Culex) annulus* Theobald, 1901
245. *Culex (Culex) barraudi* Edwards, 1922
246. *Culex (Culex) edwardsi* Barraud, 1923
247. *Culex (Culex) fuscifurcatus* Edwards, 1937
248. *Culex (Culex) fuscocephala* Theobald, 1907
249. *Culex (Culex) gelidus* Theobald, 1901
250. *Culex (Culex) hutchinsoni* Barraud, 1924
251. *Culex (Culex) jacksoni* Edwards, 1934
252. *Culex (Culex) mimeticus* Noe, 1899
253. *Culex (Culex) mimuloides* Barraud, 1924
254. *Culex (Culex) mimulus* Edwards, 1915
255. *Culex (Culex) murrelli* Lien, 1968

256. *Culex (Culex) nilgiricus* Edwards, 1916
257. *Culex (Culex) perexiguus* Theobald, 1903
258. *Culex (Culex) perplexus* Leicester, 1908
259. *Culex (Culex) pseudovishnui* Colless, 1957
260. *Culex (Culex) quinquefasciatus* Say, 1823
261. *Culex (Culex) sitiens* Wiedemann, 1828
262. *Culex (Culex) theileri* Theobald, 1903
263. *Culex (Culex) tritaeniorhynchus* Giles, 1901
264. *Culex (Culex) univittatus* Theobald, 1901
265. *Culex (Culex) vagans* Wiedemann, 1828
266. *Culex (Culex) vishnui* Theobald, 1901
267. *Culex (Culex) whitei* Barraud, 1923
268. *Culex (Culex) whitmorei* (Giles, 1904)

(iii) Subgenus *Culiciomyia* Theobald, 1907

269. *Culex (Culiciomyia) bailyi* Barraud, 1934
270. *Culex (Culiciomyia) fragilis* Ludlow, 1903
271. *Culex (Culiciomyia) harrisoni* Sirivanakarn, 1977
272. *Culex (Culiciomyia) nigropunctatus* Edwards, 1926
273. *Culex (Culiciomyia) pallidothorax* Theobald, 1905
274. *Culex (Culiciomyia) ramakrishnii* Wattal & Kalra, 1965
275. *Culex (Culiciomyia) scanloni* Bram, 1967
276. *Culex (Culiciomyia) shebbearei* Barraud, 1924
277. *Culex (Culiciomyia) spathifurca* (Edwards, 1915)
278. *Culex (Culiciomyia) viridiventer* Giles, 1901

(iv) Subgenus *Eumelanomyia* Theobald, 1909

279. *Culex (Eumelanomyia) brevipalpis* (Giles, 1902)
280. *Culex (Eumelanomyia) castrensis* Edwards, 1922
281. *Culex (Eumelanomyia) foliatus* Brug, 1932
282. *Culex (Eumelanomyia) hinglungensis* Chu, 1957
283. *Culex (Eumelanomyia) iphis* Barraud, 1924
284. *Culex (Eumelanomyia) khazani* Edwards, 1922
285. *Culex (Eumelanomyia) malayi* (Leicester, 1908)
286. *Culex (Eumelanomyia) mohani* Sirivanakarn, 1977
287. *Culex (Eumelanomyia) pluvialis* Barraud, 1924
288. *Culex (Eumelanomyia) tenuipalpis* Barraud, 1924

(v) Subgenus *Lophoceraomyia* Theobald, 1905

289. *Culex (Lophoceraomyia) aculeatus* Colless, 1965
290. *Culex (Lophoceraomyia) bengalensis* Barraud, 1934
291. *Culex (Lophoceraomyia) bicornutus* (Theobald, 1910)
292. *Culex (Lophoceraomyia) cinctellus* Edwards, 1922
293. *Culex (Lophoceratomyia) cubitatus* Colless, 1965
294. *Culex (Lophoceraomyia) demissus* Colless, 1965
295. *Culex (Lophoceraomyia) flavicornis* Barraud, 1924
296. *Culex (Lophoceraomyia) gracicornis* Sirivanakarn, 1977
297. *Culex (Lophoceraomyia) inculus* Colless, 1965
298. *Culex (Lophoceraomyia) infantulus* Edwards, 1922
299. *Culex (Lophoceraomyia) lasiopalpis* Sirivanakarn, 1977
300. *Culex (Lophoceraomyia) macdonaldi* Colless, 1965
301. *Culex (Lophoceraomyia) mammilifer* (Leicester, 1908)
302. *Culex (Lophoceraomyia) minor* (Leicester, 1908)
303. *Culex (Lophoceraomyia) minutissimus* (Theobald, 1907)
304. *Culex (Lophoceraomyia) paraculeatus* Sirivanakarn, 1977
305. *Culex (Lophoceraomyia) peytoni* Bram & Rattanarithikul, 1967
306. *Culex (Lophoceraomyia) pholeter* Bram & Rattanarithikul, 1967
307. *Culex (Lophoceraomyia) pilifemoralis* Wang & Feng, 1964
308. *Culex (Lophoceraomyia) quadripalpis* (Edwards, 1914)
309. *Culex (Lophoceraomyia) raghavanii* Rahman, Chowdhury & Kalra, 1968
310. *Culex (Lophoceraomyia) rubithoracis* (Leicester, 1908)
311. *Culex (Lophoceraomyia) seniori* Barraud, 1934
312. *Culex (Lophoceraomyia) singhbhumensis* Natarajan & Rajavel, 2009
313. *Culex (Lophoceraomyia) uniformis* (Theobald, 1905)
314. *Culex (Lophoceraomyia) variatus* (Leicester, 1908)
315. *Culex (Lophoceraomyia) wilfredi* Colless, 1965

(vi) Subgenus *Maillotia* Theobald, 1907

316. *Culex* (*Maillotia*) *hortensis* Ficalbi, 1889

(vii) Subgenus *Oculeomyia* Theobald, 1907

317. *Culex* (*Oculeomyia*) *bitaeniorhynchus* Giles, 1901
318. *Culex* (*Oculeomyia*) *cornutus* Edwards, 1922
319. *Culex* (*Oculeomyia*) *epidesmus* (Theobald, 1910)
320. *Culex* (*Oculeomyia*) *infula* Theobald, 1901

321. *Culex* (*Oculeomyia*) *luzonensis* Sirivanakarn, 1976
322. *Culex* (*Oculeomyia*) *sinensis* Theobald, 1903

II. Genus *LUTZIA* Theobald, 1903

(i) Subgenus *Metalutzia* Tanaka, 2003
323. *Lutzia* (*Metalutzia*) *agranensis* Singh & Prakash, 2008
324. *Lutzia* (*Metalutzia*) *fuscana* (Wiedemann, 1820)
325. *Lutzia* (*Metalutzia*) *halifaxii* (Theobald, 1903)
326. *Lutzia* (*Metalutzia*) *vorax* Edwards, 1921

Tribe: Culisetini Belkin, 1962

I. Genus *CULISETA* Felt, 1904

(i) Subgenus *Allotheobaldia* Broelemann,1919
327. *Culiseta (Allotheobaldia) longiareolata* (Macquart, 1838)

(ii) Subgenus *Culiseta* Felt,1904
328. *Culiseta (Culiseta) alaskaensis indica* (Edwards, 1920)
329. *Culiseta (Culiseta) niveitaeniata* (Theobald, 1907)

Tribe: Ficalbiini Belkin, 1962

I. Genus *FICALBIA* Theobald, 1903

330. *Ficalbia minima* (Theobald, 1901)

II. Genus *MIMOMYIA* Theobald, 1903

(i) Subgenus *Etorleptiomyia* Theobald, 1904
331. *Mimomyia (Etorleptiomyia) luzonensis* (Ludlow, 1905)

(ii) Subgenus *Ingramia* Edwards, 1912
332. *Mimomyia* (*Ingramia*) *fusca* (Leicester, 1908)

(iii) Subgenus *Mimomyia* Theobald, 1903
333. *Mimomyia* (*Mimomyia*) *aurea* (Leicester, 1908)
334. *Mimomyia (Mimomyia) chamberlaini* Ludlow, 1904 ssp. *clavipalpus* (Theobald, 1908)
335. *Mimomyia* (*Mimomyia*) *hybrida* (Leicester, 1908)
336. *Mimomyia (Mimomyia) intermedia* (Barraud, 1929)

Tribe: Hodgesiini Belkin,1962

I. Genus *HODGESIA* Theobald, 1904

337. *Hodgesia bailyi* Barraud, 1929
338. *Hodgesia malayi* Leicester, 1908

Tribe : Mansoniini Belkin,1962

I. Genus *COQUILLETTIDIA* Dyar, 1905

(i) Subgenus *Coquillettidia* Dyar, 1905

339. *Coquillettidia (Coquillettidia) crassipes* (van der Wulp, 1881)
340. *Coquillettidia (Coquillettidia) novochracea* (Barraud, 1927)
341. *Coquillettidia (Coquillettidia) ochracea* (Theobald, 1903)

II. Genus *MANSONIA* Blanchard, 1901

(i) Subgenus *Mansonioides* Theobald, 1907

342. *Mansonia* (*Mansonioides*) *annulifera* (Theobald, 1901)
343. *Mansonia* (*Mansonioides*) *dives* (Schiner, 1868)
344. *Mansonia* (*Mansonioides*) *indiana* Edwards, 1930
345. *Mansonia* (*Mansonioides*) *uniformis* (Theobald, 1901)

Tribe: Orthopodomyiini Belkin, Heinemann & Page, 1970

I. Genus *ORTHOPODOMYIA* Theobald, 1904

346. *Orthopodomyia albipes* Leicester, 1904
347. *Orthopodomyia andamanensis* Barraud, 1934
348. *Orthopodomyia anopheloides* (Giles, 1903)
349. *Orthopodomyia flavicosta* Barraud, 1927
350. *Orthopodomyia flavithorax* Barraud, 1927

Tribe: Sabethini Blanchard,1905

I. Genus *MALAYA* Leicester, 1908

351. *Malaya genurostris* Leicester, 1908
352. *Malaya jacobsoni* (Edwards, 1930)

II. Genus *TOPOMYIA* Leicester, 1908

(i) Subgenus *Suaymyia* Thurman, 1959

353. *Topomyia* (*Suaymyia*) *cristata* Thurman,1959

(ii) Subgenus *Topomyia* Leicester, 1908

354. **Topomyia (Topomyia) aureoventer** (Theobald, 1910)

355. **Topomyia (Topomyia) bifurcata** Dong, Wang & Lu, 1995

356. **Topomyia (Topomyia) hirtusa** Gong,1989

III. Genus *TRIPTEROIDES* Giles, 1904

(i) Subgenus *Rachionotomyia* Theobald, 1905

357. *Tripteroides (Rachionotomyia) affinis* (Edwards, 1913)

358. *Tripteroides (Rachionotomyia) aranoides* (Theobald, 1901)

359. *Tripteroides (Rachionotomyia) ceylonensis* (Theobald 1905)

360. *Tripteroides (Rachionotomyia) coonorensis* Mattingly,1981

361. *Tripteroides (Rachionotomyia) edwardsi* (Barraud, 1929)

362. *Tripteroides (Rachionotomyia) serratus* (Barraud, 1929)

(ii) Subgenus *Tripteroides* Giles, 1904

363. *Tripteroides (Tripteroides) indicus* (Barraud, 1929)

364. *Tripteroides (Tripteroides) powelli* (Ludlow,1909)

365. *Tripteroides (Tripteroides) similis* (Leicester, 1908)

366. *Tripteroides (Tripteroides) tarsalis* Delfinado & Hodges, 1968

Tribe: Toxorhynchitini Lahille, 1904

I. Genus *TOXORHYNCHITES* Theobald, 1901

(i) Subgenus *Toxorhynchites* Theobald, 1901

367. *Toxorhynchites (Toxorhynchites)* albipes (Edwards, 1922)

368. *Toxorhynchites (Toxorhynchites)* edwardsi (Barraud, 1924)

369. *Toxorhynchites (Toxorhynchites)* gravely (Edwards, 1921)

370. *Toxorhynchites (Toxorhynchites)* kempi (Edwards,1921)

371. *Toxorhynchites (Toxorhynchites) klossi* (Edwards, 1921)

372. *Toxorhynchites (Toxorhynchites)* metallicus Leicester 1904

373. *Toxorhynchites (Toxorhynchites)* minimus (Theobald, 1905)

374. *Toxorhynchites (Toxorhynchites)* splendens (Wiedemann, 1819)

375. *Toxorhynchites (Toxorhynchites) tyagii* Krishnamoorthy *et al.*, 2013

Tribe: Uranotaeniini Lahille,1904

I. Genus *URANOTAENIA* Lynch Arribalzaga, 1891

(i) Subgenus *Pseudoficalbia* Theobald, 1912

376. *Uranotaenia (Pseudoficalbia) atra* Theobald, 1905

377. *Uranotaenia (Pseudoficalbia) bicolor* Leicester, 1908
378. *Uranotaenia (Pseudoficalbia) bimaculata* Leicester, 1908
379. *Uranotaenia (Pseudoficalbia) dibrugarhensis* Bhattacharyya *et al.*, 2004
380. *Uranotaenia (Pseudoficalbia) husaini* Qutubuddin, 1947
381. *Uranotaenia (Pseudoficalbia) luteola* Edwards, 1934
382. *Uranotaenia (Pseudoficalbia) lutescens* Leicester, 1908
383. *Uranotaenia (Pseudoficalbia) maculipleura* Leicester, 1908
384. *Uranotaenia (Pseudoficalbia) mattinglyi* Qutubuddin, 1951
385. *Uranotaenia (Pseudoficalbia) maxima* Leicester, 1908
386. *Uranotaenia (Pseudoficalbia) nivipleura* Leicester, 1908
387. *Uranotaenia (Pseudoficalbia) novobscura* Barraud, 1934
388. *Uranotaenia (Pseudoficalbia) obscura* Edwards, 1915
389. *Uranotaenia (Pseudoficalbia) ohamai* Tanaka, Mizusawa & Saugstad, 1975
390. *Uranotaenia (Pseudoficalbia) recondita* Edwards, 1922
391. *Uranotaenia (Pseudoficalbia) stricklandi* Barraud, 1926
392. *Uranotaenia (Pseudoficalbia) unguiculata* Edwards, 1913
393. *Uranotaenia (Pseudoficalbia) maikalensis* Natarajan *et al.*, 2017
394. *Uranotaenia (Pseudoficalbia) dandakaranyensis* Natarajan *et al.*, 2017
395. *Uranotaenia (Pseudoficalbia) satpuraensis* Natarajan *et al.*, 2017

(ii) Subgenus *Uranotaenia* Lynch Arribalzaga, 1891

396. *Uranotaenia (Uranotaenia) alboannulata* (Theobald, 1905)
397. *Uranotaenia (Uranotaenia) annandalei* Barraud, 1926
398. *Uranotaenia (Uranotaenia) campestris* Leicester, 1908
399. *Uranotaenia (Uranotaenia) christophersi* Barraud, 1926
400. *Uranotaenia (Uranotaenia) edwardsi* Barraud, 1926
401. *Uranotaenia (Uranotaenia) hebes* Barraud, 1931
402. *Uranotaenia (Uranotaenia) lateralis* Ludlow, 1905
403. *Uranotaenia (Uranotaenia) longirostris* Leicester, 1908
404. *Uranotaenia (Uranotaenia) macfarlanei* Edwards, 1914
405. *Uranotaenia (Uranotaenia) micans* Leicester, 1908
406. *Uranotaenia (Uranotaenia) orientalis* Barraud, 1926
407. *Uranotaenia (Uranotaenia) testacea* Theobald, 1905

Chapter 6

IDENTIFICATION KEYS FOR INDIAN VECTOR MOSQUITOES - *4TH INSTAR LARVA AND ADULT*

6.1. Introduction

Identification of vectors is of prime importance in the field of medical entomology for both operational and research purposes. In the broad sense, vectors encompass any organism involved in the transmission of an infectious agent. More specifically, in medical and veterinary entomology, vectors designate any haematophagous arthropod which actively transmits an infectious pathogen from an infected host to a new host (Duvallet *et al.*, 2018). In a given zoological group, vector species are ordinarily outnumbered by other non-vector species. For instance, among the 540 species of mosquito in the subfamily Anophelinae, about 60 species are recognised vectors of Plasmodium, the agent of malaria (Cohuet *et al.*, 2010). Identification of arthropod species, both vectors and non-vectors, is a core capacity in the field of medical and veterinary entomology. Accurate identification of a species involved in transmitting a pathogen is essential for a proper understanding of the mechanisms that govern any biological system. Knowing which arthropod species transmit vector-borne diseases is also vital to providing species-specific focused control programmes and to correctly identifying disease risk and exposure (Failloux *et al.*, 2017).

Different approaches to mosquito identification are available: morphological, molecular (PCR and nucleic acids sequencing), proteomics tools, isozyme analysis, etc. (Fontenille *et al.*, 2017; Nayduch, 2009). However, for historical and technical reasons, morphological identification despite three principal limitations (the first being that it relies on expert entomologists performing the identification, especially when many species are present in the collection area; the second limitation resides in the level of preservation of the morphological characters used in the sample handling process, and the third concerns species complexes and a few morphologically similar species where identification should be based on more than one developmental stage, which is not always accessible) is still the reference method for both research purposes and operational surveillance as it requires little technical equipment, is easy to implement in the field and is inexpensive even when large numbers of individuals need to be identified.

Identification keys are provided for male adults and fourth stage larvae of the of mosquito species that transmit vector-borne diseases such as malaria, lymphatic filariasis, dengue/chikungunya/Zika and Japanese encephalitis in India.

6.2. Larval Characters and Keys

Head: Inner, outer and posterior clypeal hairs: They may be placed close together or with their bases widely separated; these hairs may be simple, very finely frayed, or may bear conspicuous lateral branches. Head setae 4-C, 5-C, 6-C, and 7-C: simple, multiple branches, distance between each hairs; hypostomal sutures well developed, incomplete or absent numbers of lateral teeth on mentum.

Antenna: Size of antenna; setae 2, 3-A as long as or longer than the proximal part.

Thorax: Pleural area of thorax speculate or non-speculate; hairs on Pro, Meso and Meta thoracic pleural hairs simple or pectinate.

Abdomen: Size of the anterior tergal plates; presence or absence of palmate hairs; number and rows of comb scales; presence or absence of sclerotized comb plate; size and shape of individual comb scales.

Siphon: Presence or absence of siphon and pecten teeth; size of the siphon; number of subventral setae; siphonal valves simple or modified.

Saddle: Ventral brush (4x) single or with many pairs.

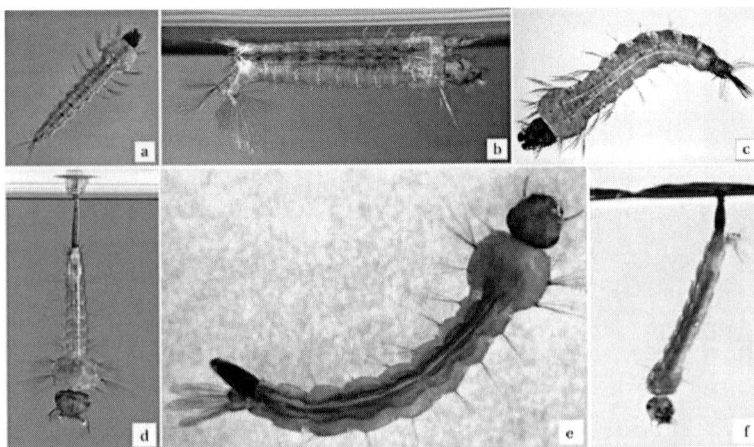

Fig. 6.1. Larvae of Anopheles (a,b), Culex (c,d) and Aedes (e,f) mosquitoes in dorsal view and while bracing water surface for respiration. (Source: Free Wikipedia)

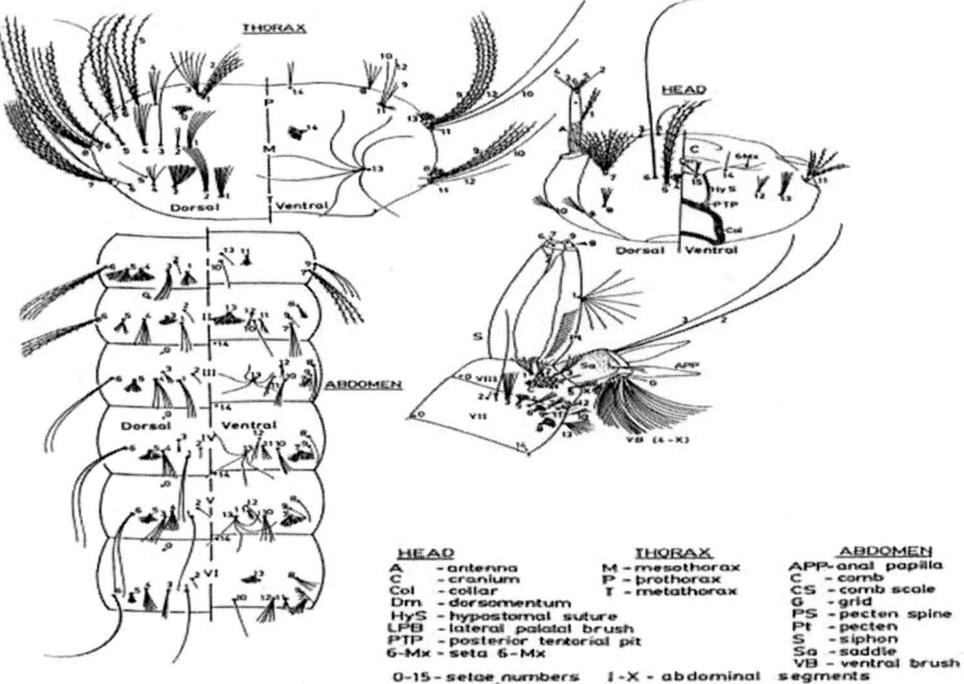

Fig. 6.2. Larval Characters for Head, Thorax and Abdomen.

6.2.1. Identification key for Genera.

1. Respiratory siphon absent; abdominal seta 1 palmate, one at least on segments IV-VII Fig.1) .... *Anopheles*

   Respiratory siphon present; abdominal seta 1 never palmate (Fig.4a) 2

2(1). Lateral palatal brush (mouth brush) with 10 thick curved, simple filaments; antenna with setae 2, 3-A arising basad of seta 1-A (Fig.2 a, b), comb scales and pecten absent .... *Toxorhynchites*

   Lateral palatal brush with numerous fine setae; antenna with setae 2,3-A distad of 1-A (Fig. 3), comb scales always present; pecten present or absent 3

3(2). Ventral bush (4-X) with a single pair of setae; siphon with two or (usually) more sub-dorsal setae as well as various ventral and sub-ventral setae; antenna short, without articulated apical segment (Fig. 4 a, b) 4

   Without this combination or characters, ventral brush (4-X) with 3 or more pairs of setae (Fig.15a) ... 6

4(3). Thoracic setae 5,6-P large, fan shaped tufts arising from a common tubercle (Fig. 5); 6-M and 7-T never stout spines; comb usually a patch of scales in two or more rows 5

   Thoracic seta 6-M and / or seta 7-T often a stout spine (Fig. 6); comb scales in single row, sometimes arising from a sclerotized plate scale .... *Tripteroides*

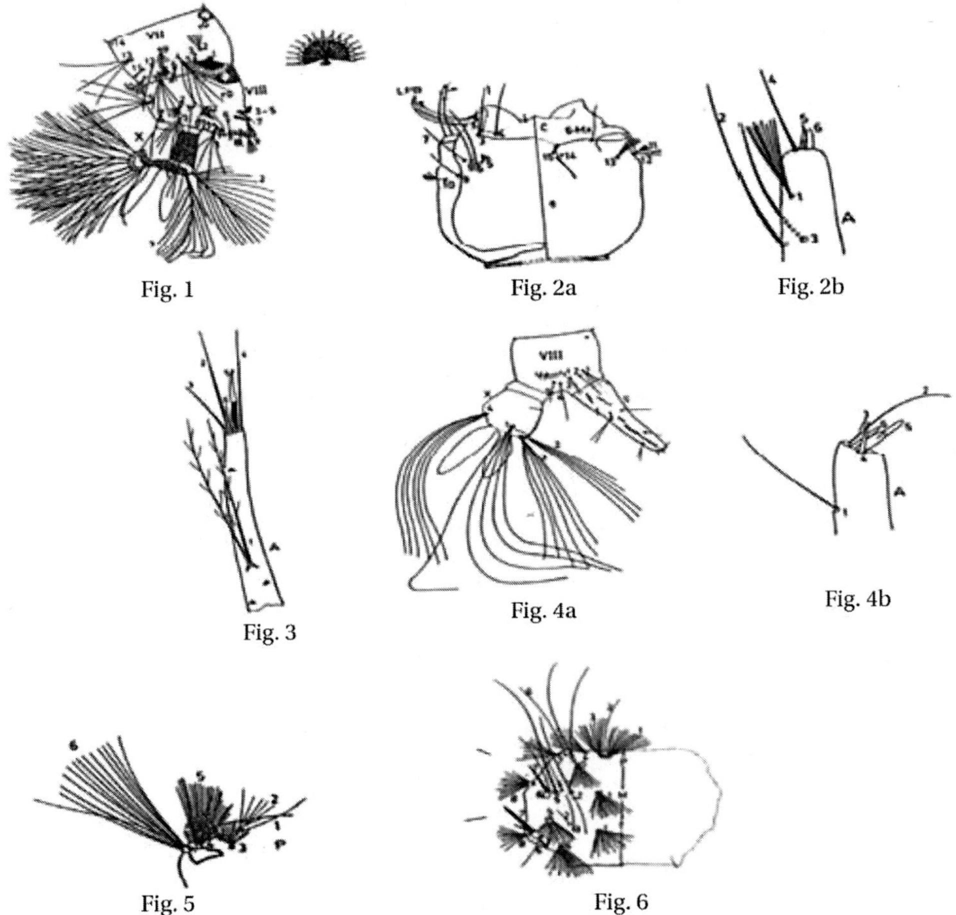

Fig. 1 Fig. 2a Fig. 2b

Fig. 3 Fig. 4a Fig. 4b

Fig. 5 Fig. 6

Legends to the Figures **(1-6) : Fig. 1**- Abdomen seta-1 palmate; **Fig. 2a**- Mouth brush; **Fig. 2b**- Setae 2,3-A position; **Fig. 3**- Setae 2,3-A position; **Fig. 4a**- Ventral brush (4-X); **Fig. 4b**- Antenna short; **Fig. 5**- Thoracic setae 5,6-P; **Fig. 6**- Setae 6-M & setae 7-T.

5(4). Abdominal segments V-VI with one or more pairs of stellate setae with numerous, short, stiff branches; maxillae with conspicuous horn; siphon at least 6 times as long as saddle (Fig. 7a,b,c) .... *Topomyia*

Maxillary horn absent; abdominal segments V-VI never with stellate setae; siphon at most 4 times as long as saddle (Fig. 8) .... *Malaya*

6(3). Siphonal valves modified for piercing tissues of Aquatic plants with sclerotized saw-toothed process at tip (Fig. 9a) 7

Siphon not so modified or, if so, without any saw-toothed process.. 8

7(6). Antenna with part distal to setae 2,3-A flexible, as long as or longer than the proximal part (Fig. 9b) ... *Coquillettidia*

Antenna with part distal to setae 2,3-A not flexible, less than 0.5 of the proximal part (Fig. 9c) .... *Mansonia*

8(6). Antenna broad flattened; some thoracic setae enormously long, others stellate; tip of siphon with paired hooks and branched setae (Fig. 10a, b, c) .... *Aedeomyia*

Antenna, thoracic setae and siphon otherwise 9

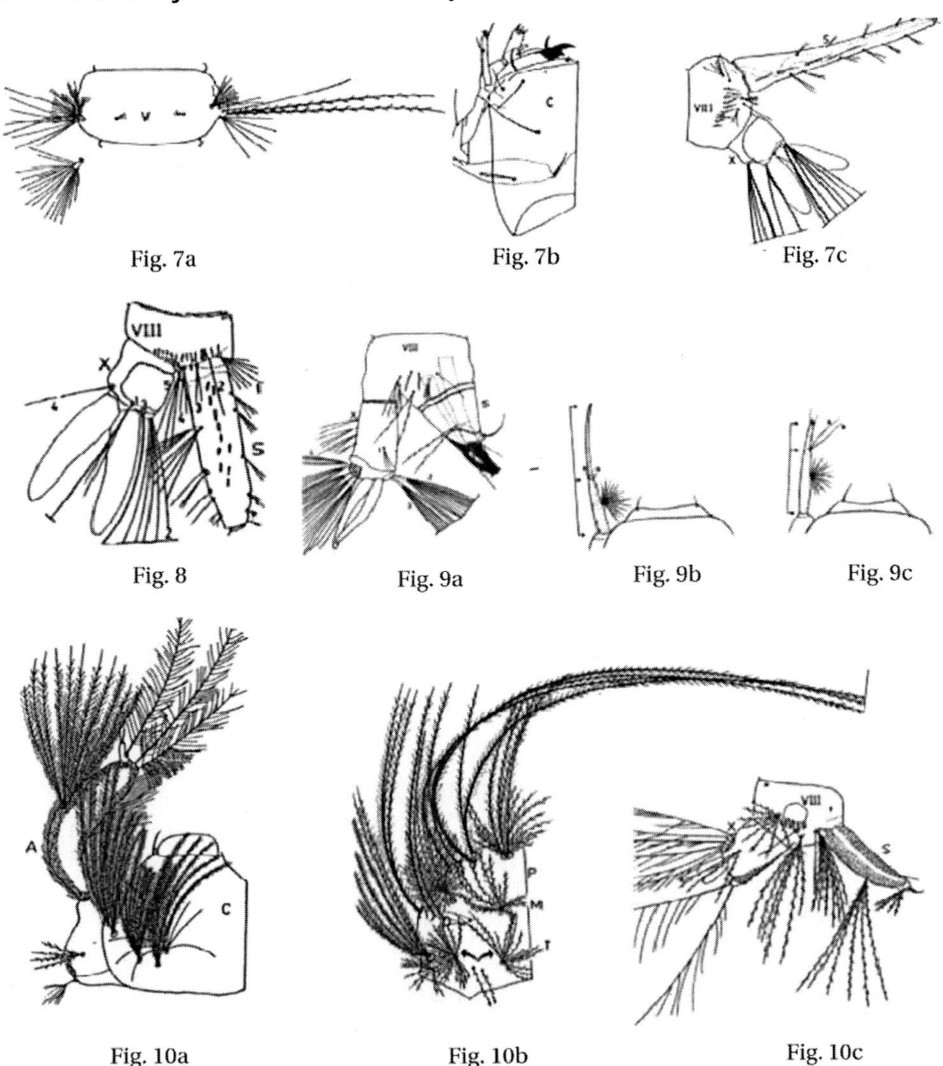

Fig. 7a Fig. 7b Fig. 7c

Fig. 8 Fig. 9a Fig. 9b Fig. 9c

Fig. 10a Fig. 10b Fig. 10c

Legends to the Figures **(7-10) : Fig. 7a**- Stellate setae in abdominal segments V-VI; **Fig. 7b**- Maxillary horn; **Fig. 7c**- Long siphon; **Fig. 8**- Short siphon; **Fig. 9a**- Modified siphon; **Fig. 9b**- Setae 2,3-A length; **Fig. 9c**- Setae 2,3-A length; **Fig. 10a**- Antenna broad flattened; **Fig. 10b**- Long thoracic setae; **Fig. 10c**- Siphon with hooks.

9(8). Siphon with a single pair of sub-ventral setae arising at not more than 0.2 of the distance from base; comb in a single row of at most 20 scales; siphon as long as saddle (Fig. 11a) 10

Without this combination of characters.................. 11

10(9). Pecten with at least 3 teeth, usually more; Seta 1-C simple; Seta 6-C single, long; Seta 4-C nearly as long as 5-C (Fig. 11a,b).. *Hodgesia*

Head seta 1-C strongly barbed; pecten with at most 2 teeth (Fig. 12a, b) .... *Ficalbia*

11(9). Distal portion of antenna freely articulated, setae 2, 3-A attached at joint; siphon with a single pair of sub-ventral setae (Fig. 13a,b); pecten absent or sometime present; ventral brush with 2-4 pairs of setae .... *Mimomyia*

Antenna never thus; siphons often otherwise; ventral brush with at least 3 pairs of setae, often with 5 or more pairs.. 12

12(11). Pecten absent; siphon very short and broad with a single pair of sub-ventral setae arising beyond half way; antennal seta 1-A minute (Fig.14a, b) .... *Armigeres*

Without this combination of characters 13

13(12). Pecten absent; siphon very long, with a single pair of sub-ventral setae; ventral brush with 6 pairs of setae or more; antennal seta 1-A arising on basal half, with 4 or more branches (Fig.15a,b) .... *Orthopodomyia*

Without this combination of characters 14

14(13). Siphon with a single pair of sub-ventral setae arising Not less than 0.25 distances from base........ 15

Siphon with several pairs of sub-ventral setae or if with a single pair then arising about 0.2 distances from the base 18

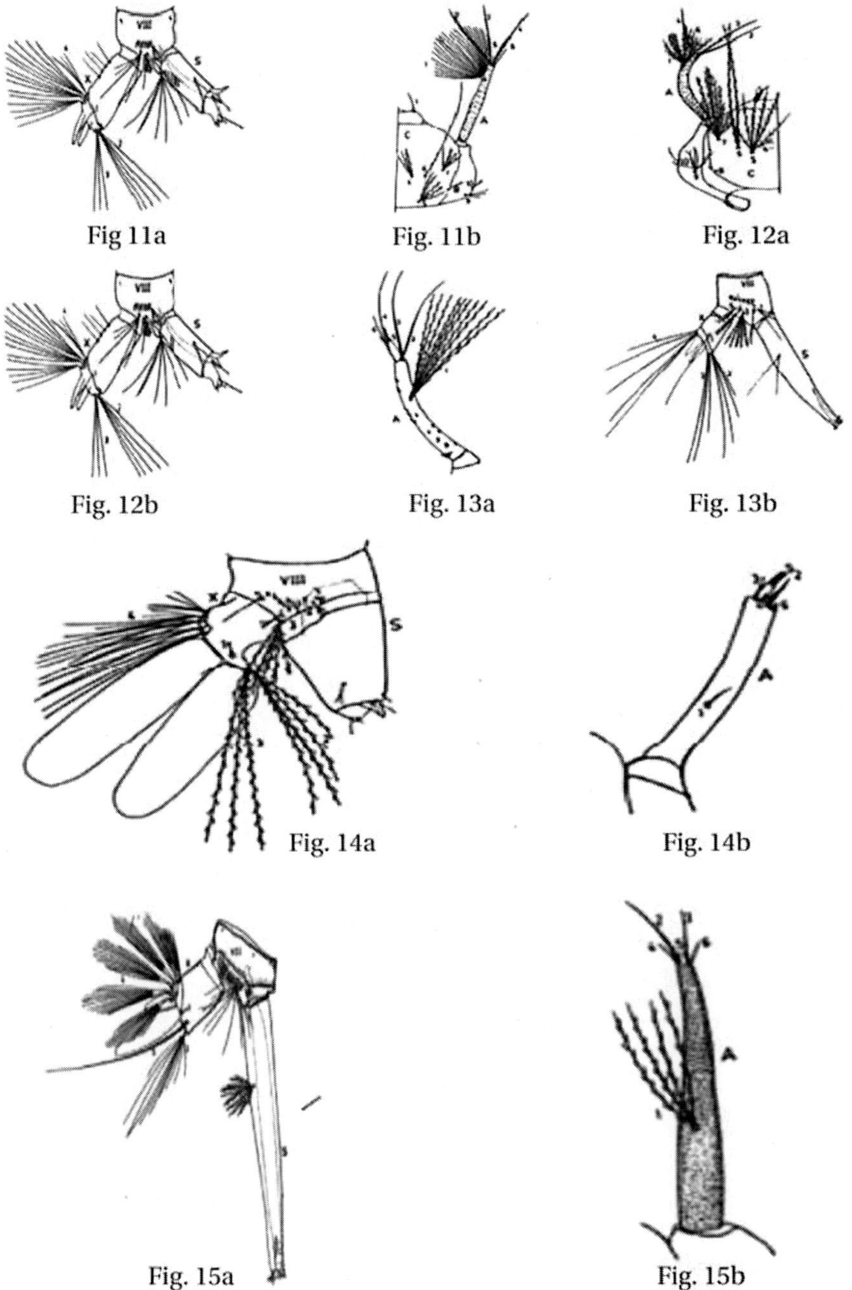

Fig 11a Fig. 11b Fig. 12a

Fig. 12b Fig. 13a Fig. 13b

Fig. 14a Fig. 14b

Fig. 15a Fig. 15b

Legends to the Figures **(11-15) : Fig.11a**- Sub-ventral setae; **Fig. 11b**- Seta 6-C; **Fig. 12a**- Head seta 1-C; Fig. 12b- Pecten teeth; **Fig. 13a**- Position of setae 2,3-A; **Fig. 13b**- Siphon with single pair of sub-ventral setae; **Fig. 14a**- Pecten absent; **Fig. 14b**- Seta 1-A minute; **Fig. 15a**- Long siphon with sub-ventral setae; **Fig. 15b**- Seta 1-A with 4 or more branches.

15(14). Hypostomal sutures absent or incomplete, not reaching posterior tentorial pit; head setae 5-C or 6-C or both sometimes spine like; comb often arising from a large sclerotized plate (Fig 16 a, b) ... *Uranotaenia*

Hypostomal sutures well developed, extending from the level of the mentum to the posterior tentorial pit; head setae 5-C or 6-C never spine like (Fig.17a); comb plate, if present, smaller 16

16(15). Comb with at most 10 teeth, in a single row; head seta 5-C single, 4,6-C shorter than 5-C, with 2 or more delicate branches, 7-C single and much longer than any of these (Fig.17b,c). .... *Udaya*

Without this combination of characters 17

17(16). Head seta 4-C large multiple branch subequal in size to 7-C; 6-C with 2 unequal branches (Fig.18a); 7-C at least 5 branched; thoracic integument without spicules; comb plate absent; saddle incomplete .... *Heizmannia*

Head seta 4-C small variously branched, less than 0.5 length of 7-C; 6-C single or with two equal branches; Seta 4-C closer to 6-C than to 5-C (Fig. 18b) .... *Aedes*

18(14). Siphon with a single pair of sub-ventral setae arising near the base (Fig.19) .... *Culiseta*

Siphon with several pairs of sub-ventral setae (Fig. 20) .... *Culex*

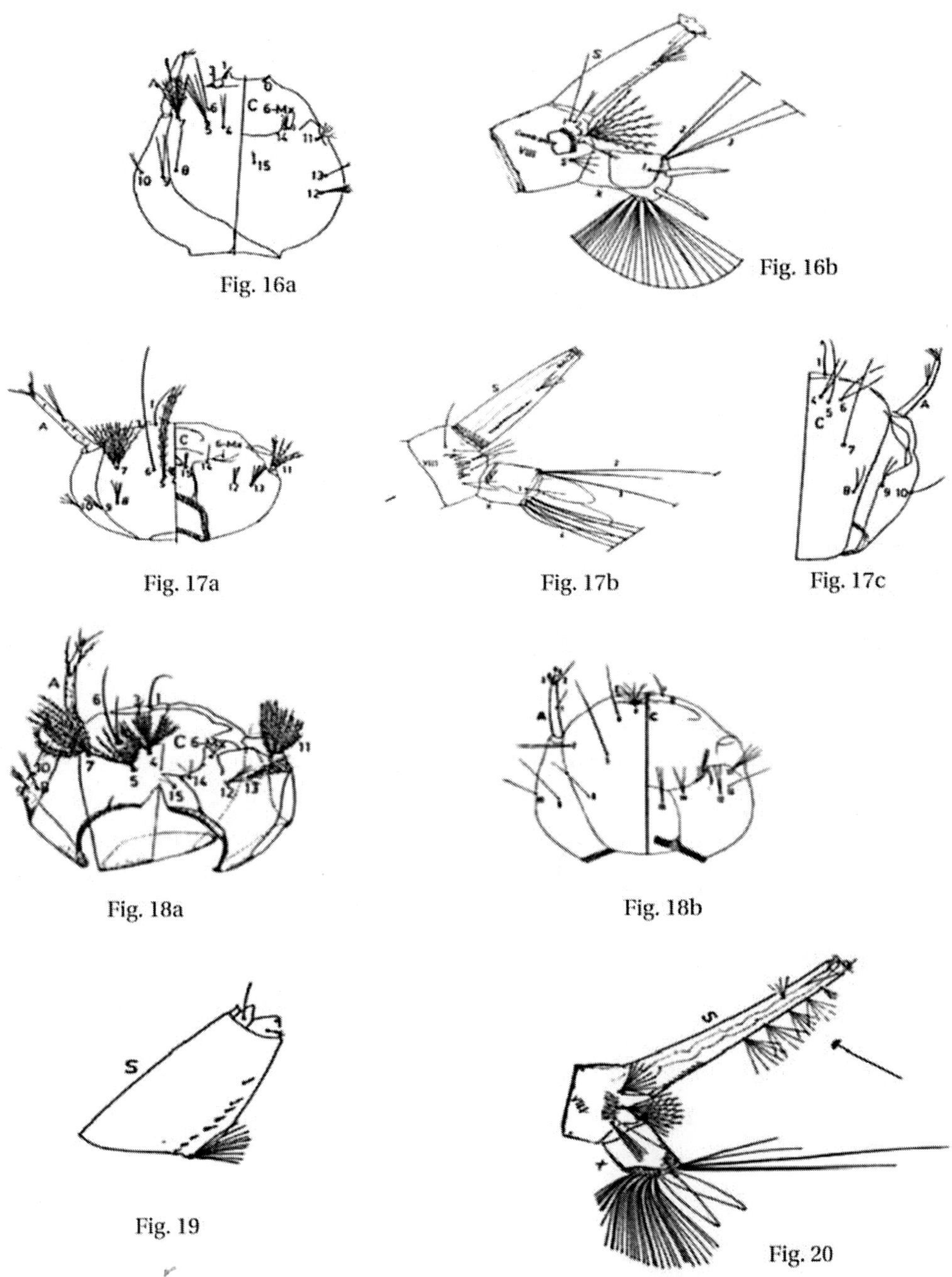

Fig. 16a Fig. 16b

Fig. 17a Fig. 17b Fig. 17c

Fig. 18a Fig. 18b

Fig. 19 Fig. 20

Legends to the Figures **(16-20) : Fig. 16a-** Hypostomal sutures absent; **Fig. 16b-** Sclerotized comb plate; **Fig. 17a-** Well developed hypostomal sutures; **Fig. 17b-** Comb scales; **Fig. 17c-** Seta 7-C long; **Fig. 18a-** Seta 6-C with unequal branches; **Fig. 18b-** Seta 4-C; **Fig. 19-** Single pair of sub-ventral setae; **Fig. 20-** Several pairs of sub-ventral setae.

6.2.2. Vectors of Japanese Encephalitis in India

Japanese encephalitis (JE) is a viral disease predominantly located in South East Asia and commonly associated with transmission between amplifying hosts, such as pigs, and the mosquito *Culex tritaeniorhynchus*, where human infection represents a dead end in the life cycle of the virus. The expansion of JE beyond an Asiatic confine is dependent on a multitude of complex factors that stem back to genetic subtype variation. A complex interplay of the genetic variation and vector competencies combine with variables such as geography, climate change and urbanization. This neurotrophic killer disease is caused by JE virus (JEV), of which five genotypes exist, two of which G3 and G1 have been reported from India. Monitoring for the presence of JE virus in mosquitoes in general can be used to estimate levels of potential JE exposure, intensity of viral activity and genetic variation of JEV throughout surveyed areas. Increased surveillance and diagnosis of viral encephalitis caused by genotype 5 JE virus is required in particular, with the expansion in epidemiology and disease prevalence in new geographic areas. Additional studies that measure the impact of vectors (*e.g.* bionomics and vector competence) in the transmission of JEV and that incorporate environmental factors (*e.g.* weekly rainfall) are needed to define the roles of Culex species in the viral pathogenesis during outbreak and non-outbreak years. The JEV can only be spread by the bite of an infected mosquito. Mosquitoes become infected soon after they bite infected pigs and other amplifying host such as birds. After a person is bitten by an infected mosquito, it usually takes 5–15 days for the first symptoms to appear, such as headache, high fever, convulsions and coma.

The following species have so far been incriminated either as primary or secondary vectors of JE.

1. *Culex (Culex) tritaeniorhynchus*
2. *Culex (Culex) vishnui*
3. *Culex (Culex) pseudovishnui*
4. *Culex (Culex) gelidus*
5. *Culex (Culex) fuscocephala*
6. *Culex (Culex) quinquefasciatus*
7. *Culex (Culex) whitmorei*
8. *Culex (Oculeomyia) epidesmus*
9. *Culex (Oculeomyia) bitaeniorhynchus*
10. *Culex (Oculeomyia) infula*
11. *Mansonia (Mansonioides) indiana*

12. *Mansonia (Mansonioides) annulifera*
13. *Mansonia (Mansonioides) uniformis*
14. *Anopheles (Anopheles) peditaeniatus*
15. *Anopheles (Anopheles) barbirostris*
16. *Anopheles (Cellia) subpictus*

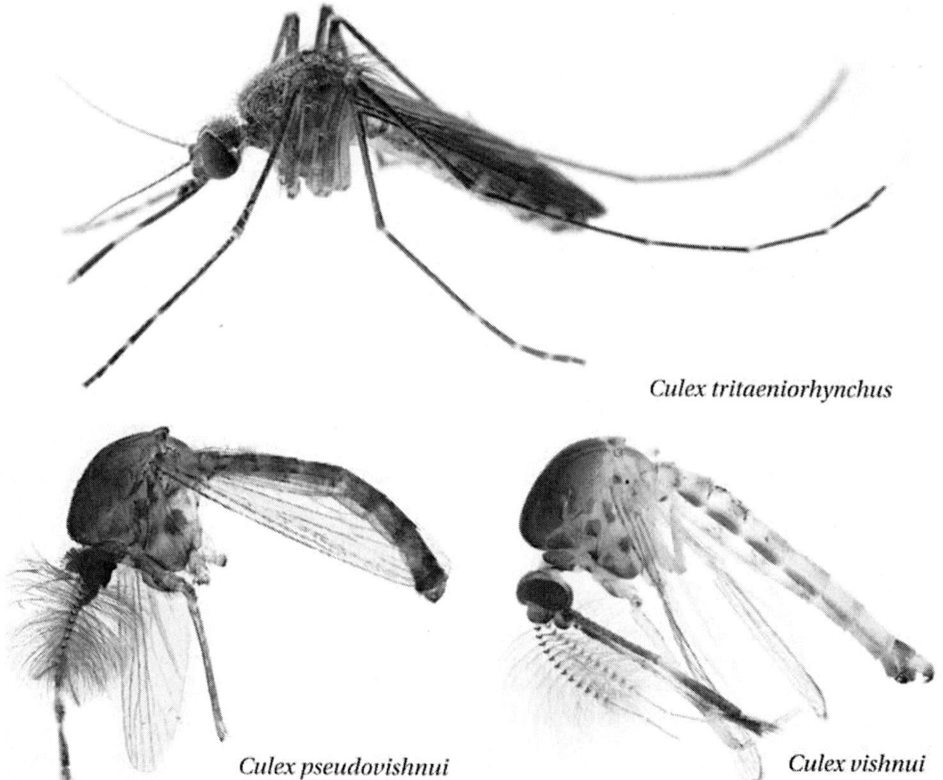

Fig. 6.3. Three major JE vectors, namely, *Culex tritaeniorhynchus, Cx. vishnui and Cx. pseudovishnui.*

6.2.2.1. KEY TO THE JAPANESE ENCEPHALITIS VECTOR MOSQUITO LARVAE (IV INSTAR)

1. Seta 1-A present at mid length; median labral plate not distinguished completely, fused with dorsal apotome (Fig 21) .... Subgenus *Oculeomyia*

   Pecten restricted to basal half of siphon; ventral brush (4-X) consists of 10 or more hair tufts; precratal setae absent; thoracic seta 3-P always single, as long and as thick as 1-P (Fig. 22a, b) ... Subgenus *Culex* (2)

2(1). Median labral plate of head capsule separated from Dorsal apotome as a distinct transverse bar (Fig. 23a) 3

   Median labral plate of head capsule not separated from dorsal apotome (Fig. 24) 12

3(2). Seta 1-C pale, slender and distally strongly tapered and filamentous (Fig. 23a) 4

   Seta 1-C dark, stout, spiniform and abruptly pointed or blunt apically (Fig. 23b) 5

4(3). Setae 5, 6-C double; mentum with 6,7 lateral teeth on each side of central tooth (Fig. 23a, c); siphon slender, more or less cylindrical, and distally tapered .... *fuscocephala*

   Setae 5,6-C with 4-6 branches ; mentum with 10-12 lateral teeth on each side of central tooth; siphon short, fusiform, and basal half strongly swollen (Fig. 25a,b) .... *quinquefasciatus*

5(3). Abdominal seta 7-I double 6

   Abdominal seta 7-I single 7

6(5). Seta 6-C double; siphon brownish, slender, ventral valve of spiracular apparatus small, siphonal tufts weak, with 2-5 branches; individual comb scales, apically rounded (fan shaped) fringed with subequal spicules (Fig. 26 a, b, c) ..... *tritaeniorhynchus*

   Seta 6-C triple; siphon yellowish, thick; ventral valve of spiracular apparatus large (Fig. 27) .... *whitei* & *barraudi*

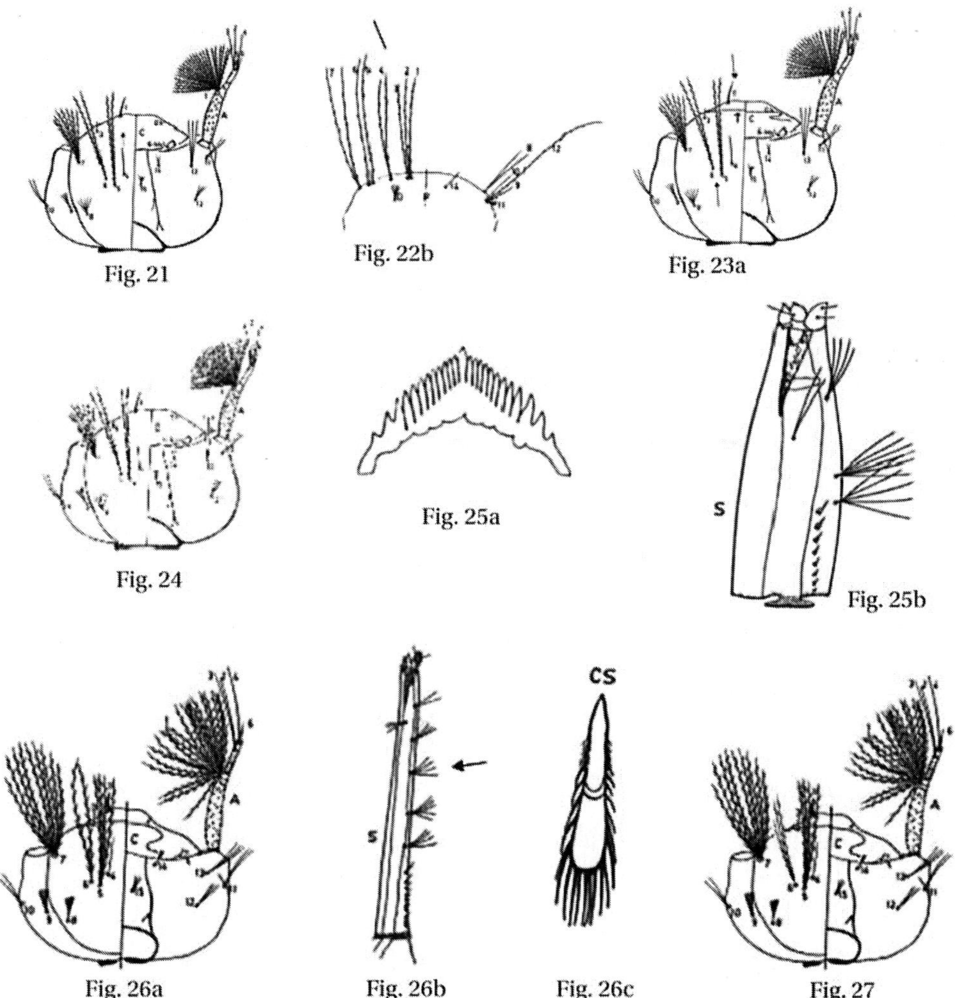

Fig. 21 Fig. 22b Fig. 23a Fig. 24 Fig. 25a Fig. 25b Fig. 26a Fig. 26b Fig. 26c Fig. 27

***Legends to the Figures* (21-27) : Fig. 21**- Pecten teeth; **Fig. 22a**- Pecten teeth; Fig. 22b- Seta 3-P-single; **Fig. 23a**- Transverse bar; **Fig. 23b**- Seta 1-C; **Fig. 23c**- Mentum; **Fig. 24**- No transverse bar; **Fig. 25a**- Mentum; **Fig. 25b**- Fusiform siphon, **Fig. 26a**- Seta 6-C double; **Fig. 26b**- Weak siphonal tufts; **Fig. 26c**- Fan shaped comb scales; **Fig. 27**- Seta 6-C triple;

7(5). Siphon short, fusiform, distinctly swollen in middle (Fig. 28) .... *gelidus*

Siphon long, cylindrical (Fig. 32a) 8

8(7). Seta 4-P double or multiple (Fig. 29) 9

Seta 4-P single (Fig. 30); siphon with some prominent spines on ventral surface of apical half ... *jacksoni*

9(8) Comb scales usually with 4-8, large spiniform scales more or less in a single row.. 10

Comb scales small, with 16-20 or more, in 3 or 4 rows . 11

10(9). Siphon strongly tapered and up-curved distally; siphonal tufts very strong with 2-3 barbed branches (Fig. 31); 4-P always double .... *whitmorei*

Siphon moderately tapered, slightly curved apical; siphonal tufts weaker with 4-6 branches; comb scale very large with long and strong apical spine, 4-P double or multiple (Fig. 32a, b).................................. *pseudovishnui*

11(9). Pleural areas of thorax spiculate; comb scales about 22, pointed apically with strong apical spine (Fig. 33a, b); seta 5-C usually double or triple; 6-C double *vishnui*

Pleural areas of thorax not spiculate; comb scales about 40 or more rounded apically with even fringe of fine spicules (Fig. 34), seta 5-C branched (4-6); 6-C triple.. *mimuloides*

12(2). Setae 2,3-A located halfway between apex and base of seta 1-A; seta 4-P short and minute, rather indistinct; Siphon with 4 or 5 pairs of sub-ventral tufts (Fig. 35a,b,c).. *sinensis* and *cornutus*

Setae 2,3-A sub-apical or closer to apex than to base of seta 1-A; seta 4-P strong and distinct (Fig. 36a, b).. 13

3(12). Siphon usually with 4 pairs of sub-ventral tufts (Fig. 37) .... *bitaeniorhynchus*

Siphon usually with 3 pairs of sub-ventral tufts (Fig. 38) .... *infula*

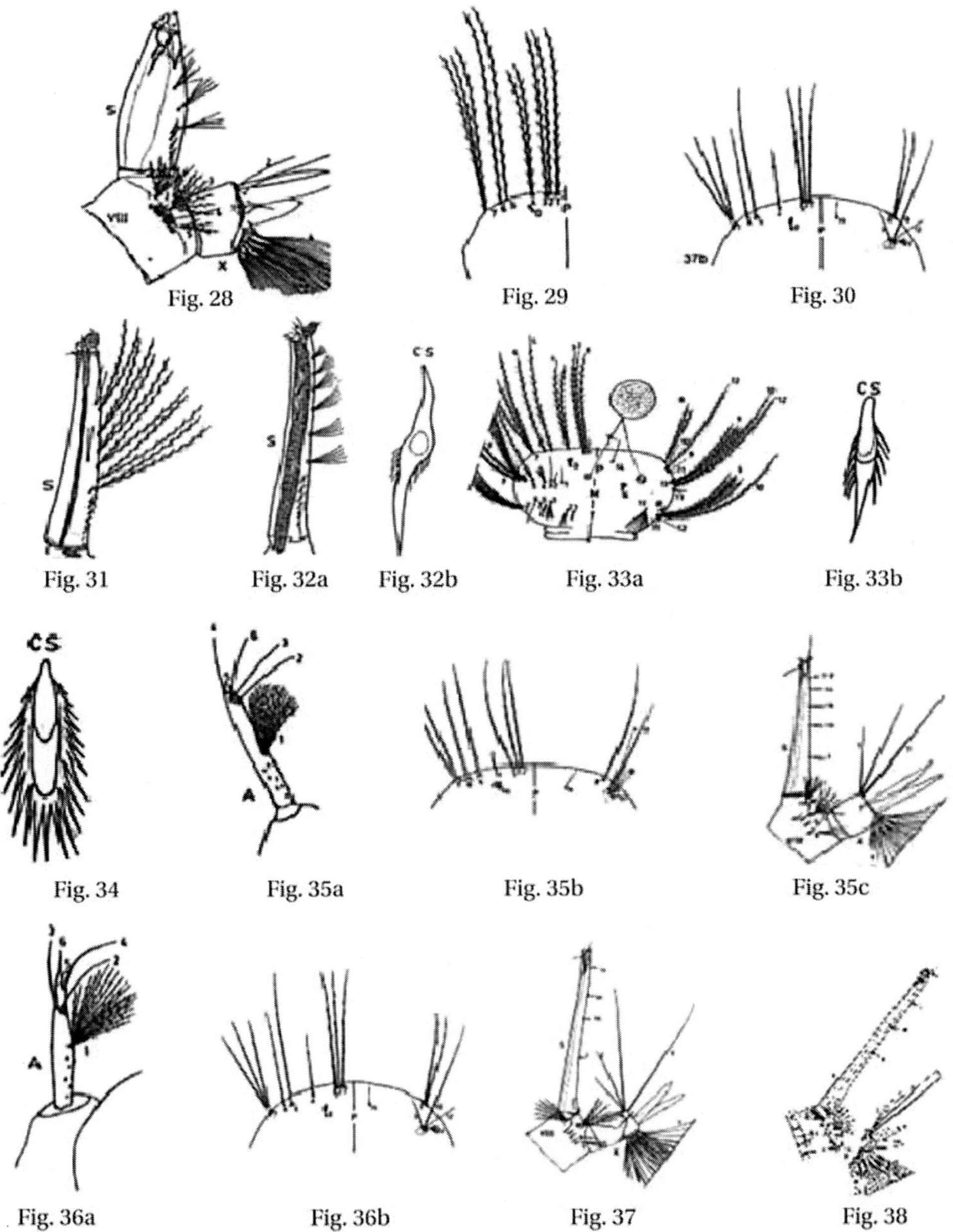

Fig. 28 Fig. 29 Fig. 30

Fig. 31 Fig. 32a Fig. 32b Fig. 33a Fig. 33b

Fig. 34 Fig. 35a Fig. 35b Fig. 35c

Fig. 36a Fig. 36b Fig. 37 Fig. 38

***Legends to the Figures* (28-38) : Fig. 28**- Short siphon; **Fig. 29**- Seta 4-P double; **Fig. 30**- Seta 4-P single; **Fig. 31**- Barbed siphonal tufts; **Fig. 32a**- Long cylindrical siphon; **Fig. 32b**- Comb scales; **Fig. 33a**- Spiculate thorax; **Fig. 33b**- Comb scales; **Fig. 34**- Comb scales; **Fig. 35a**- Setae 2,3-A; **Fig. 35b**- Seta 4-P minute; **Fig. 35c**- Sub-ventral tufts; **Fig. 36a**- Setae 2,3-A; **Fig. 36b**- Seta 4-P; **Fig. 37**- 4 pairs of sub-ventral tufts; **Fig. 38**- 3 pairs of sub-ventral tufts.

1. Bases of inner clypeal hairs much closer to one another than to outer clypeal hairs; antennal hair branched or simple (Fig. 39).. Subgenus *Anopheles* (2)

Bases of inner clypeal hairs wide apart and closer to the bases of outer clypeal hairs; antennal hair simple (Fig. 40).. Subgenus *Cellia* (3)

2(1). Outer clypeal hairs profusely branched forming a tuft; seta 1-P with many branches arising very near the base (Fig. 41a, b).. *barbirostris*

Outer clypeal hairs profusely branched forming a tuft; seta 1-P simple or bifid (Fig. 42)......................... *peditaeniatus*

3(1). Posterior clypeal hair placed not very close to inner clypeal hair; mesothoracic hair 4 most often with 2 branches (Fig. 43a,b) .... *subpictus*

Posterior clypeal hair placed not very close to inner clypeal hair; mesothoracic hair 4 most often with 3 branches (Fig. 44) .... *sundaicus*

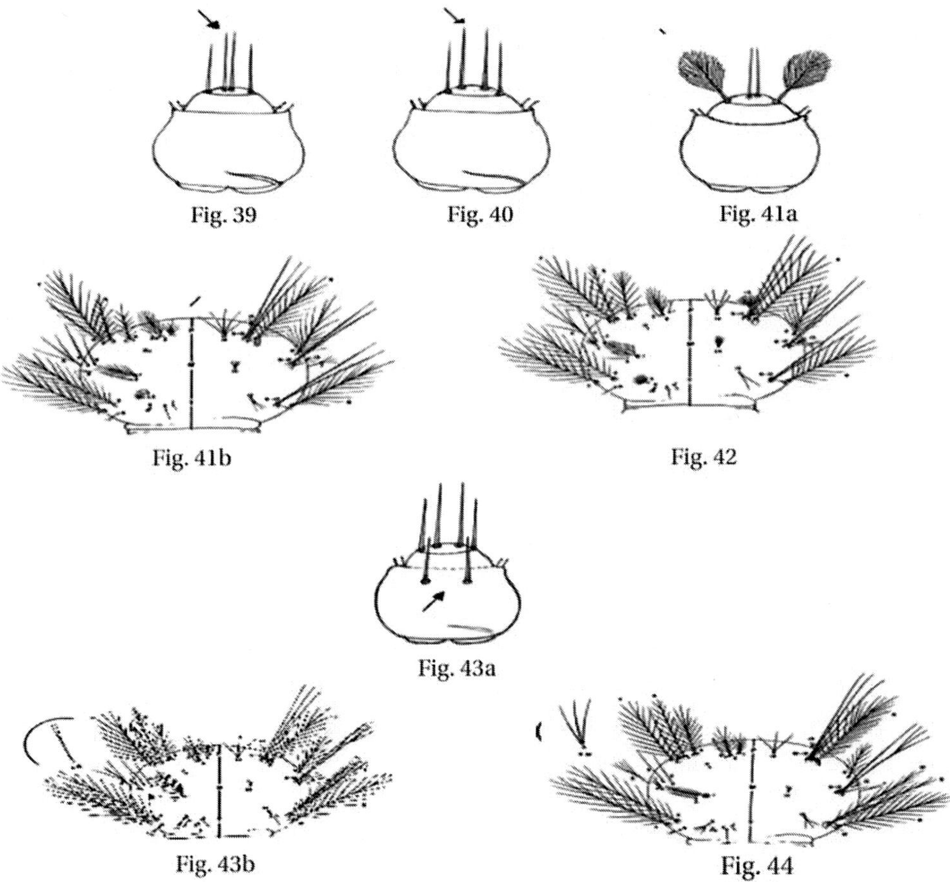
Fig. 39 Fig. 40 Fig. 41a
Fig. 41b Fig. 42
Fig. 43a
Fig. 43b Fig. 44

Legends to the Figures **(39-44) : Fig. 39**- Inner clypeal hairs- closure; **Fig. 40**- Inner clypeal hairs- wider; **Fig. 41a**- Outer clypeal hairs with tuft; **Fig. 41b**- Seta 1-P with many branches; **Fig. 42**- Seta 1-P simple; **Fig. 43a**- Position of posterior clypeal hairs; **Fig. 43b**- 4-M with 2 branches; **Fig. 44**- 4-M with 3 branches.

1. Siphon without pecten; siphonal valves modified for piercing tissues of aquatic plants with sclerotized saw toothed process at tip; (Fig. 45) .... *Mansonia*

1a. Setae 2,3-A long and not reaching very beyond end of antenna; basal half of antenna darkened; setae 6,7-C with 6-7 branches without barbed; seta 7-C distally split into 2or 3 branches (Fig.46 a,b) .... *annulifera*

1b. Setae 2,3-A long and reaching beyond end of antenna; antenna with a dark ring at base and another at level of origin of hair tuft; setae 6,7-C barbed with 7-9 branches (Fig.47 a, b) .... *uniformis* & *indiana*

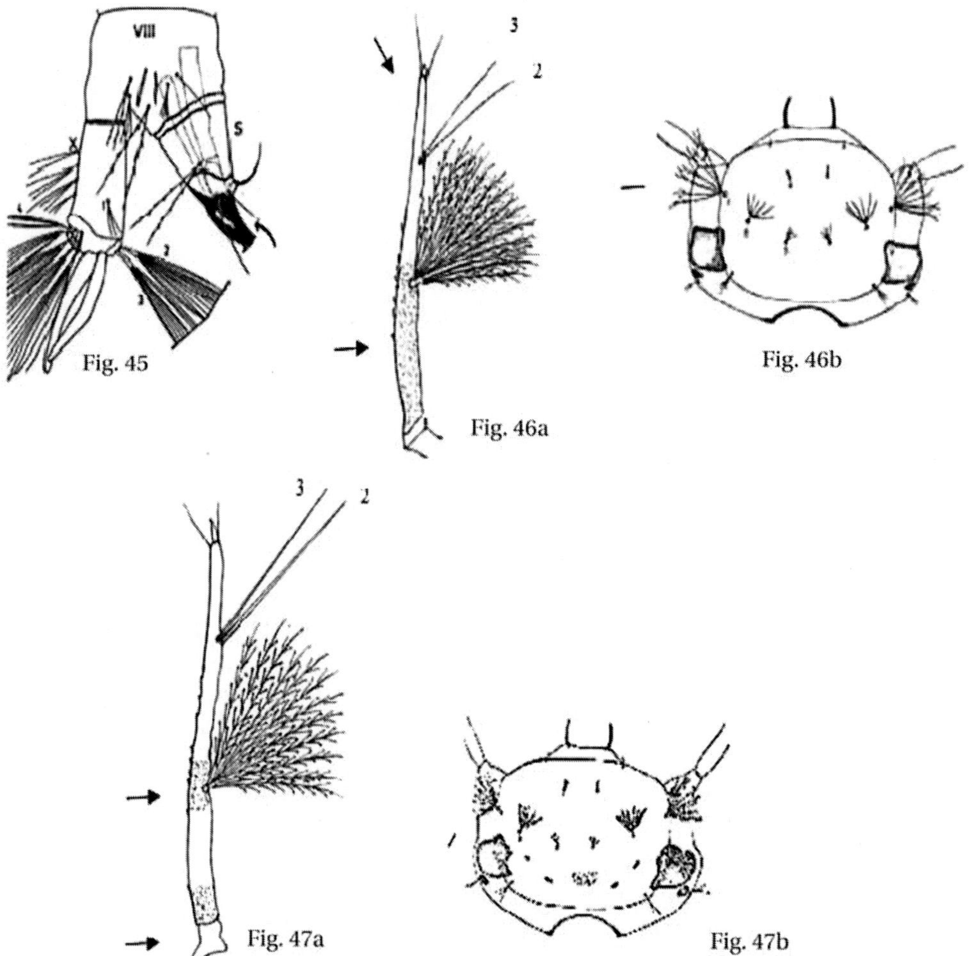

***Legends to the Figures* (45-47) : Fig. 45-** Modified siphon; **Fig. 46a-** Basal half of antenna darkened; **Fig. 46b-** Seta 7-C distally split; **Fig. 47a-** Antenna with two dark rings; **Fig. 47b-** Seta 6,7-C barbed.

6.2.3. VECTORS OF LYMPHATIC FILARIASIS

Lymphatic filariasis is transmitted mainly by the following mosquitoes:

Culex quinquefasciatus

Mansonia uniformis

Mansonia annulifera

Aedes niveus (sub-periodic filariasis in Andaman & Nicobar Islands)

6.2.3.1. *Key to the Lymphatic Filariasis vector mosquito larvae (IVth Instar)*

1. Pecten restricted to basal half of siphon; ventral brush (4-X) consists of 10 or more hair tufts; prerectal setae absent; thoracic seta 3-P always single, as long and as thick as 1-P (Fig. 48a, b) .... Subgenus Culex (2)

Siphon without pecten; siphonal valves modified for piercing tissues of aquatic plants with sclerotized saw toothed process at tip; (Fig. 49) .... Genus *Mansonia* (3)

2(1). Setae 5, 6-C double; mentum with 6,7 lateral teeth on each side of central tooth (Fig.50a, b); siphon slender, more or less cylindrical, and distally tapered .... *fuscocephala*

Setae 5,6-C with 4-6 branches; mentum with 10-12 lateral teeth on each side of central tooth; siphon short, fusiform; basal half strongly swollen (Fig. 51a, b) .... *quinquefasciatus*

3(1). Setae 2,3-A long and not reaching very beyond end of antenna; basal half of antenna darkened; setae 6,7-C with 6-7 branches without barbed; seta 7-C distally split into 2or 3 branches (Fig. 52a,b)............... *annulifera*

Setae 2,3-A long and reaching beyond end of antenna; antenna with a dark ring at base and another at level of origin of hair tuft; setae 6,7-C barbed with 7-9 branches (Fig. 53 a,b) .... *uniformis* & *indiana*

6.2.3.2. *Diurnally subperiodic filariasis vector in Andaman & Nicobar Islands*

1. Head seta 4-C small, variously branched, less than 0.5 length of 7-C; 6-C single or with two equal branches; Seta 4-C closer to 6-C than to 5-C (Fig. 54) genus Aedes

1a. Ventral brush (4-X) with 5 pairs of setae; 12-16 comb scales arranged in a single row; individual comb scales pointed, fringed almost to apex, pecten uniformly pigmented with 15-22 spines (Fig. 55a, b). *nivea**

**Downsiomyia nivea (=Aedes (Finlaya) niveus)*

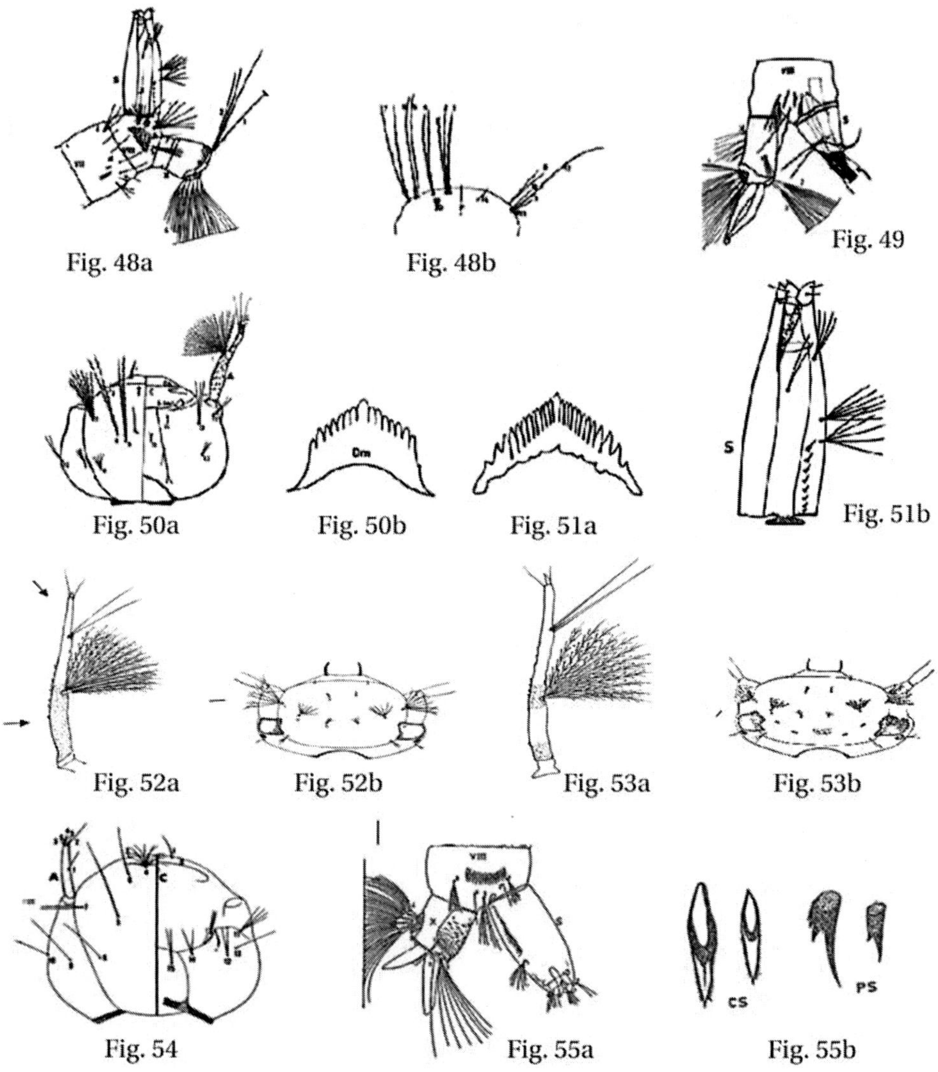

Fig. 48a Fig. 48b Fig. 49 Fig. 50a Fig. 50b Fig. 51a Fig. 51b Fig. 52a Fig. 52b Fig. 53a Fig. 53b Fig. 54 Fig. 55a Fig. 55b

***Legends to the Figures* (48-55) : Fig. 48a-** Ventral brush with 10 or more tufts; **Fig. 48b-** Seta 3-P single; **Fig. 49-** Modified siphon; **Fig. 50a-** Setae 5,6-C double; **Fig. 50b-** Mentum; **Fig. 51a-** Mentum; **Fig. 51b-** Siphon strongly swollen; **Fig. 52a-** Basal half of antenna darkened; **Fig. 52b-** Seta 7-C distally split; **Fig. 53a-** Antenna with two dark rings; **Fig. 53b-** Seta 6,7-C barbed; **Fig. 54-** Position of seta 4-C; **Fig. 55a-** Ventral brush; **Fig. 55b-** Comb scales and pecten teeth.

6.2.4. VECTORS OF DENGUE /CHIKUNGUNYA

There are two major vectors:

Aedes aegypti (= Stegomyia aegypti)
Aedes albopictus (= Stegomyia albopicta)

6.2.4.1. Key to the Dengue/Chikungunya vector mosquito larvae (IVth Instar)

1.	Head seta 4-C small variously branched, less than 0.5 length of 7-C; 6-C single or with two equal branches; Seta 4-C closer to 6-C than to 5-C (Fig. 56)	*Stegomyia*
1a.	Ventral brush (4-X) with five pairs of setae each with two branches; Pecten scales usually 8-20, each scales with 1-4 basal denticles: comb scales with distinct large median spine with strong and stout lateral denticles (Fig. 57)...	*aegypti*
1b.	Ventral brush (4-X) with four pairs of setae each with single; Pecten scales 8-14, each scales with 2 main basal denticles: comb scales with fringed denticles at the base of apical spine (Fig. 58)..............	*albopictus*

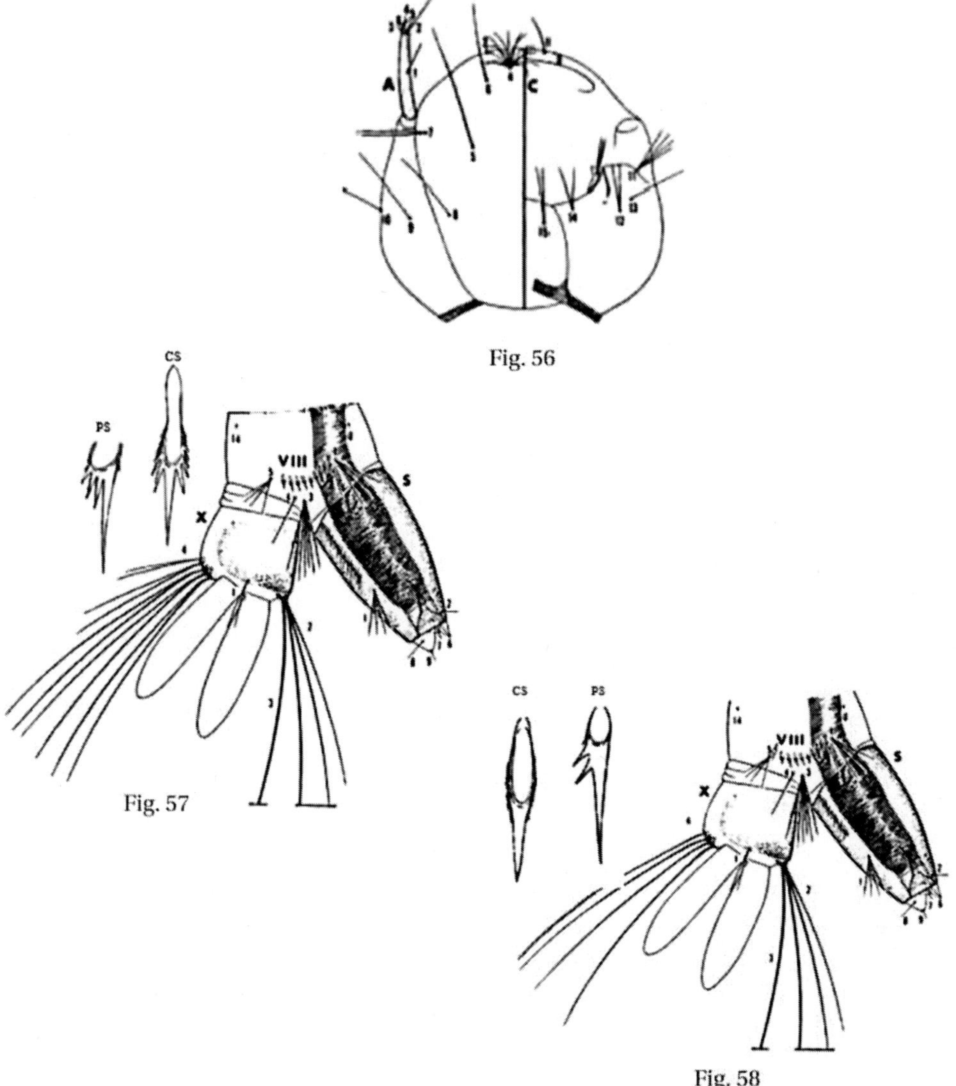

Fig. 56

Fig. 57

Fig. 58

Legends to the Figures **(56-58) : Fig. 56**- Position of seta 4-C; **Fig. 57**- Ventral brush with 5 pairs each with 2 branches; **Fig. 58**- Ventral brush with 4 pairs each single.

6.2.5. VECTORS OF MALARIA

1.	Bases of inner clypeal hairs much closer to one another Than to outer clypeal hairs; antennal hair branched or simple (Fig. 59)	Subgenus *Anopheles*
	Bases of inner cly peal hairs wide apart and closer to The bases of outer clypeal hairs; antennal hair simple (Fig. 60)	Subgenus *Cellia* (2)
2(1).	Anterior tergal plates on abdominal segments III-VII very wide, and enclose the rounded median chitinous spot posteriorly (Fig. 61)	3
	Anterior tergal plates on abdominal segments III-VII not very wide and never enclosing the rounded median chitinous spot (Fig. 62)	5
3(2).	Inner and outer clypeal hairs simple (Fig. 60)	4
	Inner and outer clypeal hairs with short scattered branches (Fig. 63)	*aconitus*
4(3).	A pair of minute hairs arising from the tergal plate on segments II-VIII (Fig. 64)	*varuna*
	The pair of minute hairs not arising from the tergal plate but lying external and a little posterior to the plate on each side (Fig. 65)...	*minimus* and *fluviatilis*
5(2).	Inner and outer clypeal hair simple or with short in inconspicuous lateral fraying	6
	Inner and outer clypeal hair with conspicuous lateral branches (Fig. 66)...	12
6(5).	Mesothoracic pleural hairs all simple, with two long hairs (Fig. 67)	7
	One of the long mesothoracic pleural hair pectinate others simple (Fig. 68)	10
	Meso and metathoracic pleural hairs all simple (Fig. 69)	11

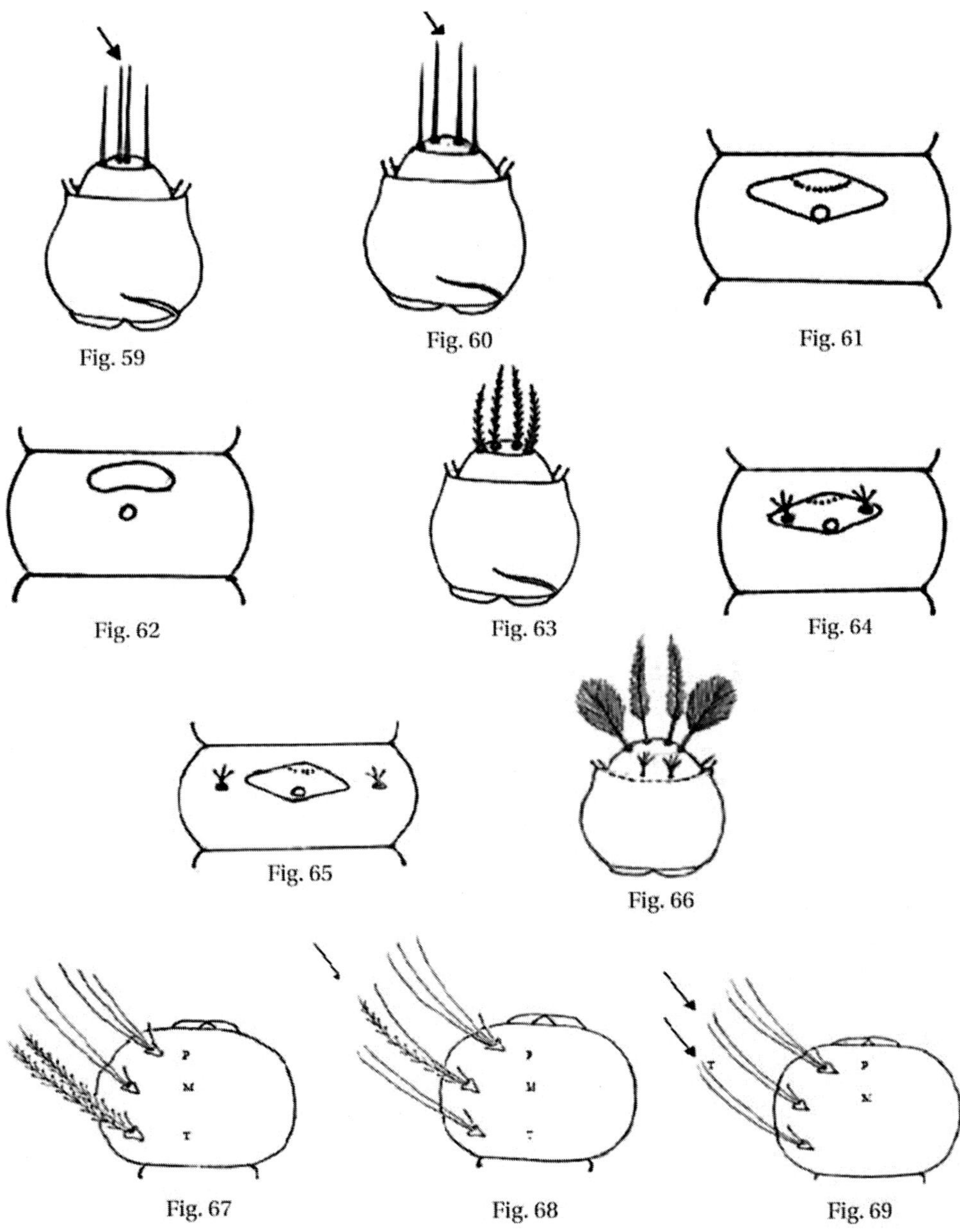

Fig. 59 Fig. 60 Fig. 61 Fig. 62 Fig. 63 Fig. 64 Fig. 65 Fig. 66 Fig. 67 Fig. 68 Fig. 69

Legends to the Figures **(59-69) : Fig. 59**- Inner clypeal hairs- closure; **Fig. 60**- Inner clypeal hairs- wider; **Fig. 61**- Anterior tergal plates; **Fig. 62**- Anterior tergal plates; **Fig. 63**- Inner and outer clypeal hairs; **Fig. 64**- Minute hairs in tergal plate; **Fig. 65**- Minute hairs not in tergal plate; **Fig. 66**- Inner and outer clypeal hairs with branches; **Fig. 67**- Mesothoracic hairs simple; **Fig. 68**- Mesothoracic hair with pectinate; **Fig. 69**- Meso and metathoracic hairs simple.

7(6).	Both the long metathoracic pleural hairs pectinate (Fig. 70); palmate hair on thorax not differentiated ...	8
	One of the long metathoracic pleural hair pectinate others simple (Fig. 71); palmate hair on thorax well differentiated	9
8(7).	Posterior clypeal hair short, and placed very close to inner clypeal hair (Fig. 72)	*vagus*
	Posterior clypeal hair placed not very close to inner clypeal hair; mesothoracic hair 4 most often with 3 branches (2-4) from near the base (Fig. 73a, b)	*sundaicus*
	Posterior clypeal hair placed not very close to inner clypeal hair; mesothoracic hair 4 most often with 2 branches (1-3) from near the base (Fig. 73a, 74)	*subpictus*
9(7).	Filaments of abdominal palmate hairs about half as long as the blades of leaflets; Seta 9-T with few branches, 10-T simple; only the end of outer submedio-dorsal caudal hairs curved to form hooks, the branches of inner hairs, fine and straight (Fig. 75a, b, c)	*culicifacies*
	Filaments of abdominal palmate hairs about half as long as the blades of leaflets; both the ends of inner and outer submedio- dorsal caudal hairs curved to form hooks, the branches of inner hairs stout (Fig. 75a, 76)	*d'thali*

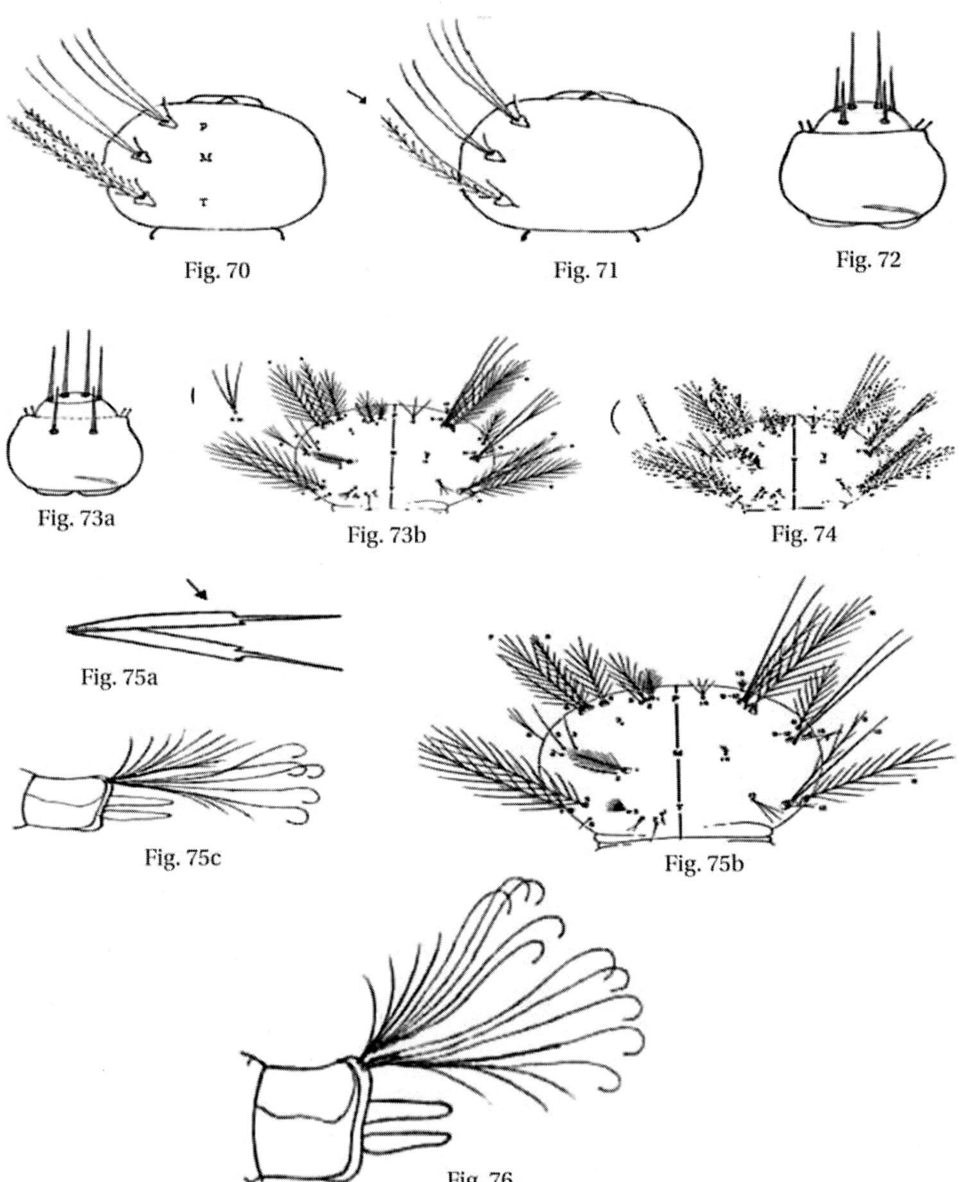

Fig. 70 Fig. 71 Fig. 72

Fig. 73a Fig. 73b Fig. 74

Fig. 75a Fig. 75c Fig. 75b

Fig. 76

***Legends to the Figures* (70-76) : Fig. 70**- Both the metathoracic hairs pectinate; **Fig. 71**- Single metathoracic hair pectinate; **Fig. 72**- Position of posterior clypeal hair; **Fig. 73a**- Position of posterior clypeal hair; **Fig. 73b**- Seta 4-M with 3 branches; **Fig. 74**- Seta 4-M with 2 branches; **Fig. 75a**- Filaments of palmate hair; **Fig. 75b**- Seta 9-T with few branches; **Fig. 75c**- Outer submedio-dorsal caudal hairs; **Fig. 76**- Inner & outer submedio-dorsal caudal hairs.

10(6). Outer clypeal hairs always simple; abdominal seta 1-I with 3-5 branches; setae 9,10-T both branched (Fig. 77a,b).. *stephensi*

Inner and outer clypeal hairs finely frayed; filamentous of abdominal seta 1 with sharp point; Seta 6-III with 20 or more branches (Fig. 78) .... *maculatus*

11(6). Seta 1-P with 2-5 branches; abdominal seta 1-II with filamentous branches (Fig. 79a,b) *tessellatus*

Innermost sub median prothoracic hair (1-P) with more than four branches arising from a large root (Fig. 80) .... *baimaii* (=*dirus D*)

12(5). Outer clypeal hairs with a large no. of long branches forming a broom like tuft; posterior clypeal hairs with 2-5 branches; filament of palmate hairs more than half or as long as blade of leaflet (Fig. 81a, b) .... *pallidus*

Outer clypeal hairs with a large no. of long branches forming a broom like tuft; posterior clypeal hairs with 7-10 branches; filament about ¼ length of leaflet (Fig. 82a,b) .... *philippinensis*

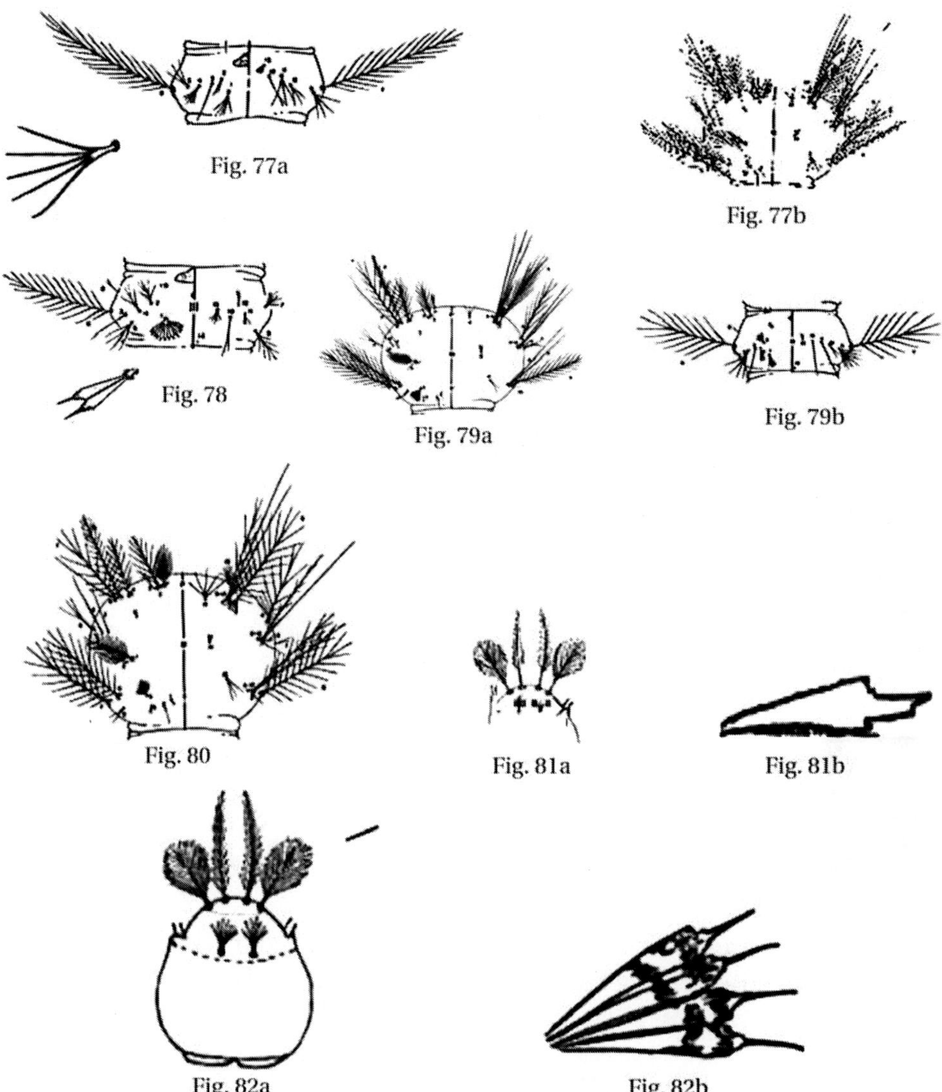

***Legends to the Figures* (77-82) : Fig. 77a**- Abdominal seta 1-I with 3-5 branches; **Fig. 77b**- Setae 9,10-T both branched; **Fig. 78**- Seta 6-III with 20 or more branches; **Fig. 79a**- Seta 1-P with 2-5 branches; **Fig. 79b**- Abdominal seta 1-II with many branches; **Fig. 80**- 1-P with more than 4 branches; **Fig. 81a**- Posterior clypeal hairs with 2-5 branches; **Fig. 81b**- Filaments of palmate hair; **Fig. 82a**- Posterior clypeal hairs with 7-10 branches; **Fig. 82b**- Filaments of palmate hair.

6.3. Adult Characters and Keys

6.3.1. *Adult Characters used in identification*

Head: Character and coloration of scaling, and arrangement of bristles; erect, narrow and broad scales on vertex.

Antenna: Coloration of torus, and presence or absence of scales; length of antenna compared with that of proboscis.

Clypeus: Shape; presence or absence of scales.

Palpi: Length, shape, ornamentation, tips of palpi dark or white; palpi with 3 or 4 pale bands; with or without speckling.

Proboscis: Ratio of palpi and proboscis, size and shape, pale rings; pale patches on ventral surface.

Thorax: Shape, with or without pale markings (ornamentation); with or without scale patches in antepronotum and postpronotum; postspiracular area with our without scales or setae, and shape of scutellum.

Legs: With or without speckling, especially presence or absence of pale rings or bands on tarsal segments, if present whether basal or apical, or both; comparative length of tibia and of tarsal segments; ornamentation of the legs is usually in connection with the tarsal segments. They may be uniformly dark, or they may be marked with white or pale bands at the joints; tarsal claws, whether toothed or simple; presence or absence of pulvilli.

Wings: Venation or arrangement of veins; venation and spotting of wing; shape of scales on veins; length of wing from base of costa to tip; presence or absence of hairs or scales on upper calypter and alula.

Abdomen: The abdomen consists of eight visible segments. The first segment usually forms a somewhat transverse bar dorsally carrying long, outstanding hairs. The dorsal surface or dorsum is called as tergites and ventral surface or venter is called as sternites; with or without basal or apical, or both pale bands.

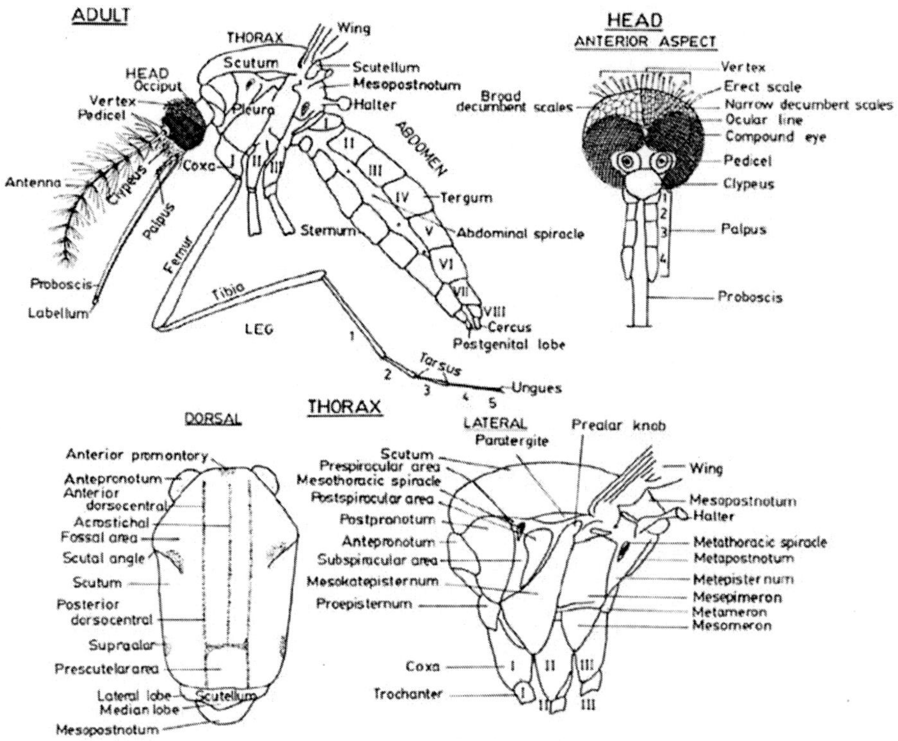

Main body parts

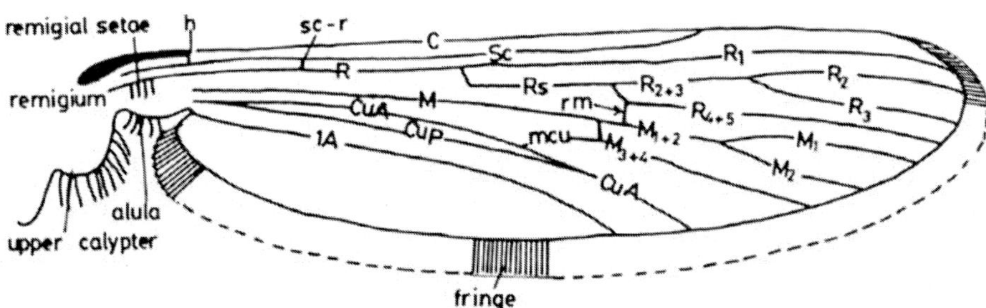

C -costa
CuA-cubitus anterior
CuP-cubitus posterior
h -humeral crossvein
M -media
M_1 -media-one
M_{1+2}-media-one-plus-two
M_2 -media-two
M_{3+4} -media-three-plus-four
mcu - mediocubital crossvein
R - radius
R_1 - radius-one
R_2 - radius-two
R_{2+3} - radius-two-plus-three
R_3 - radius-three
R_{4+5} -radius four-plus-five
Rs -radial sector
rm -radiomedial crossveir
Sc -subcosta
sc-r -subcostal crossvein
1A -anal vein

Wing

Morphology and diagnostic characters a female mosquito.

6.3.2. *Identification key for Genera harbouring species of public health importance in India*

1.	Scutellum evenly rounded; palpus about equal to the length of proboscis (Fig.1, 1a); abdomen with sterna and usually terga largely or wholly devoid of scales	***Anopheles***
	Scutellum trilobed; palpus short (Fig. 2, 2a); abdominal Sterna and terga with dense, uniform covering of scales	2
2(1).	Postspiracular setae present, or fore ungues toothed, or both (Fig. 3, 3a)	3
	Postspiracular setae absent; ungues simple (Fig. 4, 4a)	5
3(2).	Wing with broad, often asymmetrical scales; dark and pale scales intermixed (Fig. 5)	***Mansonia***
	Wing with normal scales	4
4(3).	Proboscis curved and laterally compressed (Fig. 6) ...	***Armigeres***
	Proboscis slender and usually straight, never laterally compressed (Fig. 7)	***Aedes***
5(2).	Hindtarsi with ungues very small and inconspicuous; pulvilli present (Fig. 8)	***Culex***
	Hindtarsi with ungues not very small; pulvilli bsent (Fig. 9); yellow or yellowish brown mosquitoes	***Coquillettidia***

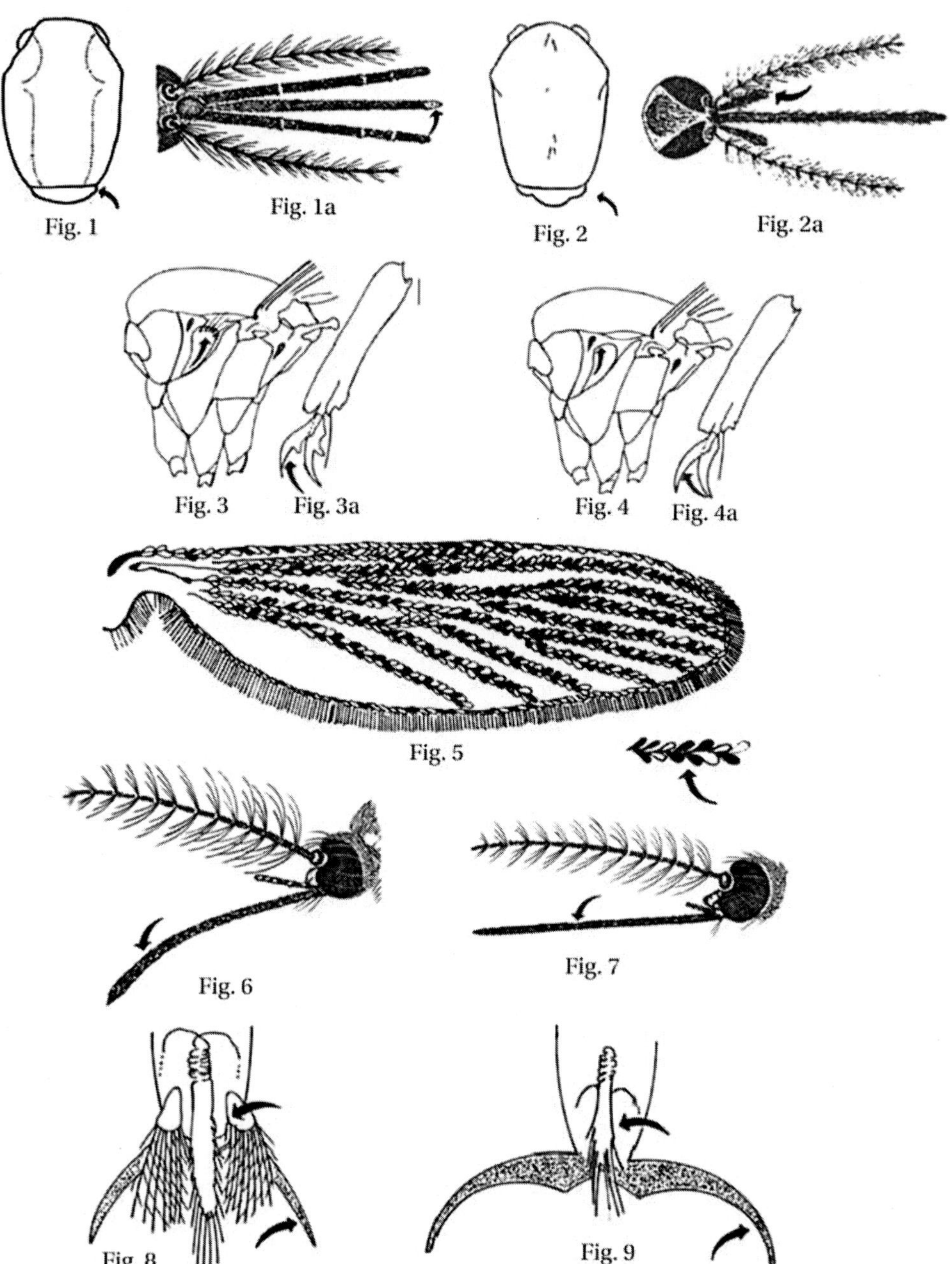

Legends to the Figures **(1-9) : Fig. 1-** *Anopheles* scutellum; **Fig. 1a-** *Anopheles* palpi and proboscis; **Fig. 2a-** *Culicine* palpi and proboscis; **Fig. 3-** pleuron with postspiracular setae; **Fig. 3a-** toothed fore ungues; **Fig. 4-** pleuron without spiracular setae; **Fig. 4a-** untoothed fore ungues; **Fig. 5-** wings with asymmetrical scales; **Fig. 6-** laterally compressed proboscis; **Fig. 7-** slender proboscis; **Fig. 8-** pulvilli present; **Fig. 9-** pulvilli absent;

6.3.3. KEY TO VECTORS OF JAPANESE ENCEPHALITS

Lower mesepimeral setae absent or 1, 2 weak ones; pleuron with distinct pale scale patches on at least upper and lower mesokatepisternum and anterior mesepimeron (Fig.10) ...... subgenus *Culex*

Wing heavily/moderately speckled with pale scale; abdominal terga II-VI with apical bands and/or apicolateral pale patches and basal pale bands (Figs. 24-26) .... (Subgenus *Oculeomyia*)

1. Proboscis and tarsi without pale rings; 1 or 2 lower mesepimeral setae present (Figs. 11, 12) 2

   Proboscis and tarsi with pale rings; lower mesepimeral setae absent (Figs. 13, 14) 4

2(1). Abdominal terga without pale bands, occasionally a few indistinct bands on posterior segments; pleuron with striking pattern of dark and pale stripes (Figs. 15, 16) .... *fuscocephala*

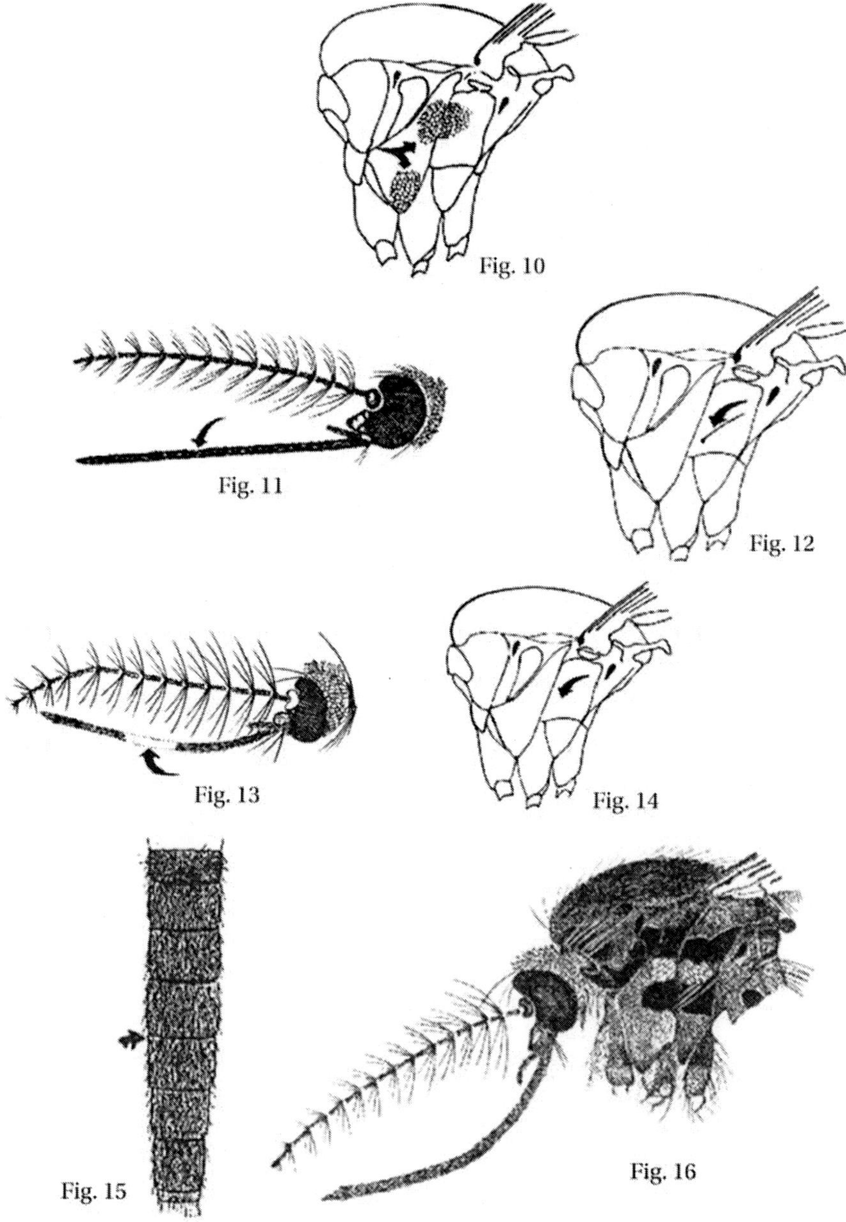

***Legends to the Figures* (10-16) : Fig. 10**- pleuron with pale scale patches; **Fig. 11**- proboscis without pale ring; **Fig. 12**- with lower mesepimeral setae; **Fig. 13**- proboscis with pale ring; **Fig. 14**- without lower mesepimeral setae; **Fig. 15**- abdomen terga without pale band; **Fig. 16**- pleuron with dark and pale stripes.

Abdominal terga with basal pale bands; pleuron without striking pattern of dark and pale stripes (Figs. 17, 18) 3

3(2). Postspiracular area with pale scale patch (Fig. 19) .... *perexiguus*

Postspiracular area without pale scale patch (Fig. 20) .... *quinquefasciatus*

4(1). Wing with pale spots on at least 2 areas of costa and 1 area of other veins (Fig. 21) .... Mimeticus Subgroup

Wing without distinct pale spots.................................. 5

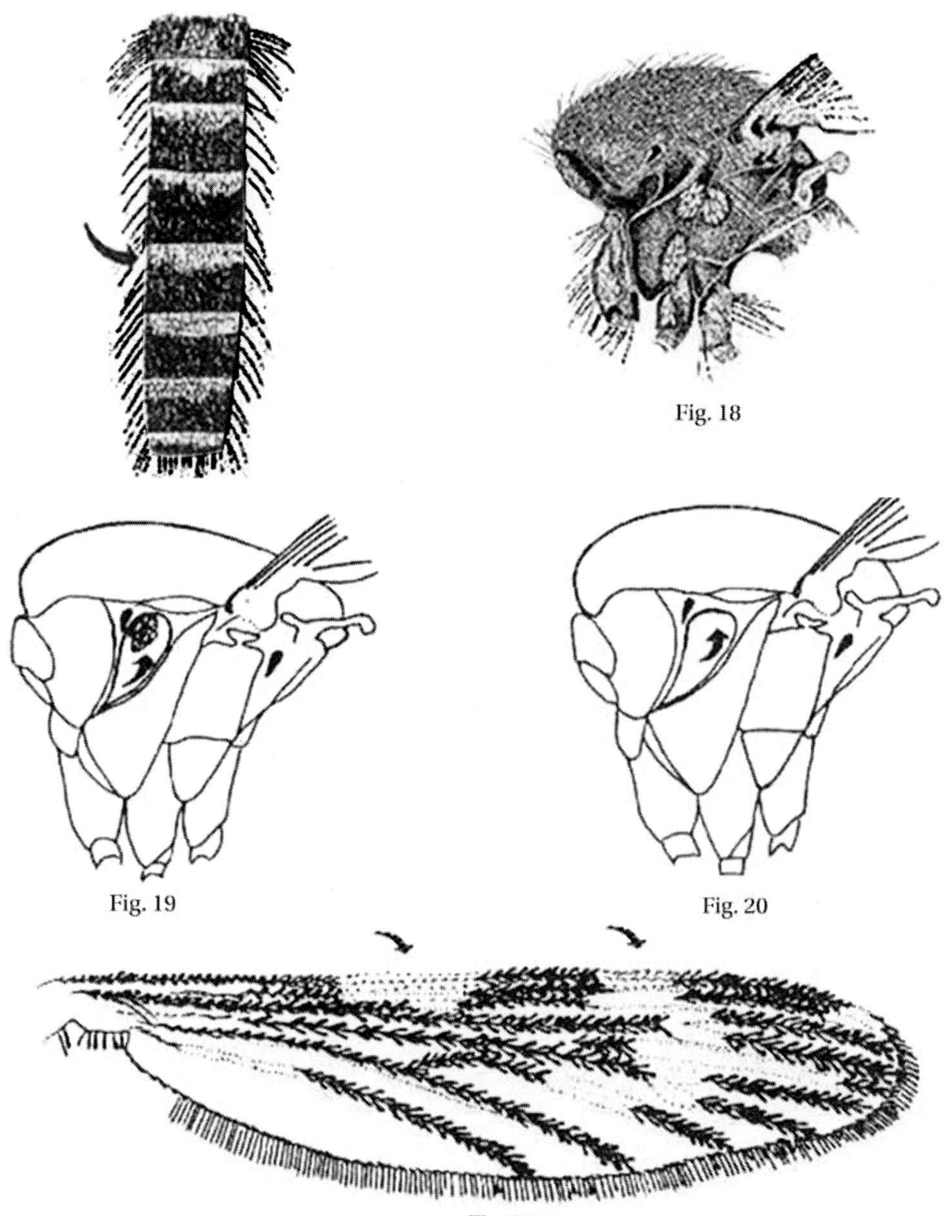

Fig. 18

Fig. 19

Fig. 20

Fig. 21

***Legends to the Figures* (17-21) : Fig. 17**- abdominal terga with basal pale band; **Fig. 18**- pleuron without dark and pale stripes; **Fig.19**- postspiracular area with pale scales; **Fig. 20**- postspiracular area without pale scales; **Fig. 21**- wing with pale spots.

5(4). Abdominal terga II- VIII largely yellowish or golden; pale yellow scales on apical portions of wing veins C, R1, R2, R3, and R4+5, forming apical pale area (Fig. 22, 23) .... *epidesmus*

Abdominal terga II- VIII with dark and pale bands, or completely dark; wing tip without definite pale area.... 6

6(5). Wing heavily / moderately speckled with pale scale; abdominal terga II-VI with apical bands and/or apicolateral pale patches and basal pale bands 7

Wing without speckling of pale scales; abdominal terga II-VI with basal pale bands 8

7(6). Abdominal terga II-VII with evenly broad apical pale bands and without apicolateral pale patches; femora, tibiae, and wings heavily speckled with pale scales (Figs. 24, 25) .... *bitaeniorhynchus*

Abdominal terga II-IV largely dark, or with narrow apical bands, apicolateral yellowish patches, and median basal pale bands or spots; terga V-VII with apical and basal bands, basal bands narrower; legs and wings lightly or moderately speckled (Fig. 26) .... *infula*

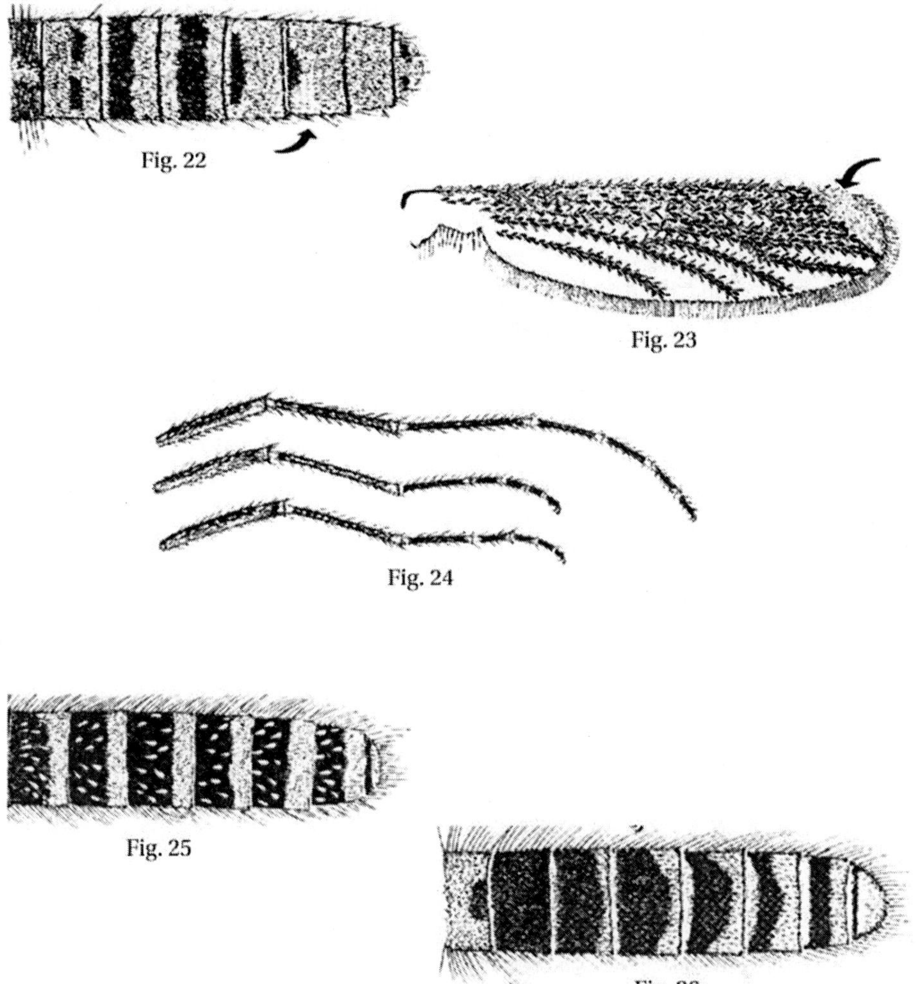

***Legends to the Figures* (22-26) : Fig. 22**- abdomen terga II-VIII with yellowish scales; **Fig. 23**- apical portion of wing with pale area; **Fig. 24**- legs with speckling; **Fig. 25**- abdomen terga II-VII broad apical band; **Fig. 26**- abdomen terga II-IV largely dark with narrow apical bands.

8(6). Anterior 0.7 of scutum covered with pure white cales . 9

Anterior 0.7 of scutum covered with beige, yellow, golden or dark scales 10

9(8). Anterior surface of fore-and midfemora without speckling of pale scales; scales on prescutellar space, behind wing base, and on scutellum entirely dark; pale band on proboscis narrow, less than length of basal dark area (Figs. 27-29) .... *gelidus*

Anterior surface of fore-and midfemora speckled with pale scales; scales on prescutellar space, behind wing base, and on scutellum mainly pale; pale band on proboscis broad, as long as or longer than basal dark area (Figs. 30-32) .... *whitmorei*

10(8). Anterior surface of fore and midfemora speckled with several pale scales, at least on apicodorsal surface 11

Anterior surface of fore and midfemora entirely dark . 12

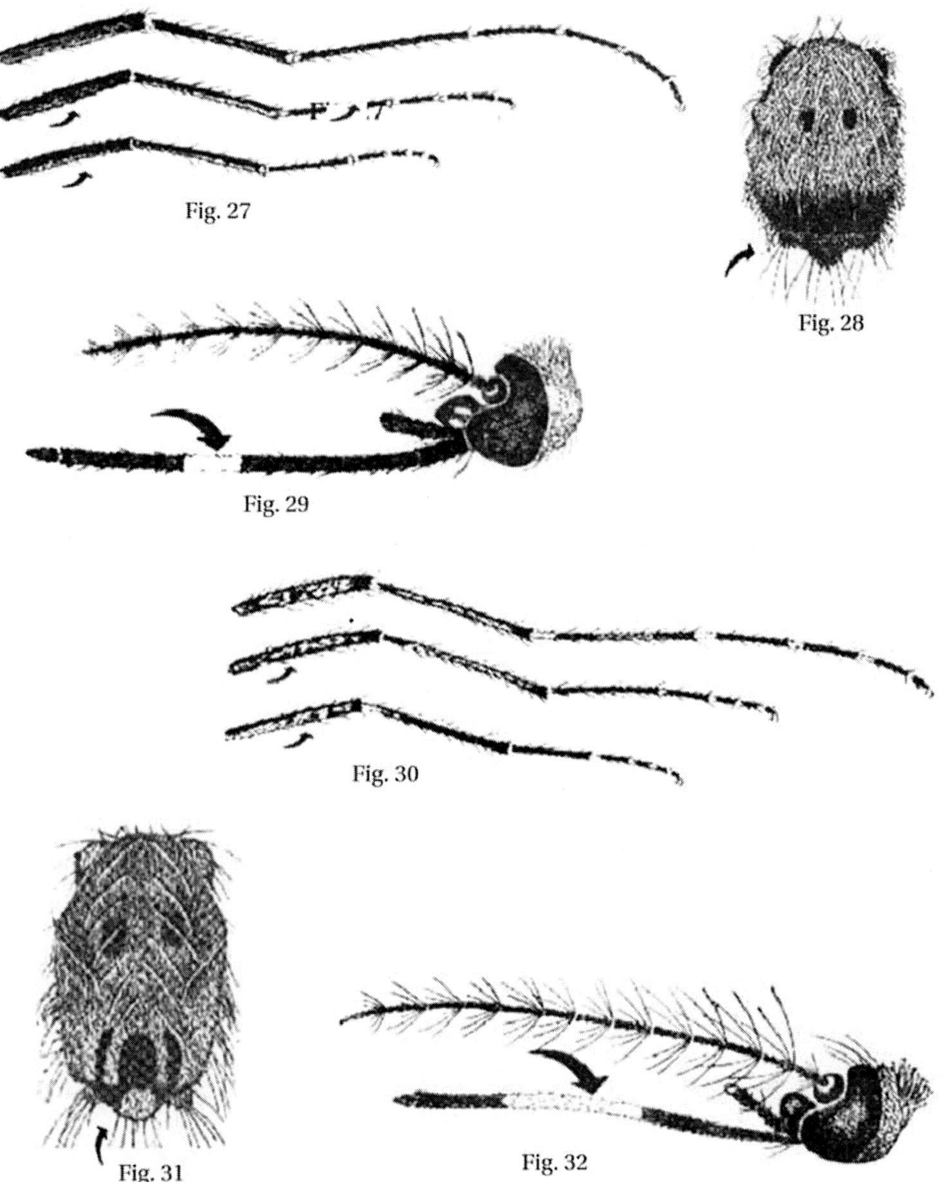

Fig. 27

Fig. 28

Fig. 29

Fig. 30

Fig. 31

Fig. 32

***Legends to the Figures* (27-32) : Fig. 27**- legs without speckling; **Fig. 28**- scutellum entirely dark; **Fig. 29**- proboscis with narrow pale band; **Fig. 30**- legs with speckling; **Fig. 31**- scutellum mainly pale; **Fig. 32**- proboscis with broad pale band;

11(10). Wing scales mainly dark; scutal integument dark; speckling of pale scales on femora contrasting sharply with dark-scaled area (Fig. 33) .... *sitiens*

Wing with few to several scattered pale scales; Scutal integument light brown; anterior surface of hind femur with pale stripe not contrasting with dark-scaled area (Fig. 34) .... *vishnui*

12(10). Proboscis usually with accessory pale patches or stripe on ventral surface: hindfemur pale with an apical dark ring (Figs. 35, 36); erect scales on vertex all dark .... *tritaeniorhynchus*

Prescutellar area entirely covered with pale scales; hind femur with pale stripe on anterior surface contrasting well with dark areas (Figs. 37, 38) .... *pseudovishnui*

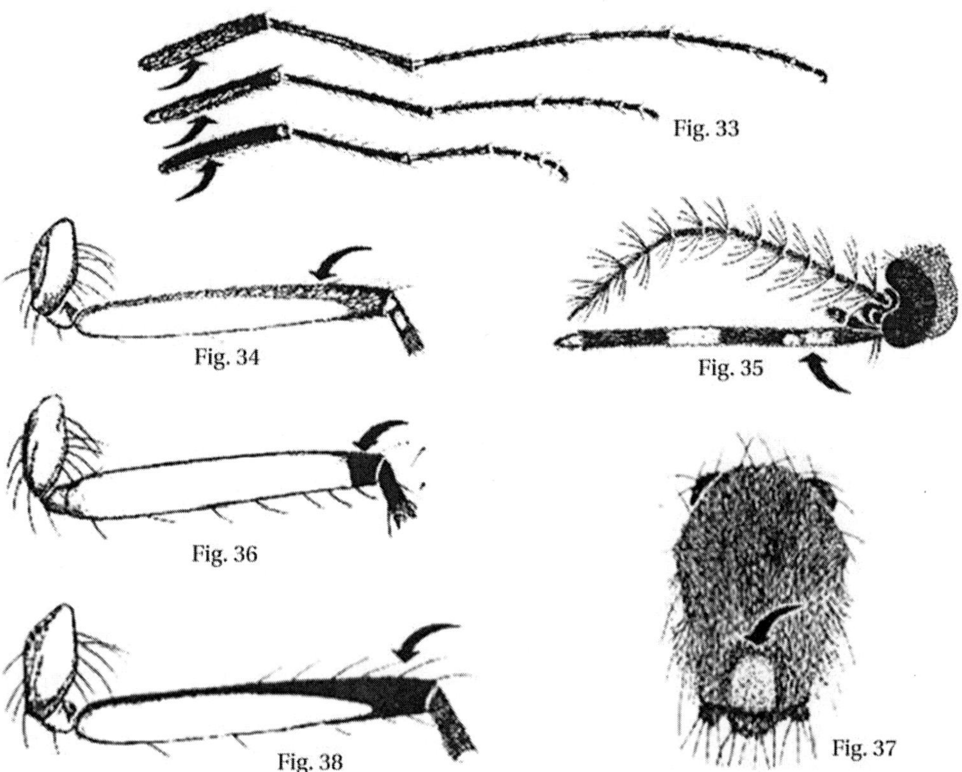

Legends to the Figures **(33-38) : Fig. 33**- legs with speckling; **Fig. 34**- hind femur not contrasting with dark scales; **Fig. 35**- proboscis with accessory pale patches; **Fig. 36**- hindfemur with apical dark ring. **Fig. 37**- prescutellar area with pale scales; **Fig. 38**- hind femur well contrasting with dark scales.

Wing with broad, often asymmetrical scales; dark and pale scale intermixed (Fig. 39) .... *Mansonia*

1. Mesonotum marked with distinct round spots of white scales.. 2

Mesonotum not so (though there may be small indefinite white spots or patches) 3

2(1). Yellowish-brown; mesonotum marked with 4 (or more) distinct round white spots; rather broad white scales on mid-lobe of scutellum (Fig. 40) .... *annulifera*

Brownish-black; mesonotum marked with 2 (or 3) round white spots; scales on mid-lobe of scutellum narrow (Fig. 41) .... *longipalpis*

3(1). Mesonotum marked with a pair of sublateral greenish stripes on a brown ground (Fig. 42) .... *uniformis*

Mesonotum dark brown, not marked with greenish stripes; some white scales, tending to form indistinct spots or patches, in some specimens (Fig. 43) .... *indiana*

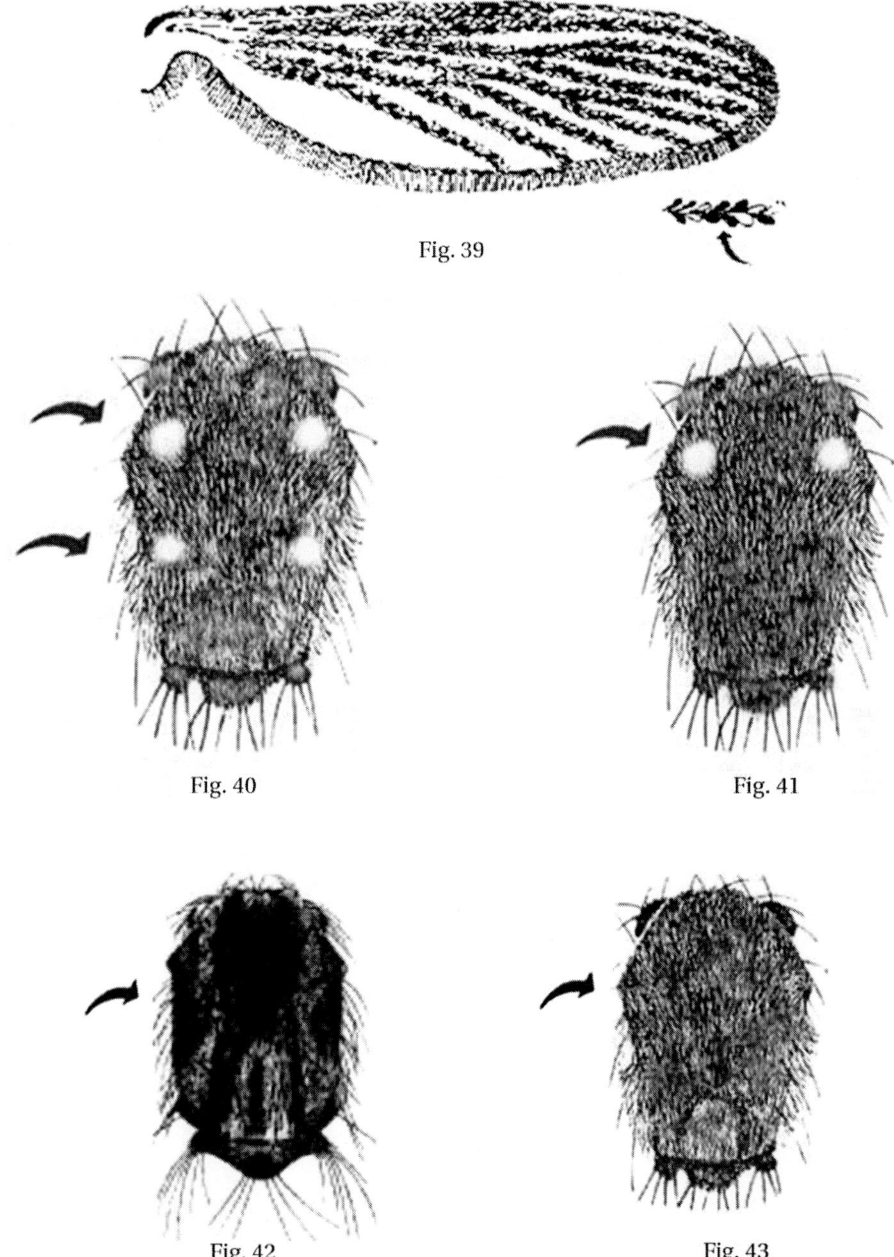

Fig. 39

Fig. 40

Fig. 41

Fig. 42

Fig. 43

Legend to the Figures (39-43) : Fig. 39- wing with asymmetrical scales; **Fig. 40**- mesonotum with 4 round white spots; **Fig. 41**- mesonotum with 2 round white spots; **Fig. 42**- mesonotum with sublateral greenish stripes; **Fig. 43**- mesonotum without greenish stripes.

1. Wings with less than three dark spots on costa, which also involve vein 1 (Fig. 44) .... (Subgenus *Anopheles*)

1a. Hind tarsomeres with broad apical pale bands; wing humeral cross vein without dark scales; remigium mostly pale scaled (Figs. 45, 46) .... *peditaeniatus*

1b. A prominent tuft of dark scales on ventral side of abdominal segment VII; inner quarter of costa with scattered pale scales, usually with a fringe spot at vein 5.2; wing with narrow lower apical fringe spot opposite to vein 3 only (Figs. 47, 48)............................ *barbirostris*

2. Wings with 4 (or) more dark spots on costa which also involve vein 1 (Fig. 49).. (Subgenus *Cellia*)

2a. Femur and tibia not speckled; hind tarsomere 5 not white; fore tarsi with broad pale bands; palpi with subapical dark band equal to the apical pale band (Figs. 50, 51) .... *subpictus*

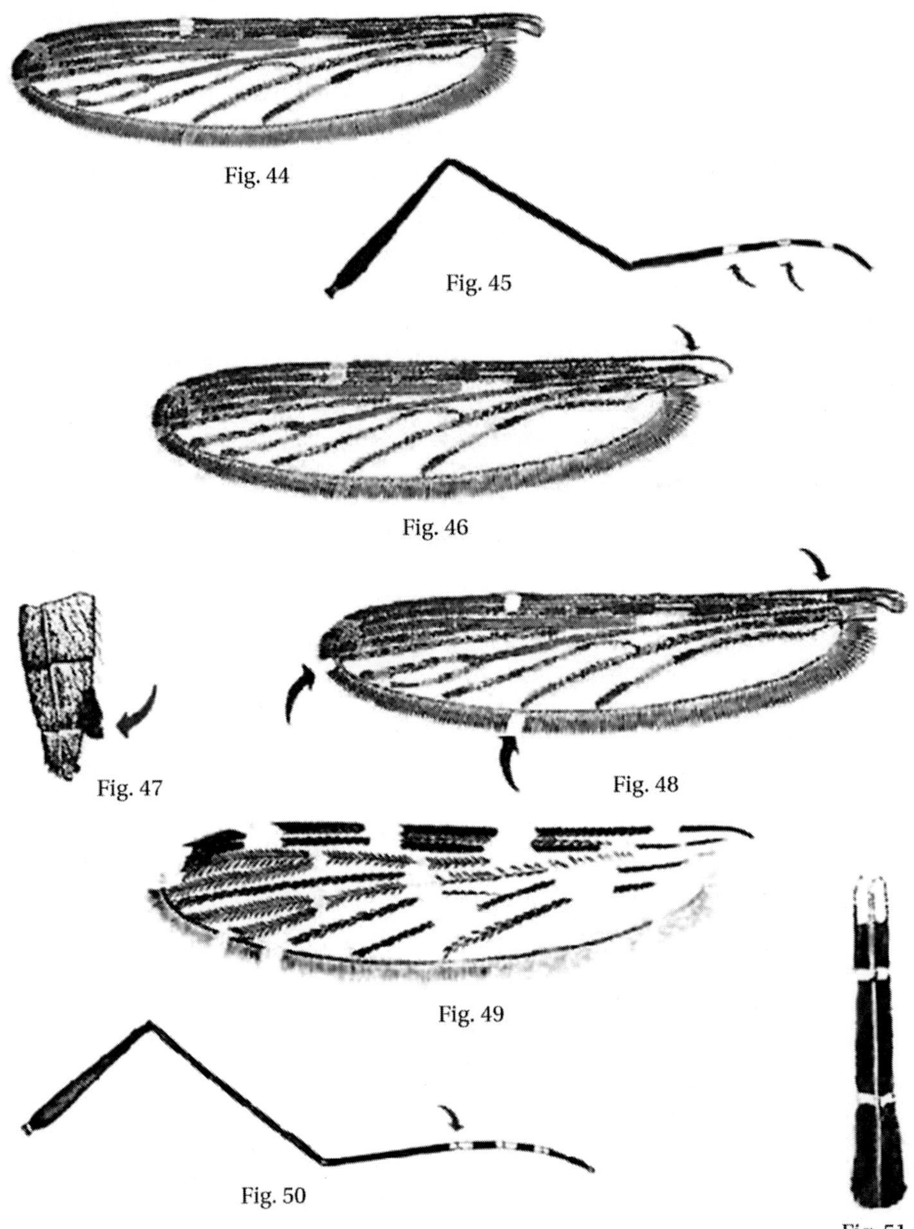

Fig. 44

Fig. 45

Fig. 46

Fig. 47

Fig. 48

Fig. 49

Fig. 50

Fig. 51

***Legends to the Figures* (44-51) : Fig. 44**- wings with less than three dark spots; **Fig. 45**- hind tarsomere with broad apical pale bands; **Fig. 46**- remigium with pale scales; **Fig. 47**- abdomen segment VII with tuft; **Fig. 48**- wing with fringe spot at vein 3 and 5.2; **Fig. 49**- wings with four or more dark spots; **Fig. 50**- fore tarsi with broad pale bands; **Fig. 51**- palpi with subapical dark band equal to apical pale band.

6.3.4. VECTORS OF DENGUE / CHIKUNGUNYA

1. Post spiracular setae present; proboscis slender and usually straight, never laterally compressed .... **Aedes**

1a. Mesonotum marked with a pair of lateral curved white lines, and usually also with a pair of submedianyellowish lines; tibiae without white rings; two dots of white scales on clypeus (Fig. 52) .... *aegypti*

1b. Mesonotum with a narrow median silvery white line; white scales on pleurae arranged in irregular patches; tibiae without white rings; abdomen with silvery basal bands on dorsum (Fig. 53) .... *albopictus*

1c. Mesonotum marked with 4-6 small white spots; all tibiae with white rings (Fig. 54) .... *vittatus*

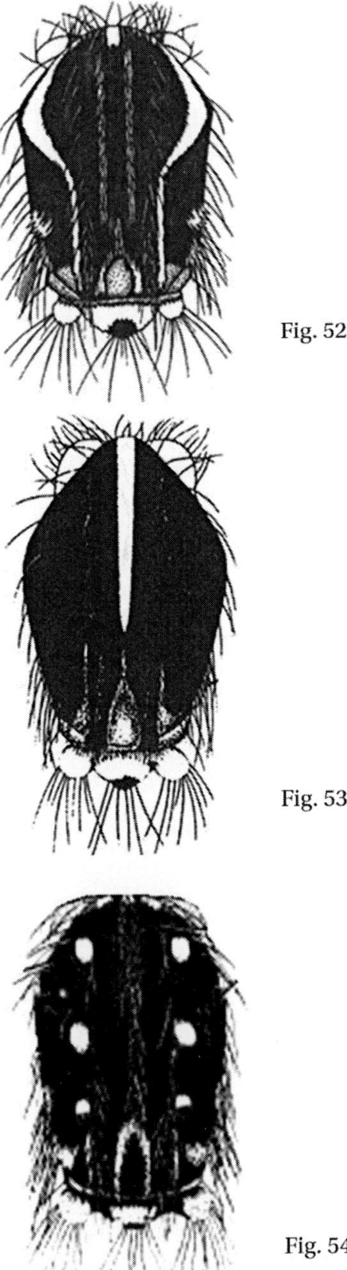

Fig. 52

Fig. 53

Fig. 54

***Legends to the Figures* (52-54) : Fig. 52**- mesonotum marked with a pair of lateral curved white lines (*Aedes aegypti*); **Fig. 53**- mesonotum with narrow median silvery white line (*Aedes albopictus*); **Fig. 54**- mesonotum marked with 4-6 white spots (*Aedes vittatus*).

6.3.5. VECTORS OF LYMPHATIC FILARIASIS

1. Lower mesepimeral setae absent or 1, 2 weak ones; pleuron with distinct pale patches on at least upper and lower mesokatepisternum and anterior mesepimeron .... Subgenus Culex

1a. Proboscis and tarsi without pale rings; pleuron without striking pattern of dark and pale stripes; abdominal terga with basal pale bands; (Figs. 55-57) *quinquefasciatus*

2. Wing with broad, often asymmetrical scales; dark and pale scale intermixed (Fig. 58) .... **Mansonia**

2a. Mesonotum marked with a pair of sublateral greenish stripes on a brown ground (Fig. 59) .... *uniformis*

2b. Yellowish-brown; mesonotum marked with 4 (or more) distinct round white spots; rather broad white scales on mid-lobe of scutellum (Fig. 60) .... *annulifera*

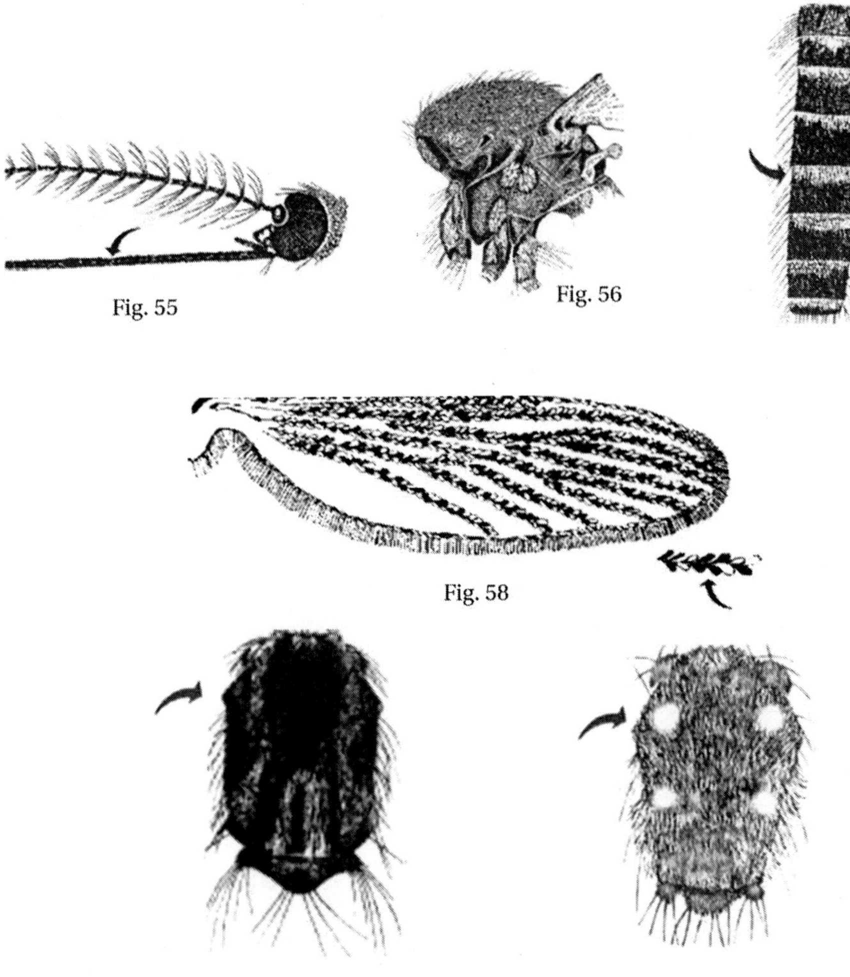

Fig. 55 Fig. 56 Fig. 57

Fig. 58

Fig. 59 Fig. 60

***Legends to the Figures* (55-60) : Fig. 55**- proboscis without pale ring; **Fig. 56**-pleuron without striking pattern of dark and pale stripes; **Fig. 57**- abdominal terga with basal pale band; **Fig. 58**- wing with asymmetrical scales; **Fig. 59**- mesonotum with sublateral greenish stripes; **Fig. 60**- mesonotum with four round white spots.

6.3.5.1. Diurnally sub-periodic filariasis vector in Andaman & Nicobar islands

1.	Postspiracular setae present; proboscis slender and usually straight, never laterally compressed (Fig. 61)	**Aedes**
1a.	Mesonotum with a large snowy white patch in front; wing scales all dark; tarsi entirely dark (Figs. 62, 63)	*nivea**

* *Downsiomyia nivea (=Aedes (Finlaya) niveus)*

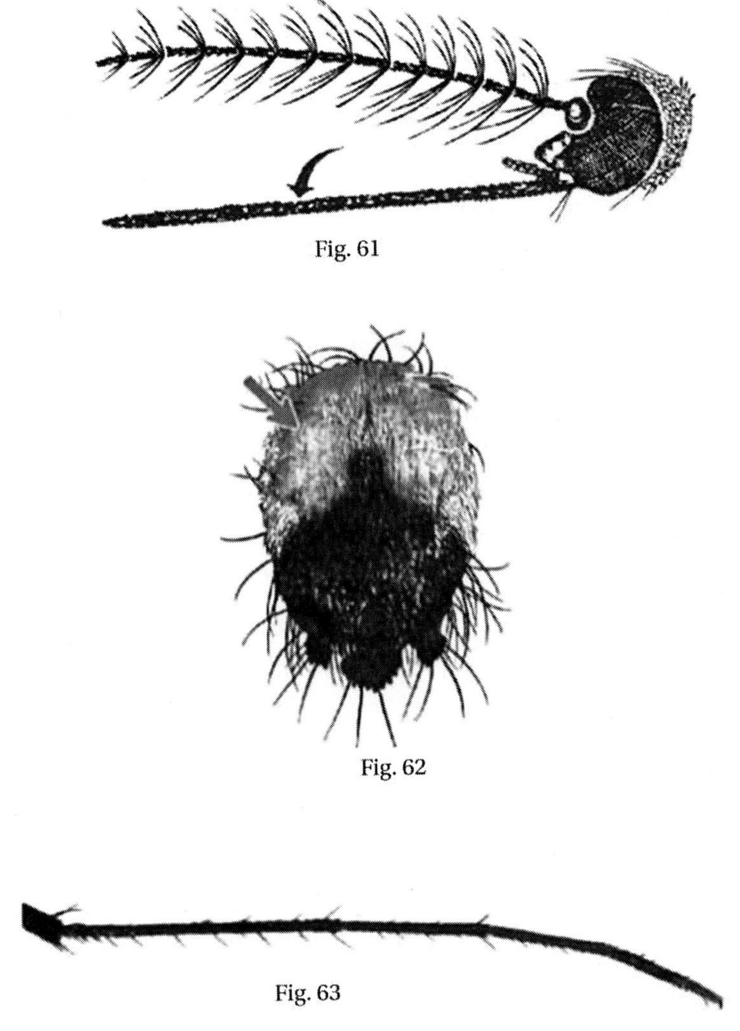

Fig. 61

Fig. 62

Fig. 63

Legends to the Figures **(61-63) : Fig. 61**- proboscis slender and straight; **Fig. 62**- mesonotum with snowy; white patch; **Fig. 63**- tarsi entirely dark.

6.3.6. KEY TO VECTORS OF MALARIA

Wings with 4 or more dark spots on costa (Fig.64) .... (Subgenus Cellia)

1. Femur and tibia speckled (Fig. 65) 2

Femur and tibia not speckled (Fig. 66) 7

2(1). Hind tarsomere 5 not white (Fig. 66) 3

Hind tarsomere 5 white (Fig. 67) 6

3(2). Palpi with three pale bands (Fig. 68) ....

Palpi with four pale bands (Fig. 69) 5

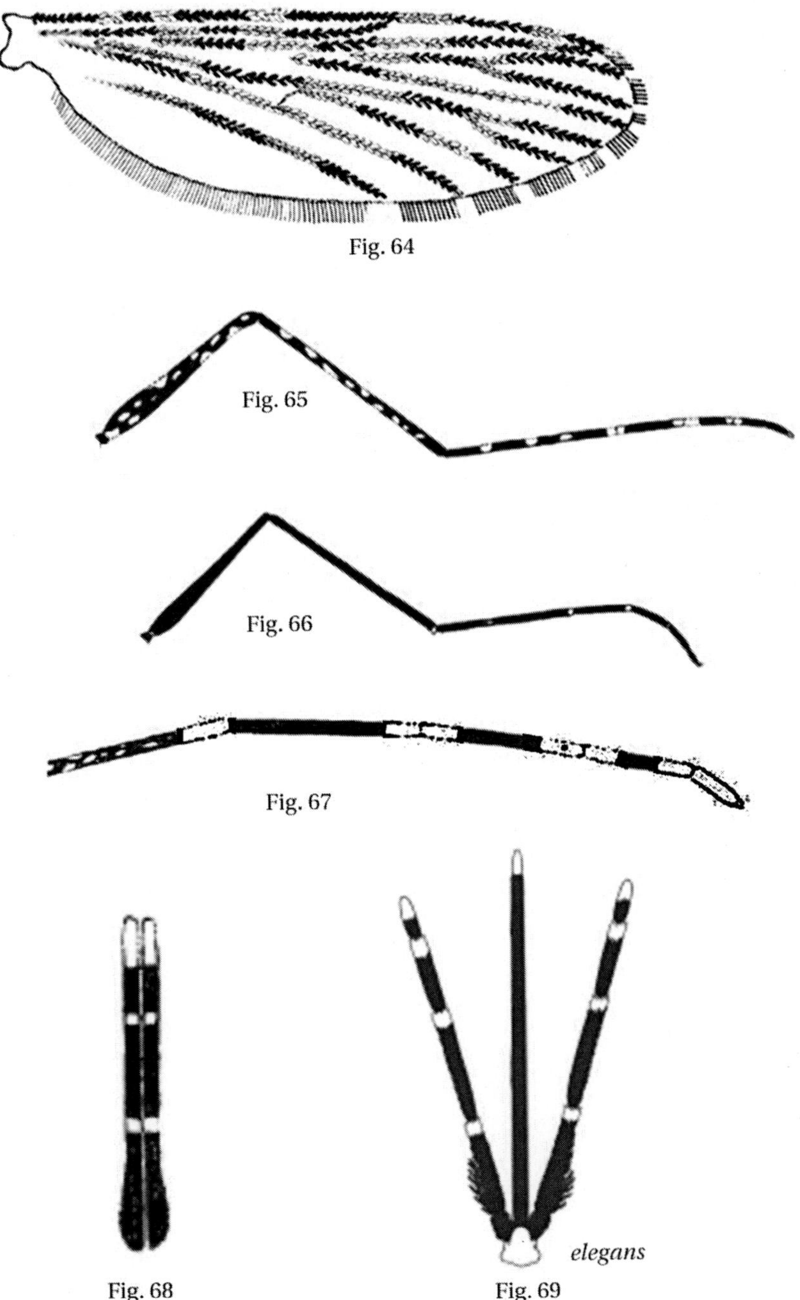

Fig. 64

Fig. 65

Fig. 66

Fig. 67

Fig. 68

Fig. 69

***Legends to the Figures* (64-69) : Fig. 64**- wings with 4 or more dark spots on costa; **Fig. 65**- femur and tibia speckled; **Fig. 66**- femur and tibia not speckled; **Fig. 67**- Hind tarsomere 5 white; **Fig. 68**- palpi with 3 pale bands; **Fig. 69**- palpi with 4 pale bands.

4(3). Apical and subapical pale bands of palpi equal; palpi speckled (Fig. 70) .... *stephensi*

Apical and subapical pale bands of palpi unequal; palpi not speckled (Fig. 71) .... *sundaicus*

5(3). Hind leg with a broad white band at the tibio-tarsal joint; presector dark mark on vein 1 without any pale interruption; apical pale band on hind tibia with a longitudinal basal dark stripe on ventral aspect (Figs. 72, 73) .... *baimaii* *(=dirus D)*

Hind leg with a broad white band at the tibio-tarsal joint; presector dark mark on vein 1 with one or more pale interruption; apical pale band on hind tibia without a ventral dark stripe (Figs. 74, 75) .... *elegans*

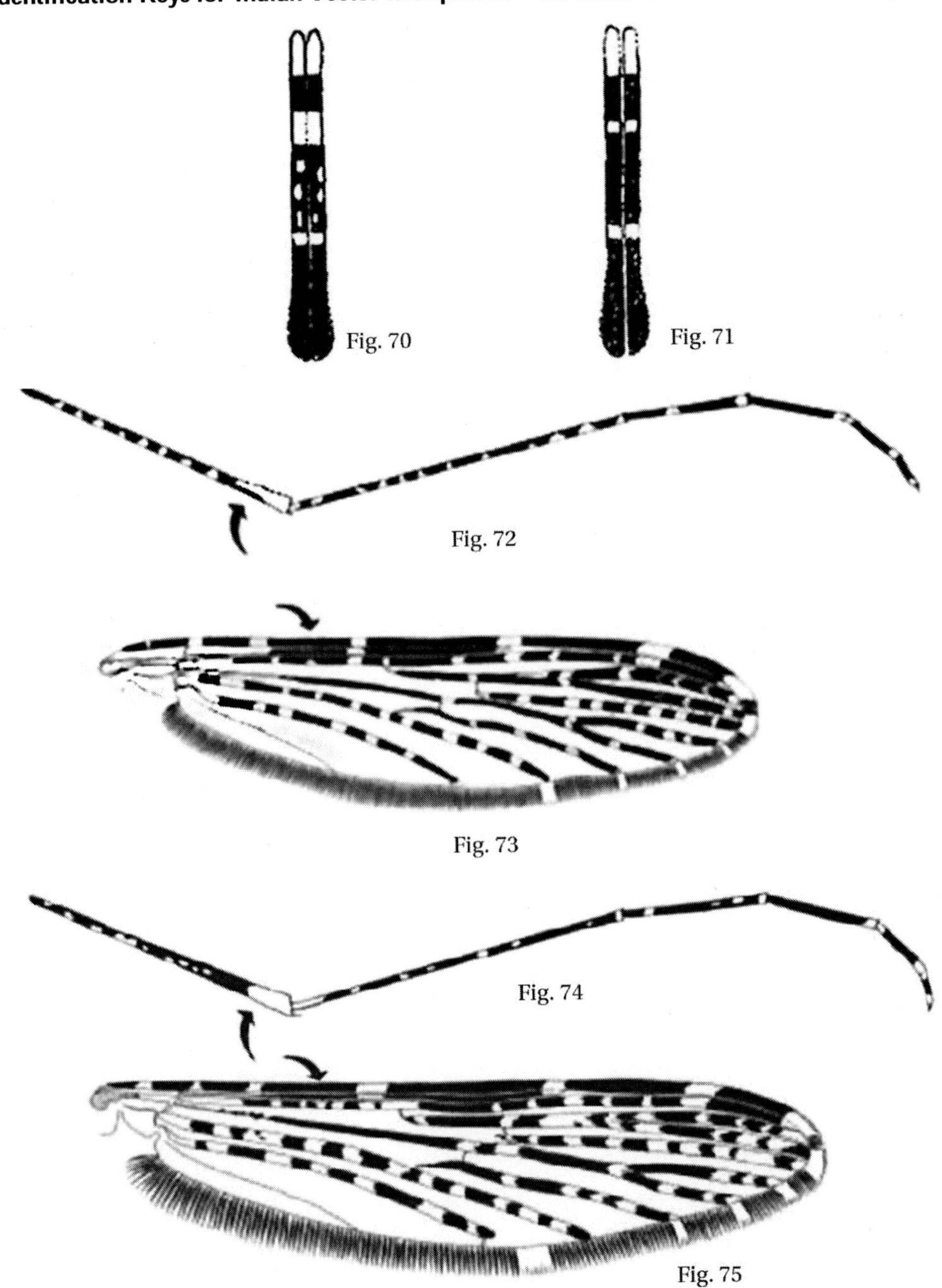

Fig. 70

Fig. 71

Fig. 72

Fig. 73

Fig. 74

Fig. 75

Legends to the Figures **(70-75) : Fig. 70-** palpi with speckling; **Fig. 71-** palpi without speckling; **Fig. 72-** hindleg with broad white band at tibio-tarsal joint; **Fig. 73-** presector dark mark on vein 1 without pale interruption; **Fig. 74-** hind tibia without ventral dark stripe; **Fig. 75-** presector dark mark on vein 1 with pale interruption.

6(2). Abdominal terga II-VIII shiny, with no pale scales (Fig. 76) .... *eudowillmori*

Abdominal terga IV-VIII with a few pale scales (Fig. 77) .... *maculatus*

7(1). Hind tarsomere 5,4 and 3 completely white; apex of hind tarsomere 1 usually with white band; wing vein 5 mainly white, no dark spot at the point of bifurcation (Figs. 78, 79) .... *philippinensis*

Hind tarsomere not white; fore tarsi with narrow pale bands (Fig. 80) 8

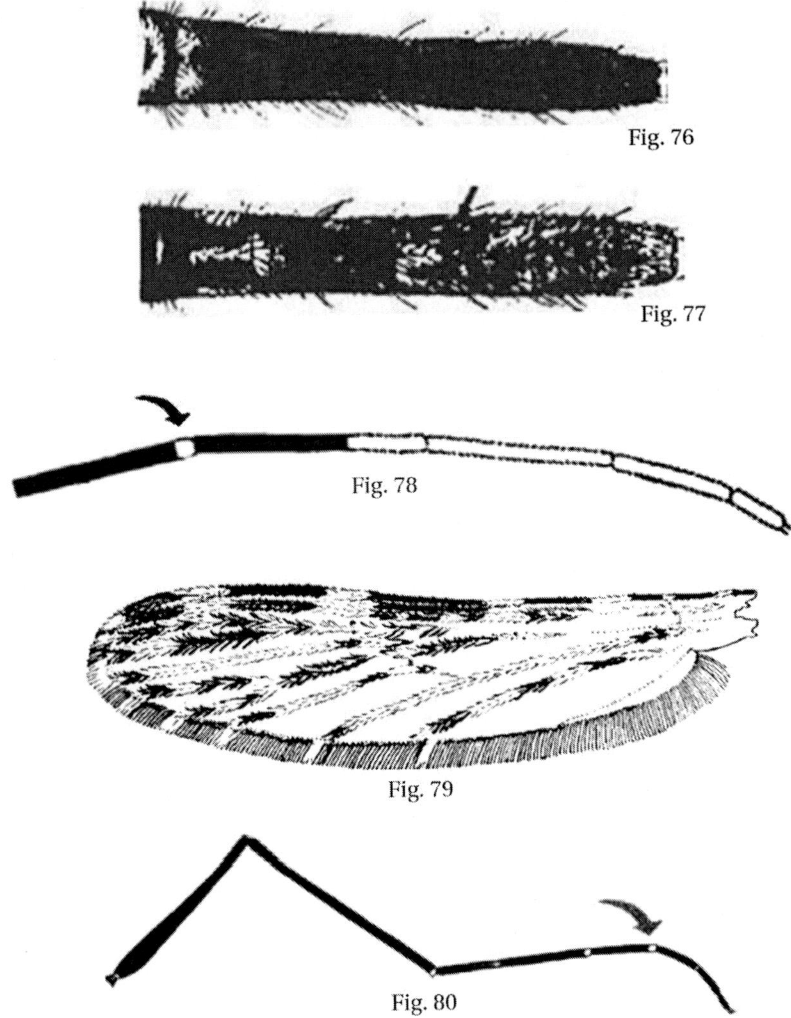

Fig. 76

Fig. 77

Fig. 78

Fig. 79

Fig. 80

***Legends to the Figures* (76-80) : Fig. 76-** abdominal terga II- VIII without pale scales; **Fig. 77-** abdominal terga IV- VIII with few pale scales; **Fig. 78-** hind tarsomere 1 with white band; **Fig. 79-** wing vein 5 white no dark spot at the point of bifurcation; **Fig. 80-** fore tarsi with narrow pale band.

8(7). Wing vein 3 mainly dark; inner quarter costa usually with pale interruption (Fig. 81) .... *culicifacies*

Wing vein 3 mainly white (Fig. 82) 9

9(8). Female palpi with both apical and subapical pale bands as broad as or broader than subapical dark band; basal third of costa with white interruption; fore tarsomere 1-4 with very small dorso-apical pale patches (Figs. 82-4) .... minimus

Female palpi with subapical pale band, narrow and subapical dark band much broader; vein 6 with two dark spots, distal half mainly dark; hind tarsomeres uniformly dark (Figs. 85-87) .... fluviatilis

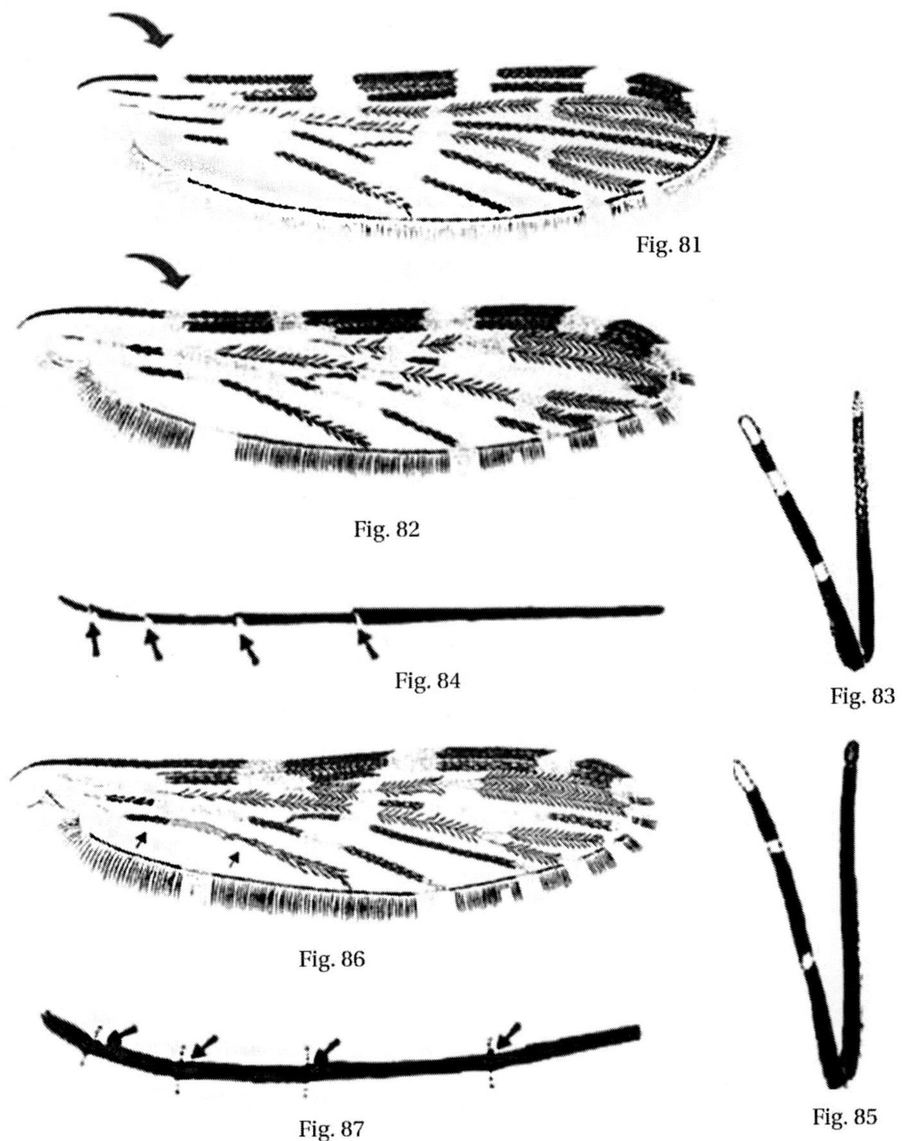

Fig. 81

Fig. 82

Fig. 84

Fig. 83

Fig. 86

Fig. 87

Fig. 85

***Legends to the Figures* (81-87) : Fig. 81**- wing vein 3 mainly dark; **Fig. 82**- wing vein 3 mainly white; **Fig. 83**- palpi- apical and subapical pale bands broader than subapical dark band; **Fig. 84**- fore tarsomere 1-4 with very small pale patches; **Fig. 85**- palpi- subapical pale band narrow and subapical dark; band broader; **Fig. 86**- wing vein 6 with 2 dark spots; **Fig. 87**- hind tarsomere uniformly dark.

Source of Fig. 6.2 to 87 is primarily Tyagi *et al.* 2012 and 2014.

Chapter 7

WORLD'S FOREMOST ANTI-MOSQUITO CRUSADERS OUTSIDE INDIA - *19TH CENTURY*

Among all the world's countries these were mainly the tropical and subtropical countries in the Central and South Americas, Africa and Asia which suffered badly due to devastating epidemics of malaria and long-term debilitating impact of filariasis, besides havocs of the Yellow Fever in Africa and the Latin America. Although most scientists today credit Dr Patrick Manson, Dr Ronald Ross, Dr Walter Reed and Dr Finlay for pinpointing mosquito's role in the transmission of various different kinds of vector-borne diseases during the late nineteenth century, yet the fact is that many researchers had cast a doubt about the critter and even professed its contribution to malaria spread (Harrison, 1978). In fact a team of great scientists like Prof. Grassi, Dr Golgi and Dr Bignami were working already on malaria in Italy, along with a possible role of mosquitoes in its transmission. Even before them Giovanni Lancisi had more or less certainly laid the seed for confidently thinking about mosquitoes in as far as malaria is concerned. This was later given weight by Dr Daniel Beauperthuy in Venezuela and Dr Josiah Nott in Alabama more than two decades before a concrete and decisive observation was put forth by the great **Dr Sir Patrick Manson** who subsequently discovered the role of *Culex pipiens* in the transmission of human lymphatic filariasis in China in 1880, which, in turn, marked the foundation of the great science of medical entomology. (Figs. 7.1-7.3).

Born at Old Meldrum, in Aberdeenshire in Scotland, UK, in 1844, Manson, the son of a bank manager, qualified in medicine at Aberdeen in 1865. Graduating from Aberdeen University, he abandoned a potentially lucrative career in industry for medicine and instead chose to sail for China to exercise in medicine - his first love! After a few years in Formosa, now known as Taiwan, he moved to Amoy on the mainland. He travelled widely in the vast stretches of the East, becoming medical officer in Formosa, and later moving to Hong Kong. During his long and tumultuous career spanning 23 years, he addressed to the problems of tropical diseases such as tinea, Calabar swelling and blackwater fever. Dr Manson, who was posted as a British medical officer in Formosa, under Chinese regime, in the 1870s, became much concerned by the various debilitating forms of gruesomeness and grotesqueness, particularly elephantiasis of legs and hydrosis of the genitals, of the disease called lymphatic filariasis that was caused by a worm and was prevalent widely across the tropical World affecting millions of people! In 1878 he observed that filariae, the worms that cause elephantiasis in man, pass part of their life cycle in the *Culex* mosquito; he thus led the way in the study of the transmission of diseases caused by parasites. All this great discovery occurred in the obscurity of Amoy where he first recognized that a bloodsucking arthropod can harbour – and presumably transmit – organisms of human filariasis: in particular, that the house mosquito is an intermediary of the filarial parasite *Wuchereria bancrofti.* Manson published his finding in 1878 and within two years discovered that the parasite's "embryos" (microfilariae) exhibit "nocturnal periodicity", i.e., they absent from blood during the day and reappear at night coincident with the vector's greatest biting activity. Unfortunately, like many of his contemporaries, Manson believed that mosquito took one blood meal and died on water a few days later. Manson theorized that the filarial larvae escaped from the mosquito into water and that people ostensibly infected themselves by drinking the contaminated water. When, however, infective larvae were found in 1900 in the mouthparts of mosquitoes, the concept that transmission occurred by bite (malaria transmission was by then understood) proved unavoidable. Although Manson was not justified in his theory, nevertheless he had shown to the world that mosquito transmitted a disease like lymphatic filariasis. Based on his conviction about role of mosquitoes in transmitting various different parasites, he made in 1894 the deduction that the parasite of malaria passed part of its life cycle in the mosquito, a theory that Ronald Ross was to verify three years later. A founder of two schools devoted to the study of tropical diseases, one at Hong Kong (1886) and the other at London (1898), Manson is often described as the father of tropical medicine. He practised medicine in China (1871) and in Hong Kong (1883), started a school of medicine that became the University of Hong Kong. He became medical adviser to the Colonial Office, and in 1899 helped to found the London School of Tropical Medicine. Manson retired to Britain in 1889 and to provide students with intellectual stimulus he wrote his '*Manual of Tropical Diseases*' which promptly became a bestseller in its field. He died in 1922. It was his natural affectionate inclination for medical education and research that he provided all that help that could have been possible on earth. It goes without mention that it was him whose advice to Ross finally led the latter to discover *Anopheles* mosquito as the carrier for malaria parasite - a work for which Ross was awarded the Nobel Prize for 1902.

While Manson was busy studying lymphatic filariasis transmission in China and, malaria in England, a superb team of scientists led by an inexhaustible American scientist, **Dr Walter Reed**, was progressing to unveil the mystery behind transmission of the deadly Yellow Fever and malaria in Central America. On September 13, 1851, Dr Walter Reed, one of the world's outstanding physicians and medical research scientists, was born in Belroi, Gloucester County, Virginia, in an affluent family of a Methodist minister, Lemuel Sutton Reed, and his Wife, Pharaba White Reed. At an early age, Walter Reed gave evidence of the intellectual brilliancy and earnestness of purpose which distinguished him in later years. After his basic education at a private school in Charlottesville, Virginia, Walter Reed matriculated at the famed University of Virginia where he completed the two-year medical course in only one year and received his degree in 1869 at the age of eighteen. He was the youngest student ever to graduate from the medical school. Since the University of Virginia had no hospital attached, he took a second degree at Bellevue Medical College in New York in 1870. During the subsequent five years he served his internship at Kings County Hospital in Brooklyn and at the Brooklyn City Hospital, and by this time he earned himself a reputation of a versatile conversant. After further internship at Brooklyn City Hospital with consultant status at Kings County, he was appointed one of the five inspectors on the Brooklyn Board of Health in 1873 at the age of twenty-two. He approached all his duties with enthusiasm and optimism, traits which contributed immeasurably to his success, both social and professional. During all these years he had developed an immense love for music and left no opportunity to attend good lectures on literature and scientific subjects, no matter where they were to be held. It was at this time that the young intern conducted his first research which was the basis of his first scholarly paper, published in 1892, called "The Contagiousness of Erysipelas." In 1874, having served on the Boards of Health in Brooklyn and New York, he returned to Virginia to visit his family, and soon got married to Emilie Lawrence, the daughter of a North Carolina planter. Because he felt the Army offered a good opportunity for research, he applied for an Army position and was accepted for an appointment of a surgeon in the Medical Department. He passed the required examinations and was appointed Assistant Surgeon with the rank of first lieutenant on June 26, 1875. So began an eighteen-year garrison life. After five years at Fort Lowell and Fort Apache, Arizona, where he served as a beloved family doctor visiting patients in the wild country surrounding his posts, he was promoted to captain and soon thereafter was transferred to Fort McHenry in Baltimore. In this city he became a student of physiology at Johns Hopkins University in his spare time during 1881 and 1882.

After serving at Fort McHenry, the ever peripatetic Walter Reed was again assigned to southern and western posts at Fort Omaha, Nebraska, and Mount Vermon Barracks, Alabama. Dr Reed returned to Baltimore in 1890 as Attending Surgeon and examiner of recruits. This assignment was quite welcome since it provided him with the opportunity to do research. He became a student of bacteriology and pathology under the tutelage of Dr William Henry Welch, head of the Pathological Laboratory at Johns Hopkins and foremost pathologist and medical bacteriologist in this country. These subjects were not taught as part of the medical curriculum of that day. It is significant that Walter Reed's career coincided with the great flowering of medical science that took place in the 1880s. The germ theory of infectious disease was now accepted as postulated and proved by Pasteur, and Robert Koch had perfected a method for studying bacteria. In the United States George Miller Sternberg, later Army Surgeon General, with whom Walter Reed had a close professional relationship, was one of the founders of bacteriology. The 1880s were critically important years in Walter Reed's life and in the destiny of the United States. It was during this period, when all areas of the central and northern part of southern Americas were plagued by yellow fever, that the mature scientific investigator was formed. He conducted his own individual research much to the delight and satisfaction of Dr Welch who had been one of Pasteur's students. His ties with Dr Welch were strengthened and Dr Welch's mutually admiring relationship with Surgeon General Sternberg was quite advantageous to Dr Reed as an Army surgeon protege. Walter Reed, then forty-two years old, was greeted with enthusiasm by an eager group of researchers at the Pathological Laboratory. Captain Reed - promoted on June 26, 1880 - learned the necessary research techniques and was soon working on his own project concerning typhoid fever.

A man of sterling character, religious by nature, prepared for practice and research, a soldier who had learned to endure hardships, a student and pathologist of the highest calibre, Walter Reed was now ready for the great achievements of his lifetime. He would live for only fifty-one years, but between 1892 and 1901, till a year before his death, he was engaged in some of the most important work in the history of medicine. This took the form of research into the etiology and epidemiology of typhoid and yellow fever. From 1891 to 1893 Walter Reed spent his last western tour in the Dakotas. He was promoted to major in 1893 and then took his post as curator of the

Army Medical Museum (now part of the Armed Forces institute of Pathology) and professor of clinical microscopy in the Army Medical School (now the Walter Reed Army Institute of Research) founded in Washington by Surgeon General Sternberg. At this same time he also held the chair in bacteriology at the Columbian University (now the George Washington University). He worked industriously for five years, teaching and working in his specialty, bacteriology. His work, clinical and academic, was accurate and original. His experience at work in the School gave him an excellent sense of judgement valuable for investigating the causes of epidemic diseases and in making sanitary inspections at military posts. Medicine was making rapid advances and military doctors had to be informed of the new techniques. An anti-toxin for diphtheria had been prepared and there was a race to find the specific agents (bacteria or viruses) responsible for communicable diseases. This was the beginning of medicine as we know it today, an era of great discovery. At the time of the Spanish-American War, however, he was casting about for a position in the field which would allow him to put his knowledge of Army routines and sanitation to use for the benefit of soldiers in Cuba.

It was not until mid-August in 1898 when the war was already over that he had an opportunity to perform this valuable service. General Sternberg was then able to secure a board of officers headed by Major Reed to investigate the spread of typhoid fever. The Typhoid Board performed its greatest service through the discovery that this dread disease, prevalent at almost all U.S. Army encampments, was spread most commonly and disastrously by contact between persons and flies soiled with human excrement containing typhoid bacilli, by human carriers who shed bacilli by the billions, and by impure drinking water. This triumph for Army medicine demonstrated for the first time the effects of intestinal disease-producing agents or the pathogens. This victory also served to dispel old notions that such diseases were caused by miasmas or foul emanations from swamps and rivers.

In May 1900 Major Reed was appointed president of a board whose purpose it was to study infectious diseases in Cuba paying particular attention to yellow fever. The other members of the board were Acting Assistant Surgeons Major James Carroll, Manor Jesse W. Lazear and Major Aristides Agramonte of Havana. As a result of this Yellow fever Board, very few people living today have any knowledge of this dread disease. The onset of yellow fever came with chills and a headache. Then followed severe pains in the back, arms and legs accompanied by high fever and vomiting. The feverish stage might last hours, days, or weeks. Jaundice, from which the fever derives its name, might then appear. Then came the so called "stage of calm" when the severity of the symptoms subsided and the fever dropped. In less serious cases this stage indicated recovery. But in the main, this stage was followed by a return of the fever accompanied by internal bleeding which caused the dreaded "black vomit" when blood released into the stomach was ejected.

Yellow fever killed more men in the Spanish-American War than did the enemy. It appeared in Central America in 1596, probably imported from Africa by slave ships. It may have been the disease from which members of Columbus' second expedition suffered in 1495. Ninety epidemics struck the United States between 1596 and 1900. In 1793 an epidemic first hit Philadelphia, then the U.S. capital, causing the Government to flee as ten percent of the population perished. Washington went to Mount Vernon while Jefferson fled the disorder caused by the onslaught of the disease. Because of frequent epidemics which destroyed ninety per cent of his expeditionary forces in 1802, Napoleon was influenced to sell the Louisiana Territory. It was chiefly because of "yellow jack," as the disease was nicknamed from the pennant which was flown during quarantine, that the French were unable to complete the Panama Canal. The danger of contaminating the southern states was considered to be a major factor in the annexation of Cuba. Reed and Carroll had estimated that there were 300,000 cases in the United States between 1793 and 1900, which cost the nation almost $500,000,000 with a mortality rate usually at forty per cent but sometimes as high as eighty five per cent. The scourge of yellow fever had plagued the south-eastern United States for two hundred years, but nowhere was it more prevalent than in Havana. Surgeon General George Miller Sternberg was this country's leading expert on yellow fever. Because neither he nor other researchers had been able to pinpoint the specific cause, he was astounded to hear the claim of Dr Giuseppe Sanarelli, who in 1897 stated quite conclusively that the fever was caused by *Bacillus icteroides*.

On June 25, 1900 Walter Reed arrived at Columbia Barracks in Quemados about six miles from Havana. Major Reed attended Major Jefferson R. Kean, Chief Surgeon of the Department of Western Cuba, who had himself contracted the disease and fortunately recovered. Then Doctor James Carroll and Major Reed set out enthusiastically to prove Sanarelli`s theory. By August, 1901 however they had found no causal relationship between *Bacillus icteroides*, a member of the hog-cholera group, and yellow fever. The Board then turned its attention to the theory of **Dr Carlos Juan Finlay** and examined it more carefully. For nineteen years this resident of Havana had contended that yellow fever was carried in the body of a common house mosquito which at that time was called *Culex fasciatus,* later *Stegomyia fasciata,* and is now known as *Aedes aegypti.* This theory had been expounded even earlier, but it was Finlay who was its staunchest exponent. However, after some 100 experimental inoculations had failed to produce any cases of the disease under strict laboratory control, Finlay was scoffed at and even ridiculed as the "mosquito man". There was evidence, however, that tended to lend credence to this theory which even the Yellow Fever Board, optimistic though it was, had doubted. First of all, the disease skipped erratically from house to house, jumping around corners. One member of a household might contract the disease while others in close contact never became ill or did so after a period of about two weeks had elapsed. This was quite unlike any other infectious disease except malaria which had already been shown to be spread by the Anopheles mosquito, by Ronald Ross in India (*vide infra*). The Yellow Fever Board then decided that the best way to study yellow fever was not by searching for a specific agent but rather by identifying the means by which the fever was transmitted. For this purpose Major Reed organized the Board in the following manner: Major Reed himself was in-charge of the entire project, **Dr James Carroll** was in-charge of bacteriology; **Dr Jesse W. Lazear** was in-charge of the experimental mosquitoes, and **Dr Aristides Agramonte** was in charge of pathology. After several weeks, the Board determined that the alleged agent causing yellow fever was not *Bacillus icteroides.* Carroll experimentally showed that the agent could pass through a Berkefeld filter and, therefore, was not bacterial. Dr Lazear had recently been working with malarial mosquitoes and attended his duties with great enthusiasm in view of the information he read concerning the observations of Dr Henry R. Carter of the Marine Hospital Service (now the Public Health Service). Dr Carter had observed that it took two or three weeks for the first case of yellow fever to produce the next case in a community. On the basis of this observation, he suspected that an insect might be the intermediary since this would account for the delay in transmission as the disease ripened in the mosquito. Dr Finlay had given some of the black, cigar-shaped eggs to the Board, and Lazear allowed them to hatch. It wasn't difficult to maintain a supply since the mosquitoes bred in any clean, still water. Dr Carroll then volunteered to be bitten by *Aedes aegypti* on August 27, 1900, and promptly developed a successful case of yellow fever which was experimentally defective since there may have been other sources of contamination. Carroll luckily recovered and went on with his work in bacteriology. Next, Lazear got Private William Dean of Ohio bitten by the female *Aedes aegypti.* He developed the first successful case and recovered. Finally, Dr Lazear allowed himself to be bitten and, after several days of delirium and black vomit, died - a true martyr to science. Major Reed, although grieved at Lazear's death, did not lose heart and picked up meticulously the threads of Dr Lazear's highly exciting experiments (minutest details of which he scrupulously noted in a diary). Tracking down the secret of the fever Walter Reed found that in order for a mosquito to become infected, it had to bite a yellow fever patient during the first three days of his illness as only during that time was the agent present in the bloodstream. Further, it required at least twelve days for the agent to ripen in the female mosquito (only the female *Ae. Aegypti* draws blood) and migrate to her salivary glands before the fever could be passed to another person. He solved the puzzle on 13th October 1900, announcing on his results to the annual meeting of the American Public Health Association that "the mosquito serves as the intermediate host for the parasite of yellow fever." Thus, by an independent move, Dr Reed also disproved the fallacious belief that yellow fever could be transmitted and induced from clothing and bedding soiled by the excrement of yellow fever sufferers. These articles were known as fomites and were commonly thought to carry the disease. Just as "everybody knew" that the mosquito theory was foolish, so "everybody knew" that fomites were dangerous. These two cases, although sufficient to convince the Yellow Fever Board that they were at last experiencing some success, were not enough for the thorough scientific mind of Walter Reed, nor would they be for a public which the press had instructed in the "foolishness" of the mosquito theory. With the express permission of General Leonard Wood, Governor General of Cuba, Camp Lazear, was established on November 20, 1900. Moreover, General Wood authorized the Board to use and pay American and Spanish volunteers for the experiments since at that time yellow fever was thought to be a disease afflicting only humans. Dr Carroll had exhausted the list of experimental animals, rats and the like normally used for scientific research, failing to produce any cases of the fever in them. In addition to the mosquito theory, "bacillary theory" was also contemplated. Had it not been for Major Reed's fair and thoroughly scientific approach to the problem and misconceptions concerning the disease, especially the whole contagion theory, yellow fever might have continued for years. His efforts were providentially assisted by the inestimable courage of a large number of volunteers, just like Hussein proved for Ross's malaria research.

In February, 1901 Walter Reed returned to the United States Where he was an instant success in medical circles. All through the acclaim he remained modest and reserved. His constant hope of doing something to relieve the suffering of mankind had been fulfilled; his dedication to duty, sound judgment, and thorough scientific methods was an inspiration to the deans of medical research. Soon, he took up an assignment of professor of pathology and bacteriology at the Columbian (George Washington] University Medical School and as professor of bacteriology in the Army Medical School (the present Walter Reed Army Institute of Research). In the summer he was awarded two honorary degrees: a Master of Arts from Harvard University and a Doctor of Laws degree from the University of Michigan, He was appointed librarian of the Surgeon Generals library on November 1, 1902.

Dr Reed died of peritonitis on November 23, 1902, following an appendectomy. He was buried in Arlington National Cemetery. On his simple monument is inscribed the following epitaph, taken from the remarks of President Eliot when Harvard University conferred the Master of Arts degree; "He gave to man control over that dreadful scourge, yellow fever." Today a great hospital and medical centre stand in constant tribute to Walter Reed. In 1945 he was elected to the Hall of Fame at New York University, the first physician to be so honoured. On November 21, 1966, a bronze bust of Major Reed was unveiled at Walter Reed Army Medical Centre by the Walter Reed Memorial Association, an organization which, since its inception in 1903, had resolved to erect a memorial in Washington to perpetuate his fame and memory. During his career in army Dr Reed travelled extensively mostly due to frequent transfers. Truly, one of the several marvels of his life is that his relegation to frontier garrisons, unfavourable for intellectual contacts, did not ruin him. On the contrary, they provided him infinite opportunities to accomplish the rarest tasks of great human values.

Dr Reed's researches on yellow fever, in particular, and that of Dr Ronald Ross's in India as well as Dr Giovanni Battista Grassi in Italy ushered in new hopes in controlling the vector mosquitoes involved. **Surgeon General William Crawford Gorgas** (October 5, 1854 - July 3, 1920) was the son of General Josiah Gorgas, a native of Pennsylvania who graduated from the United States Military Academy in the class of 1841. Assigned to the ordnance service, he was in command of Mount Vernon Arsenal, near Mobile, Alabama, in 1853. The junior Gorgas was married in 1885 to Marie Cook Doughty in Cincinnati. They had one child, Aileen Tyster. His varied duties took him to all parts of the world and he received honours too numerous to list. A person of initiative, vision, purpose, tenacity and energy, he was probably best known for translating the discoveries of Dr Walter Reed concerning yellow fever and malaria to practical application. Gorgas' work conquered these diseases by destroying the disease-bearing mosquitoes of the tropics.

Gorgas' early education had been quite irregular. He had had the advantages of a private school in Richmond but the times were too distracting for satisfactory progress. He spent six years at the Sewanee school and in 1875 graduated with the degree of bachelor of arts at the University of the South, and his MD from Bellevue Medical College in New York. From the experiences of his childhood Gorgas had acquired a strong desire for a military career. Every effort was made to obtain for him an appointment to the West Point academy but without avail. Much against his father's wishes he determined to enter the service by way of a medical degree. He entered Bellevue Hospital Medical College in New York in 1876 and was graduated in 1879 after three years of financial difficulties. Following an internship in Bellevue Hospital he was appointed, on June 16, 1880, an assistant surgeon in the medical corps of the army. For nearly two decades thereafter, Gorgas' life was that of the average medical officer of the period. Following several years in Texas posts and a tour of duty in North Dakota he spent practically the entire decade preceding the Spanish-American War at Fort Barrancas in Pensacola Bay, Florida.

Shortly after the beginning of his army career he went through an epidemic of yellow fever at Fon: Brown, Texas, and was himself stricken with the disease. Thereafter, as an immune, he was frequently drafted for service where yellow fever existed. This accounts for his long service at Fort Barrancas, a post in a section notorious for its epidemics and itself frequently subject to visitations of the disease. He was promoted to the rank of captain on June 16, 1885, and then to major on July 6, 1898. To Gorgas, as to others, yellow fever was an enigma. Its suddenness of appearance, its puzzling choice of victims, and the inutility of ordinary means of disease prevention were quite beyond understanding. With the Spanish-American War in progress, Gorgas arrived at Siboney, in Cuba, on July 7, 1898. While negotiations for the capitulation of Santiago were in progress yellow fever broke out in the American forces. Gorgas was assigned to the yellow fever section of the Siboney hospital and shortly thereafter took over the command of the hospital from Major Louis A. La Garde. It is significant of the view of the disease then currently held that it was recommended that the village of Siboney be destroyed by fire. Not only was this recommendation carried out, but in September the Siboney group of hospitals with much of their equipment was similarly destroyed.

Major Gorgas was returned to the United States in September, convalescent from typhoid fever, Later in 1898 he returned to Cuba and early in 1899 when he became chief surgeon of the Department of Havana Following the appointment of General Leonard Wood as military governor of Cuba in December 1899, Gorgas was made chief sanitary officer for the city of Havana. Though yellow fever at this time showed only as sporadic cases it was, as always, the chief concern. Gorgas applied to the city the generally accepted methods of disease control. He cleaned up the city, segregated the sick, and quarantined infected localities. Though a friend of Dr Carlos J. Finlay and familiar with his theory of the mosquito transmission of yellow fever Gorgas placed no confidence in the idea. Despite the greatly improved sanitary condition of Havana the yellow fever situation, instead of improving, became much more serious. It was not until the board, of which Major Walter Reed was the head, furnished proof that the *Stegomyia* mosquito (which we now know as *Aedes aegypti*) was the carrier of the disease that truly effective methods could be instituted. *Aedes aegypti* was the common mosquito pest of the city. It is a highly domesticated insect, breeding in all kinds of water containers in and around habitations. The surest control of the insect was deemed to be the elimination of its breeding places. This plan was adopted and, though the task had many difficulties, Havana was not only freed of its mosquitoes but was permanently rid of yellow fever. The results obtained by his work in Havana brought him an international reputation as a sanitarian. The years from 1900 to 1904 brought the gradual development of plans for digging the Panama Canal. There was early recognition of the necessity for expert sanitary advice upon the project and in 1902 Gorgas was transferred from Havana to Washington and assigned to this work.

Fig. 7.1. (a) Dr Sir Patrick Manson, (b) Dr Walter Reed, (c) Dr Carlos Juan Finlay, (d) Dr J.C. Carroll, (e) Jesse William Lazear, and (f) Dr Aristides Agramonte. (*Source:* Free Wikipedia)

Fig. 7.2. (a) Dr William Crawford Gorgas, (b) Dr Robert Koch, (c) Camillo Golgi, (d) Dr Ettore Marchiafava, (e) Dr Amico Bignami, and (f) Dr Angelo Celli. (*Source:* Free Wikipedia)

In March 1903 Congress raised him to the grade of colonel in recognition of his services in Havana. For two years he studied the canal problem, reviewing the experience of the French on the isthmus and making visits to the Suez Canal and to Panama. Actual work upon the canal commenced in 1904, and Gorgas with his staff of assistants arrived in June of that year. He early encountered administration difficulties. Despite the positive knowledge that the French failure had been due to disease the American administration was disinclined to support adequate measures for preventing a repetition of that experience. The first Canal Commission, headed by Admiral John C. Walker, had strongly in mind the prevention of graft and extravagance that, with of course the scourges of malaria and yellow fever, had caused much of the disaster of the French effort, particularly of De Lesseps. Expenditures for sanitary improvements were regarded as falling under the latter head. It required a visitation of yellow fever, starting in November 1904, to obtain for Gorgas any substantial support for his work. He began in the Canal Zone the measures which had been highly successful in Havana. Again the mosquito was to be deprived of breeding places and cases of yellow fever segregated and protected from mosquitoes. The situation in Panama presented more difficulties than that in Havana and results were far less prompt. It was well into 1905 before yellow fever had been eradicated, and in the meantime determined efforts were being made to discredit Gorgas' work and to supplant him. It is probable that these could have been successful but for the interest aroused by a report made by Dr Charles A. L. Reed of Cincinnati to the American Medical Association in March 1905, in which the obstructive hand of Commissioner Carl E. Grunsky was so largely featured. The discharge of the Walker Commission at about this time and the appointment of another headed by Theodore P. Shonts did little to mitigate Gorgas' troubles. Yellow fever was still prevalent and the new commissioners were dissatisfied that the first interest of the sanitary service was the elimination of mosquitoes rather than the general improvement of the cities of Panama and Colon. They recommended the removal of Gorgas which not only drew the disapproval of President Roosevelt but caused an order for active support of his work. In November 1906, the President paid a visit to Panama and shortly thereafter Gorgas was made a member of the canal commission. For a time he had practically a free hand, but after the reorganization of the commission in 1908, with Colonel George A. Goethals as chairman and chief engineer, his troubles began anew. Goethals, given unusual powers by executive order, ruled the Canal Zone with a despotic control. He was free in criticism and centred his attacks upon the expense of the sanitary service. Despite the difficulties thrown around his work, due to lack of cooperation from the chief commissioner, Gorgas not only freed the Canal Zone from yellow fever but he made the cities of Panama and Colon models of sanitation comparable with any city of the United States. In the meantime his reputation had extended until he was generally regarded as the world's foremost sanitary expert. In 1913 he was asked by the Transvaal Chamber of Mines to visit South Africa and make recommendation for the control of pneumonia among the negro mine workers. It was while engaged in this work that he received the notification of his appointment as Surgeon General of the Army, with the rank of brigadier general, on January 16, 1914. He returned to the United States in April to take up his new duties, and on March 4, 1915, he was advanced to the grade of major general. The recently organized International Health Board enlisted him as an advisor, and in 1916 sent him with a staff of assistants for a tour of South and Central America with a view to continuing the fight on yellow fever in these sections. Following this trip, a plan for the elimination of yellow fever was adopted and Gorgas was made director of the work. General Gorgas had always been more interested in disease prevention than in office administration. In January 1917 he informed the Secretary of War that he wished to retire. During war days, General Gorgas' time was largely taken up by consultations with representatives of the Secretary of War, the Council of National Defence, the General Medical Board, and with Congressional committees. General Gorgas was retired on account of age on October 3, 1918, shortly before the Armistice and again became available for work with the International Health Board.

General Gorgas was commissioned to investigate the yellow fever situation on the West coast of Africa and in May, 1920 he sailed with his staff for London. After attending the meeting of the International Hygiene Congress in Brussels he returned to London where he experienced a stroke of apoplexy, and died a month later, on July 3, 1920, in the Queen Alexandria Military Hospital at Millbank. The funeral was held in St. Paul's Cathedral and the body returned to the United States to rest in the Arlington National Cemetery. He was survived by his wife, Marie Cook, and a daughter. Many books and articles were written about Dr Gorgas recognizing him as a benefactor of all mankind. General Gorgas was memorialized at the University of Alabama, in Mobile, and in many other places around the world. His statue is in the New York University Hall of Fame. General Gorgas had been the recipient of many honours. Honorary degrees had been bestowed upon him by the University of the South, by Harvard, Brown, Pennsylvania, Johns Hopkins, and by Oxford Universities. He had been decorated by a number of foreign countries. The University of the South and the University of Alabama, both of which his father had served, offered him their presidencies. In 1908, the American Medical Association elected him its president. During his last illness he was visited by King George and knighted. A lesser man than Gorgas might have failed before the opposition he encountered even after the success at Havana gained world-wide publicity and acclaim. United States officials in the Canal Zone refused to believe in Gorgas. They pigeon-holed his requests for equipment, refused him authority to deal with the people, discredited him to higher authorities and frequently clamoured for his dismissal. His greatest gift to the world was a dengue/mosquito-free Panama. Someone thus truly said of him as, "His application of sanitary measures eliminated yellow fever as a blight on the health of the people of the world and made possible the construction of the Panama Canal." Throughout his extraordinary anti-mosquito feats in central and southern Americas General Gorgas was much helped by the quests for factors behind malaria transmission and by the epochal discoveries by a battery of dedicated malariologists, particularly Dr Ronald Ross in India and Professor Giovanni Battista Grassi in Italy.

While the works of the above referred latter two great men of science, i.e., Walter Reed and General Gorgas, were of utmost practical importance, the seed for need of basic research was actually sown by the great Dr Patrick Manson. Undisputedly many contemporary scientists had been inspired from Manson's irrefutable examples and lessons, directly or indirectly. Such is the case of the Italian scientist, **Professor Dr Giovani Battista Grassi** (1854-1925) who uninterruptedly contributed to the enrichment of diverse scientific knowledge especially in the fields of biology and parasitology, till his last breath at a ripe age of 71. He helped conquer malaria disease, and the practical application of his scientific contributions has brought health and happiness to many communities throughout the world. He may therefore be considered as one of the great benefactors of mankind. Besides being an extraordinarily hard worker, Grassi also had an arduous zeal for the unknown. He devoted himself with great passion to scientific research, and as soon as he discovered a truth and laid the foundation for its far reaching practical implications, that same passion drove him toward new adventures and explorations of unknown realms. Grassi was an example of the total consecration of man to the higher ideals of learning and the worship of truth, which he upheld throughout his life. Originality and thoroughness were the main traits of his scientific personality. Moreover, he as a great master founded a scientific school which is still being carried on by his disciples in Italian universities and other countries. In Chile, his ideals were spread through one of his outstanding disciples, the late Professor Doctor Juan Noé (1877-1947), who

Fig. 7.3. (a) Prof. Dr Giovanni Battista Grassi, and (b) Dr Ronald Ross. (*Source:* Free Wikipedia)

founded the Biological Institute of the University of Chile, now bearing his name, and which comprises the Departments of Parasitology, Biology, Histology, and Embryology.

Giovani Battista Grassi was born in the small town of Rovellasca in Lombardy, Italy on March 27, 1854, where he spent his early childhood. At a very early age he showed signs of an extraordinarily vivid intellect and an intense application to work. In elementary and high school he was always head of his class. Because he had a marked talent for mathematics, one of his high school teachers urged him to study this discipline further. But young Grassi felt an intense calling for the natural sciences. He could always be found roaming the countryside and forests collecting plants, flowers, and rocks to take home. It was therefore not a sudden decision when in 1872 he entered the Faculty of Medicine at the University of Pavia. His primary aim was to simultaneously enter research and the practice of medicine, although later on he had to give up the latter, thereby foregoing the opportunity of a splendid financial situation. He obtained his doctorate in 1878 and became Professor of Zoology at the University of Catania till 1896. An established taxonomist of animals, he soon became an expert on the life history and societies of the Termitidae, and on the developmental relationship between Leptocephalus and the eel and other Muraenidae. But, above all, his epochal researches in the field of malariology will be ever remembered in golden letters in the annals of medical history of malaria. Independent of all his contemporary researchers, particularly Ross, and yet most thoroughly, he demonstrated in 1898 that the *Anopheles* mosquito carries the *Plasmodium* parasite of malaria in its digestive tract. They both competed with each other throughout their life, also in the case of the decision for the 1902 Nobel Prize which, after a humongous trail of wrangling and bitter war of words and letters, and even involvement of an independent (that was in fact not without a bias) arbitrator, Dr Robert Koch, finally went to Dr Ross in 1904 (Tyagi *et al.*, 2020)! Even though the balance of pioneering research work on Anopheles mosquito as the carrier for malaria weighed heavier on Ross's side, nevertheless no one could ever deny the genius of Professor Grassi, whom many considered as the most authoritative zoological scientist of his time. Dr Patrick Manson, who mentored Ronald Ross and played a key role in his achieving the most significant discovery of the 19th century, was even in favour of Grassi when a decision about the Nobel Prize was to be taken! The Italian Government had duly recognized the genius, however, and issued a stamp (25 L) in his honour on November 19, 1955, on the occasion of his 30th death anniversary, bearing portrait of Battista Grassi, mosquito and microscope. The people of Italy also dedicated a boulevard in Milan in his memory and named it as Giovanni Battista Grassi Street 74, harbouring the famous Luigi Sacco Hospital. Professor Grassi received many honours and distinctions among which the Darwin Medal presented to him in 1896 was very dear to him. He died in 1925.

Chapter 8

MOSQUITO TERMINATORS OF INDIA - *AN EXTRAORDINARY LEAGUE OF CULICIDOLOGISTS*

8.1. Introduction

'*Mosquitoes v/s man*' fracas created a timeless war wherein entomologists, bacteriologists, medical doctors and technicians etc. discovered *de novo* the connection between mosquitoes and various deadly and/or debilitating infections such as malaria, yellow fever, dengue, chikungunya, Zika and encephalitides of different kinds, innovated insecticides with varied pathways of action, and invented vaccines to counter their onslaughts. At present these discoveries, inventions and innovations might appear simple but, on delving deeper into the intricacies of their development, they would prove to be really fundamental necessities of science - for instance, how a malaria parasite (oocysts) was first viewed by Dr Ronal Ross, a medical officer in British Indian Army, in a mosquito (possibly *Anopheles stephensi*) without a sophisticated microscope with high resolution lenses; how, for the first time ever, Dr Manuel Elkin Patarroyo, a self-employed immunologist in an impoverished and slowly developing nation Colombia, discovered the world's first malaria vaccine in an ordinary laboratory saving millions of lives in his country and elsewhere; and how Dr Vijay Veer, a DRDO entomologist, an entomologist, fought tooth and nail in the court of law for seeking patent in favour of Defence Research & Development Organization and thereby commercializing of the first ever DEET-equivalent Indian repellent of international standard, DEPA (N,N, -diethyl phenyl acetamide), to save thousands of army personnel posted in difficult frontier terrains particularly in the northeastern States of India and, in the process, saving trillions of Indian rupees every year otherwise wasted in importing DEET, a US-patented repellent? The list of these extraordinary scientists is very long, originating from many countries like Great Britain, Italy, France, United States of America, Russia, Australia yet perhaps nothing more important was ever discovered anywhere in the world than in India - the country where mosquito was decisively pinned down to demonstrate its role in transmitting the human malaria – truly the greatest discovery of the millennium! In fact, Dr Patrick Mansion had once advised Dr Ronald Ross that 'India was the best place to conduct malaria research,' following which his protégé hit the bull's eye by pinning down mosquitoes to demonstrate their role in transmitting malaria parasites – the starting point of Ariadne's thread that would eventually lead to the exit from the labyrinth of human malaria and lay the foundations for scientifically based disease eradication through vector control in future.

Surprisingly, most of the British India scientists engaged in vector taxonomy were medicos and still, with their inherent spirit of self-help, perseverance and patience, cultivated meticulously the habit of learning and mastering other disciplines, such as taxonomy, systematics, zoogeography and evolution, thereby converting apparently the most insuperably proved obstacles in way to their advantage. In fact, in many instances they turned these very difficulties their best helpers, by evoking their powers of labour and endurance, and stimulating into life faculties which might otherwise have lain dormant: *Where there is a will, there is a way!* By the turn of 20th century native Indian entomologists, parasitologists, microbiologists and medico-epidemiologists have also emerged to give their own highly appreciable contribution in various fields of vector-borne disease control. For reason of convenience in describing, therefore, all these scientists who principally worked in India, whether of Indian origin or otherwise, are unquestionably considered "mosquito hunters of India". This chapter features about five dozens of Indian culicidologists who by sheer dint of hard work, dedication and application have contributed significantly toward the control of mosquitoes. From Dr Ronald Ross in the late nineteenth century to Dr I.M. Puri, Dr T.R. Rao, Dr V.P. Sharma and Dr P.K. Rajagopalan, Dr V. Dhanda, Dr A,K. Hati and scores of many others, in the twentieth and twenty-first centuries, their stories represent each researcher's individual style and intellectual involvements ranging from breakthrough novelties such as the demonstration of connection of human malaria with mosquitoes (read anophelines) to the pioneering research underlying the discovery of potential to transmit dengue viruses (DENV) by the mosquito *Aedes albopictus*, as a primary vector, and in the absence of *Ae. aegypti*, in the Kerala State and the numerous pieces of extraordinary investigations into mosquito taxonomy, systematics, biology, ecology, chorogeography, embryology, physiology, genetics and control reflect an unparalleled example of scientific feats that were vigorously emulated often in other countries across the world as well. Thus, this section brings together innovatively a sea of biobibliographic accounts, otherwise no longer easily accessible, from the lives of so many dedicated and focused culicidologists - like a garland of flowers of different blooming times yet so debonairly arranged systematically to offer it a magnificent and meaningful look, a truly international collection on the Indian soil!

8.2. The mosquito ambushers

These are the culicidologists who by their energetic character not only waged a decisive war against their deadliest foe, mosquito, but by the example of diligence and perseverance which they set before future generations of scientists, largely influenced the current scientific activity in all directions and contributed in a great degree to form the national character. Their individual industry, energy and uprightness as well as leadership skills are a glorious example of focused vector-borne disease control research for the entire scientific fraternity engaged in defeating an enemy who is better equipped biologically by more than 200 million years in evolution! (Figs. 8.2, 8.3 and 8.5-8.10).

In as far as India is concerned any scientific research involving mosquitoes commenced with the malariological investigation on 'trial and failure' basis by **Dr Ronald Ross** in the late 19th century which eventually won him the 1902 Nobel Prize for his discovery of mosquito-malaria connection in 1897 while serving the British Indian Army as a Major Surgeon in Secunderabad, India (Tyagi *et al.*, 2020). Dr Ronald Ross (1857-1932), eldest among the eight children to General Ross, also in British Indian Army, was born in Almora (now in Uttarakhand) on 13th May, 1857. He had early schooling at home under the watchful eyes of his parents and later moved to England to complete both schooling and college as well as medical education. Ross nurtured various different kinds of interests such as poetry, drama, art, music and, above all, mathematics, but his father was adamant to send him to the Indian Army with a high prospect of personal comfort and professional advancement in rank being enjoyed by all Britishers in the Indian Army in those days. Ironically, during the course of his leisure he began to take interest in research. Following a long tumultuous career in Army, he hit the bull's eye not only by creating his *de novo* discovery but also bagging the coveted Nobel Prize. For such unassumed happening of the highest order most people in science generally mockingly said that he became a medical scientist almost by accident, which was in fact completely unfounded. He was an utterly indefatigable genius *nonpareil* with indelible scientific temperament and one of those rare self-taught scientists who by their energetic character not only practically educate the budding researchers in habits of industry, but by the example of diligence and perseverance which they set before them, largely influence the scientific activity in all directions and contribute in a great degree to form the research character. He was such a scientist who turned turtles on the face of worst adversaries and moved ahead undaunted to secure what he wanted. History is witness that accident does very little towards the production of any great result in life. Though sometime what is called "a happy hit" may be made by a bold venture, the common highway of steady industry and application is the only safe road to travel. "A happy hit" certainly did not happen in Ross's instance because he persevered hard and applied most meticulously and systematically all his mental and physical faculties to the extreme limits to bear the unique fruits of his labour – the discovery of mosquito being responsible to mediate transmission of malaria parasites from man to man! He toiled through thick and thin during almost entire length of his active career in the British India Medical Service, between 1881 and 1899, and never rested until achieved his target. The capability of putting all his energies in unravelling the Ariadne's thread in decoding the labyrinth of malaria transmission by mosquitoes is acquired only by the labour of a life, else the probability is that the scientist who has not carefully trained himself beforehand, in attempting to produce a brilliant discovery at a dash, will only produce a blotch. Sedulous attention and painstaking industry always mark the true worker. The greatest achievers or scientists are not those who "despise the day of small things," but those who improve them the most carefully. Although there are discoveries which are said to have been made by accident, if carefully inquired into, it will be found that there has really been very little that was accidental about them. For the most part, these so-called accidents have only been opportunities carefully improved by genius. It is the mind that sees as well as the eye. Where unthinking gazers observe nothing, men of intelligent vision - the scientists - penetrate into the very fibre of the phenomena presented to them, attentively noting differences, making comparisons, and recognizing their underlying idea: *"Wise haveth their eyes, in the head; Fools waketh through forest, and see no firewood!"*

History of malaria research reveals that many great scientists before Dr Ronald Ross had come close to make a link between the mosquito and the malaria transmission to humans, and in fact some of them were better equipped both parasitologically and entomologically including also the identification of various different species of mosquitoes, yet, as if the destiny kept it all for Ronal Ross, a hyper-energised, combating and thinking scientist, who would later solve the riddle puzzling the world for many centuries. One major factor behind the epoch-making success of

Dr Ronal Ross is that he sincerely followed the advices of his mentor, Dr Patrick Manson, "*Look for the mosquitoes*" and "*India is the best country for malaria research*".

When he was appointed as a medical officer in the Madras Regiment of the British Indian Army in 1881, he anyway himself nurtured the thinking that this would end before long and he would return sooner to England for a much comfortable future than in India which he then knew as 'a country full of dust, heat and mosquitoes', never ever realizing in the remotest sense that the same country will one day raise him to reign the world of malariological science and be honoured with the world's loftiest recognitions and distinctions including the enviable Nobel Prize!! The road ahead was however not smooth and he had had to pass through a battery of litmus tests time and again! Ross's first tenure to India lasted eight years, from 1881 to 1888 during which he, as for most of the young British officers, with only a couple of hours duty every day, had plenty of leisure time; golfing, fishing and hunting, and only disinterestedly performing his medical duties. He had finished his first novel, as well as writing some poems and short plays, and was developing a long-standing interest in mathematics. All these being close to his heart he nonetheless began to feel a big void in his strong desire to apply life to fruition. The thought "It was not this that he had come to this far, and he needed to do something great, perhaps greatest" continuously sounded in his head and he was getting restless. With a pinch of salt in his day-to-day medical duties at the hands of his superiors on one or another reason, he was exacerbated to the point of being a paranoid, but never confronted seniors. He began to think to do something extraordinary which others could never do and what could carry him to higher pedestal than others. Amidst the pressure of his everyday medical duties, Ross spent three years studying mosquitoes and their parasites. In doing so, he naturally faced enormous problems since he was unable to identify mosquitoes and had never seen a malaria parasite earlier. He realized that it was inevitably indispensable for him to have firsthand knowledge about the malaria parasites if he had to succeed in his objective. He was restlessly casting about with an underlying volcanic pressure to learn more practically. He searched for the parasite which Dr Laveran, a French army surgeon, had found in 1880 in the blood of victims of malaria in Algeria, but he was unsuccessful. Taking cue from the lives of great scientists like Pasteur and Koch he enrolled in England on his own costs for obtaining training in public health, bacteriology and microscopy. There he also got married to lovely Ms Boxom. At the end of his leave, with wife and holding a microscope that he only had assembled, he sailed back to his posting, only to find him packing again and again to almost a dozen new and distant places, manipulated as punishment postings during the next ten years of service in India under highly precarious and offensive professional environments, which he bravely turned in his favour.

In India Ross continued to brace failures in his attempts to see a malaria parasite or to understand any role that mosquitoes could play in malaria transmission, all this happening under the constantly increasing pressure of humiliation and indifference rendered to him by his seniors. He had begun to acquire a tarnished reputation of a useless medical officer who was more interested in research on malaria rather than executing his mandatory military medical assignments for which he was paid. Ross was however least interested in anything else but malaria. Frustrated he took leave for the second furlough due by then and went to England in 1894, not knowing about his next steps. As if destiny had it in store to give his life's greatest turning point, he providentially met the great Dr Patrick Manson, who had already shown a 'mosquito-filariasis' connection in 1880s while serving in China and was a highly respected expert on tropical diseases as well as well-versed with malariology and, what's more, he had seen Laveran's malaria parasites. Ross had heard about Manson, and got a common acquaintance's advice to see him. Ross took proper appointment and on the fixed day and time they met at the clinic of Manson. Realizing a very calm and fragrant evening breeze outside his clinic, Manson took Ross on a campus walk. During the stroll Manson knew Ross's interests but also that he did not know anything about either mosquitoes or the Laveran's malaria parasite. Ross was obviously confused about his future and in research arena he was comparable to a ship sailing without knowing its destination. Manson further knew that he was also burning an inferno within to do something lofty, and only needed encouragement at the right time!

From this point on Ross spent many hours tagging along behind Manson on ward rounds at the Seamen's Hospital and in Manson's private laboratory. In course he had impressed Manson by his inquisitiveness, ceaseless energies and the fiery desire for unravelling the centuries-old mystery of malaria transmission. Manson, who then took Ross as his student, showed to Ross malaria parasites for the first time, and demonstrated to him how to find the parasites in blood smears. Manson had inferred, based on his earlier experiences in China in respect of filarial worm, *Wuchereria bancrofti* being transmitted by *Culex pipiens* that mosquitoes could be responsible for transmitting malaria, too, although his hypothesis remained merely a conjecture as he had himself proved unsuccessful in bringing his point home. One afternoon in November that year Manson began to expound upon his ideas about the plasmodia. "Do you know," he remarked to Ross while the two were on the way to the hospital, "I have formed a theory that mosquitoes carry malaria just as they carry filaria." Though this nicely rhymed observation might have been made into another new poem, Ross was initially unimpressed. The mosquito had been suggested as a carrier of malaria many times before, he responded. But then the two men discussed the facts that favoured this argument. As Ross began to be swayed, Manson brought in his own theory: "The filaments in the crescents, believed by everyone to be evidence of a death spasm, were actually living bodies. The mosquito, feeding on the blood of a malaria patient, sucked the filamented crescents into its stomach. The filaments proceeded to travel through the stomach into the insect's tissues. After the mosquito died laying its eggs, the 'flagellated spores' emerged into the water, ready to infect anyone who came to drink." This theory, which had earned for Manson the titles "pathological Jules Verne" and "Mosquito Manson," sent the young Ross into raptures of ecstasy. But suddenly, somewhere in the back of his mind, the fame that had eluded him despite years of writing poems, music, plays, novels, and equations seemed to fall within his grasp. He had but to prove what Manson had presented to sound like gospel truth. Overwhelmed by the conviction that Manson was entirely correct, Ross made only the most superficial attempt to familiarize himself with the life cycle of the mosquito, despite Manson's entreaties. After having failed to examine books on the mosquito in the library of the British Museum, the only place in England at the time with such books, Ross found himself in a quandary and did not have much hope to solve this conundrum. With a dream in his eyes, Dr Ross, equipped with the resolution to prove Manson's "Grand Induction" of mosquito-malaria parasite-human linkage, set on sailing back to India once more. His ignorance was offset by an almost manic enthusiasm: on the ship to India, Ross rushed among passengers and crew members, frantically pricking fingers and examining blood. At the ship's ports of call, he besieged the local hospitals for blood specimens of malarious patients. He even tried to prepare himself for anatomical studies of Indian mosquitoes by dissecting the ship's cockroaches, even fish! He truly fell in love with mosquitoes!

Once again back in India, now at a new posting awaiting him in Secunderabad, Ross continued to dart around like a frenzied dragonfly looking for cases of malaria. His preoccupation quickly incurred the hostility of his superiors who spoilt his Annual Confidential Report to make him suffer professionally. Colleagues thought him so mad that they kept authenticated malaria cases away from him. The local doctors would often turn over fever cases to him just to rid themselves of his annoying presence.

Even he offered the malarious natives he discovered a rupee a prick, yet many still fled. The natives, their fingertips inflamed from his repeated pinpricks, would dash to hide from him frustrating him further. But, as a true army personnel and a great personal family legacy of valour, he refused to give up. Spurned in his garrison hospital, he haunted the municipal hospitals and the other regiment infirmaries looking for cases of malaria. After he finally managed to obtain the cooperation of one patient with a documented case, fellow doctors challenged him to demonstrate the parasite. But somehow, when the moment came, the plasmodia mysteriously vanished from the blood. Poor Ross had to endure gales of laughter.

Ross felt broken internally beyond comprehension and doubted if he was moving on the right pathway of Manson's "Grand induction". He vigorously communicated with Manson and narrated everything in detail both about his research status on mosquitoes and professional bottlenecks. Manson cautioned him to focus on his objective, and promised to help him out of his professional difficulties. Ross captured the ubiquitous mosquitoes and tried to induce them to bite malaria patients. But they obdurately refused to bite anyone, even Ross. The mosquitoes caught were probably too disturbed to bite, Ross reasoned. So he raised new mosquitoes from grubs. Still no luck. He baked the patients in the hot sun "to bring their flavour out." Nothing happened. On May 13, 1895, Ross' birthday, a heavy rain soaked the bed and netting of a malaria patient. Made ravenous by the moisture, the mosquitoes attacked the patient with alacrity. Ross grabbed four of them, expressed their ingested blood on a glass slide, and peered at it with his microscope. Just as Manson had professed, there were the parasites. To be certain of the results, Ross tried the same experiment with six more mosquitoes the next day. Abjuring food and sleep, he peered through the microscope so intently that his eye was blackened by repeated contact with the eyepiece. He immediately wrote to Manson about the veracity of his "Grand Induction" theory and that every point that he had predicted seemed to come true. Ross thought that the environment of mosquito's stomach likely supported a greater degree of proliferation of parasites present in the blood of the mosquito than in blood from the finger, although the development of the crescents and the formation of the flagella seem to be favoured by conditions in finer-prick blood. The crescent-sphere-flagella metamorphosis does go on inside the mosquito to a much greater degree than in control specimens of finger blood. At one point, Ross became so enthralled watching a white blood cell try to engulf a flagellum that in his letter he compared the battle to a swordfight. "I shall write a novel on it in the style of *The Three Musketeers*," he added joyously. Ross apparently drew energies from his artistic skills!

Dr Ross, typical in his military-disciplined daily routine of maintaining minute-to-minute account of malaria-mosquito explorations, extensively corresponded with Dr Manson in England and sought his advice. Manson immediately wrote back with more instructions. Ross meticulously followed the whole process; he once collected four mosquitoes biting a patient named Abdul Kadir sick with malaria, put them in a bottle of water and allowed them to become gravid and lay eggs from which hatched out tiny larvae. At this point of events Manson adhering to his dogma rather wrongly advised him to give that mosquito-water to people to drink, thinking that the pathogen was passed to water by larvae/eggs. Of course, Ross obediently followed his advice, and himself got drifted to a wrong way; he allowed the mosquito-water kept in a bottle to be drunk by his native servant, Lutchman, who was promised a suitable financial reward for swallowing the liquid in the bottle. Day after day for ten days, Ross waited anxiously for Lutchman to develop fever. On day eleven, Lutchman complained of a headache and was found to have a slight temperature elevation. Now thoroughly excited, Ross admitted his experimental subject to the hospital and impatiently sat by him, measuring his temperature every thirty minutes. Lutchman probably only had the flu. Not a parasite was to be seen in his blood. A few days later, he had recovered completely. Ross repeated the experiment with other volunteers, but these men were likewise totally unaffected by the mosquito water. The volatile Ross was plunged into deep despair. Nothing seemed to be going right. Careful dissection of the mosquitoes revealed that the flagella Manson believed to be spores remained active only a few moments in the insect's tissue. Perhaps, thought Ross, ingestion of the parasites by the mosquitoes was only incidental, unrelated to human disease. Thoroughly discouraged, he began, as usual in his moments of dismay, writing poems and novels, thinking perhaps this was a better way to succeed.

Manson being a gracious mentor and a responsible teacher understood from the various correspondence getting exchanged between them the mental state of Ronald Ross at this point of time. He desperately tried from England to re-energize his slacking protégé pushing him to never give up. He prophetically announced that the malaria germ went into the mosquito for a purpose and not for fun. Ross was certainly feeling reinvigorated from Manson's encouragement but failed to convince his bosses in the Indian Medical Service who completely disagreed with Dr Ross's intention to pursue such a research. They felt that Ross had possibly developed into more of a nuisance than a genuine researcher. To put him to something more challenging and useful, his Commandant in-Charge sent him to cholera epidemic-hit Bangalore. This interrupted Ross's malariological investigations and he was infuriated and exasperated completely with the orders of movement to Bangalore particularly when he felt he was very close to the solution. In Bangalore Ross was neck-deep busy for the next eighteen months and had little time to do anything but visit cholera cases and fight with civilian and military authorities over sanitary reforms.

After attending to Bangalore's cholera epidemic Ross returned to Secunderabad – disgruntled and confused as to his future malaria explorations. In England, Manson too was distressed for the official indifference towards Ross's works on malaria. He explained to Ross about the progress being made by French and Italian scientists and that he had very little time to complete his experiments. Ross took two months' leave at his own expense and tried to continue his malaria work. But within a few days he himself contracted malaria and spent his vacation too ill to do anything. When he finally recovered, the weather had turned hot and dry. A stiff wind filled his laboratory with mounds of dust. The sweat-rusted microscope, its one remaining eyepiece cracked, sat idle as Ross sulked, brooded, and wrote, his all-time accomplice and refuge, poetry. Inside him, however, a spate of thoughts rushed in and he worked hard to pick them up one-by-one to judge the whole logic behind Manson's theory of 'Grand Induction'. This counter-reasoning was the first sign of evolving original ideas of his own and a glimmer of hope shone in his eyes: "*God give me unclouded eyes and freedom from haste. God give me a quiet and relentless anger against all pretence and all pretentious work and all work left slack and unfinished. God give me a restlessness whereby I may neither sleep nor accept praise till my observed results equal my calculated results or in pious glee I discover and assault my error. God give me strength not to trust God*!"

As he was struck by the crucial inspiration, Dr Ross suddenly sprang up and emerged from the bleak period of depression. Re-reading Manson's original article on filariasis, he was reminded of the fact that only one species of mosquito among the four found in Amoy, *Culex pipiens*, was capable of carrying filariasis. Manson had also suggested that each form of the malarial plasmodia might require a particular mosquito species. Ross suddenly realized he had used the wrong species of mosquito. Most of his cases had been falciparum malaria, and he had consistently employed the common house mosquito, a grey insect with plain wings that sat with its tail pointing down. The larvae of these mosquitoes floated head downward in water, breathing by means of a long tube. After a little searching, Ross was able to come up with another species of mosquito. This "dappled winged mosquito," as he called it, had no breathing tube and floated parallel to the water surface in its larval form. The adult sat with its tail pointed upward. A malaria patient named Husein Khan was the first experimental subject. Ross let loose a dozen of the brown mosquitoes under the mosquito netting of Husein Khan's bed, and trapped each one, after it had fed, in a separate bottle. Two mosquitoes were then examined immediately. Two days later others were dissected. More of the insects died during the night and quickly decomposed. In all of them Ross found nothing. Worse, search as he might, no more of the same species of insect were to be procured but his dissection of one mosquito on 19th August appeared different in that there were clearly visible round bodies with black pigment on the wall of the stomach of the mosquito. He had almost jumped in excitement but feared lest it would be mere his exceptional view and wanted to demonstrate it again in other left out mosquitoes first. He feared this to inform even Manson and did not write any letter that day to him. On August 20, 1897, a watchful but tired and discouraged Ronald Ross dissected one of the two remaining brown (dappled wings) mosquitoes. Just as he was about to give up, he renoted the same quaint structure within the cells lining the insect's stomach: an almost perfectly circular cell containing a group of black pigment granules very similar to those seen in the malarial parasites found in blood smears. Nearby were identical circular cells. Ecstatic with joy, Ross wrote, poetically of course, "Those circles in the wall of the stomach of the mosquito, those circles with their dots of black pigment, they can't be anything else than the malarial parasite, growing there...." In a blitzkrieg he shot out a note to his mentor, Dr Manson, on his achievement as customary for him in past but, at the same time, expressed his many scepticisms over the same (Fig. 8.1). Manson, on receiving his report, felt greatly delighted and quickly responded to Ross about his desire to present these observations in a scientific meeting of the Royal Society for everyone to know in the first instance.

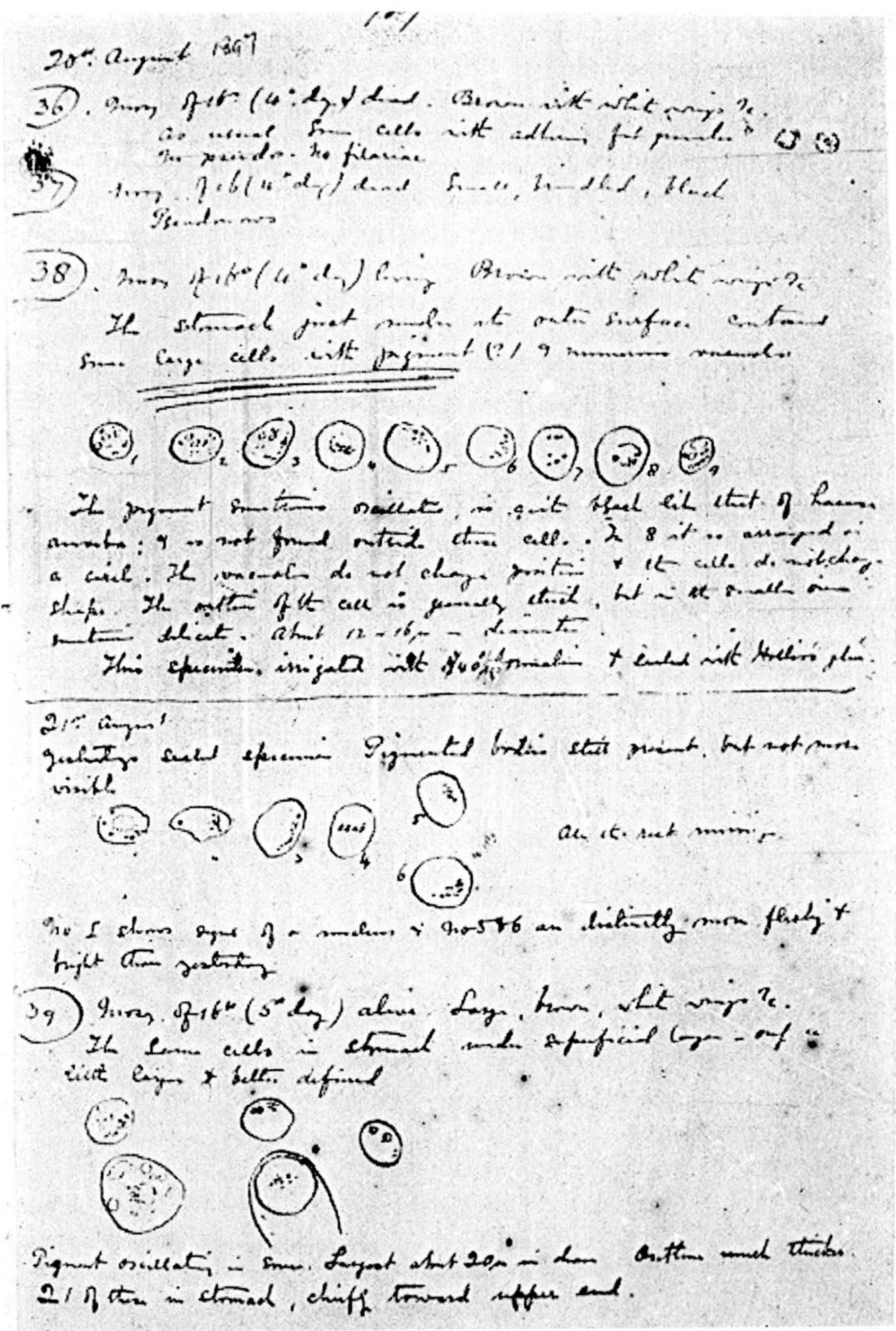

20th August 1897

36). Mosq. of 16th (4th day) dead. Brown with white wings &c.
As usual some cells with adherent fat granules [illegible]
No [illegible]. No filariae

37) Mosq. of 16 (4th day) dead. Small, brindled, black
[illegible]

38). Mosq. of 16th (4th day) living. Brown with white wings &c.
The stomach just under its outer surface contains some large cells with pigment (?) & numerous vacuoles.

The pigment sometimes oscillates, is quite black like that of haemamoeba; & is not found outside these cells. In 8 it is arranged in a circle. The vacuoles do not change position & the cells do not change shape. The outline of the cell is generally thick, but in the smaller ones sometimes delicate. About 12–16μ in diameter.
This specimen irrigated with [illegible] formalin & sealed with Hollis's glue.

21st August
Yesterday's sealed specimen. Pigmented bodies still present, but not more visible

[illegible]

No 1 shows signs of a nucleus & Nos 5 & 6 are distinctly more fleshy & bright than yesterday

39) Mosq. of 16th (5th day) alive. Large, brown, white wings &c.
The same cells in stomach under superficial layer – [illegible] thick layer & better defined

Pigment oscillating in some. Largest about 20μ in diam. Outline much thicker.
21 of them in stomach, chiefly toward upper end.

Fig. 8.1. Facsimile of the page in Ross's diary No. 1 (*Source:* PPT slides of Col. A. Kumar and SRRIP archive).

Dr Ross was overwhelmed by his feat just as any other scientist in another discipline would be but he hardly conjectured the global significance of his own masterpiece. Well, as expected, he commemorated his triumph by writing instantaneously the now much famous ode:

This day designing God
Hath put into my hand
A wondrous thing.
And God Be praised.
At his command
I have found thy secret deeds
Oh million-murdering Death.
I know that this little thing
A million men will save
Oh death where is thy sting?
Thy victory, oh grave?

On August 21 he dissected and examined the last brown mosquito, and again saw the pigment-filled circles. Dr Ross jumped to his feet with an overwhelming feeling of great happiness and excitement. His joy saw no bounds! Now, the major question nagging him all the times was whether the granule-containing circular cells were to be related to the malarial parasites at all?

Preceding the above great discovery, Dr Ross had begun thinking independently along certain novel research lines and his approach to the entire issue of 'mosquito-malaria' connection differed fundamentally. He built his argument on the basis of the fact that mosquito feeds on human blood and while doing so she might ingest malaria parasites as well, just as the mosquito was always injecting a small quantity of fluid with her bite. To test this eminently reasonable idea, he allowed mosquitoes that had fed on a malaria patient to bite a healthy man. Nothing happened. The experiment was repeated again and again but no fever developed. Still enthusiastic, he communicated his new notion to Manson who, however, had quite an imperfect understanding of mosquito behaviour. Believing that the insect bit only once during its life, he was convinced that Ross could not be correct. "Follow the flagella," he wrote back, adding, "and forget this crazy idea". Ross obediently went back to dissecting mosquitoes and in February 1897 was able to observe the true fate of the flagella. Within a blood smear he saw two parasites near each other. The first was giving off flagella, while the second, which was spherical and unsegmented, had a single flagellum wriggling slowly inside. As we now know, had Ross begun a moment earlier, he might have watched the flagellum emitted by the first parasite penetrate the second and so perceive the true nature of the process of fertilization. But from his vantage point, he surmised, quite oppositely, that the single wriggling flagellum was trying to escape the sphere rather than fertilize it. A few weeks later when Dr McCallum in Baltimore, USA, correctly interpreted the process, Ross felt deeply humiliated and disgraced as a man of science for incorrectly interpreting his own observations.

Dr Ross still believed that research in malaria was to be his path to fame and fortune, although he did not receive any advancements by his employer for all his labour and pains so far in malariology. He pestered the military hierarchy incessantly to allow him time for research but got no favours whatsoever. Apparently he had stepped on too many toes to expect any favours from his superiors. In June 1897 he was sent back to the steamy, depressing hospital in Secunderabad. Yet, he refused to give up. He persuaded Manson to intercede with high government figures on his behalf. Manson thereupon approached Lord George Hamilton, the secretary of state, and even petitioned Lord Elgin, the Viceroy of India, but of no avail! As luck would have it, swarms of brown mosquitoes suddenly appeared in Secunderabad. Now able to repeat his experiment, Ross confirmed that only those insects that had bitten malaria patients carried the circular cells. He submitted the results to the British Medical Journal, and in December 1897 his report, "On Some Peculiar Pigmented Cells Found in Two Mosquitoes Fed on Malarial Blood," was published. Then, just at his moment of triumph, his superiors foiled Ross by serving on him the wired orders: “UNDER INSTRUCTIONS FROM COMMAND HEADQUARTERS, SURGEON MAJOR R. ROSS, I.M.S., WILL PROCEED IMMEDIATELY TO BOMBAY FOR MILITARY DUTY.” Infuriated, Ross bombarded the commanding officers with angry telegrams. Even Manson’s intervention did not help. His new station, to his dismay, was totally unsuited to his research. Yet the resourceful Ross did not quite languish. He knew of Danielewsky's studies of bird malaria, and verified for himself that some types of pigeons carried the disease. At the time, many biologists believed that mosquitoes did not attack birds. After studies of pigeons, sparrows and crows, Ross verified that birds were indeed bitten by mosquitoes, as well as by other insects. In the meantime, Manson reached the right bureaucrat and on January 29, 1898, after weeks of futile pleading for a furlough or a change of station, Ross was notified by wire of his attractive new assignment in Calcutta (now Kolkata) where he found no shortage of mosquitoes and the disease. He was assigned the dilapidated laboratory of a recently retired physiologist. Dr Ross engaged two assistants, namely Mohammed Bux and Purboona, on his own expenses. Purboona disappeared after the first payday. But Mohammed Bux was a particularly fortuitous choice. He became quite devoted to Ross and was singularly adept at the work. He would even sleep on the laboratory floor at night to keep stray cats from killing the experimental animals. Ross directed Mohammed Bux to capture mosquitoes, hoping again to find the circular, pigment-filled cells within the insects' stomachs. Mohammed Bux climbed through the sewers, the drains, the stinking tanks that abounded in Calcutta and brought back all kinds of mosquitoes to feed on patients. But the work had begun between malaria seasons. All the human experiments were a failure. For this reason, Ross turned to the study of bird malaria. His laboratory soon became filled with the live sparrows, larks and crows that Mohammed Bux had snared. Into the cages covered with mosquito netting went the mosquitoes. In almost no time Ross could demonstrate that the plasmodium passed from bird to mosquito, just as it did in humans. Moreover, the same pigmented, circular cells would form in the wall of the insect's stomach. Only one difference was noted. The common grey mosquito (*Culex*) was the carrier of bird malaria. The brown dapple-winged vector (*Anopheles*) or the brindled mosquito (*Aedes*) could not be infected by the bird parasite. These studies were carried on in desperate haste. Ross was forever haunted by the fear that hostile superiors would once again interrupt his work. He was forbidden to publish any of his results, though compelled to send long, meticulously detailed summaries or reports to headquarters. The only outsider aware of the research was Dr Patrick Manson.

But how did the parasite pass from the mosquitoes to the birds? According to Manson's theory, the parasites were ingested with water in which the mosquitoes had died while laying eggs. Ross easily tested this theory by feeding infected mosquitoes to healthy sparrows. The result: The birds remained free of malaria. The true answer finally emerged as Ross continued to study infected mosquitoes. As the circular cells within the mosquito's stomach enlarged, the pigmented granules grew into little rod-shaped bodies. And soon, Ross discovered, these rods appeared in the insect's thorax. Since the stomachs of some mosquitoes showed only ruptured capsules where circular cells had been, the implication was clear: The rods migrated. On July 4, 1898, Ross discovered their destination. Examining the insect's head, he noted the salivary gland to be so loaded with rods that it quivered. Here then was the answer. Malaria was passed back to the birds through the mosquito's saliva during the act of biting. As a final verification, he sent Mohammed Bux to capture a group of healthy sparrows. Mosquitoes that had fed on infected birds were allowed to bite these healthy ones. Within a few days the blood of the new birds was loaded with malarial parasites. Ross communicated his results to Manson in a state of intense excitement. Manson received the news at a meeting of the new Tropical Diseases Section of the British Medical Association. When he read Ross's report to the assembled delegates, the excitement generated was intense. "I am sure you will agree with me," Manson added, "that the medical world, I might even say humanity, is extremely indebted to Surgeon Major Ross for what he has already done, and I am sure you will agree with me that every encouragement and assistance should be given to so hard-working, so intelligent, and so successful an investigator to continue his work." The audience may have agreed with Manson, but the Indian Medical Service did not. For Ross's original assignment was also supposed to include the study of another tropical disease, Kala-azar, Ross, just when at the peak of his success, thus received instructions to abandon the malaria work and report to a new post in Assam to do research on Kala-azar.

In a state of deep despondency, Ross was forced to close his small laboratory in Calcutta and leave for his next destination. In fact, the research still to be done was more complex than that for, as Manson remarked in his address to the Tropical Diseases Section, "One can object that the facts determined for birds do not hold necessarily for men." Ross could not continue to answer this question but in Italy Professor Grassi succeeded to unravel the mystery in an irrefutable scientific manner. Ross had a tumultuous career throughout since many of his medical superiors and colleagues did not see eye-to-eye. When his service bosses were unsympathetic to his requests for leave, at times he threatened to take early retirement from the I.M.S. and return to his first love, namely, literature. It was the rock-like support of Manson, his mentor till that time, who, on one hand, encouraged him to continue his research during these periods of discouragement and despair, but, on the other hand, also endeavoured hard to convince his I.M.S. bosses to think favourably about his efforts.

For his pioneer work, Ross was awarded the 1902 Nobel Prize in Physiology or Medicine. He subsequently held various positions notable amongst which were (i) Professor of Tropical Medicine at University College, Liverpool (from 1902), (ii) Physician at King's College, Hospital, London, (iii) Consultant on malaria to the War Office, and (iv) Director of the Ross Institute and Hospital for Tropical Diseases, London (from 1926). He was knighted in 1911 (for more details see Tyagi *et al.*, 2020).

The events of the closing years of the 19th century connected with the disclosure that malaria was transmitted by mosquitoes initiated an outburst of scientific interest in this group of insects that continues unabated to the present time. Quickly realizing that species identification was essential to the research and control of mosquitoes, administrators of the British Museum (Natural History) in 1899 obtained the services of **Dr Fred V. Theobald** (1868-1930), an English entomologist, to begin the taxonomic study of this hitherto greatly neglected family of Diptera. Theobald proved to be a rapid and indefatigable worker, with the result that between 1900 and 1914 he issued at least 65 separate publications dealing with mosquitoes. This included his mammoth five volume work entitled "A Monograph of the Mosquitoes of the World (1901-1910)" comprising a total of 2459 pages, with support of the British Museum. Before completing his magnum opuses he studied Indian mosquitoes in Indian Museum, Calcutta and described several new taxa. Theobald was the first individual to work extensively with the systematics of mosquitoes of the world. Although many of his concepts of relationships within the Culicidae have subsequently had to be readjusted, the value of his pioneering contributions to culicidology should not be overlooked or underestimated. His research publications are masterpieces of authoritative explorations and every Indian scholar needs to know about these, and must appreciate going through them as the gems of scientific collection with enormous historic significance.

One of the greatest exponents of greatness of Dr Ronald Ross's scientific feats was the famous **Dr Sir Rickard Christophers** whose contributions to malaria and Yellow Fever vectors' taxonomy as well as the irreplaceable volume on Culicidae under the Fauna of British India, among several other great scientific memorabilia, are good testimony to have his name written in golden letters in the annals of mosquito taxonomy. Brevet Colonel Sir Samuel Rickard Christophers, CIE, OBE, FRS was a British protozoologist and medical entomologist specializing in mosquitoes. The son of Samuel Hunt Christophers and Mary Selina Christophers née Rickard, Christophers was born and raised in Liverpool and, after attending a kindergarten school, he went at the age of thirteen to the Liverpool Institute where he did particularly well in mathematics and hydrostatics but subjects such as history, grammar and languages held no appeal, possibly because the subjects were badly taught. To counterbalance this defect he, on his own volition, read widely every form of literature he came across, making the best use of public libraries. As a result he passed the London Matriculation at the age of 16. He entered University College Liverpool as a medical student in 1889 and qualified M.B. Ch.B. in 1896.

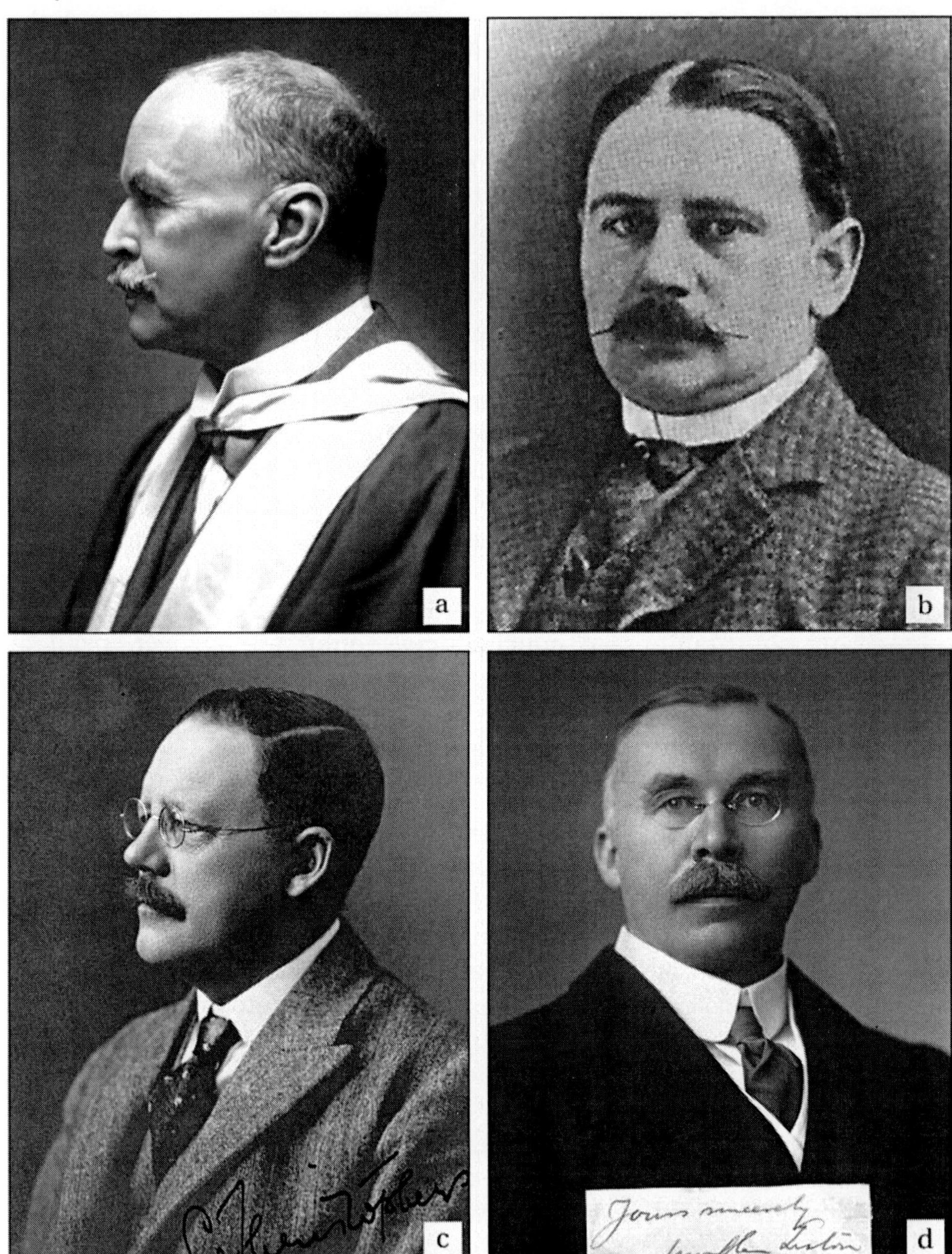

Fig. 8.2. (a) Sir Dr Ronald Ross (1857-1932), (b) Dr Fred V. Theobald, (c) Dr Rickard S. Christophers, (d) Dr William Glen Liston. (*Source:* Free Wikipedia)

Fig. 8.3. Dr John Alexander Sinton, (b) Dr T. Ramachandra Rao, (c) Dr C. F. Curtis, and (d) Dr G.B. White. (*Source:* a,b - Free Wikipedia; c,d - Dr B.K. Tyagi's personal archive)

In 1897, he took part in an Amazonian expedition and in 1898 went to Italy as part of the Malaria Commission, followed by a trip to Africa to study malaria. In 1901, the Malaria Commission moved to India. On his return to England in 1902, he became a Lieutenant in the Indian Medical Service, moving back to India in 1904. In 1910 he was appointed the first Director of the Central Malaria Bureau, coordinating anti-malarial training and research throughout India. He spent World War I on anti-malaria duties in Iraq and in 1919 returned again to India as Director of the Central Research Institute at Kasauli in the foothills of the Himalayas.

Christophers's earliest scientific work was on serology and as early as 1898 he studied the effect of normal human serum on *Bacillus coli communis*. This was before his introduction to the tropics as medical officer on a steamer of the Booth Line which was the first ocean steamer to traverse the upper reaches of the Amazon river. The young medical man was fascinated by this experience which undoubtedly influenced him in his decision to adopt a career in the tropics. An opportunity soon arose; he was appointed a member of the Malaria Commission of the Royal Society and Colonial Office in 1898-1902 and this appointment gave him five years of experience of the tropics in Africa and India with his senior colleague the late Professor J. W. W. Stephens, F.R.S. He was so impressed with the opportunities offered by India for research in tropical problems that on the termination of the Malaria Commission he decided to return to that country and enter the Indian Medical Service by the usual examination. This service offered an almost unlimited scope for work in many scientific disciplines besides medicine and surgery and Christophers's previous experience inevitably marked him out for appointment in the Research Department of the service. Although the Indian Medical Service was primarily a military service, and junior officers commenced as medical officers in the Indian Army, Christophers's talents were so obvious that this normal procedure was waived. After a short period as medical officer of the 1st Brahmins he was put in charge of antimalarial operations at Mian Mir, but was soon placed on special duty to investigate the findings of Colonel Donovan in Madras of the newly described cause of Kala-azar, in patients suffering from the disease. At this time, 1904, he was appointed to the Directorship of the newly constituted King Institute of Preventive Medicine in Madras, an appointment he held for four years. In 1908 the occurrence of blackwater fever among planters in the tea gardens of the Duars led to an urgent call for investigation of the causes responsible for this serious condition. To study this Christophers and Dr C. A. Bentley were placed on special duty but in 1909 a more urgent matter led to the recall of Christophers to investigate a serious epidemic of malaria in the Punjab. His appointment to this investigation was to lead to important developments in the organization of malaria research and measures to control the disease in India.

Between 1910 and the outbreak of World War I a great deal of activity took place in the organization of malaria research, including measures to be taken in implementing active anti-malarial measures in the field. Each province now had its own malaria officer with organization and facilities for field work. Under Christophers there was formed a Central Malaria Bureau with a laboratory in the grounds of the Central Research Institute, Kasauli, with provision for workers, literature, collections and other facilities. As officer in-charge of the bureau, Christophers was now responsible for all matters relating to malaria in the whole of India and especially for the operation of 'malaria surveys' in order to gain an idea of the distribution and severity of the disease nationwide. In this period Christophers published his standard book on the anopheline mosquitoes of India. He was undoubtably the most outstanding scientist of the Indian Medical Service. In spite of his scientific achievements and the world-wide renown in which he was held, he never lost the modesty which was inborn in his character. To his scientific peers and to humble students alike he was equally approachable and by many of the latter the time and care he took in helping them over their difficulties were always a source of encouragement and true inspiration to them.

The years 1916-18 saw Christophers in an entirely different environment and his adaptation to this was so rapid and so complete that he seemed to assume almost a different personality. The war in Mesopotamia (Iraq) was in progress and, apart from the military operations, many situations arose which posed problems not only for the military personnel but others, affecting their health and welfare directly and indirectly from the close contact between troops and the civil population. It was obvious that some organization capable of dealing with a diversity of problems bearing on the health of the troops had to be provided and to meet this demand a central laboratory was established in Basra under the command of Christophers who was appointed D.A.D.M.S. (San.) to the Mesopotamian Expeditionary Force. This laboratory was accommodated in the Indian naval ship, the Elphinstone, which was moored in the Shatt-el-Arab near the left bank and immediately opposite the city of Basra. The entire ship and its Lascar crew were under the command of the Director of the laboratory and the scientific staff and laboratory personnel all had quarters in the ship. Under Dr Christophers, the senior staff comprised W.D.H. Stephenson, I.M.S. senior bacteriologist,

Dr Miskin, a Lebanese bacteriologist trained at the Pasteur Institute, Paris, Dr Bassett, a highly qualified chemist, Dr A.J. Grove, an entomologist, the Reverend Aitken, an amateur entomologist, P.J. Barraud, an entomologist, and H. E. Shortt, I.M.S., a protozoologist and malariologist. The work of the Central Laboratory was extremely varied; much of it was not strictly medical as many aspects affecting the troops indirectly had to be studied, such as the quality of the rice supplied for the Indian troops and the components in the blankets as regards the amount of 'shoddy' in their composition. Christophers now showed his extraordinary adaptability. As one knew him in India his reaction to any problem was a strictly scientific one. Before forming a conclusion or giving an opinion on any matter he would consider it from every point of view, almost giving an impression of indecisiveness. The scientific side remained, but its expression was completely altered. He quickly realized that the military wanted quick and decisive answers to problems on which to base actions and this became the *modus operandi* with emphasis on practicability. One of the main operations of the laboratory was to carry out malaria surveys in all areas where troops were operating. The area of operations throughout the whole campaign covered a very large area from Basra on the Shattel-Arab to Kirkuk in Kurdistan, well beyond Baghdad, and the whole of Persia up to the Caspian Sea. These surveys were chiefly to ascertain the endemicity of malaria by assessing the spleen rates of children, and the results provided information on areas where there was special need for anti-malaria operations.

Another activity, which was of lesser importance in that it did not have a direct impact on military operations, was the study of 'oriental sore' or 'Baghdad boil'. There was a high incidence of this condition, and the phlebotomine flies which were the transmitters rivalled by mosquitoes as agents of irritation. In little things, as in great, Christophers took an equal interest.

In 1921 Christophers returned to India to rejoin his previous post as Assistant Director of the Central Research Institute, Kasauli, until 1925 when, on retirement of the Director, Colonel Harvey, I.M.S., he took over as Director, a post he held until his retirement in 1932. Parallely Dr Christophers was also an honorary physician to King George V from 1927 to 1930. Earlier he was awarded CIE in 1915, OBE in 1918 and knighted in 1931. He retired from the Indian Medical Service on 27th November, 1930 as a Brevet Colonel. On his retirement, he joined the London School of Hygiene & Tropical Medicine where he became Professor of Malaria Studies in the University of London and Leverhulme Fellow of the Medical

Research Council in-charge of the Malaria Unit at the LSHTM (1932-38). In 1944, Christophers was awarded the Manson Medal by the Royal Society of Tropical Medicine and Hygiene for his significant contribution to the fields of tropical medicine and hygiene. He was elected a Fellow of the Royal Society in May 1926. He was the sixteenth president of the Royal Society of Tropical Medicine and Hygiene from 1939 to 1943. An expert on tropical medicines, Christophers studied many diseases, particularly malaria. His work on the research of this disease won him the Royal Society's 1952 Buchanan Medal for "outstanding research" on the Anopheles mosquitoes that transmitted malaria. In his career he also contributed to the taxonomy of other ectoparasites. Among the several classical writings, his magnum opus *"The Fauna of British India, including Ceylon and Burma.* Diptera. Volume IV (1933)" deserves a special mention.

Dr Christophers was distinguished for his work as an entomologist, protozoologist and sanitarian. His immense mastery of histological techniques played a vital part in his researches both on insects and protozoal parasites of man and animals. In all these disciplines he made many discoveries, but probably his greatest originality lay in his synthetic approach to the field of 'malariology' (this word was introduced into medical literature late in the nineteenth century following the visit in October 1899 of Sir Patrick Manson and Dr L. Sambon to the laboratory of Grassi in the University of Rome; the Roman 'malariologists' demonstrated their results on the transmission of human malaria parasites by local anopheline mosquitoes and Manson was greatly impressed by Grassi's beautiful drawings of the anatomy of Anopheles and of specific features of the larvae. The word 'malariology', however, did not come into general use until 1915 when the specialist journal *La Malariologia* first appeared in Naples under the editorship of Ernesto Cacace. Christophers's introduction to malaria began a year before the visit of Manson to Italy. He and J. W. W. Stephens were appointed, in 1898, jointly by the Royal Society and the Colonial Office, as members of a Malaria Commission to study malaria in tropical Africa. They were requested to travel to Italy first in order to learn about the most recent developments in malaria research. At that time, the discoveries of the cycle of the parasite in the mosquito, made by Surgeon-Major R. Ross in India in 1897, had scarcely been recognized; in fact Professor M. Foster, Secretary of the Royal Society, in a letter dated 2nd August 1898 to the Colonial Office, stated 'the view that the mosquito is an important agent in propagating the disease has at present advanced very little beyond the stage of hypothesis, and even if proved to be a true view, the important problems to be solved are pathological in nature'. They spent a month in Italy, first in Pavia to consult Golgi about his recently described periodic cycles in tertian and quartan malaria and secondly in Rome to see the work which Grassi, Bastianelli and Bignami were doing in the San Spirito Hospital on the transmission of human malaria by mosquitoes. Stephens and Christophers left England in December 1898 and eventually arrived at Blantyre on 1st February, 1899. They investigated malaria in the surrounding country, then known as British Central Africa, and later successively as Nyasaland and Malawi. After eight months, they decided that conditions for work were unsuitable and they returned to England, with the proposal that they should transfer their operations to West Africa, which they reached in December 1899. The first report of Stephens and Christophers dealt with the protozoological aspects, their second (and later ones) with the entomological problems, and their fourth, fifth and sixth with sanitation. These two investigators found numerous sporozoites and 'zygotes' in the mosquito, *Anopheles costalis* (= *gambiae*), caught in Sierra Leone. These observations must represent some of the earliest confirmatory evidence of Ross's discovery of the role of the mosquito in the transmission of the human disease in the tropics. In 1902, Stephens and Christophers moved to Bengal and carried out work on much the same basis as their researches in tropical Africa. But they added a further dimension - the problem of malarial endemiology which was to occupy much of Christophers's time in later years in India. Christophers's interest became extended to other protozoa and he described sporozoan parasites in the salivary glands of *Anopheles rossii* (= *subpictus*) which he had previously found in *A. costalis* (= *gambiae*). Half a century later he gave a list of parasites in mosquitoes, a subject of increasing importance today in biological control. Christophers and Stephens embodied the knowledge they had acquired in Africa and India in a little book '*The practical study of malaria and other blood parasites*' which first appeared in 1903, was translated into French by the Sergeant brothers in 1906 and had three editions. A similar, practical work appeared in 1928 with the title of '*How to do a malaria*' which went through six editions and was under the joint authorship of S.R. Christophers, J. A. Sinton and G. Covell. These books may be out of date today, in view of the development of residual insecticides and the powerful synthetic drugs, but up to the time of World War II they were the constant companions of most malaria workers in the tropics. Their highly scientific approach was responsible for leading many of the latter into the 'research life'.

Their careful analysis of the possible aetiological factors of blackwater fever in India and Africa demonstrated that three factors are involved: *Plasmodium falciparum* malaria accompanied by very few parasites in the blood, previous attacks of the disease which had been inadequately treated with quinine, and the administration of quinine shortly before the onset of blackwater fever. The exact mechanism of the intravascular haemolysis was not discovered in these researches; it is still not precisely known, though the trigger is now thought to be due to an autoantigen-antibody reaction. As a result of these observations they were able to establish proof of the aetiology of blackwater fever, which previously had been mainly conjecture. The nature of malaria endemicity in areas of intense transmission of malaria was studied in both continents. They appreciated the different effects, noting that in Africa, once young children had recovered from their first infections, symptoms of the disease disappeared and the population became almost immune; in India, however, the immunity took longer to develop and was less complete. The nature of such immunity resembles the 'salting' of animals with trypanosomiasis and piroplasmosis. As a corollary to these observations, Christophers reported that gametocytes ('crescents') were only to be found in any number in the young children, and recommended that European houses should be sited well away from native villages. Robert Koch, about the same time, and the Dutch workers in the East Indies showed that a similar situation prevailed elsewhere in the tropics, while later in Africa the concept of holoendemicity was developed by Bagster Wilson (1950) as a result of his tutelage at Kasauli in India. Christophers appreciated the varying degrees of endemicity, and sought to establish statistical standards for its classification. These were based primarily on the incidence of splenomegaly and on parasite rates in the blood in the different populations. His methods for measuring the size of the spleen in relation to the size of the individual (the 'average enlarged spleen') have been widely adopted and add precision to the records. The malaria survey was then completed by investigating the mosquito factor in all its manifold aspects, *e.g.* species, habits, malaria infection rate and relationship to the various meteorological conditions. The analysis of these data enabled Christophers and others to predict epidemic years. George Macdonald developed these methods with a high degree of sophistication, and made them the basis of modern malaria control and eradication. As a result of his observations of over 25 years in India, particularly in Mian Mir, the Duars, throughout the Punjab and into Assam, Christophers, around 1948, came to some surprising conclusions. In regions of hyperendemic malaria, where the population was saturated with the infection, the pathogenicity was low and he considered that malaria was a serious cause of mortality neither in adults nor in children. On the other hand, he found that regional epidemic malaria, as in the Punjab, was accompanied by an enormous degree of morbidity and a very heavy mortality especially in infants and young children. In hyperendemic malaria, infection is nearly

100% throughout the year. In epidemic malaria the initial rate is negligible, but rises to a high level in the course of the epidemic. The explanation of the paradox is to be found in the immune status of the population—almost complete in the hyperendemic region, but practically absent in epidemic conditions.

Early in 1909, mounting records of heavy mortality from the epidemic areas in India began to show the importance and extent of the disease, and when Christophers found that under the official system accurate records of deaths had for many years been kept, he was able to map not only the present epidemic but those that for many years in the past had been occurring at intervals of some six or seven years. He was able from studying these records to confirm that one of the most important causes of heavy mortality from malaria was flooding from the many large rivers crossing the country. The importance of these and other findings led to a meeting under the Viceroy in Simla at which important questions in organization were decided upon. It is difficult to convey any idea of how devastating these epidemics were and how completely they destroyed living conditions in a country of villages dependent on local activities in agricultural pursuits for their very existence. His associate Captain E. C. Hodgson, I.M.S., described the ordeal as "hardly a soul was moving about; many of the villagers lay on charpoys (native beds) racked with fever, others were lying on the ground incapable of any effort and there were dead bodies lying unburied, small naked children were neglected and many died; few had the energy to cook and help by neighbour to neighbour was at an end".

Christophers naturally took a great interest in the malaria parasites themselves, and he and Stephens were the first investigators to describe the diagnostic changes which *Plasmodium falciparum* induces in erythrocytes (these changes are usually referred to as 'Maurer's clefts' from the name of the German worker who described them two years later but the term, 'Stephens's and Christophers's clefts', should have had precedence). The correct names of the different species (and genera) of malaria parasites have been debated from the time of their discovery in 1880 by Laveran, and in the course of the succeeding years the situation became increasingly complicated until 1938 when Christophers and Sinton showed conclusively that the name, *Plasmodium falciparum* in common use for the parasite of malignant tertian malaria, was zoologically invalid and that the correct name was *P. malariae* (Marchiafava & Celli 1885). The latter, however, was the name commonly applied to the parasite of quartan malaria and it was abundantly clear that the reversal of these names would cause great confusion in the literature of tropical medicine. Accordingly in 1943 a case was submitted to the International Commission on Zoological Nomenclature, based on the analysis by Christophers and Sinton and supported by malariologists in the U.S.A., for the stabilization of the nomenclature of these important human parasites. A full investigation was then completed by the Secretary of the Commission (Frank Hemmings) in which Christophers played a leading part and in 1954, Opinion 283 was issued which validated the generic and specific names in common use.

Christophers's interest in taxonomy no doubt stemmed from his systematic work in entomology. His old colleague and friend, J.W.W. Stephens, F.R.S., described two other species of human malaria parasites (*P. ovale* and *P. tenue*), about which there has been much debate. Christophers supported the validity of the former from the start. He also accepted the existence of *P. tenue* which he had observed himself in the semi-aboriginal races of Central India. In Christophers's last years in India, he became familiar with the malaria parasites of Asian monkeys, including *Plasmodium knowlesi*, and on his return to England in 1932 he began a new phase in his career, by studying the biochemical properties of this parasite. With J. D. Fulton, he continued a series of equally innovative studies on the respiratory metabolism of *P. knowlesi* and *Trypanosoma rhodesiense*.

From the start Christophers revealed a flair for the study of mosquitoes, sandflies, ticks and other arthropods in all their aspects. His first contributions dating from 1901 were illustrated by meticulous drawings showing histological details of the internal organs and tissues of anopheline and culicine mosquitoes. These drawings were based on the examination of stained sections and of fresh tissues. The excellence of such illustrations were a feature of all his subsequent entomological writings and finally culminated in his masterpiece, "*Aedes aegypti*: the Yellow Fever Mosquito", published in 1960. Christophers, Shortt and Barraud made a detailed study of the sandfly, *Phlebotomus argentipes*, and especially the mouthparts, while his assistant, Barraud, developed a breeding technique for all stages from oviposition to the emergence of the adult fly. Observations on the detailed histology of specimens enabled these observers to trace the anatomical course of experimental infection from the midgut to the proboscis. Dr Christophers was always aware of the importance of zoology in his work, and in particular of taxonomy, and knew that without correct identification of material he could not hope to achieve a scientific perspective of the problems. When he began his studies on mosquitoes their identification was very largely based on the external characters of the female adult; he soon added the criteria of the larval and, later, of the pupal characters. He also made a point of examining the interior of the insect and studied the various organs with great precision.

His work on the structure of the ovary is of prime importance, especially the development of the ovarian follicles and the nurse cells; the appearance of the five different stages was found to indicate the age of the mosquito, at least during the first 24 hours after emergence, and also whether it had already oviposited. Christophers placed less importance on the deterioration of the wing fringe as a clue to the age of the insect. In his early studies on mosquitoes in India with Stephens, Christophers devoted much attention to the species of *Anopheles* and classified them into natural groups, based on the appearance of the adult, larva and egg and their respective habits. At that time (1902) only 16 species had been identified, and no formal subdivisions had been made; by 1933, the number had reached 52 and a classification into three subgenera had become possible. The system of subgenera originated with the work of Christophers, who in 1915 described three subgenera: Anopheles (widely distributed), Myzomyia (later renamed Cellia) (Old World) and Nyssorhynchus (Neotropical). These mosquitoes were described in considerable detail in Christophers's volume on Anophelini under *The Fauna of British India* (1933). Of great importance are his observations on zoogeography and phylogeny which confirmed the grouping of the subgenus, *Myzomyia*, into five (Indian) sections: Neomyzomyia, Myzomyia, Paramyzomyia, Neocellia and Pseudomyzomyia. It is interesting to compare the synonymy of these Indian anophelines in the course of this interval. Dr Christophers himself was responsible for the sinking of many of the well known but incorrect names, though *Anopheles christophersi* (= *minimus*) and *Anopheles rossi* (= *subpictus*) were sunk by Edwards. It is unfortunate that the names of these two, most illustrious, malariologists can no longer be attached to anopheline mosquitoes.

Christophers carried out original work, either alone or in conjunction with his able assistant P. J. Barraud, on the structure of mosquitoes; *e.g.* their pilotaxy in systematics; male and female genitalia including the evolution of the terminal abdominal segments and hypopygium, and a new terminology for the structures; wing venation; characteristics of the egg. He devoted much time to broader studies of mosquitoes and described in detail their bionomics, relations to disease, susceptibility to various parasites and distribution, In this way, he was one of the first entomologists to realize the importance of specific differentiation of the vector in malaria control, *i.e.* by the introduction of species sanitation, in which the attack is made solely on the known malaria carrier, breeding in what might be a unique situation (*e.g. An. stephensi* in wells), and not on mosquitoes in general. As early as 1900, Christophers recommended practical measures for control of malaria by reducing the breeding places of *Anopheles* spp. in

West Africa, *e.g.*, in Lagos by filling swamps with sand (9), but he was sceptical at that time if any anti-anopheline operations would do much to diminish the prevalence of the disease, and even added that to stamp out malaria from the native population was 'chimerical'. Instead he concentrated on the segregation of Europeans from the illimitable native reservoirs of infection. In India, however, he became more optimistic and in conjunction with S. P. James, F.R.S., he reported success in mosquito destruction operations by treatment of the irrigation canals, especially in the notoriously unhealthy conditions of Mian Mir. In these early years, Christophers stressed the necessity of differentiating prophylaxis of malaria into two groups: as applicable to 'native communities' and as applicable to 'Europeans'. He could not emphasize too strongly the importance of personal precautions, and stated that in over three years of residence in Africa and India, neither he nor J.W.W. Stephens had contracted the disease. He placed the proper use of a mosquito net as 'by far and away the greatest means of individual protection'.

During the Mesopotamian campaigns in World War I, Christophers and Shortt quickly realized the danger that malaria represented and directed their attention to two problems. The first was personal prophylaxis for the troops dispersed in small units, and for this purpose quinine was given daily and mosquito nets were used when practicable. The second problem was to reduce the mosquito population around the Basra base, an area of about 10 square miles. Anopheline breeding places comprised water in excavations and waterlogged ground. Silt was dredged from the river bed and bunds were constructed; the water table sank to 12 ft and mosquito breeding was largely eliminated; the incidence of malaria fell dramatically. After World War II, Christophers's advice was sought throughout India, particularly in Bombay, Delhi, Assam, and some of his last papers summarized his views on malaria control in India. Naturally, by this time, he emphasized the paramount importance of malaria in the rural populations in its different degrees of endemicity, but also he laid great stress on the continuing importance of research; fortunately this aspect is still receiving much attention. He concluded that organization is a very essential part of antimalarial work, and the resurgence of malaria in recent years in the Indian sub-continent illustrates how terrible can be the results of the discontinuation of organized control. The use of mosquito repellents was investigated by Christophers and he introduced strict methods for the determination of the efficacy of various compounds, and for their proper application. One of the best preparations was devised or modified by him and has the following composition: (DMP cream) dimethyl phthalate 12.5 ml, cera alba (B.P.), 9 g and arachis oil 27.5 ml. An equally magnificent contribution to the study of mosquitoes was made by Christophers in the last years of his active life, when he produced his monograph on the life history, bionomics and structure of Aedes , 'the yellow fever mosquito'. This work was carried out in the Zoological Laboratories of Cambridge University and was published with addenda in 1960. Although 'yellow fever' figures prominently on the title page, fewer than a dozen pages are devoted to it in the text; nevertheless, it is clear from these pages and other writings that he was intrigued by the complicated life history of the infection in the monkey reservoirs in tropical Africa and America and in a wide variety of curious arboreal mosquitoes (aegypti plays only an accidental role and that outside the forest). Perhaps the most remarkable feature about this book is that it epitomizes the character of Christophers himself; it is thoroughly practical, it is based largely on structure (*i.e.* the visual aspect of the mosquito) and it contains many original features, including the first ever description of the brain of the mosquito in both larval and imaginal stages. Nearly all the 86 line drawings which accompany the text were made by the author and are of the finest calibre.

Even as late as 1960, Christophers was still fascinated by the Rules of zoological nomenclature, and was only too delighted to take up the challenge of Sir William MacArthur who objected to the use of a diaeresis in the 'generic name Aedes'; it was typical of him also that at the age of 87 years he was still capable of changing a lifetime's habit and accepting Aedes as the correct form (Article 27 of the 1964 Rules validates the omission of a diaeresis and all diacritical marks). In spite of his advanced age, Christophers was able to include in this monograph, most of the important new developments up to its date of publication, including the mode of action of, and resistance to, insecticides, genetics and chromosome structure, insect hormones and other recently discovered physiological processes. During the course of Christophers's observations on haemogregarines in the erythrocytes of dogs in Madras, he noticed that these animals were heavily infested with ticks, in which arthropods he found that the haemogregarine *Leucocytozoon canis* (= *Hepatozoon canis*) underwent sporogonic development. In order to follow the latter in detail, he had to ascertain the anatomical and histological structure of the internal organs of ticks. He took as an example of argasid ticks, the desert-inhabiting Ornithodorus and of ixodid ticks, *Rhipicephalus annulatus* and *Hyalomma aegyptium*. He found a great similarity in their internal structure, but reported profound differences in their physiological characters. His observations on embryology and the ovum are of much importance in view of the hereditary transmission of parasites in ticks, which process he was one of the first to demonstrate.

Dr Christophers was personally much impressed with Dr Ronald Ross's prowess of malaria research and personally he regarded him as a great malariologist. Dr Ross's ideology prompted him, like many others across the globe, to undertake many campaigns all over the world including India, such as the one in Mian Mir cantonment of Punjab where malaria was targeted for control through environment sanitation under his charge in 1901-03. One of the greatest advances made in connection with malaria was the introduction in the Punjab of malarial forecasts, year by year, by Lieutenant-Colonel C.A. Gill, based on his study of the meteorological conditions under which malaria spreads. For four years in succession malarial forecasts have been made for the Punjab, and have turned out to be remarkably accurate. Given such an accurate forecast, it was possible to mobilise personnel at the danger areas in advance and to take measures beforehand, and these forecasts were then acted upon by the public health organisation of the province, and by the Director of Medical Services, India. Colonel Gill's book on the role of meteorology in epidemic malaria was considered a masterpiece. Lieutenant-Colonel S.R. Christophers worked extensively and reported epidemic malaria in Singhbhum where during the first two years of life every child in the community suffered almost continuously from acute febrile infestation, with parasites to the number of some 10,000/mm^3 being continuously present in the blood, and the spleen enlarged some 5 cm below the costal margin. Among the adults the spleen rate was 50%, the number of parasites only about 100/mm^3 of blood, and the adult become immune and afebrile. Christophers and Rai Sahib Kazan Chand set up a standard method of measuring enlarged spleens in doing survey work, the measurement being expressed in cm, and a table was given for correction in accordance with the sitting height, age or nipple-umbilicus line. Measurements were recommended from the lower pole of the spleen to the umbilicus and to the mid-line of the body, and a standard chart given together with a notation which gives standard results of effectiveness and precision. Colonel Christophers also studied the frequency distribution of measurements of enlarged spleens in a malarious child community, and raised the question of the relationship of the size of the spleen to the number of attacks and relapses suffered. Hill-station malaria attracted a good deal of attention of malariologists and Lieutenant-Colonel T.C. McCombie Young studied malaria in Shillong where he showed *Anopheles maculatus* to be the important local carrier, and that it bred in the river pools in the Um Kra stream, the rice fields being less important. He surmised that if the rainfall in the early phase of the monsoon were heavy, these pools could have become scoured out and the year would be relatively non-malarious. He recommended a system of contour drainage. The same subject was also dealt with by Major H. E. Shortt in Shillong where the malarial season was from June to October, and of late years the disease had become increasingly prevalent. The spleen rate among the indigenous Khasis was very low, but the Pasteur Institute patients constituted a continuous reservoir of the infection in the neighbourhood of the Um Kra and Um Shirpi streams where *An. maculatus* bred. He recommended for these streams to be canalised. On Ambootia Tea Estate, Kurseong, Dr C. Strickland showed that *An. maculatus* and *An. willmori* were the carriers, and there was a close relationship between rainfall and malaria. He explained this relationship: should the early rains be heavy, the breeding places were scoured out and a year of low malarial incidence followed. In fact the curve of severity of the monsoon and of malarial incidence in the next year were in an inverse relationship. He studied opposite conditions present in Kurseong and Shillong, respectively.

Colonel Christophers was invited to visit Bombay in 1923 in connection with the Back Bay reclamation scheme, and advised the services of a wholetime officer. The organisation instituted had grappled very successfully with the problem, the whole area being divided into sectors, each of which was covered twice a week by the search and larvicide parties, respectively. The chief carriers wre *An. stephensi, An. culicifacies* and *An. vagus*, with *A. rossi* also prevalent; and the chief breeding places which had required attention were the mouths of the sludge pipes, joints of pipes where leakage occured, fissures in the drying sludge, and depressions in the new land surface. Work had also been necessary in the adjacent areas in Colaba, and much public health propaganda work was carried out. Admissions to Colaba hospital for malaria among British troops had fallen from 973 in 1921, to 418 in 1923, and to 169 in 1924. In Assam, Dr Strickland incriminated *An. umbrosus*, which bred in cut bamboo stumps, as an important vector, whilst in Madras, Dr D. A. Turkhud reported, that special measures had to be taken to deal with epidemic malaria in the work of the Forest Department at Sappal Hill and Top Slip. Drs Ramaswami Iyer and K. P. Menon were specially trained and sent to the two areas, respectively, and had submitted detailed reports and recommendations for the consideration of the government. Major J. A. S. Phillips, in connection with five towns in the U.P., asserted significance of even partial measures undertaken. He reported results; in Saharanpur a reduction of the spleen rate from 79 to 7%, in Nagina from 79 to 13%, in Kosi from 81 to 42%, in Meerut from 6 to 1.3%, and in Lucknow from 2.2 to 1.6%, together with greatly reduced fever rates.

Majors H. W. Acton, and R. Knowles advocated a standard treatment for malaria with Indian cinchona febrifuge. The timing of the dose was considered important. Major O.A.R. Berkeley-Hill reported that of 35 cases of benign tertian malaria treated orally with cinchona febrifuge on the above lines, and carefully watched over a successive period of same months, only four had relapsed at intervals of from 2 weeks to 2 months after completing treatment. Dr T.H. Bishop advocated the following prescription, which became in use use all over the E.B. Railway system, and which, according to him, was pleasanter and far more palatable than a quinine mixture. Dr U.N. Brahmachari dealt with the treatment of difficult cases of malaria. Quinine salts being markedly hemolytic, he used an anti-hsemolytic solution for intravenous administration which needed to be administered slowly at the rate of not more than 10 c.c. per minute and might prove suitable for cases of blackwater fever. Majors R.N. Chopra and Knowles, with Assistant Surgeon J.C. Gupta recorded the absolute failure of *Vitex peduncularis* in the treatment of malaria, whilst Dr Bayley-de Castro recorded the case of a woman six months pregnant, suffering from mixed benign and malignant tertian malaria, together with typhoid fever, who recovered after treatment with intravenous quinine.

Sir Rickard Christophers received many tributes on the occasion of his centenary in November 1973. The Royal Society of Tropical Medicine and Hygiene (of which he had been President from 1939 to 1943) published in its Transactions messages from 17 of his friends and past colleagues from India, Nigeria, United States, Venezuela, France, Portugal, Italy, Germany and the United Kingdom. The Royal Entomological Society held a special meeting, honouring him as one of the foremost entomologists of the world (1974). The year before his death, a symposium to commemorate the 'Birth of Medical Entomology' was arranged by Dr Michael Service; its proceedings (published in the Transactions of the Royal Society of Tropical Medicine and Hygiene 1968) were dedicated to Christophers who had done so much to nurture the 'Offspring'. Since 1979 a 'Christophers Medal' had been endowed by the Royal Society of Tropical Medicine and Hygiene to commemorate his life and achievements.

Lieutenant-Colonel S. R. Christophers worked mostly together with Dr J.W.W. Stephens and Captain Barraud on many projects. With Barraud, he determined the tracheation and wing venation of the mosquitoes. **Dr Philip James Barraud** (1879–1948) was an English entomologist who specialised in mosquitoes. He worked in Iraq and India where he invented the Barraud cage (a box of made of muslin or organdie suspended on a wire frame placed inside insulators containing wet towels) for the transport and study of live mosquitoes. Barraud described many new genera and species of Culicidae. His collection, which included Palearctic Lepidoptera, is in the Natural History Museum. The genus *Barraudius* Edwards, F.W., 1921 is named for him.

Prof. Dr John William Watson Stephens, FRS (1865–1946) was a British parasitologist and expert on tropical diseases. After a term at Christ College, Brecon and then completion of secondary school at Dulwich College, Stephens matriculated in 1884 at Gonville and Caius College, Cambridge, graduating there with B.A. in 1887. He received his medical education at St Bartholomew's Hospital, receiving there M.B. and B. Chir. in 1893 and D.P.H. in 1894. Some of the important publications by Dr Stephens are listed below:

1. Two new species of Culex (Diptera, Culicidae) from Assam. *Indian J. Med. Res.* 11: 507-509 (1923).
2. A revision of the culicine mosquitoes of India, Part XXIII. The genus Aedes (sens. lat.) and the classification of the subgenus. Descriptions of the Indian species of Aedes(Aedimorphus), Aedes (Ochlerotatus), and Aedes (Banksinella.), with notes on Aedes (Stegomyia) uariegatus. *Indian J. Med.Res.* 15: 653-69. (1928)
3. A review of the culicine mosquitoes of India. Part XXIV. The Indian species of the subgenera Skusea and Aedes, with descriptions of eight new species, and remarks on amethod for identifying the females of the genus Aedes. *Indian J. Med. Res.* 16: 357-75. (1928).
4. 1934. *The Fauna of British India, Including Ceylon and Burma.* Diptera V, family Culicidae, tribes Megarhinini and Culicini. Taylor and Francis, London. 463 p.(1934).

Many others were influenced by their influential teamwork, *e.g.*, Dr M.O.T. Iyengar who described a new species of anopheline common in Bengal, *Anopheles varuna* which is annectant to *An. listoni* and *An. minimus*. Strickland on the other hand considered that there are no definable points of distinction between the African *An. funestus*, its Indian variety *An. listoni*, and *An. minimus*, and that therefore all these forms could be considered merely geographical races of *An. funestus*; *An. aconitus* on the other was considered by him a distinct form entitled to specific rank.

Major J. R. Adie, I.M.S., has been a great resource person for Dr Sir Rickard S. Christophers who admirably spoke about him as having helped him with mosquitoes from difficult areas in India. Major Adie had interests in plants' effects on mosquito growth and development. He brought considerable evidence to bear on the benefit of *Lemna minor* as a means of keeping mosquitoes from laying their eggs on water. He showed that tanks covered with this green flat weed never contain larvae of *Culicidae,* whilst others at the same time of year are full of them. As a test he cleared certain areas near the banks of all *Lemna* and enclosed them with light floating structures, which were fixed enough to resist the winds-in fact made experimental pools. He found in due time plenty of *Anopheles* larvae in these pools. This seemed to prove that *Lemna* acts as a mechanical obstruction to the process of egg laying, and a very obvious method of prevention. This small yet widely distributed genus of floating plants evidently has a very marked effect upon the frequency of culicid larvae in natural and artificial collections of water. The little *Lemna arrhiza,* or the Rootless Duckweed, occurs in Asia, Africa, South America and Europe, and apparently has the same effect as the larger *L. minor.* His work on anopheline mosquitoes in Kolkata remains one of the foremost references (Adie & Alcock, 1905).

Taxonomy of mosquitoes reached a point of zenith with the dedication of **Lieutenant-Colonel Sydney Price James** (1870–1946) who was also an officer in the Indian Medical Service where he relentlessly worked on malaria control in India. Although his period of service briefly overlapped that of Giles, another giant on mosquito research in those days (James, 17 years younger to Giles, entered the I.M.S. in 1896 and Giles retired in 1901). James hardly ever mentioned the name Giles in his writings whether personal or professional. James published 15 references dealing with mosquito taxonomy; the last of which was published in 1934. After retiring from the Indian Medical Service in 1918, he went on to complete a long and distinguished career in several aspects of tropical medicine but discontinued any research on mosquito taxonomy on which he long joined with Dr W.G. Liston.

Lt Col, CIE, William Glen Liston (30th July, 1872–18th October, 1950), a distinguished member of the Indian Medical Service and best known for his sterling research on elucidation of the transmission of plague from rat to man by the rat flea, was borne in Secunderabad, India where his father was an Army Chaplain. He was educated at George Watson's College, Edinburgh, and at the Albany Academy, Glasgow. He studied medicine at Glasgow University where he was one of the leading students of the year graduating in 1897. He did his MD from the University in 1902 with a thesis titled *'Aetiology of malaria'.* In the following year he entered the Indian Medical Service. During the training course at Netley he first came under the influence of Almroth Wright, then Professor of Pathology at the Army Medical School. Liston arrived in India in October 1898 and was posted to

Secunderabad for duty at the Station Hospital. Though a significant member of the plague research in India, he also had a brief but important encounter with mosquitoes. Liston was another member of that extraordinary group of British medical officers who were serving in India at the turn of the century and who so quickly moved to elucidate the mosquito species linked with the transmission of malaria. In collaboration with James, he made important contributions to the knowledge of anopheline taxonomy, biology and control, as well as to the vectorial capabilities of the Indian species. In his taxonomic explorations he received unstinted artistic support from his colleague Dr Turkhud.

The famous mosquito artist and semantic scholar **Dayaneshwar Atmaram Turkhud** was born in 1862 in a highly respected and knowledgeable family. His illustrious father Dr Atmaram Pandurang Turkhadekar (or just Turkhad in English publications) was, in the first place, an Indian physician and social reformer who founded the *Prarthana Samaj* and was one of the two Indian co-founders (the other being Sakharam Arjun) of the Bombay Natural History Society (Gupta, 2010). He was an entomologist, too, as he devoted much of his later years of service to anopheline mosquitoes in Kodaikanal near Madurai and malaria vectors in southern India. Dr D.A. Turkhud worked at the Haffkine Institute and served as a director of the King Institute of Preventive Medicine at Guindy and worked in Kodaikanal on *Anopheles* mosquitoes (Gupta, 2010). It is irony that Ronald Ross who rather struggled with the identification of mosquitoes later in mid-1890s did not know anything about the existence of Dr Tarkhadkar, although a new species was justifiably erected in his honour, i.e., *Anopheles turkhudi* Liston, 1901. The world famous coloured drawings of mosquitoes in the 'Anopheles of India' monograph by James & Liston (2008; revised 2011) were made by Dr Turkhud, in whose praise they had eloquently said that the beautiful and convincing coloured drawings of mosquitoes were a great asset for the book and would certainly help the entomologists in their field identification.

Even though some taxonomists did not directly work on Culicidae yet their dedication to taxonomy was deep that mosquito taxonomists had felt honoured in remembering their contributions in one or another way. One such great scientist was **Dr Thomas Nelson Annandale** (1916-1924) who was born on 15th June, 1876 at Edinburgh city of United Kingdom. He came to India in 1904 as Deputy Superintendent of the Natural History Section of the Indian Museum and in 1907 was appointed as Superintendent on the post vacated by Lieut-Col. Alcock. Under his strong and wise administrative ability, the Natural History Section of the Indian Museum flourished both with human and faunal resources. Two scientific journals, *viz., Records of the Indian Museum* and *Memoirs of the Indian Museum* were initiated in 1907. Dr Annandale was the first Director of Zoological Survey of India, a great responsibility which he performed till his last breath. He had published a series of scientific reports on zoology as well as on anthropology. He also worked on Indian freshwater Mollusca and their possible role in acting as vectors of human Schistosomiasis. Dr Annandale became a legend in his life time having published a lot almost on every group of zoology. He was an emphatic believer in the necessity of field work. He travelled all corners of the Indian Empire and acquired a personal knowledge of its fauna. Within a very short life span of only 48 years he produced an enormous number of publications in almost all groups of zoology and anthropology also. Altogether he published 538 scientific reports and findings, including 72 articles on anthropology and only one on Botany. He started working on the aquatic fauna of India, an untouched field of research and gave much of his attention to the fauna of freshwater and began a series of study on Hydra as well as freshwater sponges and Polyzoa, which he summarized and published in '*The Fauna of British India and Ceylon*' in 1911. A few mosquitoes, *Aedes annandalei* (Theobald, 1910) and *Anopheles annandalei* Baini Prashad, 1918, named in his honour, area a clear testimony to his authority and the reverence with which he was held by his colleagues and successors in scientific arena.

John Alexander Sinton (1884–1956), the third of the seven children of Walter Lyon Sinton and his wife, Isabella Mary, née Pringle, was born on 2nd December 1884 at Victoria, British Columbia, Canada. He was a very famous malariologist. His parents were both from the north of Ireland, of Scottish lowland origins, Quakers on his father's side, Presbyterian on his mother's. The family returned to Ulster in 1890 and that remained Sinton's base throughout his much travelled life. Academically gifted from an early age, Sinton attended the Nicholson Memorial School, Lisburn, and the Royal Belfast Academical Institution before graduating MB BCh BAO (Belfast) in 1908, first of his year, adding the diploma in public health (Belfast and Cambridge) in 1910 and the diploma in tropical medicine (Liverpool) in 1911. He was first in the entrance examination for the Indian Medical Service (IMS) and was gazetted lieutenant IMS in July 1911. However, before being posted to India he was seconded for a year as Queen's University research scholar to the Liverpool School of Tropical Medicine where his contact with Sir Ronald Ross may have influenced his later career as a malariologist.

Sinton served with great distinction in the military branch of the IMS from 1912 to 1921, when he transferred to the civil branch in which he served until his retirement in 1936. In the former he saw action in the turbulent North-West Frontier Province, in Mesopotamia, Afghanistan, Waziristan, and Turkestan, and was decorated many times, receiving the Victoria Cross in 1916, the OBE, and the Russian order of St George, and was mentioned in dispatches six times. Although shot through both arms and through the side, he refused to go to hospital, and remained as long as daylight lasted, attending to his duties under very heavy fire. In three previous actions Captain Sinton displayed the utmost bravery. During the transcaspian campaign of 1918–19 Sinton commanded an Indian field ambulance which marched through Persia to Turkestan where he became senior medical officer of the Turkestan military mission. Heavy fighting along the Trans-Siberian Railway resulted in casualties among troops already weakened by cholera, typhus, and malaria. The situation was exacerbated by the 1918 influenza pandemic which left Sinton's unit with no fit men to collect rations. The military hospitals were overwhelmed and sick and wounded overflowed into Russian hospitals. Without Sinton's energy and expert medical knowledge the death rate in this campaign would have been substantially higher.

Although military matters dominated those years, Sinton found time for research, publishing papers on cholera, leishmaniasis, mosquitoes, and sand flies. Following transfer to the civil branch he was posted to the government of India's principal centre for medical research and vaccine production at Kasauli, in the Punjab, where his academic potential blossomed. Initially Sinton divided his time between the Pasteur Institute and the Central Research Institute, but in July 1921 he was put in charge of the quinine and malaria inquiry under the newly formed central malaria bureau. At Kasauli, Sinton met Eadith Seymour Steuart-Martin (1894 - 1977), daughter of Edwin Steuart-Martin and Ada May Martin (née Martin), whom he married on 19 September 1923. Their daughter, Eleanor Isabel Mary Sinton, was born at Kasauli on 9 December 1924.

In 1925 Sinton became the first director of the Malaria Survey of India at Kasauli, a position he held until he retired from the IMS. He was a dynamic leader and established an international reputation for himself and for his unit. Sinton made important contributions to the enumeration, identification, pigmentation, and nomenclature of parasites. In the field of treatment he discovered the metabolic effects of quinine, notably hypoglycaemia. By controlled research he distinguished between parasite suppression, which could be followed by recrudescence, and cure which involved parasite elimination, and he defined the principles for the assessment of the efficacy of the antimalarial drugs then being introduced. Drug resistance rendered many of the early synthetic antimalarials useless, and quinine returned to favour. The principles defined by Sinton remain highly relevant in the continuing search for effective antimalarial therapy. The *in vivo* tests for resistance defined by the World Health Organization are based on his ideas. On monkey malaria, Sinton showed that multiple infections and super-infections occurred and that longitudinal studies could detect the presence of more than one plasmodial species in the same host. These observations contributed significantly to an understanding of the mechanisms of immunity in malaria and were still cited in research papers at present. A highly disciplined and affable officer, Sinton worked with many native geniuses. With Dr R.B. Lal he dealt with albuminuria in malaria. The incidence in 64 malignant tertian cases treated between August and mid-September was 18.7% and in 110 cases treated between mid-September and mid-October was 12.4%. The combination of quinine with alkalies in the doses given had no deleterious effect on the kidneys, and the results show a slight diminution in the incidence of albuminuria in the cases treated with alkalies. Major Sinton introduced a most ingenious method for enumerating the parasites and leucocytes in malarial blood. A standard suspension of the nucleated red corpuscles of fowls was made in 1 in 3,000 perchloride of mercury solution, so that 1 mm^3 = 20,000 cells. The malarial blood was then mixed with an equal volume of this suspension, and the mixed fluid spread in a thick smear, when, as the number of nucleated cells was known, the number of parasites or leucocytes could easily be arrived at. Major T.A. Hughes found a condition of latent hemolytic jaundice to be present in four cases of hepatic cirrhosis with enlargement of the spleen and ascites by the van den Bergh test, and considered that a similar condition might be present in individuals continually resident in malarial districts, who showed no clinical signs of active malaria. Together with Major Sinton, he studied the functional capacity of the liver in malaria, as measured by the laevulose tolerance test. There wes some disturbance of the glycogenetic function of the liver in a certain number of ordinary acute cases of malaria, but the degree of inefficiency found was not marked. In cases showing pernicious symptoms the inefficiency was probably greater. The ingestion of glucose or laevulose was indicated in all cases of malaria, more especially in severe cases, to protect the liver cells and prevent further damage to the other functions of the liver, which were protected by the glycogen reserve of this organ. Major Sinton with sub-assistant surgeon J. D. Baily worked on the " acidosis" - or rather lowered alkali reserve - in malaria. The amount of sodium bicarbonate required to render the urine of healthy Indians alkaline to litmus was found to be on an average 6.5 gm whereas in a series of ordinary cases of malaria the corresponding amount was found to be 14 gm.

After his first retirement Sinton continued malaria research as Manson Fellow at the London School of Hygiene and Tropical Medicine and at the malaria laboratory of the Ministry of Health at Horton Hospital, near Epsom. He also became adviser on malaria to the Ministry of Health. With the outbreak of the Second World War, Sinton was recalled as an IMS reservist and commanded a hospital in India. At the age of fifty-five he was again retired, but was appointed consultant malariologist to the east African force and later to Middle East command, retiring with the honorary rank of brigadier in August 1943. Two weeks later he was yet again recalled and appointed consultant malariologist to the War Office, travelling widely to Assam (India), Australia, Burma, Ceylon, New Guinea and the Solomon Islands, where his expertise in malaria was invaluable. Further military decorations resulted from this period, after which Sinton returned to Northern Ireland and retired to Cookstown. He was elected FRS in 1946. In his final retirement Sinton's compelling sense of duty was manifested in service to his university and to the local community. At Queen's University, Belfast, he sat on the senate, was pro-chancellor and president of the University Association, and held a number of other offices. Sinton published over 200 scientific papers, many on malaria including a monumental bibliography of 2200 malaria research papers in India from Ross's discovery of the parasite in 1898 up to 1928. Another forty of Sinton's papers were on sand flies and included descriptions of twenty-seven Phlebotomus species, sixteen of which were new to science; Sinton named one *Phlebotomus eleanorae* after his daughter, and another *Phlebotomus eadithai* after his wife. Others honoured Sinton by naming three mosquito species, *Aedes sintoni, Anopheles sintoni,* and *Anopheles sintonoides,* one sandfly species, *Sergentomyia sintoni,* and one subgenus Sintonius of the genus Phlebotomus, after him. Besides, he received many awards and was Arnot memorial medallist of the Irish Medical Schools' and Graduates Association (1917); MD (honoris causa) of Belfast (1919); honorary fellow of the Ulster Medical Society (1920); DSc of Belfast (1927); Chalmers memorial medallist of the Royal Society of Tropical Medicine and Hygiene (1928); Bisset-Hawkins medallist of the Royal College of Physicians, London (1944); Robert Campbell memorial orator and medallist of the Ulster Medical Society (1946); Mary Kingsley medallist of Liverpool School of Tropical Medicine (1949); and Manson medallist of the Royal Society of Tropical Medicine and Hygiene (1956). He died at his home at Slaghtfreedan Lodge, Cookstown, co. Tyrone, on 25th March 1956 and was buried with full military honours on 28th March at Claggan Presbyterian cemetery in Cookstown. His name is commemorated in a University hall of residence, Sinton Hall, and in Sinton Court, a British Legion home. Sinton is unique as the only man ever to have been both a holder of the Victoria Cross and a Fellow of the Royal Society. In the words of Colonel H.W. Mulligan in an obituary in the British Medical Journal: *'Sinton had an exceptionally quick, receptive, and retentive brain, but his greatness sprang not so much from his unusual intellectual gifts as from the simple qualities of absolute integrity and tremendous industry'.* His name and fame crossed borders and malaria research got the central attraction in many western countries even though the northern America was largely free from malaria, and one such genius was Dr P.F. Rusell, the true global malariologist working in India.

Dr Paul Farr Russell was born in Boston on August 12, 1894. He received his bachelor's degree from Boston University in 1916 and his medical degree from Cornell University in 1921. In 1929, he earned a master's degree from the Harvard University School of Public Health. Dr Russell, who lived in Richmond, was a research specialist and administrator with the Rockefeller Foundation for 36 years until he retired in 1959. In World War II, when he served as a colonel in the Army Medical Corps, he was assigned to General MacArthur's headquarters in Australia. He extensively worked on malaria and mosquitoes studying disease epidemiology and insecticidal control of vectors in erstwhile Madras Presidency's Ennore-Nellore coastal areas, Nilgiris district, Thanjavur and many southeastern areas. His works on *Anopheles culicifacies* in irrigation fields, mostly in collaboration with the famous malariologist and mosquito specialist Dr T. Ramachandra Rao, are classical examples of quality research extensively published in *Indian Medical Gazette* and *Journal of Malaria Institute of India*. His *magnum opus* 'Practical Malariology' (1963) has a chapter by another eminent Indian culicidologist, Dr B.L. Wattal (Wattal, 1963). Another great mosquito scientist who was associated with him in India was Dr B.N. Mohan (On experimental malaria infections in certain Anopheles of southeastern Madras. *Jour. Mal. Inst. Ind.* 1939; 2: 425–431). Paul Russell's 'Man's Mastery of Malaria', written in 1955, opens with these wistful words: *"While keeping in mind the realities, one can nevertheless be confident that malaria is well on its way toward oblivion. Already as a malariologist, I feel premonitory twinges of lonesomeness, and in my own organization I am now a sort of 'last survivor."* However, Russell was misguided in thinking that the possibilities that had opened up with Ronald Ross's discovery of the life cycle of the malaria parasite in 1897 (not to mention his remarkable attempts thereafter to understand the dynamics of infection through a simple mathematical model) were to be realized so soon. We are still struggling to conquer malaria, but now have at our disposal some very sophisticated methods of attacking the mosquito vectors, such as with bacterial larvicides and entomopathogenic fungi; new drugs like artemisinin have replaced those to which the parasite has become resistant, and candidate malaria vaccines are reaching a new stage of maturity. Malaria has a complex life cycle and vaccines are now being developed to attack it at both its pre-erythrocytic stages and erythrocytic (blood) stages, as well as those that are specifically tailored to prevent onward transmission (Fig. 8.4). It is broadly envisaged that blood-stage vaccines will mainly impact on disease severity rather than infection, whereas pre-erythrocytic vaccines are being held to the higher standard of preventing infection as well as disease. A handful of pre-erythrocytic vaccines have shown partial efficacy, and others that target either the asexual blood-stage parasites or the transmissible forms (gametocytes, and also the zygotes within the mosquito) are in various stages of development. When it comes to elimination, a clear case can be made for synergistic interactions between pre-erythrocytic and transmission-blocking vaccines and bed nets. The vaccines and bed nets may act in concert to diminish the availability of susceptibles as well as to reduce the per capita rate of infection, thereby tipping the system into negative growth. In contrast, vaccine-induced immunity against the blood stages may neither prevent infection nor impact on onward transmission. Fortunately a malaria vaccine (RTS,S, known by the trade name Mosquirix; RTS,S/AS01 is a recombinant protein-based malaria vaccine), a low efficacy one, is given in four injections. A pilot project for vaccination has been launched on 23th April 2019 in Malawi, on 30th April 2019 in Ghana, and on 13th September 2019 in Kenya. RTS,S (developed by PATH Malaria Vaccine Initiative (MVI) and GlaxoSmithKline (GSK) with support from the Bill and Melinda Gates Foundation) is the most recently developed recombinant vaccine.

Dr Paul F. Russell, a specialist in malaria and tropical diseases who served on Gen. Douglas MacArthur's senior staff in World War II, died at St. Luke's Hospital in Richmond at a mature age of 89 years. For his multifarious investigations on malaria he was awarded the Walter Reed Medal by the American Society of Tropical Medicine in 1946 and the Darling Medal by the World Health Organization in 1957.

The Ross and Russell legacies embodied finest quality of malaria research in India furthered in all directions of mosquito and malaria research by **Dr Henry Edward Shortt**, CIE, FRS, MD, DTM&H, who worked for the Indian Medical Service in his initial years. Shortt was a great physician-naturalist and a devoted protozoologist known for many seminal contributions to medical science, especially malaria. He had varied interests, from single-celled parasites, tigers, trout to houseflies, was a untiring protozoology man and yet generous and caring to his associates and students. Drs Christophers and Shortt carried out the first scientific malaria survey of war during 1914-1918, modelled on Christophers' earlier malaria surveys in India. Shortt later worked with Indian Medical Service as Director, King Institute of Preventive Medicine, Madras from 1935-1939; Inspector-General of Civil Hospitals and Prisons in Assam during World War II and later as Professor of Parasitology at London School of Hygiene and Tropical Medicine from 1945-1951. For his seminal research on malaria he was awarded the Darling Foundation Prize of WHO in 1951.

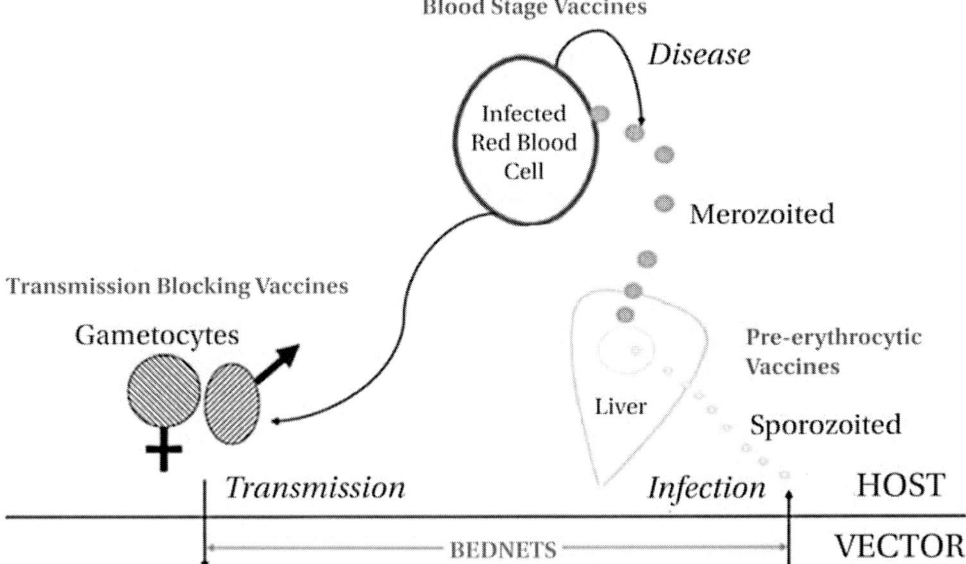

Fig. 8.4. The malaria parasite, Plasmodium falciparum, is injected by the mosquito vector into its human host in the form of sporozoites. These sporozoites quickly hone in on the liver and set up intracellular infections of hepatocytes. Vaccines targeting these stages are known as "pre-erythrocytic" vaccines. After an incubation period of ~1 wk, merozoites burst out of the liver and go on to invade RBCs (erythrocytes). Infected RBCs produce further merozoites that can invade uninfected RBCs: this highly proliferative stage of the life cycle is specifically associated with malarial disease, and vaccines targeting the blood stages could conceivably have an effect on disease severity without impacting on infection or transmission. The differentiation of infected RBCs into sexual stages or gametocytes completes the life cycle, as these can be taken up by the mosquito to mate within its gut. Vaccines targeting these stages can block transmission without interfering with either infection or disease and, therefore, be of no immediate personal benefit to the vaccinee but afford significant long-term benefits. Bed nets can prevent both infection of the human host and transmission of parasites from an infected host to the mosquito vector.

One of the most powerful entomologists and influential malariologists in the post-independence India was **Dr I.M. Puri**, who gave to the Indian fraternity engaged in controlling malaria in the country several important research works notable amongst which is his 1960 monograph, "*Synoptic tables for the identification of the full-grown larvae of the Indian anopheline mosquitoes.* Health Bulletin No. 16, Malaria Bureau No. 7., Government of India Press, Calcutta. pp.104.

Although not a mosquito scientist, in a direct sense, nevertheless, because of his overall influence on malaria control in India in the early decades of operation **Dr A.P. Ray** was a monumental figure in the Indian malariological success story during mid 1960s and will be respectfully remembered for his multi-directional influence on the disease control. Dr Amar Prasad Ray (1913-1996) was born in Kolkata on 26th February 1913. He belonged to a Jamindar family of Shri Hemant Kumar Ray and Smt. Nirmal Nalini from Sugandha in Hoogly district, West Bengal. Amar's early education was in Patna where he attended Raja Ram Mohan Roy Seminar School, and subsequently completed intermediate in science from Patna Science College. He joined the Prince of Wales Medical College and Hospital, Patna University obtaining MBBS with distinction in all subjects. He completed the internship and house job and later in 1957 did his PhD in Medicine from Patna University. His doctoral thesis was on "*Cirrhosis due to Malaria Infection in Monkeys*". Dr A.P. Ray joined as the Chief Medical Officer in the state of Maharaja of Hathuwa (an independent state under the British rule) in 1938. He married Miss Kalyani in 1942. He was selected in the Indian Medical Service and joined the armed forces in 1939 as Captain. He was posted in Middle East, travelled extensively in war torn countries. He participated in the 2nd World Biographical Memoirs and trekked the most difficult terrain in Middle East serving the Allied Army Regiments at the war front. In 1945 he was deputed by the Army Medical Headquarters to undergo training in malaria control methods at the Malaria Institute of India, Delhi. In the initial training courses in malariology, he attended in 1945, Dr S.P. Ramakrishnan and Dr N.G.S. Raghavan were his co-participants. The three stalwarts joined Malaria Institute of India within a gap of six months. In 1946 Dr Ray joined the Malaria Institute of India as Assistant Director (rechristened to National Institute of Communicable Diseases, NICD; presently National Communicable Disease Centres, NCDC) after demitting office as Major in the Army Medical Corps of British India. At that time Major General G. Covell was Director, Malaria Institute of India. He was highly impressed by Dr Ray's brilliant career, enthusiasm and his commitment to research. Dr A.P. Ray was promoted to the post of Assistant Director at the first opportunity superseding many senior officers in the Institute. Because of his brilliant performance at the Institute, he was promoted to the post of Deputy Director in 1956. The former two colleagues superannuated as Directors of NICD while Dr Ray retired as Director of National Malaria Eradication Programme (NMEP) in 1968. At the Malaria Institute of India he devoted all his time and energies in research on malaria. Dr AP Ray's first research paper was on prophylactic effect of Paludrine in tea estate which was published in 1948 in the Indian Journal of Malariology. In the next two decades (1948-1966) Dr Ray published more than 125 scientific papers mostly on therapeutic and prophylactic effect of antimalarial drugs and the development of drug resistance in animal models. He worked with antimalarial drugs such as Acridines (Mepacrine), Pyremethamine (Daraprim), 4-aminoquinolines (Chloroquine and Rasochine), antifolate compounds (Proguanil), Amodiaquine (Camaquine), Pamaquine, Stabrin, Bromoguanide and used experimental models of *Plasmodium knowlesi, P. inui, P. gallinaceum, P. cynomolgi* and *P. berghei.* He pioneered the work on parasite resistance and cross-resistance against Paludrine in *Plasmodium knowlesi,* and induced chloroquine resistance in *P. knowlesi* in chicks. He successfully cultured *Plasmodium gallinaceum* in vitro in 1951. He isolated a new strain of *Plasmodium knowlesi* (Nature 172 p. 122, 1953). He again reported his work on two fold resistance to chloroquine in *Plasmodium gallinaceum* (Nature 176 p. 1291-1292, 1956). The same year he discovered the synergistic action of quinine and pyrimethamine in *P. falciparum,* and reported the development of resistance to pyrimethamine in *P. cynomolgi.* His work on chemotherapy of malaria and drug resistance was acclaimed internationally. His studies on human malaria infections of *P. vivx* and *P. falciparum* helped in formulating Anti Malarial Drug Policy in India. Dr Amar Prasad Ray spent time in laboratory cutting sections, screening slides and recording data for communication. It is notable to mention that he was himself conducting experimental work with little help. He was interested in pathology of the disease and invariably he could be seen examining liver sections of infected monkeys. When the discovery of hypnozoites was announced by Dr W.A. Krotoski in 1980, he recalled having seen the hypnozoites in the liver sections. He wanted to re-visit these slides but all his slides were lost during the period of his long absence from the country. He regretted that a great discovery in malaria had slipped away from his hands. In 1959, Dr A.P. Ray took over as the Director of the National Malaria Eradication Programme (NMEP), a centrally sponsored activity and one of the biggest health programmes in the world. During this period at the Malaria Institute of India he was pioneer in many fields of research and field operations. Before undertaking the community based field trials with antimalarial drugs, he used to personally monitor experimental studies in confined communities like police hospitals, jail hospitals etc. After fully satisfying himself with preliminary results, he used to extend the field operational studies in endemic areas. He laid the foundation of field studies related to drug trials and testing of insecticides. His work in UP Terai on chloroquine efficacy is valid even after six decades. He carried out many pilot studies in difficult terrain throughout the country to control malaria by combination of anti-parasitic and anti-vector measures. All his work was directed to benefit the community by reducing malaria related morbidity and mortality using different strategies in space and time. He blossomed in full fragrance in the conduct of his work and enjoyed visits to tribal and inaccessible areas like Bastar district in Madhya Pradesh, the entire Andman and Nicobar Island. His blue book entitled "Administrative and Technical Guidelines of NMEP" is a bible for all malaria workers concerned with the eradication of malaria. In the field every malaria worker had area-wise time bound programme of movement schedule for each working day of the week. The field worker was supervised by the malaria inspector who was directly responsible to guide any work-checking visitor to the village area with exactness. A saying was common during Dr Ray's time: "*The Sun rays may miss a village but not the visit of a malaria field worker.*" He himself used to wade through a rivulet and flood waters, and muddy terrain taking his trousers and shoes in hands to inspect the surveillance and DDT spraying in houses. Dr Ray used to walk long distances of more than 20 km in a single stretch to verify the spray of insecticides in the remotest parts of the difficult terrains. He introduced the surveillance at fortnightly interval to collect the blood smears of all the fever cases and administer radical treatment to m positive cases. His dynamism, hard work, knowledge and objective assessment, and at times harsh comments, encouraged the staff to perform better with full dedication. As a result there was healthy competition in the performance of malaria control activities among the districts. He was a man with indefatigable energies. Dr A.P. Ray has the credit of providing capable leadership to the country's most prestigious programme of malaria eradication. He treated his entire staff as one family and always admired his good hard working staff, at every level, but at the same time he was merciless to the non-performing delinquent staff. His emphasis was on total coverage with DDT of the indoors of houses, thorough search of fever cases and strict supervision of all field work to ensure operational efficiency.

Dr Ray had the art of putting his point sarcastically even to the VIPs with a pinch of salt. Once he met the Director of Health Services and the

State Finance Secretary Punjab, and during discussions Dr Ray was told that Punjab would not be able to spend the required money for malaria eradication as it had become a poor state. The same evening, Dr Ray happened to meet the State Finance Minister and narrated the story of state poverty to the Finance Minister. The Finance Minister took Dr Ray to meet Mr Pratap Singh Kairon, the most powerful Chief Minister in those days. When Mr Kairon asked Dr Ray how Punjab was doing in malaria eradication work, he responded by saying that Punjab was at the top of the country a few month back; but now the state has become poor and therefore the state officials want malaria work to be carried out by the central government. On hearing this Mr Kairon appeared upset and exclaimed "Dr Ray are you trying to threaten me?" Then Dr Ray narrated the discussion he had with the state officials. Mr Kairon immediately transferred both the Director of Health Services and Finance Secretary and requested Dr Ray to call on him whenever he visited Punjab. Later he used to brief Mr Kairon personally of his field visits to the endemic areas of Punjab. A similar incidence happened in West Bengal when the famous Mr B.C. Roy was the Chief Minister. Dr A.P. Ray narrated to the Chief Minister the sad state of affairs of malaria work in the state and the Chief Minister took prompt action to rectify the errors in the programme. When Dr Ray landed at Palam Airport after meeting him, he received a message from Mr B.C. Roy that he had already taken action as suggested by him for improving malaria work in the state. Such was the hold Dr Ray had on the top leaders of the state and the Centre alike. In the field of malaria, he was close friend of some of the most noted malariologists of his time such as Dr P.C.C. Garham, Dr G. MacDonald, Dr E.J. Cohen, Dr D. Clyde, Dr S.R. Christophers, Dr L.J. Bruce-Chwatt, Dr R. Senior-White, Dr R.H. Black, Dr Paul F. Russel, Dr G. Covell, Dr M.F. Boyd, Dr E.Y. Pampana, Dr W. Peters, Dr J.F.B. Edison, Dr G. Davidson and many others. Among the Indian names that had association with him are Dr B.A. Rao, Dr T. Ramachandra Rao, Dr G.K. Sharma, Dr Jaswant Singh, Dr S. Pattanayak, Dr I.M. Puri, Dr M.O.T. Iyengar, Dr M.K. Afridi, Dr D.K. Vishwanathan, Dr S.P. Ramkrishnan, Dr V.P. Sharma and, of course, a large number of his colleagues from the state services and research institutions. Dr Ray as a person was very compassionate and considerate. He would instantly come for help of his staff physically, financially and morally.

Dr A.P. Ray was elected Fellow of the National Academy of Science, Fellow of the Indian Society for Malaria and other Communicable Diseases, Fellow of the National Academy of Medical Sciences and Fellow Royal Society Tropical Medicine and Hygiene. In 1967, he was honoured by the President of India with Padma Shree. In 1968 he was offered the post of

Dean and Director, International Malaria Training Centre under the World Health Organization at Manila, Philippines. In 1970 he was promoted as the Senior Malaria Adviser, Western Pacific Regional Office, WHO, Manila. He remained on this post up to 1972. After his superannuation from WHO, he became Director, Malaria Research Centre, Ingleburn, Sydney. In 1974 WHO honoured Dr A.P. Ray by awarding him the Darling Foundation Prize. He was honoured with Geoffrey Harkness Medal in Australia. He left Australia in 1978 due to illness of his wife, and back home he served as the Chief Co-ordinator of the *Plasmodium falciparum* Containment Programme (*Pf*CP) from 1979-1987. He was the longest serving Member of the WHO Expert Committee on Malaria for 32 years. He passed away on 24th September, 1996 after a brief illness at his home. Dr A.P. Ray had become a legend in his life time as an incomparable malariologist.

One of the stalwarts in the science of medical entomology, malariology and arbovirology in post-independence period, **Dr T. Ramachandra Rao**, son of Shri Thammajee Rao and a member of the Mysore State Civil Service, was borne on 9th October, 1907. Having completed his early education and graduation in science from the famous Central College, Bangalore, young Ramachandra Rao moved to Calcutta University where he earned his M.Sc. in Zoology, with distinction. Subsequently, he took up lectureship at the Mysore Medical College for a couple of years, however all the while concentrating on further higher education. As a result, in 1951, he was bestowed with the degree of Doctorate in Science from his alma mater, the Calcutta University. After completing education in Bangalore, Karnataka he earned a Rockefeller Foundation Fellowship in 1936 and worked under the guidance of Prof. P.A. Buxton at the famous London School of Hygiene and Tropical Medicine. He started his research career withy interests in the insect cytogenetics, mostly working on grasshoppers. Soon, however, with the ever mounting importance of the medical parasitology or entomology, he before long joined Dr Paul Russell, a world renowned malariologist, in 1932, and undertook investigations on mosquitoes' bioecology and control. The team of Dr Rao and Dr Russell conducted a very large number of innovative and pioneering investigations, particularly on the main malaria vector, *Anopheles culicifacies*, in various different States in the country endemic or hyperendemic for malaria. His love for malaria control brought him as an entomologist to the newly established Malaria Organization of Bombay State (1942), where he amply shone his remarkable mosquito/malaria-based knowledge and got promo-ted to the rank of Deputy Director of Public Health of the Maharashtra State, and in-Charge of Malaria and Filariasis. Here, he had another marvellous opportunity to prove his grip on mosquitoes, and, with Dr D.K. Viswanathan, carried out numerous studies on the bionomics and control of *An. fluviatilis*, another important vector of malaria, in the hilly and sylvan environs. The early breakthrough came in his life when he demonstrated effectively the usefulness of DDT in controlling malaria through indoor residual spraying in rural India. These investigations on DDT application against malaria mosquitoes became the hallmark for both the National Malaria Control Programme (1953-1958) and the National Malaria Eradication Programme (1958-1999), besides fetching him international acclaim.

A man of great natural instincts, he never seized to learn new things in biomedical research. Therefore, when Virus Research Centre (now the National Institute of Virology) at Pune was established, under joint collaboration of the Indian Council of Medical Research and the Rockefeller Foundation, he was naturally the first choice to head this premier institution. There he was the Director between 1961 and 1970, a position that he held till his retirement. An able medical biomedist and an astute administrator, Dr Ramachandra Rao developed the VRC into an international reference institute on virologic research. A vastly travelled scientist, he was a member to several committees of the World Health Organization, particularly on matters relating communicable diseases such as malaria, insecticides, vector biology and control, and virus diseases. While, he served as a consultant and an advisor to the Indian Council of Medical Research for 35 long years, the WHO also availed his vast knowledge and expertise as the Team Leader in Afghanistan and Tanzania. He also served as the Officer on Special Duty with the ICMR. He was a Fellow of some highly reputed international societies, besides having chaired as President of Zoology and Entomology Section of the Indian Science Congress Association in 1972. His magnum opus, "*The Anopheline of India*" (1984), is a major reference work on the subject for the Indian researchers. Dr Ramachandra Rao was an epitome of hard work and dedication throughout his life, albeit failing eye-sight in the wee years. He was ever vigilant and keenly interested in all malariological researches in the country till his last breath, on 8th November, 1984. I have a vivid memory of Dr Ramachandra Rao, whom I met in the then ICMR-Malaria Research Centre's Director, Dr V.P. Sharma's Chamber, who introduced me to Dr Rao and the latter took the opportunity to give me my first malariological lesson. "Think, Dr Tyagi, how you can eradicate this menace, malaria." This, of course, became my guiding force all these years and my own malaria research in the tribal southern Gujarat, the Thar Desert in the north-western India and the difficult terrains of tribal Koraput, Kalahandi and Malkangri districts of Odisha (earlier Orissa) in eastern India instilled in me an invaluable knowledge of the subject. During past nearly thirty five years I have incessantly worked on malaria, filariasis, dengue, Japanese encephalitis, leishmaniasis, chikungunya and scrub typhus in more than a dozen States and Union Territories in the country, but all along my long, chequered and cumbersome journey I had never forgotten the 'essence' of his advice.

India is blessed with an array of highly luminescent entomologists many of whom rose to the enviable international fame and virtually became most dependable authority on their subject of study and **Dr M.O.T. Iyengar** was such a shining star in our galaxy of outstanding entomologists who by sheer virtue of their rare and energetic character not only practically educate the budding researchers in habits of industry, but by the example of diligence and perseverance which they set before them, largely influence the scientific activity in all directions and contribute in a great degree to form the national character. The national progress, it is aptly said, is after all the sum of individual industry, energy and uprightness – traits which Dr M.O.T. Iyengar most profoundly exhibited in the glorious annals of Indian entomology. His life presents a glimpse of the rare human attributes of perseverance and application which invariably make an entomologist a great scientist in both the personal and professional life. He was an entomologist of international repute and contributed to the development of medico-entomological disciplines a very great deal. He had studied entomology both in laboratory and field, extensively working in far off and difficult conditions – often setting up field laboratory in make-shift arrangements with minimal tools and assistance. His greatest accomplice was his determination and whatever he resolved to achieve in science he had left no stone unturned and kept making efforts till he got the desired results from his long and tiring experimentations! He never left on tomorrow the work of today and would see to it that all data are properly recoded on data-sheets for analysis. Such was the devotion of Dr M.O.T. Iyengar to medical entomology that even the best of world's entomologists conformed, without an iota of skepticism, to his observations.

Dr M.O.T. Iyengar had many great virtues but one that can be quoted here for the youngsters to emulate is that he always first accomplished his task, with all information duly obtained through rigorous path of examining and re-examining the datasets, then would speak of his achievement – a characteristic that the budding scientists should invariably emulate in their own scientific pursuit. As a principle, all genius men and women of science do not necessarily do extraordinary things right away, but they certainly do things, even the most trivial and insignificant looking

Fig. 8.5. (a) Dr M.O.T. Iyengar, (b) Dr Jaswant Singh, (c) Dr T.N. Ananthakrishnan, and (d) Dr S. Pattanaik. (*Source:* Free Wikipedia/Dr B.K. Tyagi's personal archive)

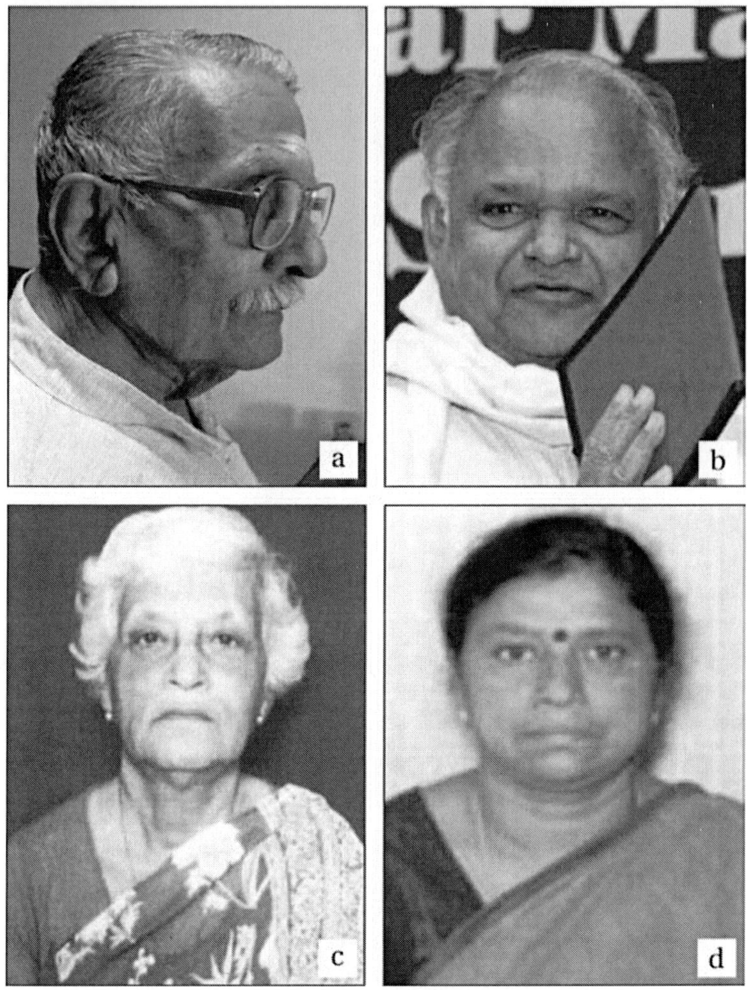

Fig. 8.6. (a) Dr P.K. Rajagopalan, (b) Dr V.P. Sharma, (c) Dr R. Reuben and (d) Dr Sarala K. Subbarao. (*Source:* Free Wikipedia/Dr B.K. Tyagi's personal archive)

matter, in a most extraordinary way. That is what makes them genius, and Dr M.O.T. Iyengar was such an extraordinary genius entomologist. Dr M.O.T. Iyengar (affectionately known as "Thinnani" to his close friends and relatives) was born in Madras (now known as Chennai) on 6th February 1895. He came from a very illustrious family. His father was a senior judge in the then Madras Presidency High Court of British India. His elder brother, Professor M.O.P. Iyengar (1886 - 1963), a prominent botanist and physiologist at the Presidency College, University of Madras, who researched different types of algae in relation to their structure, cytology, reproduction and taxonomy of major groups, and, for his excellent research contribution to algal science, was also referred to as the "Father of Algology in India".

After his early education at Hindu High School, Triplicane, Madras, and his graduation from Presidency College, Madras, Dr M.O.T. Iyengar distinguished himself as a medical entomologist. His main interest was the study of mosquitoes, but he was equally interested in other insects of medical importance. These included infestations of human intestines by corpid beetles, a study of the life history of the tinged bug and the systematics of rat-fleas in India. He was also a keen botanist. His studies on dimorphism of the flowers of two species of *Monochoria* as an adaptation for cross-pollination by insects is still very favourably regarded by many botanists.

Dr M.O.T. Iyengar began his career as an Entomologist in-Charge of Bengal Malaria Research Laboratory, Calcutta, India (1918-1931 & 1935-1947), but intermittently also took up to the post of the Instructor on Medical Parasitology and Entomology to D.P.H. students (1920-1922), Professor of Medical Entomology, School of Tropical Medicine, Calcutta (1922-1923), and Medical Entomologist to the Government of Travancore under the auspices of the Rockefeller Foundation (1931-1934). Of particular mention is his vocation in Travancore State (now known as Kerala) where he studied the incidence and distribution of filariasis. The results of his studies showed (i) the distribution of *Wuchereria bancrofti* and *W. malayi* infections in this southern state of India, (ii) their endemicity, (iii) factors that determine this endemicity, and (iv) methods for their control. The correlation between the infection rate and the disease rate, and the supposed antagonism between malaria and filarial infections are discussed in his ICMR *Memoir* published in 1938. This publication is still regarded as a text book on filarial epidemiology. At the young age of 33, he was awarded the prestigious Earl of Minto Gold Medal for medical research in India, in 1929. By 1934, his fame as a medical entomologist was well established and he was invited to many universities and research organizations in Europe. During the two years he spent in Europe he was associated with leading entomologists and parasitologists like Prof. P.A. Buxton (of London School of Hygiene and Tropical Medicine), Prof. W. Schuffner and Dr S.L. Brug (of Institut voor Tropische Hygiene, Amsterdam, Holland), Prof. Emile Brumpt and Dr Henri Galliard (of Institut de Parasitologie, Faculte de Medecine, Paris), Prof. Swellengrebel (of Institut fur Schiffsu. tropen-krankheitan, Hamburg, Germany) and Mr P.G. Shute (of Malaria Reference Laboratory, Horton Hospital, Epsom, England). In 1939 he obtained his Degree of Doctor of Science from the University of Madras.

Dr M.O.T. Iyengar was with the Government of West Bengal for nearly three decades, initially as Entomologist, Bengal Malaria Research Laboratory, and later as Malaria Officer. After highly celebrated works of Stephens and Christophers in the early 1900s, it was Dr Iyengar who was responsible for the surveys of malaria in Calcutta and its environs. His papers on the incidence of malaria in deltaic Bengal and methods for the control of the disease and problems related to it, still hold good to this day. He was an ardent advocate of biological control of mosquito larvae. Two fungi of the *Coelomomyces* group described by him have been successfully used in the control of anopheline larvae in India. The fact that one species of a mermithid nematode, *Romanomermis iyengari,* and a culicine mosquito, *Culex iyengari,* have been named in his honour, is a good testament of his huge popularity. *Romanomermis iyengari,* in particular, has been shown to be an effective parasite of mosquito larvae in Brazil.

During his long and challenging career Dr M.O.T. Iyengar shouldered several positions of great responsibilities. Between 1941 -1944, he served as Honorary Adviser on malaria control to the US Army, British Land Forces and Royal Air Force in Bengal, and as Malaria Officer to the Government of West Bengal from 1947 to 1950. From 1949 to 1952, he was a Consultant on Filariasis to the World Health Organization and had assignments in Ceylon (now Sri Lanka), Maldive Islands, Thailand and Afghanistan. His love and dedication for tropical medicine was unmatched and would never give a second thought whenever such an opportunity came along his way. Therefore, when soon after he was awarded a decoration by the Public Health Services of the Government of Thailand in recognition of his researches on filariasis in Thailand in 1952, he was asked by the South Pacific Commission (SPC), in 1953, if he would accept the position of Research Specialist on Mosquito-borne Diseases, Dr Iyengar instantly agreed whole-heartedly for his love for research knew no bounds. He also knew that while in the South Pacific he would be associated with some great French epidemiologists like Dr Guy Loisson, Dr Emile Massal and Dr J. Kerrest – a judgement which later proved to be a great boon for Dr Iyengar in his own scientific pursuits.

The South Pacific Commission was founded in 1947 by six countries, *viz.*, Australia, France, The Netherlands, New Zealand, the United Kingdom and the United States of America (Netherlands and the United Kingdom have since withdrawn). It is a regional inter-governmental organization whose membership includes both nations and territories. The Commission's purpose is to advise the participating Governments on ways of improving the wellbeing of the people of the Pacific islands in health, economic and social matters. Altogether twenty-two governments and territories are involved with approximately five million people scattered over some thirty million square kilometers. Less than two per cent of the region is land. Dr Iyengar's task was to compile an annotated bibliography on filariasis, elephantiasis and related aspects and circulate to member Governments, territories and others concerned with these problems. This involved the collection of papers, either as reprints, photocopies or micro-films, published in journals or periodicals throughout the world but pertaining to filariasis in the South Pacific Region. That this was achieved by Dr Iyengar who reviewed all the papers and completed the abstracts will be seen from his publications in Part-I: Epidemiology of filariasis, Part-II: Studies on mosquitoes in the South Pacific Region, Part-III: Symptomatology, etiology, pathology and diagnosis of filariasis, Part-IV: Treatment of filariasis, and Part V: Prophylaxis and control of filariasis. Over 2000 published articles by over 1000 authors were reviewed and abstracted in his 7-year stay with SPC – a truly remarkable piece of work! In addition, Dr Iyengar carried out filariasis surveys in 8 of the 22 territories and recommended to the governments concerned measures for the control of the disease. In these investigations he was ably assisted by his technical assistant, Mr M.A.U. Menon.

Even after his return to India, Dr Iyengar continued to write about filariasis in the South Pacific - such was his passion for research! In a paper published in 1965, he wrote about the two biological races of *Wuchereria bancrofti*, the periodic and the non-periodic. Each of these two races is further divided into two zones depending on the vectors but no difference in morphology or in pathogenecity was found. The evolution of the Micronesian strain of periodic *W. bancrofti* was also discussed. Dr Iyengar was an epitome of devotion, determination and discipline. His sense of promptness and sincerity was superlative. In the early fifties he was asked by the World Health Organization to study and report on the incidence of filariasis in the Maldives Islands. He was accompanied by his two assistants, Dr M.I. Mathew and Mr M.A.U. Menon. A 60-ton schooner was placed at their disposal by the ruler of Maldives Islands and it served as their home and laboratory. After completing his investigations in four months, they set sail to return to India. Halfway through, a cyclone hit the schooner. Tossed around for two days, the schooner was swept ashore in the island of Seychelles in East Africa. This near-fatal incident did not deter Dr Iyengar and after hardly four months elapsed he soon was set on accomplishing another assignment to Thailand for WHO! Dr Iyengar loved teaching and interacting with young scientists was his great motto. As a teacher and mentor, he guided students through their DPH and DTM&H

courses at the Calcutta Medical College and the Calcutta School of Tropical Medicine. He was a tutor. mentor and inspiration to many entomologists, parasitologists and epidemiologists, and his work on malaria and filariasis will always remain as a monument. It is felt that this publication would have served its purpose if it inspires students, teachers and research workers to follow in his footsteps and excel in their vocation. Dr M.O.T. Iyengar passed away on 16 September 1972, and in him the world lost a great scientist. His wife, Mrs. Rukmani Iyengar, instituted through the Indian Council of Medical Research, "The Dr M.O.T. Iyengar Memorial Award" which is given annually to scientists who have made outstanding contributions in the fields of malaria, filariasis, plague or medical entomology. *I was one of those few Indian scientists who were honoured with this prestigious award, in 2008.*

Before independence, medical and health services in the country were basically managed by the British officers. After independence, in 1947, the entire responsibility fell on the shoulder of Indian medical and health personnel. During British rule, curative and preventive services ran separately. Due to resource and personnel constrains, preventive services were at disadvantage. With the growing influence of specialisations in clinical subjects, manpower development for preventive medicine was hampered. There was hardly any organized body of scientists/professionals in the field of Public Health. In this scenario, the 'Indian Society for Malaria and Other Communicable Diseases' was born in 1950 with the efforts of **Lt. Col. Jawsant Singh**, the then Director of Malaria Institute of India, Delhi. Lt Col Jaswant Singh (*b.* 4th December, 1901) did his MB and ChB from University of Edinburgh, DTM&H from University of London and DPH from England. He remained Director, Malaria Institute of India (now National Institute of Communicable Diseases, Delhi). Dr Singh made some significant contributions to malaria studies in India, which included malaria survey, laboratory techniques used in malaria diagnosis and control, and chemotherapy of malaria with antimalarials such as atebrin, paludrine, resochin and camoquin. In conducting a 'screening test' for malaria by using the plasmodial infection of monkeys, Singh opened up a new line of study of chemotherapy with antimalarial drugs. He also did some work on synthetic insecticides in relation to vector control. He is immortalised by the introduction of 'JSB Stain' used to colour the malaria parasites in blood taken for microscopic slide preparation and observation. As Director of Malaria Eradication Training Centre (WHO) in Kingston, he trained senior professionals and advised the Indonesian government in setting up a Malaria Institute in that country. He was a Fellow (1956) of the Indian National Science Academy. He was also Member of the American

Society of Tropical Medicine, and Hygiene WHO Expert Committee on Malaria.

There are always some great scientists who were though not directly working on any mosquito or mosquito-borne diseases public and/or animal diseases yet they, by sheer their guidance and mentoring nature, imprinted indelibly a gigantic impression over the minds of brooding mosquito scientists for their infallible future principles of systematic and thorough investigations which became the hallmark of quite many Indian culicidologists of the current generation. In this context, he did carry out some very interesting studies on interaction of mosquito larvae and various plant species with reference to understand plants' role either as a repellent or attractant. He is the greatest ever Indian entomologist **Professor Dr T.N. Ananthakrishnan**, an utterly indefatigable entomological genius *nonpareil*, whose life truly reflects the spirit of exceptionally diverse and deep knowledge of nearly all the biological sciences, though entomology has been his greatest forte. He made innumerable outstanding contributions to the understanding of insect (or, in view of vast phyletic coverage, the arthropod) taxonomy, insect-plant interactions and the application of such knowledge for developing crop protection strategies. One of his many phenomenal contributions following four decade-long epoch making researches on thrips (Class: Insecta; Order Thysanoptera) has been to the basic understanding of 'insect-weed-crop interactions' which has been extended to the field for increasing agricultural productivity all over India. The diversity, versatility and magnificence of Prof. Ananthakrishnan's scientific work can be measured, if any indication, by his more than 400 research papers in a large number of national and international journals plus 30 books written by him which are significant landmarks in the evolution entomological sciences. He described new species in many invertebrate groups of animals, but most of these (>200) come from his favourite, Thysanoptera or thrips - the tiny insects that cause severe damage to plant tissues often by developing galls or other deformities, but some have been also recorded incidentally biting humans! He impressed the world authorities by the scope of his contributions and the depth with which he has analysed an almost infinite variety of the thrips. His steady investigative activities have resulted in remarkable eclectic publications that scan the biological kingdom and include taxonomy, systematics, morphology, bionomics, behaviour, distribution, reproductive biology, chemical cues and ecology *etc.* It is no exaggeration, therefore, when the world famous entomologist Professor Douglas Whitman of the Illinois State University, USA referred to him as *"a brilliant biological treasure."* A great visionary he exhibited a comprehensive world-view of entomology even at that time when little was known about thrips' impact on agriculture entomology wherein he has made a conspicuous and indelible mark to be emulated by the thousands of young entomologists in India and elsewhere.

Prof. T.N. Ananthakrishnan was born on 15th December, 1925, at Mangalore. He did his high school classes at St Aloysius College High School from 1938 to 1941 and intermediate studies at Government Victoria College, Palghat, between 1941 and 1943. He then moved to Madras Christian College, Tambaram for his BSc (Hons.) degree course in Zoology during the tenure of Dr Alexander John Boyd as principal. He won the Buckie Fellowship for pursuing his Honours course. It was at Madras Christian College that his passion for intensive and extensive field work began to take shape under encouragement of Dr John P. Joshua who took him out periodically for field work, in addition to annual trips to the Gulf of Mannar. He passed his BSc (Hons.) in 1946 and Dr Boyd offered him Demonstrator's post on a monthly salary of Rs. 47.50. He has by this time resolved to research, which ultimately remained with him as a mission and passion throughout his career, and even today. Only a few months ago he, with the support of his numerous colleagues, students and scholars, laid foundation of the country's first such entomological society, the *ENTOMOLOGY ACADEMY OF INDIA*, which is truly represented democratically by all the major disciplines of the day, including 'medical arthropodology' – all for the cause of creating a healthy and sustained entomological research activity in India. In his early research days, Prof. Ananathakrishnan was acquainted to thrips by Dr T. V. Ramakrishna Iyer, the great Indian entomologist and an international expert on thrips. Impressed by the ceaseless energy, perseverance and inquisitiveness of the ever agile youth in Dr Ananthakrishnan, Dr Ramakrishna Iyer gave him his lifelong collection of identified and invaluable material for reference. Dr Ananthakrishnan became a temporary lecturer in zoology at the Madras Christian College in 1947-48, during which period Mr. Samuel Raj, his senior colleague, guided him in the field work. During weekends both of them used to walk from Tambaram to St. Thomas Mount and back, collecting insects, so much so that Dr Ananthakrishnan became conversant with the more common insects. Together they spent late hours every day studying the collected material. They also used to visit the nearby Vandalur Hills which was known for its exquisite flora and fauna in those days. His interest in botany was kindled and augmented by two very brilliant teachers Prof. K. R. Venkatasubban and Dr K. R. Rangaswamy Iyengar. In 1948, Dr Boyd recommended the name of Dr Ananthakrishnan to Fr Jerome D'Souza who was then the principal of the Madras Loyola College. Thus at the age of 23 he began his career as lecturer in Zoology and this was the beginning of the Zoology section of the Natural Science Department in Loyola College with Fr Rapinat, a great botanist who interacted with Dr Ananthakrishnan in many ways. In this phase of his career, Dr Ananthakrishnan was greatly influenced by Fr Lourdu Yeddanapalli, Prof. Ekambaranathan, Dr Rao Bahadur Y. Ramachandra Rao and Prof. M. S. Mani. He had the good fortune of assisting Prof. Ekambaranathan in bringing out two volumes of 'Manual of Zoology' which became the only classic Indian work in zoology for the BSc students for nearly two generations. Thanks to the British Council, he visited the British Museum of Natural History, London and several overseas universities, besides his memorable visit to the Mecca of Thysanopterists, LINZ in Austria where he had occasions to study thrips under Prof. H. Priesner, the greatest among Thysanopterists.

In 1957, at a young age of 32, Dr Ananthakrishnan became a professor of Zoology at Loyola College. Growing under the careful sight of the great academician Dr Lakshmanasamy Mudaliar, who was then the Vice Chancellor of Madras University, Dr Ananthakrishnan was able to complete his M.Sc., Ph.D. and D.Sc. degrees from the University of Madras within a short span of 6 years! In 1960 he went to England for studying thrips at the British Museum of Natural History, London. This was the beginning of his association with experts in his field from different parts of the world. A versatile and dynamic orator, Dr Ananthakrishnan was always the most sought after entomologist to offer lectures - in as many as 22 universities in the USA and in UK, France, Germany *etc.* In 1964 he relinquished his position as Professor and Head of the Department of Zoology to start the 'Entomology Research Institute', his brainchild and a future *alma mater* of a large number of entomological geniuses. From then on the Centre developed into the Entomology Research Institute in 1985 under his outstanding scientific leadership. By his tireless exertions Dr Ananthakrishnan has developed this Institute into a research Institute of national and international significance. He has successfully guided 48 students to their Ph.D. degree in as diverse subjects as chemical ecology of insect-plant interactions, insect-weed-crop interactions, cecidiology, bioecology, insect taxonomy and systematics, and biodiversity. With so much of academic and scientific responsibilities and commitments, Dr Ananthakrishnan, being himself an avid cricket lover, never misses an opportunity to buck up the Team India against their opponent at the famous Chepauk stadium!

An internationally reputed entomologist and zoologist, Professor Ananthakrishnan has a highly chequered professional career, both as a scientist and a teacher, spanning over a period of nearly 60 years during which time he successfully had a host of coveted positions including, to mention a few just to illustrate, the Professor and Head of the Department of Zoology, Loyola College, and Founding Director of the Entomology Research Institute located at the Loyola College campus at Chennai, and the Director of the Zoological Survey of India at Kolkata. Of many significant advancements that Dr Ananthakrishnan had brought about in the scientific area of various institutes and organizations he has been associated with, perhaps the most glittering example hails from his tenure as the Director of Zoological Survey of India where he introduced ecologically oriented programmes through initiating four new research stations, the Freshwater Research Station at Hyderabad, the Western Ghats Station at Calicut, the Estuarine Research Station at Berhampur and Sunderbans Station in West Bengal. Efforts were made during his period as Director to have an integrated approach through collaborative efforts with other national institutions. He is also a keen observer of Nature and, therefore, the woold's highest nature conservation body, the International Union for Conservation of Nature & Natural Resources (IUCN)/Species Survival Commission (SSC) conferred on him the *Certificate of Appreciation* in recognition of his years of important work, in 2004. As a member of the Menon Committee of the Government of India on the Silent Valley, he visited various parts of the valley on a fact finding mission and was involved in several scientific discussions relating to the ecological and biosystematics aspects of the area, which was subsequently saved.

He was a Fellow of the Indian National Science Academy, Indian Academy of Sciences and the Indian Academy of Agricultural Sciences. He, along with Prof. M.S. Swaminathan, is also the patron of the recently founded Entomology Academy of India. He has won many awards, most notable of which are Kidwai Memorial Award of ICAR (1972), Jawaharlal Nehru Fellowship (1980), INSA Senior Scientists' Award (1987), Pitambar Pant Award of the Ministry of Environment and Forests (1986), Indian Ecology Society (1998), Plant Protection Association of India (1999), J.C. Bose Award of Indian Science Congress (2000) and Sundarlal Hora Medal of INSA (1999) and the life time achievement Award of Monsanto Research Centre, Bangalore (2002). For several decades he had been on the International Advisory Board of the Annual Review of Entomology, USA; Bulletin of Entomological Research, London, besides being on the editorial boards of leading biological journals in India including *Entomon* and *Oriental Insects.* In addition he had served on national research advisory committees/Task Force of ICMR, ICAR, CSIR, UGC, DST, DBT and Ministry of Environment and Forests. For many years during 1980s, Prof. Ananthakrishnan had been the Chairman of the Scientific Advisory

Committee of the Centre for Research in Medical Entomology (ICMR). A widely travelled scientist across the world, he had by his visiting lecture assignments enriched entomologically the varied audiences with his vast knowledge and matchless articulation. In spite of his international stature of a scientist, Prof. Ananthakrishnan was foremost a teacher.

He passed away in peace in 2015, with all the family members by his side, in the USA. Generations of students and younger scientists hold him in veneration and admiration as a 'father figure' who has rare qualities of mind and heart. Many owe their career in the field of science to his personal example as a scientist and teacher. For the world, he was a living academic statesman of global significance in the field of entomology. For him aiming for achievements has been everything and never even once the thought about success could penetrate his mind. Medical entomologists got his advice on their wish and were always encouraged by him.

Born 23rd October 1926 as the third child among seven siblings to the illustrious parents, Shri Madhusudan Pattanayak, a leading legal practitioner at Cuttack High Court, and Smt. Durgabati Devi in the holy city of Puri, **Dr S. Pattanayak** graduated in medicine in 1950 from SCB Medical College, Cuttack. He acquired Diploma in Tropical Medicine from STM, Kolkata in 1951-52 and MPH from the School of Hygiene, John Hopkins University, Baltimore, USA. He underwent training in insect-borne diseases from US Public Health in 1957 and in the following year, he had the basic grinding in malariology training under eminent teachers like Dr A.P. Ray, Dr S.P. Ramakrishnan, Dr I.M. Puri etc. at Malaria Institute of India (presently NCDC). His initial posting was Assistant Research Officer, ICMR, Delhi from 1952-56. He moved to Malaria Institute of India in 1956 and steadily climbed the rungs to the post of Deputy Director (Epidemiology) which he held from 1966-74. He developed a strong team of epidemiologists on communicable diseases. He became Director of NMEP (presently NVBDCP) in 1974 and continued till 1981. Under his able stewardship, Modified Plan of Operation was introduced to contain morbidly and mortality due to resurgent malaria. Although he was a medic yet his control on associated subjects such as medical entomology and vector-borne diseases was infallible. After successful stint in the national programme, he moved to international arena when WHO appointed him as team leader to contain malaria in Burma (presently Myanmar) from 1981-84. The world body adroitly utilized his expertise in the containment of communicable diseases in different capacities such as Regional Advisor to the Regional Director, SEARO, Short Term Consultant for many research projects covering AIDS, Epidemiological Surveillance System, Hepatitis B, etc, and Acting Director (PCD). He was Member of W O Global Advisory Committee on Malaria (1978), Member of 17 WHO Expert Committee on Malaria (1979) and many other expert committees and assignments spanning a period of over two decades. He was a towering personality among stalwarts in the field of public health and communicable diseases in the Southeast Asia Region. He contributed many scientific publications in journals of high repute on epidemiology and various communicable diseases. He was chairman for many national committees with far reaching recommendations in the containment of malaria, filarial, plague and many other communicable diseases. He was elected as Fellow of ISMOD, IAE, MAMS, ISME and IMSA. He was a great mentor, proficient moderator and superb manager in guiding activities of ISMOCD and IAE. He was a constant source of inspiration and encouragement to the young scientists. He died on 26th Nov 2012. In recognition of his valuable contributions in the field of communicable diseases, he was awarded WHO's Darling Foundation Prize and Medal. He was also the first recipient of ISMOCD Award and Citation in 1996 in view of his contributions in public health and control of malaria in particular.

In modern times if there would be a recounting of rare entomologists/ vector-borne disease specialists, then, in addition to fewer like Dr T.R. Rao, names of Dr V.P. Sharma and Dr P.K. Rajagopalan will appear foremost. Though of contrasting characters, they are both genius *nonpareil* in every sense. **Padma Bhushan Dr Vinod Prakash Sharma** was unquestionably the greatest malariologist so far in the independent India, standing tall in the same extraordinary league of as great malariologists as Dr Ronald Ross, Dr McDonald, Dr T.R. Rao, Dr C.F. Curtis, who convincingly put his malariological research through 'Gandhian Philosophy' using environmental means without harming the environment on the international front! Dr Sharma was a scientist of international repute known for his path-breaking work and contribution in the field of basic and applied research of malaria and vector biology. An astute nature conservator and a leader *par excellence*, he demonstrated to the world as to how the malaria could be controlled without resorting to the toxic insecticides. He was a scientists' scientist and would think not an iota less than perfection in all his undertakings. Dr Sharma was active till his last day taking care of his duties as the Meghnad Saha Distinguished Professor at the Centre for Rural Development and Technology, Indian Institute of Technology, Delhi. He was also spearheading the "Safe Water" campaign of the National Academy of Sciences, India launched during the Year of Science. His work on popularization of science, creating awareness and science education programmes of NASI are widely appreciated including in very remote areas.

Born on April 6, 1938 in Bulandshahr district of Uttar Pradesh in India, Dr Sharma did his early education from his native place. He subsequently joined Allahabad University from where he secured M.Sc. in 1960 and D.Phil. in 1964. In 1965, he left to South Bend, Indiana, USA and joined the University of Notre Dame as a postdoctoral research associate in the laboratory of another Indian great, Prof. Karamjit Singh Rai, but later shifted to Purdue University, USA. Sharma returned to India in 1968 and again joined Allahabad University to complete his D.Sc. in 1979. Dr Sharma started his career at the world famous Forest Research Institute & Colleges, Dehradun, as a Pool Officer in the Entomology Branch headed by Dr P.K. Sen-Sarma, another one of the greatest entomologists India ever had, in 1969. A year later, in 1970, he joined the WHO sponsored project under the aegis of Indian Council of Medical Research, as a Senior Scientist on genetic control of culicine mosquitoes, where he worked till 1975. When this project was abruptly aborted he joined ICMR's newly established Vector Control Research Centre and soon after his *'karma-kchetra'* Malaria Research Centre (rechristened recently as National Institute of Malaria Research) in 1976, as Deputy Director. He became its first Director, a post from where he retired, in 1998, with a lofty decoration of Additional Director General of Indian Council of Medical Research. With over forty years of research and field experience, Dr V.P. Sharma was considered to be an authority on malaria and vector biology. He has been credited with many scientific and research achievements. His studies on the chemo- and radio- sterilization of male mosquitoes are well documented. He is credited with developing a new technique for sex separation of mosquitoes. He is also acknowledged for his contributions to vector biology. A versatile orator and writer (>400 scientific papers in reputed national and international journals, about two dozen books, both in Hindi and English), he also extensively contributed to the popularization of science, through his books, journals and education programs, TV and radio talks. But, possibly his greatest contribution has been the unfailing demonstration of bio-environmental malaria control as a certain eco-friendly, community driven and economical possibility which led to many innovations in the subject. It was he who conceptualized a novel bioenvironmental malaria control strategy as an alternative to spraying chemicals and successfully demonstrated malaria control in various eco-epidemiological settings. As a result of his indefatigable efforts, today we have technologies to fight malaria which are environmentally safe, cost effective, produce sustainable impact on the society and are free from ecological hazards that are often associated with DDT and other insecticides.

Dr Sharma held many positions of importance, such as the Chairman of the WHO's Roll Back Malaria initiative, by way of recognition and those of responsibility like Member, WHO Expert Committee on Malaria, throughout his career. Dr Sharma was a fellow of all the three prominent Indian science academies, *i.e.*, Indian National Science Academy, New Delhi; Indian Academy of Sciences, Bangalore; National Academy of Sciences, India (Allahabad), besides several other learned national as well as international societies such as Royal Asiatic Society, Entomology Society of India, New Delhi, Fellow of National Environmental Science Academy, New Delhi, Fellow of Indian Society for Parasitology, Fellow of Indian Society of Malaria and other communicable diseases and Zoological Society of India. He was revered by one and all for his authenticity and clarity of mind, and therefore also acted as the President of some of the prestigious associations like National Academy of Vector borne Diseases and National Academy of Sciences (India) Allahabad. Dr Sharma dominated the malariological researches in India for more than three decades so much so that he was crowned with the numerous and the most prestigious honours and awards that any other contemporary scientist in the field of vector-borne diseases can hardly imagine of; Padma Bhushan, Gujar Mal Modi Award, Meghnad Saha Distinguished Fellowship, Padma Shri, WHO Darling Foundation Prize, Government of India Inventions Award, Chancellor's Prize, Dr M. O. T. Ayengar Memorial Award, B. R. Ambedkar Centenary Award, Om Prakash Bhasin Award, Ranbaxy Award, M. L. Gupta Trust Award, FICCI Cash Award, Vaswik Award, Green Scientist Award, Best Scientist Award, G. P. Chatterjee Memorial Award, Dr R. V. Rajaram Oration Award, Dr U. S. Srivastava Memorial Lecture Award, Professor L. S. Ramaswami Memorial Oration Award, Life Time Achievement Award by the Indian Society of Malaria and other Communicable Diseases, B. N. Singh Oration Award, B. K. Srivastava Oration Award, Distinguished Parasitologist-World Parasitologist Federation (WPF), Gold Medal lifetime Achievement Award – Indian Academy of Environmental Sciences.

Dr V.P. Sharma died in 2015, after a brief illness. Dr Sharma will be always remembered for his innovativeness, liking for indigenous methodologies and emphasis on involving community participation at all steps of translational malaria research so as to reduce the gap between research and their end users! *To me, he was always very special, and highly respected, as he was the one who brought me to the Indian Council of Medical Research via a WHO/TDR funded research project at the ICMR-Malaria Research Centre, in 1981.*

Parallel to this fame and brilliance in medical entomology, malariology and filariology is **Padma Bhushan Dr Payyalore Krishnaier Rajagopalan**, the former Director of the ICMR-Vector Control Research

Centre, Puducherry, who has spent all his professional life fighting the mosquito menace in India and striving towards eliminating vector-borne diseases. He was born as the second son to P.R. Krishnaier and T.A. Gowri in Mukteswar, a hill station in the Kumaon Hills (now in Uttarakhand), on October 27, 1930. Young Rajagopalan grew up in present-day Palakkad under the care of his maternal grandfather. After secondary schooling, Rajagopalan moved to Varanasi in 1945 to study at the Banaras Hindu University. He completed an MSc in zoology with entomology as specialisation in 1951. In 1953, he joined the Rockefeller Foundation's Virus Research Centre (or VRC, which was renamed the National Institute of Virology in 1978), Pune, as a research assistant. He rose through the ranks to become Assistant Director in early 1970. He was appointed Deputy Director of the Research Unit on the Genetic Control of Mosquitoes (RUGCM), New Delhi, which was an initiative of the World Health Organisation (WHO)/Indian Council of Medical Research (ICMR). He was at this post until 1975. He was then appointed Director of the Vector Control Research Centre (VCRC), which was newly started under the aegis of the ICMR. For 15 years, Rajagopalan led the VCRC on a trail-blazing path in the battle against mosquitoes.

While Dr Rajagopalan was at the VRC, the Government of India deputed him for advanced training in medical epidemiology to the University of California Berkeley where he pursued a master's in public health. While at the university's Department of Epidemiology in the School of Public Health, Rajagopalan was trained by William C. Reeves, who considered it critical for him to get training on malaria from Lewis W. Hackett of the International Research Division of the Rockefeller Foundation and on zoonotic diseases from Karl F. Meyer of the Hooper Foundation for Medical Research because he came from a developing nation. Reeves also arranged for Rajagopalan to gain experience at his encephalitis field laboratory in Bakersfield, California. Rajagopalan cherishes what he learned while in the United States about the value of hard work, honesty and research ethics. In 1965, Rajagopalan completed his PhD at the University of Poona (now the Savitribai Phule Pune University). He was fortunate to come under the tutelage of the pioneering Indian medical entomologist Dr T.R. Rao, who trained him in sampling mosquito ecology and generously enabled his scientific encounters with the mosquito and its biology in the 1950s. At the VRC, Rajagopalan was charged with the task of dealing with the then newly recognised viral disease Kyasanur Forest disease (KFD) in Hassan district of Mysore State (now Karnataka). Whether the virus was transmitted by migratory birds from Siberia was a daunting question that had to be answered. Over the course of a two-year investigation, Rajagopalan examined about 8,000 birds and extracted some 10,000 ectoparasitic ticks but, curiously, found no exotic ticks. Tests of the blood sera from the birds at the VRC disclosed a cross reaction to the Russian Spring-Summer encephalitis complex, which indicated that KFD was indigenous. Salim Ali of the Bombay Natural history Society trained Rajagopalan in the handling of birds and understanding migratory ecology. The distinct absence of yellow fever in India was a key factor behind the establishment of the VRC and the evolution of a committed interest in medical entomology and acarology in the country in the late 1950s. Rajagopalan was one of the beneficiaries of this interest. With JE being recognised in 1954 and KFD being reported in 1957, interest in arboviruses and their transmitting arthropods grew significantly in India. Rajagopalan studied the ecology of several medically relevant arthropods of Devimane Ghat, a virgin tropical evergreen forest in what is now Uttara Kannada district of Karnataka. This study included extensive sampling of arthropods and collection of blood samples from a large number of humans and other mammals for serology and characterisation of possible viruses. Devimane Ghat was Rajagopalan's first field site, where he learnt about the spatial distribution of mosquitoes in a tropical forest. The theory uppermost in Rajagopalan's mind was that these arboviruses had zoonotic origins. He and his team had an unequivocal demonstration of this while working on KFD. The construction of a dam across the Sharavathi river in 1964 inundated an extensive natural forest area and displaced the wild animals that lived there. This forced the ectoparasitic ticks present on the animals to move into human suburbia and eventually live by parasitizing livestock. Rajagopalan's studies at the Devimane Ghat site enabled him to clarify vital but unknown aspects of the zoonotic cycles of various viruses. He was posted to Vellore, Tamil Nadu, in the mid-1950s when there was a breakout of JE. He and his team sampled thousands of culicine mosquitoes (*Culex vishnui* group) and were able to isolate the JE virus. The work on JE was extended to Akividu, Andhra Pradesh, to explore the role of migrant birds in introducing the JE virus to an area since pelicans and similar birds nested in the Akividu lake. He realised that operational research methods were sorely needed to manage the populations of medically relevant arthropod vectors. He was interested in exploring what the most practical and viable methods to save human lives would be and how one could predict an epidemic.

While at the RUGCM, he studied the ecology of *Cx. quinquefasciatus*, the mosquito that transmits filarial nematodes. The RUGCM's mandate was to manage malarial and filarial vectors through genetic manipulation. The RUGCM folded up in 1975 because of a sudden decision of the government. The sterile-insect technique trialled in the Midwestern U.S. to manage populations of screw-worm flies was considered a turning point in the management of nuisance insects. The females mated once in their lifetime, which is also the case with mosquitoes. Because of a distinct size difference between the pupae of male and female screw worms, after successfully mass rearing them, American scientists were easily able to separate the male pupae and sterilise them with radioactive exposures, and release them into natural populations. This practice reduced the screw worms that affected American cattle. A few foreign and Indian medical entomologists were keen on trialling sterile-insect technique to regulate mosquito populations. But it was not possible to segregate the sexes in mosquito pupae. A 2-5% contamination, which is a large number considering the size of mosquito populations in India, always prevailed during the segregation of male pupae from females. Moreover, the sterilised and genetically manipulated male mosquitoes were not as competitive as the males in the wild, and therefore the results were disappointing.

Dr Rajagopalan's leadership qualities came to the fore after he took over the reins of the VCRC in 1977 as its first Director. The then Director General of the ICMR, Colathur Gopalan, saw the need to establish a centre to manage the exploding populations of mosquitoes in southern India, and the VCRC was born in Puduchery in 1975. Over the next few years, it grew into a premier organisation offering cutting-edge knowledge on the ecological management of different species of mosquitoes. Today, it has grown into a centre of excellence for research and training in vector-borne diseases and control and a WHO collaborating Centre for Research and Training in Lymphatic Filariasis and Integrated Methods of Vector Control because of Rajagopalan's vision and unstinting efforts. It is due to him that a two-year-long master's programme in medical entomology, unique in India, commenced at the VCRC in 1986. Close to 20 students got their PhDs with Rajagopalan as their principal supervisor in various sub-disciplines at the VCRC. A key strategist, Dr Rajagopalan designed and launched a locality-specific management of human lymphatic filariasis in both Pondicherry (Puduchery) and Kerala, which was based on a comprehensive understanding of mosquito biology and the manner in which the insect transmits protozoan or nematode parasites. The Filariasis Control Demonstration Project, which used large-scale, integrated vector management (IVM) tactics, was launched in 1980. It reduced filarial parasite (*Wuchereria bancrofti*) infections in Puduchery. The IVM tactic entailed the reduction of the population density of *Culex quinquefasciatus*, the agent that transmits *W. bancrofti*. He established a research project in Cherthala (Kerala) to manage *C. quinquefasciatus*, which transmits the nematode *Brugia malayi*. The Cherthala project involved the physical and mechanical removal of the water plants that served as breeding sites for *Cx. quinquefasciatus*. Another VCRC project aimed at mosquito management at Rameswaram Island by targeting the unusual breeding sites of the *Anopheles culicifacies*, the local malarial mosquito. The issue was complicated by periodical movements of local human residents, mostly fisherfolk, who passively transmitted the parasite to nearby islands and mainland India. The malarial parasite confined to this island was resistant to common anti-malarial drugs.

Dr Rajagopalan investigated the breakouts of JE in Tirunelveli, Tamil Nadu (1978), and in Burdwan and Bankura, West Bengal (1976). He demonstrated that JE was seasonal and that its epidemiology varied subtly in different regions. *Cx. tritaeniorhynchus* breeds profusely in rain- and floodwater pools and feeds on livestock. Rajagopalan was of the opinion that forecasting an epidemic and initiating proactive measures of vector control was more significant than implementing management strategies later. He vehemently argued that this strategy necessitated long-term studies on vector populations in JE-prone areas so that an epidemic breakout could be prevented. Even a 50 per cent reduction in the vector population, Rajagopalan argued, would prevent an outbreak. Since *Cx. tritaeniorhynchus* was zoophilic, only the odd mosquito would bite humans. His argument that the human was an incidental link in the endemic infective cycle between *Cx. tritaeniorhynchus* and wild birds deserves a thorough investigation today. Under certain specific circumstances, larval control measures could be practical and economically justifiable, but Dr Rajagopalan was of the strong opinion that they would not work in the context of JE virus transmission.

To investigate tribal malaria, he opened a field station in Jeypore, Koraput district, Odisha, where he encountered difficulties with the terrain and the hostility of the local people. Malaria due to *Plasmodium falciparum* and *P. vivax* had been persistent in this region for several years. Rajagopalan's wise decision to appoint a woman doctor to deal with the health needs of the tribal women in particular and to have the VCRC distribute medication free of cost enabled it to gain the trust of people in Jeypore. The VCRC team found that cases of *P. malariae* and P. ovale also occurred there. The Jeypore field station was a jewel in the crown of the VCRC because it demonstrated how quality work could be achieved in adverse, and even dangerous, conditions. The VCRC was able to carry out long-term ecological studies on mosquitoes. For two years the author of this book worked on malaria vectors and malaria epidemics in the most difficult forested terrains of Koraput, Kalahandi and Malkangiri.

He worked closely with various national and international agencies, such as the WHO, identifying evidence-based solutions for operational challenges faced in various vector-management programmes. The Royal Society of Tropical Medicine & Hygiene, London, recognised him as a Fellow in 1990. The Government of India conferred the *Padma Shri* on him in 1990. All through his professional life, Rajagopalan fought the mighty mosquito and strove towards achieving the goal of better health for his fellow citizens. He may not have been 100 per cent successful in his efforts to keep mosquito populations within reasonable threshold levels, but his efforts were always genuine. Retired from service on 31st October, 1990, after serving the ICMR for 38 years, he served for nearly a quarter of a century as member of many scientific committees, including the Scientific Advisory Committee of the VCRC for nearly a quarter of a century. He retired in 1990 and is based in Chennai. On October 27, 2020 he turned 90, and is as highly active as half-a-century ago! Around that time, I was in correspondence with him, and he stated that he felt week in walking and general health, but his brain was still very strong and could speak and memorize all things with precise details. No doubt, therefore, that he is still publishing articles and scientific papers on regular basis in complete defiance of advanced age and/or physical infirmity, whatso ever! Kudos !

There are scientists who always work with a ruggedly defined objectivity - the excellence in scientific pursuit come what may, and are always remembered in history. One among them was **Dr Rachel Reuben**, a world renowned medical entomologist who sacrificed her entire life to the development of science of medical entomology. Her original publications on the vectors of Japanese encephalitis continue to be essential reference works on the study table of a serious scientist. Rachel Reuben was born in Bombay (now Mumbai) on 12th May, 1934. Her father, David Ezra Reuben, a member of the ICS, had opted for the judiciary and was posted to Bihar. Consequently, Rachel's childhood was spent in Gaya, Ranchi and Patna, where the sprawling old-style houses and rambling gardens gave her ample opportunity to explore the natural world. Her early childhood education was entirely conducted by her mother, Sophie nee Israel, herself a trained teacher. Rachel was also influenced by her father's deep interest in nature and by the age of nine had already made up her mind to become an entomologist and was very soon making notes and observations of her own. The years 1944 to 1947 were spent in boarding school at Loreto Convent in Shimla, where the mountain scenery, flora and fauna of the Himalayan foothills were a continual source of delight. This is clear in an essay on stag beetles written at the age of 13 which displays her enthusiasm, curiosity, sense of observation and keen desire for scientific accuracy. After her final year of schooling at St Joseph's Convent in Patna, she enrolled at the Patna Science College where she completed her BSc, absorbing herself with such eagerness in her studies in zoology that she set about teaching the subject to her younger sister, making her participate in such practical exercises as the dissection of a frog and the observation of the laws of heredity in cross-pollinated sweet-peas in the family garden.

On her father's retirement from the Patna High Court in 1953, the family moved to Bombay (Mumbai) while Rachel enrolled at Delhi University, completing her MSc in 1955. By this time, she was determined to work in the field rather than in the laboratory, an unusual choice for a young woman in those days. Pursuing this goal, her PhD thesis on "*The ecology of biting midges Culicoides impunctatus in relation to climatic factors*" in Scotland, prepared under the guidance of Dr D.S. Kettle of Edinburgh University, was based on field work carried out on the shore of Loch Fyne, an idyllic setting in Argyllshire (Scotland). Although a first class student throughout, Rachel found that academic standards were much higher in Britain and she had a struggle to catch up. On the other hand, she found the atmosphere among her peers stimulating and exciting. Indeed, her return to India after completing her PhD in 1959 was planned to be no more than a short visit before she continued her career in this congenial setting abroad. However, given the social commitment imbibed from her parents, it was not long before Rachel came round to the view that she would be able to make a more meaningful contribution in her field if she remained in her own country. She never regretted her decision and looking back in later years, often remarked that she was deeply thankful that she had made this choice. In 1960 she joined the Virus Research Centre (now National Institute of Virology), India, as Rockefeller Foundation Fellow. Between 1960 and 1970 she worked as part of a team in Tamil Nadu, south India, on entomological and epidemiological aspects of the mosquito-borne virus diseases Japanese encephalitis and Dengue. Vector species was incriminated, natural cycles elucidated and the biology of the vectors was studied for the first time in India. It was during this period, she re-described in detail the three vector species responsible for transmitting Japanese encephalitis virus, *viz., Culex tritaeniorhynchus, Culex vishnui* and *Culex pseudovishnui,* which was published in *Bulletin of Entomological Research.* Later, a series of papers (6 parts) on the detailed bionomics of these vectors were published in *Journal of Medical Entomology* during 1971. In the year 1970 she was appointed to the post of Senior Scientist in the WHO/ICMR collaborative project on *Genetic Control of mosquitoes,* New Delhi. Between 1970 and 1975 she carried out intensive studies on the ecology of the urban dengue vector, *Aedes aegypti* in Rajasthan, Haryana and Delhi. Studies were made of dispersal, survival, absolute population estimation and population dynamics, to provide essential baseline data for development of a control strategy by mass release of sterilized males of this species. Due to certain misunderstandings and other reasons this project had to be terminated.

In 1975 she joined as Deputy Director at Vector Control Research Centre (VCRC), Pondicherry and conducted studies on the biology and control of anopheline vectors of malaria from 1976 to 1985 at the VCRC field unit in Salem. All aspects of the ecology of the urban vector *Anopheles stephensi* were studied in relation to malaria transmission in the town of Salem, Tamil Nadu. The day time resting places of the adults were defined, and the correct dose of the organophosphorous compound, temephos, was determined to control its breeding in the wells of the town. Intensive investigations in the riverine villages of the Sathanur Dam area showed that persistence of malaria at a high level was due to outdoor resting habits of the vector *An. culicifacies* as well as to poor spray coverage. The use of ULV/thermal fogging with malathion as a supplementary control measure, successfully reduced the incidence of malaria to a very low level. In 1985 she served as Temporary Advisor to World Health Organization on a committee to make an in-depth evaluation of the National Malaria Eradication Programme of India. The Committee's report was presented to the Government of India. Between 1976 and 1986 she was a Member of the Expert Advisory Panel on Vector Biology and Control of the World Health Organization.

The Centre for Research in Medical Entomology was created by the Indian Council of Medical Research in mid-1985, and she was appointed as its first Director. The Centre initiated its work on the epidemiology, laboratory diagnosis and control of Japanese encephalitis (JE). Dr R. Reuben was the pioneer contributor to the understanding of the epidemiology of Japanese encephalitis through her original approaches to the study of vector behaviour. Under her able leadership the Centre worked in collaboration with the Tamil Nadu Agricultural University to develop simple, cost effective technology for mosquito control in rice fields (where the vectors of JE breed). In the course of the research trial two promising methods were identified which were acceptable to the farmers and therefore suitable for community use – (i) Intermittent irrigation, in which there was 20% saving in water and increase in grain yield, with 70% reduction in mosquito breeding (ii) Use of neem (*Azadirachta indica*) coated over urea as fertilizer to provide higher grain yield with suppression of vector breeding as a side benefit. In addition, the use of cheap pyrethroid-impregnated door and eaves curtains for reduction of man-vector contact in JE endemic villages were also experimented and found to be successful. She was instrumental in taking up extensive faunistic surveys in the Western Ghats, which initially started with special reference to the prevalence of malaria vectors. During these collections many new species and undescribed stages of various mosquito species were found. Among these one of the newly found mosquitoes was named after Dr R. Reuben as *Aedes reubenae* (now renamed as *Tewarius reubenae*). While at Madurai she was a keen supporter of the Palani Hills Conservation Council, and was instrumental in setting up the "*Reuben Arboretum*" in her father's memory. Dr Rachel Reuben had been associated with Bombay Natural History Society (BNHS) of which she had been a life member since the 1960s, and was engaged in nature conservation research. She had a keen interest in natural history with a focus on birds. She had family friendship with Humayun Abdulali and Dr Sálim Ali (renowned ornithologists), and compiled and edited books such as a biography on ornithologist Humayun Abdulali titled "*Humayun Abdulali - Naturalist*". Dr Rachel Reuben passed away on 1st November 2010 after a brief illness in hospital.

Medical entomology was on its zenith during the mid-nineteenth century and many experts inventorized and/or produced significant taxonomic works. **Drs B.L. Wattal and N.L. Kalra's** (1961) *Region-wise keys to female Indian Anopheles* was extremely popular among laboratory and field workers.

West Bengal has been a seat of many groundbreaking researches, and many internationally reputed malariologists and medical entomologists were born and brought up here. One of the most outstanding medical entomologists in this golden land is **Professor Dr A.K. Hati** who was born in the erstwhile undivided Bengal State of India on 10th June, 1935. His early education was completed locally. Being a brilliant student in biology, he opted for medicine and completed his MBBS from Calcutta University in 1959). He worked as a Research Scholar during 1960-1962 and obtained his PhD degree on the dissertation titled on a medico-entomological subject from the same university in 1966. His heart continued to beat for dierse academic excellence and, therefore, he mastered in Bengali language with a MA degree (1967), and the LLB (1968), all from his *alma mater*, the Calcutta University. He began his career as a Demonstrator (1962-1966) and soon became the Lecturer (1966-1971), Assistant Professor (1971-1973), Reader (1973-1975), Associate Professor (1975-1977), and full Professor (1977 till retirement). He was Chairman of the Division of Parasitology since 1981, and Deputy Director (1994) and finally the

Director of Calcutta School of Tropical Medicine and the Chairman of the Division of Parasitology (Feb 1995 - June 1996). He retired in July 1996 from the post of Director, CSTM.

He is a prolific writer. He has not only published a large number of research publications, his various textbooks are permanent fixtures on the study table of a serious medical entomologist, a medical graduate and a PG student of Zoology. His books are read all over the country and occupy a place in all the major university libraries. He has to his credit more than 250 research publications which are a testament of his genius in the science of vector-borne infections, particularly Japanese encephalitis which he has mastered and is an authority on it. Many a monograph penned by him are a necessity for every day's consultation for a serious student of Medical entomology or parasitology. Some of these include:

(i) Studies on vectors of Japanese Encephalitis in West Bengal, 1986.

(ii) J E Entomology, a training module, 1990.

(iii) Studies on snake venoms and epidemiology of snake-bites in West Bengal, 1994.

(iv) Studies on vectors of some public health importance, 1995.

(v) Kala-azar in India. Report of a contractual service agreement 1st May - 31st May, 1997; WHO Project: SE ICP CTD 001 RB 96.

He received several honours and awards during his professional career, most important of which include the Rabindra Puraskar for publication of a science writing in Bengali on a book entitled 'Cancer' in 1985, Government of West Bengal; Coate's Gold Medal, University of Calcutta, 1987, Gold Medal, W.B. Medical Council, 2001; Partha Chatterjee Memorial Award, Bangiya Vignan Parisad, West Bengal, 2001; Gopal Chandra Bhattacharyya Memorial Prize, Government of West Bengal, 2002; Indian Association of Epidemiologists' Annual Oration Award, 2006; Dr B.C. Roy Memorial Oration, R G Kar Medical College (1.7.2014); IAE Annual Oration, Agra (11.02.06); Professor Hirendra Kumar Chatterjee Re-union Oration (19.12.05); R G Kar Medical Collage; Arindam Datta Chaudhuri Memorial Oration AIASP (All India Association of General Practitioners) (05.011.05); Dr Maniklal Maitra Oration, IMA, Howrah Branch (03.09.05); J N Moitra memorial Orator 92nd Indian Sc. Cong. Assoc. Ahmedabad (04.01.05); Professor Dilip Banerjee Memorial Oration. IMA, Baranagar - Cossipore Branch. (30.11.03); Prof. S. C. Seal Memorial Oration IPHA, West Bengal Branch, Sarat Sadan, Howrah (26.07.03); R G Kar

Memorial Oration at R G Kar Medical College, Kolkata (23.08.01); Calcutta School of Tropical Medicine Re-union Oration (24.02.01); Harbans Kapoor Medallion Oration (1987); Elected as Sectional President, section for Medical and Veterinary Sciences in the 81st Indian Science Congress Association, Jaipur, 1994; and Elected President, Indian Public Health Association, West Bengal Branch, 1995. He was elected Fellow of the Indian Public Health Association and the West Bengal Medical Council. He was also a distinguished and active Life Member of a large number of scientific societies such as, *e.g.*, Indian Science Congress Association, Indian Association for the Study of Sexually Transmitted Diseases, Indian Science News Association, Indian Public Health Association, Asiatic Society, Your Health, People's Relief Committee, and Janaswastha Committee.

Prof. Hati is now but an "eighty years-old young man" who can even now beat any young man in the force of agility and industry for long hours. What is most startling to most young entomologists is that his hands are always full with several research activities. He was awarded the highest honour of the Society of Medical Arthropodology - The SOMA Life Time Achievement Award - in 2017, during the Tenth National Conference of Medical Arthropodology (SOMA) held in Kolkata.

The dynamism of Dr V.P. Sharma in developing erstwhile Malaria Research Centre (now National Institute of Malaria Research) was discernible in grooming young and energetic scientists of the institute. **Dr Sarala K. Subbarao,** born and brought up in Secunderabad, joined Malaria Research Centre following her brief stint in the GCMU. After doing her MSc in Zoology, with specialization in helminthology, she went to USA and had a chance to work under Professor John R. Laughnan, a distinguished corn geneticist, on her thesis entitled, "*The nature and sensitivity of breaksites in the long duplication of X-chromosome following X-irradiation*" and obtained her doctoral degree in Genetics in 1973 from the University of Illinois, Urbana-Champaign. In 1974, upon her return to India she secured a insect's post in a World Health Organization (WHO)-Indian Council of Medical Research (ICMR) joint research project on "Genetic Control of Mosquitoes", in New Delhi. There she worked on *Culex quinquefaciatus*, a filaria vector, to resolve the variations seen in the cytoplasmic incompatibility/compatibility between Indian and European strains of this species. This was also her first exposure to medical entomology. She began research in malaria in 1975 after foundation of the Malaria Research Unit (later called Malaria Research Centre) under ICMR at Delhi. She gained complete freedom of work and sharpening her skills in mosquito genetics under the watchful eyes of her mentor and Director, Dr

V.P. Sharma, where upon she investigated deeply into mainly the sibling species complex of primarily the *Anopheles culicifacies* complex, but also other vectors of malaria (WHO SEARO, 2007). In those days distinguishing cryptic and indistinguishable taxa among the complexes was a great challenge and it required urgent attention. Dr Subbarao took upon herself this unique opportunity and developed a team of extremely hard working scientists with the help of which she was able to develop a clear picture of identification of at least five sibling species of the *Anopheles culicifacies* complex, besides other vectors. This, of course, was her greatest contribution to mosquito vector science which continue to fascinate a serious researcher even now.

While Dr P.K. Rajagopalan had given a solid foundation to ICMR-Vector Control Research Centre, it was Dr P.K. Das's uncanny style of working through 'self-help' or personal supervision in every research project undertaken at the Centre that attracted me most. He had that unique style of forming a communication bridge with the workers and civil communities that always helped him accomplish the task more effectively. He is a great field scientist, with whom I went on establishing many scientific explorations in the difficult terrains of Jeypore hills in the western Odisha (=Orissa) and urban agglomeration's mapping in Bengaluru (=Bangalore) for developing a Master Plan for Mosquito Control. Dr P.K. Das subsequently became the Director of Vector Control Research Centre, Pondicherry supervising several research programmes of national importance. After retirement he entered into developing a eco-friendly village nearby Pondicherry township. He keeps providing his guidance to his Centre,

North-eastern States always occupied a centre-stage in expressing constraints in implementing various mosquito-borne diseases, particularly malaria, for their rather difficult physio-topography and distantly located human settlements. In such a ecosystem rose a scientist, **Dr J. Mahanta**, the former Director of the ICMR-Regional Medical Research Centre, Dibrugarh who, besides his medical responsibilities, offered equal weightage to the vector mosquitoes of the region. Born on March 1, 1952, he is the eldest son of Late Sri Jadav Chandra Mahanta and Smt. Basanti Mahanta of Puran Burka Satra near Guwahati. Elder brother to two brothers and two sisters, Dr Mahanta completed his schooling with flying colours before taking admission in Cotton College, Guwahati for his Pre-University education in 1974. After that, he qualified for medical studies at the Gauhati Medical College, Assam, where too his academic brilliance was highly appreciated. Dr Mahanta had much earlier resolved of putting his life enriched by medical education in the service of people as had witnessed during his childhood in fifties and sixties the havoc caused by the deadly and disfiguring small pox disease which often cursed an entire family! Dr Mahanta took special interest in the discipline of surgery and was moved by the magnitude of the maladies with which people suffered. The more he say peoples' surgical complication, the more he got resolved to put himself in the service of man, truly in the line of what the Father of Nation, Mahatma Gandhi, has preached, "*I cannot imagine anything nobler or more national than that for, say, one hour in the day, we should all do the labour that the poor must do, and thus identify ourselves with them and through them with all mankind. I cannot imagine better worship of God than that, in His name, I should labour for the poor even as they do.*"

Always very keen to observe and study with minutest details any case of human infection or illness that encountered him, Dr Mahanta believed in a phlebotomist's thoroughness and precision of the matter in hand rather than trifling casual touch, however brilliant, recalling the famous maxim by the great Robert Koch, "*Infection of wounds was common and I saw the suffering of patients. I wanted to help not in fire-fighting manner but rather delving deeply on the subject and discovering the cause and its prevention.*" Dr Mahanta is like the Pacific Ocean – vast and quite, but deep and full of life, never ever pretending to exhibit his huge knowledge and, at the same time, impacting extraordinarily everyone who comes in contact with him! In his making what he is today he always take pride in acknowledging the encouragement of two of his teachers, Dr J.N. Sarma Bordoloi and Dr P.C. Mahanta, who were a great source of inspiration for Dr Mahanta choosing to specialize in Medical Microbiology. "*During those days, there was dearth of specialists in the subject in Northeast India. My professors wanted someone trained in Medical Microbiology since the problem of infectious diseases in Northeast was rampant and there was an urgent need to develop the discipline there. They inspired me to take it up*", recalls a gratefully beaming Dr Mahanta. The road to success for Dr Mahanta was not simple. He had to take head on all the challenges like treading on a double edged sword. To accomplish his mission he had to first ensure his admission for higher medical education in the country's one of the most prestigious institutions, the Post Graduate Institute of Medical Education and Research (PGIMER), in Chandigarh, to further specialize in Medical Microbiology. The PGIMER truly provided him the opportunity to look at infectious agents firsthand and deal with them meaningfully. There he taught for three years before returning to Assam. No surprise that, with such a backing of masterful experience and inexhaustible energy, he was soon offered a faculty position in Gauhati

Medical College, ephemerally though, and subsequently in Regional Medical College (now Regional Institute of Medical Sciences) at Imphal. He later joined Oil India Limited (OIL) at Duliajan in Dibrugarh District and joined as the Head of the combined Department of Pathology, Microbiology, and Biochemistry in Oil India Hospital.

Never the one overvaluing the financial dividends however lucrative such as that at the OIL, his heart was yearning for something more meaningful whereby he could fulfil his goal of serving his people. The opportunity came sooner as in 1993 he joined as a senior scientist (Deputy Director) the recently established ICMR's new institute, Regional Medical Research Centre which aimed at researching the regional maladies and provide knowledge-based guidance for their control. But, as the destiny would have decided, none of the directors in recent past was willing to stay any longer and directors kept on changing with a regular frequency. The ICMR finally sighted Dr Mahanta who not only stayed solidly for past 17 years as its Director since 1996 but brought his institution at the highest apostle in scientific arena. Dr J. Mahanta developed a very strong department to study mosquitoes and mosquito-borne diseases under which they discovered three new species. To honour Dr Mahanta's contributions in varied disciplines and evolving the institute of RMRC to greater heights, even a mosquito was named after his name, *Armigeres mahantai*. The RMRC, along with Centre for Development of Advance Computing (CDAC), Pune had won the Mosquito-millionth SAARC Countries Award for 2013 for their collaborative effort in developing a surveillance system for real time reporting of malaria from remote areas. The Centre also developed a module for security forces to fight against malaria in forest fringe areas. During the time of his Directorship the RMRC was recognized as a national Centre for Excellence by the Government of India, thereby partnering in the International Centre of Excellence for Malaria Research by the National Institutes of Health, USA.

Dr Mahanta built an enviable reputation both as a scientist and an astute administrator. He published more than 325 research papers in peer reviewed international and national journals. There are also books to his credit. He is a Fellow of Royal College of Tropical Medicine and Hygiene, London, Fellow of Indian Society for Parasitology, INSA Fellow under scientist exchange programme. He is a member of National Academy of Medical Sciences of India and International Vaccine Action group. His has been conferred several honours, *e.g.*, the Annual Oration Award (2004) for Indian Society for Malaria and other Communicable Diseases, ICMR Award (2008) for his research work in North east India, Dr S.C. Parija award (2011) of Indian Academy of Tropical Parasitology, and Dr W.P. Lyton

Oration Award of Northeast Microbiology Society (2013). Dr Mahanta holds membership in several international scientific bodies and societies like International Society for Leptospirosis, American Society for Microbiology, Royal Society for Topical Medicine and Hygiene, besides being an expert member to review the Millennium ecosystem Assessment of United Nations. He was the member of World Federation of Parasitologists from 2002 to 2006 and a life member of several national scientific societies. He is President and/or Chairman for many societies. After retirement as Director, RMRC, Dibrugarh he joined the institute as Consultant and continues to work from there.

Born in 1951, **Professor Dr Aditya Prasad Dash** is a versatile medical entomologist and vector-borne specialist, and one of the rare scientists who have shown an uncommon blend of character of hard work and focused application coupled with unflinching religious devotion; all churned to yield a future many in the contemporary world would envy. Credited with Ph. D. and D. Sc. degrees in Zoology, he is at the moment of writing this account the Vice Chancellor of the Central University, Tamil Nadu since 6th August, 2015.

Dr Dash began his career as a senior scientist on an ICMR project in Odisha during early 1980s. He later joined the ICMR mainstream as Senior Research Officer in late 1984. His trajectory of success was such that as early as 2000 he became the Professor Director of the Institute of Life Sciences Bhubneshwar and the momentous occasion was when the then Prime Minister of India, Shri Atal Bihari Vajpayee, dedicated the institute to the nation on 15th July 2003. In late 2003 he became the Director of ICMR-RMRC in Jabalpur and subsequently that of the National Institute of Malaria Research in 2005. In 2007 he was appointed Regional Advisor, Neglected Tropical Diseases, WHO-South East Asia Regional Office at New Delhi. After superannuation from the World Health Organization, he occupied the ICMR-Distinguished Scientist Chair at the Institute of Life Sciences (Department of Biotechnology, Government of India), Bhubaneswar on 11th May, 2015. All through he contributed extensively to the control of malaria and other VBDs in India as well as other southeast Asian countries.

Prof. Dash has significantly contributed in the field of biomedical science, particularly on transmission biology of tropical diseases affecting the poorest of the poor. He has devoted his career in developing various tools / technologies / strategies in biomedical science. His notable contributions include: establishing *Anopheles annualris* as an important vector in rural areas and identification of telomerase activity in gametocytes of *Plasmodium falciparum.* He demonstrated efficacy of drug combinations. He was instrumental in proving the efficacy of Mass Drug Administration (MDA) in India. His group has also shown that supplementing MDA with Integrated vector management can prevent lymphatic filariasis (LF) transmission more efficiently than MDA alone. This strategy has now been included in the LF elimination strategic plan by the World Health Organisation. He has contributed in developing an animal model for chemotherapeutic and immunological studies for parasitic diseases.

Dr Dash and his colleagues also developed a simple technique for detecting dengue virus antigens in desiccated mosquitoes which serves as an important tool for surveillance. His contributions on climate change impact and on infectious diseases are noteworthy. He has played an important role in involving community in control of diseases like malaria, lymphatic filariasis, visceral leishmaniasis, dengue and chickungunya, as well as soil transmitted helminthiasis. His contribution in the field of ecological succession of mosquito species distributed in India is most striking. Specially, the research output correlating the change in malaria, dengue, chickungunya and filariasis epidemiology to the change in climate parameters are exemplary. His contribution on modern biology of disease vectors is especially significant and meaningful. For example, he was a part of research that sought answers to the question on the molecular taxonomy of Indian mosquito vectors. Employing the multiple genetic fragments with DNA sequencing studies he has contributed in inferring the phylogenetic interrelationships among Indian malaria vectors. Interestingly, the findings corroborated the inferences based on the traditional cytotaxonomical approaches. Besides research on malaria vectors, he researched on malaria parasites in identifying the telomerase activity in gametocytes. His contributions include elucidation of uniqueness of the evolutionary patterns of the chloroquine-resistant gene in *P. falciparum.*

Dr Dash has contributed in comparative genomic studies of genomes of malaria parasites, *P. vivax* and *P. falciparum.* The results, for the first time have revealed significant genomic similarities between the two genomes in comparison to essential differences between these two species in terms of epidemiology and pathogenicity. He has also contributed in developing novel genomic markers to understand the population structure and demographic history of Indian *P. vivax.* The contribution of his group in scanning the whole genome of the African malaria vector, *An. gambiae* has been remarkable. Through whole genome scan of this species of malaria vector, which is a model organism for other species of Anopheles has revealed interesting genomic organization of this species. To be noted that such information has been widely used in inferring many evolutionary phenomena of the genomes of Indian malaria vectors. The genomic study of Indian isolates of *P. falciparum* and *P. vivax* has become a benchmark. He has also contributed in developing novel genomic markers to understand the population structure and demographic history of Indian *P. vivax*. In this context, he has contributed into unravelling the interesting genomic organization functional gene in the malaria parasite, *P. falciparum*. Such information is unique and has provided baselines for the study of genetic pattern of drug resistance and virulence associated genes in field populations of this species. Along with his colleagues, he has also contributed in understanding the genomic organization of many human genes that are responsible for providing resistance to malaria infection. Using comparative genomic approaches he has contributed in the basic understanding that how such genes are evolving in humans. One of the major contributions the group has made in the field of molecular epidemiology is the determination of high proportion of mixed-species malaria parasite infections in India. Applying an array of molecular protocols and using PCR diagnostic approaches, his group has contributed to the understanding that mixed infections of the two principal malaria parasites, *P. falciparum* and *P. vivax* can be very high. Apart from laboratory level research activities, he also undertook field trials of various intervention measures. These activities were a unique experience to translate science into deliverable products and testing them for actual use to monitor the performance under field conditions.

Dr Dash has also made significant contributions on epidemiology of malaria in India and the challenges faced by the Indian programme to control malaria. He also studied along with his colleagues the histopathology of fatal respiratory distress caused by *P. vivax* malaria in India. He has published more than 250 papers in reputed peer reviewed journals and guided many Ph.D. and post-doctoral students, who have become successful scientists.

Dr Dash demonstrated exceptional organizational ability, accomplishing many complex tasks in complicated situations in many countries including India, through diligence, foresight and sound judgment. Through his able leadership, he inspired all personnel who worked with him to such an extent that all assigned missions were performed consistently in an exemplary manner in management of neglected tropical diseases in marginalized communities.

Fig. 8.7. (a) Dr B.K. Tyagi, (b) Dr A.P. Dash, (c) Dr J. Mahanta, and (d) Dr Vijay Veer. (*Source:* Free Wikipedia/Dr B.K. Tyagi's personal archive)

Fig. 8.8. (a) Dr R. Rajendran, (b) Dr R.S. Sharma, (c) Dr P.K. Srivastava and (d) Dr Kaushal Kumar. (*Source:* Free Wikipedia/Dr B.K. tyagi's personal archive)

The indomitable pair of field entomologists whose research works in the field of human lymphatic filariasis will remain in the memory of scientists across the world are **Drs R. Rajendran** and **I.P. Sunish**, old colleagues of the author at the ICMR-CRME at Madurai. Dr Rathinasamy Rajendran was born at Port Novo, near Chidambaram in Southern India on November 22nd, 1953 & left this materialistic world on June 26, 2010. His father was a local successful business man and proud owner of agricultural lands. Since he was the eldest child in the family of seven children, the entire responsibility of the household was on him and he managed his household responsibility with equal weightage as that of his official duties. He got into wedlock in the year 1980 with Smt. Bhavani, and both were blessed with an intelligent and obedient son *viz;* Manikandan. He was able to bring up his son in a manner which all the others envied, finally making his son to obtain post-graduate in medical profession, to become well qualified medical doctor to serve the humanity at large. He did basic school education at his native *viz;* Govt. Boy's High school, Portonovo, and entered Annamalai Univarsity, Chidambaram in 1970 to pursue Higher Secondary (PUC) education. In the year 1974, he passed B.Sc. examination with Zoology major; in 1976 he gained his M.Sc. degree also in Zoology (with specialization in Entomology), obtaining the highest distinctions. In the same university, he joined for the doctoral programme, under the inspiring guidance of Prof. N. Ramalingam and obtained his doctorate (Ph.D.) on insect reproductive physiology, in the year 1983.

He joined as Research assistant in 1981 in a malaria project at Tiruvannamalai field station of Vector Control Research Centre based at Pondicherry (now Puducherry); an organization of Indian Council of Medical Research, New Delhi. On 24th November, 1983 he entered into council's regular service. In the year 1985, the field station at Tiruvannamalai was winded up and the entire staff of this unit moved to Madurai to set up yet another ICMR institute *viz*; Centre for Research in Medical Entomology (CRME), under the able leadership of the founder Director, Dr R. Reuben. CRME at Madurai was started with staff strength of seventeen, in May 1985. During the establishment of CRME, he was Assistant Research Officer, and contributed to the larval ecological studies pertaining to the mosquito vectors responsible for transmitting Japanese Encephalitis virus. Dr R. Rajendran was inducted into ICMR research cadre on 6th March 1991. Through the ICMR assessment he was able to go up the ladder of scientific cadre and during his last years of his life he was holding the post of Scientist "E" (=Deputy Director).

He carried out extensive field studies in rice field agro-ecosystem along with the scientists at the Tamil Nadu Agricultural University, Madurai. In this instance, it should be highlighted that, even though he was accompanying the field team as a supervisor, in the field he worked as a member of the team carrying out field works equal to those in the team without any inhibitions. After his basic ecological studies on rice field breeding mosquito vectors, which are the main agent responsible for transmitting the deadly JE virus; he went on to develop novel control strategies to prevent these vectors breeding in rice fields. Along with the agricultural department, he conceived the idea of using a water fern Azolla in rice fields which had dual benefit to the agricultural farmers, *viz*; control mosquito vectors as well as obtain better grain yield. He carried out small scale and large scale studies on the usefulness of Azolla in rice fields for the vector breeding in rice fields. Another control strategy focused was water management (alternate dry and wet irrigation), which also was highly effective in reducing the breeding of mosquito vectors in rice fields. Apart from the field based research, he had also laid his hands equally on laboratory activities. The important contribution of CRME, in developing a suitable methodology to detect JE virus by inoculation through Toxorhynchitis sp. larvae was first standardized at CRME laboratory by

Dr Rajendran. In view of his hard work and dedication to scientific studies, in the year 1995, he was posted as the Field in-Charge to one of the field station of CRME at Vridhachalam to carry out in-depth evaluation studies on the control of JE. During this period, CRME had set up another field unit at Tirukoilur, where the control trial on lymphatic filariasis had initiated, for which also he was made responsible. For the next 10 years, he managed both the field units with much enthusiasm and vigour. He often shuttle between the two units without caring whether it is day or late night. He was a good advisor to his subordinates and was always there whenever his staff was in need, both officially as well as personally. In the control of JE, a new technology was successfully experimented in the villages near the field unit of Vridhachalam, *viz*; pyrethroid impregnated curtains, in order to reduce man-vector contact and thus helps to curtain transmission of JEV. This technology was later taken up by the state health department to carry out the trial on a larger scale. He was responsible in developing an index (Infected Vector Abundance =IVA) for estimating the load of JEV infection in mosquito vectors, which was later used widely. In the field of rural filariasis control, he leaded a team to develop a low cost technology to control the vectors of bancroftian filariasis in rural settings. This technology was highly appraised by the filariasis scientific community, which was recommended by various expert committees to replicate this technology at other places, and is the need of the hour, when annual mass drug administration was unable to effectively bring about the expected objective in the Global Elimination of Lymphatic Filariasis. Additionally, when the Indian Government could not come to a conclusion, whether to include the drug, albendazole to DEC or not in the annual MDA, it was CRME team under the able leadership of Dr R. Rajendran which proved that inclusion of albendazol was superior to DEC drug alone in reducing the infection load in the community. This changed the policy of the National Programme to Eliminate Lymphatic Filariasis to implement the two drug regimen of DEC + albendazole in all the endemic areas covered under the LF elimination programme.

In the year 2005, he returned back to the CRME Head Quarters, Madurai where, apart from managing the field station activities, he took responsibility in the overall scientific and administrative matters of the centre. He was involved in almost all the scientific projects at CRME, and critically analysed the data collected from various field sites, to device new control strategies. He guided many M.Sc. and M. Phil. students from various colleges and universities for their projects. He acted as a member in the Board of examiners for evaluating M. Phil degree examination. He was a life member of various scientific societies/academies. He had a good rapport with his colleagues and was always willing to involve in the scientific and technical details of the works. He was a person who was endowed with noble human qualities. He toiled for science like a saint, selflessly. This great scientist is a great example to all, young and old, and Dr I.P. Sunish carries on his legacy of perseverance, sincerity and discipline in research.

Born on January 1st, 1956 in a tiny village family near Dhampur in north Uttar Pradesh, the life of **Dr Vijay Veer** has been a meandering journey full of ups and downs which gave him a great sense of achieving something high and rich in future on his own. As a village urchin, he grew up under strict watchful solicitude of his well-to-do farmer parents in close association with nature which inherited him a strong sense of resilience and perseverance at a tender age of ten as he had got to walk for 12 km every day to attend his middle school classes. He passed his Intermediate from Raja Ranjit Singh Memorial College, Dhampur and graduated from the Hindu College, Moradabad in Uttar Pradesh. After completing his Masters from the Government College, Kotdwar (Uttarakhand) he moved to Dehradun only to meet his *alma mater*, the DAV Post Graduate College! During those days, in 1970s, there used to be very few government research fellowship programmes, and, as a result, most research scholars had to undertake their PhD research studies on their own which in every imagination was a daunting task. As in vogue in those times, young Vijay Veer came to Dehradun with a recommendation from his Professor in Kotdwara to try out his luck if he could be accepted as a research scholar in the Department of Zoology of the famous DAV College. The young Vijay Veer was taken as a research student by Prof. S.K. Kulshrestha. Days spent in the DAV College founded a solid base for Dr Vijay Veer. The school of difficulty is the best school of moral discipline. When difficulties have to be encountered, they must be met with courage and cheerfulness. Did not Aristotle, the great Greek teacher and philosopher, say that happiness is not so much in our objects as in our energies? Grappling with difficulties is the surest way of overcoming them. The determination to realize an object is the moral conviction that we can and will accomplish it. Our wits are sharpened by our necessity, and the individual man stands forth to meet and overcome difficulties which stand in his way. Young Vijay Veer, too, faced a string of difficulties but he surmounted them bravely. He mastered the art of articulation and writing good research projects. As if providentially set, at that point of time around mid-1976, Dr T.N. Ananthakrishnan, a professor from Loyola college, Chennai took over the Directorship at the Zoological Survey of India, at Kolkata. Earlier the young Vijay Veer had got a chance to visit him in Loyola College to see his collection of thrips. So, when Prof. Ananthakrishnan visited Dehradun on his maiden visit to the valley town, the first thing he did was to call Vijay Veer and gave him the much needed advice and encouragement to write a research project for funding by the Department of Science & Technology where Prof. Ananthakrishnan apparently had a say in deciding grant of funds to genuine research projects! This opportunity changed the future course of pursuit of research by Vijay Veer, and he for next 20 years was virtually submerged in the thrips bio-ecology, taxonomy biodiversity. From this point onwards he did not look back and persevered hard to make his own path, however meandering and difficult it was. He was awarded PhD degree in 1980 under the newly carved out Garhwal University.

Dr Vijay Veer was mentored and guided by some of the world's most celebrated entomologists like Prof. T.N. Ananthakrishnan, Loyola College, Chennai, Prof. J.S. Bhatti, Delhi University, Delhi, Prof. L.A. Mound, Natural History Museum, London, Prof. T. Lewis, Dr D.G.H. Halstead, Slough (UK), Dr G.S. Robinson, London, Prof. R.S. Beal Jr, California, Dr R. Kzur Strassen, Germany, and Dr I. Kudo, Japan. He met some of them and other eminent scientists such as Prof. M.F Claridge, Prof. Mark Jervis and Dr David Pinniger during his visit to Cardiff, UK and had fruitful discussions with all of them pertaining to thrips and other insects' eco-biology. There in UK he befriended several entomologists like Mr David Edwards, a freelance entomologist, Dr John Deeming of Wales Museum and Dr Hassan Ali Dawaha of Cardiff University, with whom he made field trips on insect collection in and around Wales State. Needless to over-emphasize, therefore, that when Dr Vijay Veer, co-editor with Dr B.K. Tyagi, invited Dr Mound to contribute an article for their upcoming book in the commemoration of Prof. Ananthakrishnan, he immediately accepted the offer and wrote a highly thought-provoking taxonomic paper, at a mature age of 80 years! To our immense delight, Prof. Dr M.S. Swaminathan, Chairman, MSSRF, and the 'Father of India's Green Revolution' has written the 'Foreword' for the book. titled, “Pests of forest importance and their management” (B.K. Tyagi & Vijay Veer, 2008).

Even though Dr Vijay Veer explored many insect orders in his long career with the Defence Organization, thrips could never be completely forgotten simply because these tiny insects offered him a vast opportunity to learn insect ecology and biology in the most weird manner (sexual behaviour including haplo- and diploploidy, parthenogenesis, sex selection, gall formation, plant tissue destruction, flight range, plant disease transmission), all that with only one mandible developed to full scale. Following truly the principles of collecting, preserving and identifying these tiny, yet agriculturally significant pests, Dr Vijay Veer created possibly one of the largest collections of slides of mounted specimens of the thrips from all over India, next only to those of Prof. T.N. Ananthakrishnan and Dr J.S. Bhatti. Many of these insects still remain to be fully described taxonomically, though these may form a versatile source of knowledge for any young thysanopterologist.

Dr Vijay Veer did not restrict himself to the thrips research alone as he sharpened his explorative inquisitiveness by delving into fungi-insect (dried dragonfly specimens) relationship, white grubs (scarabeids), dermestids, tineid moths, tabanids, mosquitoes etc. After completing his PhD dissertation he joined the world famous Forest Research Institute & Colleges, Dehra Dun (now rechristened as the Indian Council of Forests Research & Education, being also the headquarters of the Council) where he worked with Dr P.K. Sen-Sarma, Director, Biological Research, on the toxicity of insecticides against forest insect pests, larval taxonomy of white grubs and on diagnosis of nursery and plantation pests, and took part in several insect pests surveys in the Himalayan and Assam forest expeditions. He reported several new pests of poplar tree from WIMCO plantations in West Bengal where the pest caused huge financial loss to the cultivators. For his long association with Dr Sen-Sarma, he, along with Dr B.K. Tyagi and Dr Shri Prakash, wrote a commemorative volume titled, "*Pests of Forest Importance and their Management*' (Tyagi, Vijay Veer & Shri Prakash, 2009). On tabanids Dr Vijay Veer is an authority as much as he has a towering impact on mosquito research, as evinced by two of his recent monographs; one on Tabanidae and another on malaria mosquitoes, *Anopheles*. His forte is taxonomy and there his imprint is quite discernible: 20 new species, and many other treatments. His diversity in scientific publications can be gauged simply by having a look on the quantum of papers published (>90 in reputed journals), in addition to 6 patents, 2 Indian Standards, a monograph on tabanid flies and three books.

Dr Vijay Veer moved to Defence Research & Development Organization (DRDO) in 1984 and posted at Defence R & D Establishment, Gwalior, Madhya Pradesh where over the years he has successfully completed various projects on the pests of defence importance and their management, namely, pests of woollen textiles, semio-chemicals for rodent control, bedbugs, mosquitoes, etc. as well as diagnosis and taxonomy of dermestids, horse flies and the thrips. In the process he brought a lot of advancement in entomological researches under the aegis of his parental department, the Defense Research & Development Establishment at Gwalior, now a pioneer in vector mosquito control through innovation and invention. Dr Vijay Veer's dedicated and focused researches have taken his entomology division to new heights fostering both basic and applied research programmes. By now almost three decades in the DRDE, Dr Vijay Veer has left an indelible imprint in researches pertaining to pests of agriculture, forest, storage, veterinary, medical and forensic importance. All along he has advocated a practical utilization of all kinds of researches, keeping in mind the necessity of defense *jawans* in the frontier and/or difficult areas. The result of this well ingrained philosophy by him has paid repeatedly rich tributes to his entomology team's hard work. A few examples can be worth mentioning here: (i) Successful commercialization of the mosquito repellent, DEPA (N' N' diethyl phenyl acetamide), (ii) Invention of a hormonal control of vector mosquitoes, (iii) Discovery of a new sensillum for chemical cues, (iv) Development of an insecticide impregnated mosquito net, (v) Indigenous development of *Wool Care*, for protection of woolen textiles, (vi) Invention of *Ratox*, for rodent control, (vii) Production of *Roachtox* & *Roachline*, for cockroach control etc. His initiatives truly had a bearing on translational research since he did not cease working with the discoveries and inventions of products, but determinedly followed their possible commercialization for the benefit to all, but the *jawans* in the first place. He has also served as an expert to the 'DRDO Think Tank', better known as a Group for Forecasting and Analysis of Systems & Technologies (GFAST) situated in New Delhi during 2003-06. With all these and other great traits it is quite unlikely that his vast knowledge on defense-related entomological science can be surpassed by many of the extant entomologists in the country. Taxonomic feats, bejewelled by over a dozen *species nova* and *genera nova,* is an ample proof of his unmatched genius.

He is one of the rare defence medical entomologists who have described several *species nova* in various different orders of insects/ arthropods and/or won so many patents and copyrights etc., such as *Tabanus hisarensis* Vijay Veer, Parasahar & Rao, 1999 (Order: Diptera, and *Trogoderma halsteadi* Vijay Veer & Rao, 1994; *Thaumaglossa indica* Vijay Veer, 2004; *Anthrenus umbellatarum vijaii* Vijay Veer, 2011 (Order : Coleoptera). Dr Vijay Veer had even described a bacterium: *Chrysobacterium culicis* Kämpfer, Chandel, Prasad, Shouche, Vijay Veer, 2009.

Dr Vijay Veer is recipient of several awards including prestigious DRDO Technology Spin-off awards. He is a Fellow and Member of many prestigious national and international scientific societies and advisory council of scientific institutions and university. He is referee to many journals, and also edits *Annals of Entomology* from Dehra Dun, India. About half a dozen PhD students are working under his stewardship. He has received patents, copyrights and appreciation for a large number of developments, such as the following:

1. PATENTS:

(i) A method of defluoridation of water from fluoride contaminated water using natural waste materials, and to bring down the fluoride level below the permissible limit fit for drinking purpose (M, Lohborlang, Bodhaditya Das, Vijay Veer, Kochi Ismail, Uttam Thapa, Teiborlangmukhim, Shakira Sultana Islam),

(ii) An improved process for preparation of a composition of n,n diethyl phenyl acetamide (DEPA) and herbal essential oil for mosquito control (Vijay Veer, Reji Gopalakrishan, Prasant Raul),

(iii) A male odorant and irritant formulation for personal protective and mob control and an improved process for preparation of the same (Gogoi, BJ, Chattopadhyay, P, Chaurasia, AK, Vijay Veer (3540/DEL/2014),

(iv) Mosquitocidal trapping Device (Gopalakrishanan R, Vijay Veer & Baruah I (Design patent 264649),

(v) Oligonucleotide primer sequences for cercosporin toxin biosynthesis (CTB) genes and method of detection of cercosporin species (Meghavansi MK, Khan MH, Gupta R, & Vijay Veer (2991/DEL/2014),

(vi) A biotoilet system for complete digestion of human faecal matter with purification of effluent water for reuse (Chatterjee S, Vairale MG, Prakash A, Gogoi HK, Vijay Veer, Singh L. (No. 1654/DEL/2014),

(vii) Preparation of herbal based wound healing formulation and evaluation thereof (Chattopadhyay P, Upadhyay A, Goyary, D., Vijay Veer (No. 660/DEL/2014), (viii) Photobioreactor for growth of microalgae (Purkayastha J, Gogoi HK, Bora A, Singh L, Vijay Veer (2559/DEL/2013)

(ix) Novel Oligonucleotide Primers for Cercosporin Toxin iosynthesis CTB Genes and a Related Method for the Detection of *Cercospor*a (Meghvansi MK, Khan MH, Gupta R, Vijay Veer)

(x) Delivery of eserine and pralidoxime chloride through matrix type transdermal path formulations for prolongs release and a process for the preparation thereof (Chattopadhyay P, Banerjee S, Ghosh A, & Vijay Veer (3033/DEL/2013),

(xi) Herbal composition for snake repellent spray and a process for the preparation thereof (Chattopadhyay P., Banerjee S., Agnihotri, A, Karmakar S, Goyary D, Rudragoud P, Vairale, MG, Chaurasia Vijay Veer (No.3217/DEL/2013).

(xii) A Spine Column system and a method of solid phase extraction of target compounds involving the same (Gupta, VK, Naglot, A. Chattopadhyay, P. Chaurasia, AK, Shrimali, DD, Kalita MC, Gogoi, H.K., Vijay Veer (No. 2560/DEL/2013).

(xiii) Synthesis and development of polymeric nano composite for removal of bacteria from contaminated water and process of removal thereof (Devi RR, Raul PK, Policegourda R, Banerjee S, Vijay Veer ((No.1108/DEL/2013),

(xiv) Insect repellent nanogel formulation for clothing impregnation (Patanjali PK, Raza SK, Vijay Veer, Shweta, Kala S, Gopalakrishanan R, Baruah I. (No.2053/DEL/2013),

(xv) An improved thermoplastic for development of multi - insect repellent polymer yarn, net, fabric and tarpaulin and development process thereof (Vijay Veer, Selvamurthy W. (No.1949/DEL/2013),

(xvi) Autocidal ovitrap for monitoring dengue and chikungunya vectors (Gopalakrishnan R, Vijay Veer, Baruah I, Das M. (No 1016/DEL/2013),

(xvii) Insect repellent spray formulation for treating cotton fabrics (Gopalakrishnan R, Vijay Veer, Baruah I, Chaurasia AK, Kumar D, (No. 512/DEL/2013),

(xviii) Natural mosquito repellent lotion formulation and process for the preparation thereof (NG Das NG, Talukdar PK, Rabha B, Baruah I & Vijay Veer. (No. 1085/DEL/2013),

(xix) Bio-Tank: A subsurface process for the human waste disposal and treatment system (Singh L, Chatterjee S, Vairale M G, Kamboj DV, Pandey A, Gogoi HK, Vijay Veer (Aplication No. 2249/DEL/2012),

(xx) Bio-toilet (Singh L, Chatterjee S, Vijay Veer, Kamboj DV. Design patent No. 247200.

(xxi) Wool care composition (United States Patent No.US 8,524,770 B2, Date of Patent: Sep. 3, 2001 (Vijay Veer, Chandel K, Mendki MG, Shri Prakash, Vijayaraghavan R, Sekhar K (PCT). (Granted. In UK, France, The Netherlands, USA),

(xxii) Attractant composition for bed bugs (Parashar BD, Ganesan K, Rao YVS, Sukumaran D, Mendki MJ, Suryanarayana MVS, Vijay Veer, Shri Paraksh, Vijayaraghavan R, Sekhar K. (545/DEL/2009).

(xxiii) An Improved formulation N, N diethyl phenyl acetamide (DEPA) - a multi insect repellent protection of humans from flying, biting and blood sucking insects (Mendki MJ, Tikar SN, Gautam A, Pravin Kumar, Parasahar BD, Vijay Veer, Shri Prakash, Vijayaraghavan R, Sekhar K. (1676/DEL/2010),

(xxiv) An improved process for preparation of insecticide spray for protection of woollen items from insect pests (Vijay Veer, Shri Prakash & Rao KM. (Grant No. 189729),

(xxv) Process for preparation of rodenticide improved bait. Rao KM, Shri Prakash, Vijay Veer, Purnanand, Kumar S & Gopalan N. (Grant No. 189767).

(xxvi) A process for preparation of an improved rodenticide bait (Rao KM, Shri Prakash, Vijay Veer, Purnanand, Kumar S, Gopalan N, Malhotra RC, Pandey KS. (Grant No. 196997).

(xxvii) A process for preparation of bio-degradable controlled release insecticide matrix composition for protection of woollen fabrics (Chauhan RS, Vijay Veer, Ramana KV, Gopalan N & Rao KM. (Grant No. 190832).

2. TRADEMARK REGISTRATION:

(i) DrlPure™ (Grant No.2574253),

(ii) Defender Net™, (2391126),

(iii) Capsispray™, 2676308,

(iv) Capsigrenade™ (2831116),

(v) Wormipower™ (2826675),

(vi) Toxmos™ (2826677),

(vii) Dentrap™ (2826676).

3. BIS STANDARDS/ SPECIFICATIONS PREPARED:

(i) Method of Test for Evaluation of insectproofness of woollen textiles. First Revision IS: 7250. 1995. The Bureau of Indian Standards, New Delhi, and

(ii) Textiles- Estimation of common preservatives -part -2, IS: 3522, 1989, amendment no.1, 1992.The Bureau of Indian Standards,

New Delhi.

Dr Rakesh Kumar Jauhari (*b.* 28th December, 1955-d. 2019), Head of the Department of Zoology, D.A.V. (PG) College, Dehradun (Uttarakhand), was a self-taught medical entomologist, and a lonely crusader against mosquitoes in the Garhwal Himalayan region. He wrote a manual on "Application of Remote Sensing in Vector-borne diseases".

Almost like a tradition, in the first hundred years following the discovery of mosquito-malaria connection the entire focus for malaria control in India has been in rural settings and urban malaria, despite clear-cut policy, did not attend much attention. For the first time a due emphasis was laid on urban malaria control by the efforts of **Dr Rajander S. Sharma**, born on 10th October 1955 in Mandkola town of Faridabad District. He passed High School from Govt High School Faridabad 1971 and Intermediate from MB Intermediate College, Dadri. He obtained B.Sc. (Honours) in Zoology from M.M. Modi College, Modi Nagar and M.Sc. from D.A.V. College, Muzaffar Nagar. Having won a junior research fellowship from Council of Scientific and Industrial Research (CSIR), New Delhi he carried out his Ph.D. from Meerut University on the topic entitled "*The studies on Ecological characteristics in relationship to malaria transmission and control of vector Anophelines in Gurgaon urban*" – a subject that remained his lifelong mission.

Dr Sharma first joined as Biologist/Anti Malaria Officer under National Malaria Control Programme in 1980 and established a new department of Urban Malaria Scheme where he conducted studies on larval ecology including breeding habitats, larval and pupal density and seasonal prevalence of malaria and dengue vector, epidemiological malariometric surveys in different sectors in relation to vector density and abiotic factors. He conducted studies for the evaluation impact of different larvicides, *viz.*, fenthion, abate, Mosquito Larvicidal Oil (MLO), Pyrethrum and Paris Green in relation to Anopheline density and malaria incidence. He also carried out susceptibility status of anophelines to fenthion and Abate larvicides used for vector control, observed biting behaviour of malaria vector *Anopheles culicifacies* and *Anopheles stephensi* and environmental control for reducing the breeding places to check malaria. He continuously sensitized the community to check the man-made breeding places and environmental control to control malaria. He was key manager for management of drugs and insecticides under urban malaria control scheme. Overall responsible for monitoring of urban malaria control activities and planning, organization and executive supervision of Urban vector borne diseases. After serving for over 8 years in Haryana he joined National Malaria Eradication Programme (now National Vector Borne Disease Control Programme, as a Senior Entomologist. He accomplished most significant works here at the HQrs, NVBDCP, Delhi where he supervised nationwide monitoring of malaria and other vector-borne diseases and undertook various field investigations for mosquito-borne diseases of known/unknown aetiological agents. He later joined the department as Assistant Director. As a top entomologist of the country he was instrumental in preparing various guidance documents on management of vector control programme at national level, Plan of Action (POA) of malaria for WHO Biennium, POA for Roll Back Malaria (RBM) and their implementation in the country for the biennium 2000-2001. Besides malaria Dr Sharma also participated in outbreak investigations Japanese encephalitis and dengue in the country. He especially emphasized on dengue and *Aedes aegypti* surveillance in National Capital Territory, Delhi. He developed training material on biology of anophelines, bionomics of *Aedes aegypti* in relation to Port Health and Dengue DHF surveillance, bionomics of Japanese encephalitis and spray technique for vector control personnel in India, prevention and control of outbreaks of vector borne diseases, and operational review of entomological research projects. Because of his significant contribution malaria investigations he was appointed as the Project Co-ordinator with National Institute of Malaria Research on *"Studies on ovitraps for Aedes surveillance and control in MCD areas of Delhi"* and *"Monitoring of insecticide resistance in India and rotation of insecticides"*.

Dr Sharma again promoted to Director level scientist in 2012 at the NVBDCP, Delhi. Being senior most entomologist in the country he shifted to National Centre for Diseases Control Delhi as HOD, Centre for Medical Entomology and Vector Management and served there between 2012 and 2015 whereupon he superannuated in October, 2015 as Director-Grade-Scientist. Dr Sharma is a great colleague and a mentor; many of his colleagues notably Dr P.K. Srivastava, Dr Kalpana Baruah, Dr Kaushal Kumar, Dr S.N. Sharma and Dr Roop Kumari have, in a true sense of a great legacy of their predecessors like Dr I.M. Puri, Dr A.P. Ray etc., have earned a great reputation in the respective field of their medico-entomological and vector-borne disease control research.

From the beginning of anti-malaria times a major attention was on the control of vectors and the disease. Controlling vectors needed identification in the field. Earlier Wattal & Kalra (1961) had offered a useful guidance document to identify the vectors, but many of these mosquitoes continued to pose challenge to field workers. **Dr Jagbir Singh Kirti** (*b.* 6th December, 1956) studied electron microscopy of the vectors and resolved many issues related to their identification. Having passed his intermediate from the Government Randhir School, Kapurthala and pre-medical from DAV College, Jalandhar, he subsequently joined B.Sc. (Hons. School) at Punjab University, Chandigarh and studied in the Department of Zoology for BSc (H.S.) and MSc (H.S.) degrees, with entomology as specialization, submitting his master's thesis on taxonomy of 60 species of noctuid moths under the able guidance of Prof. Dr H. R. Pajni, a highly renowned entomologist of India. In course he researched on Curculionidae as a JRF under a USPL-480 project and the pyralid moths under a UGC project in Punjabi University, Patiala.

Having an insatiable hunger for seeking more entomological knowledge, he visited many reputed scientists' laboratories and institutes. During this period he came in contact with Prof. T.N. Ananthakrishnan, the entomological genius *non pareil*, who not only impacted a great deal on the future development of Dr Jagbir Singh but also selected him as one of the 18 young scientists for a three week-long workshop in his Entomology Research Institute at Loyola College, Madras (now Chennai) in 1990. He also visited the insect museum in the famous School of Entomology founded by the great entomologist Dr M.S. Mani, whose two highly dedicated colleagues, Prof. Santokh Singh, and Prof. H.N. Baijal, have carried out an extensive entomological exploration in the high alpine region in northern India, describing a large number of new insect species, including those of Lepidoptera. His meetings with Dr Mani, Prof. Jairaj and Prof. David left a great impact on his mind and Entomology became his first love. During a conference in Bhubaneswar in 1994, he came in contact with, besides many others, also Dr V.P. Sharma, Director, Malaria Research Centre, Delhi and Dr R. Reuben, Director, Centre for Research in Medical Entomology, Madurai, the two noted vector scientists of the country, and his interactions there led him to undertake *de novo* research on mosquito taxonomy and pioneered a new interest in medical entomology in northern States of Punjab, Haryana, Himachal Pradesh and Jammu & Kashmir etc. Though he worked on many aspects of various insect groups, yet his first love was undoubtedly the insect taxonomy using scanning electron microscopy, and offering new dimensions to lepidopterous and dipterous species identification. Inspired by the advice of Prof. Santokh Singh of St. Johns College, Agra, Dr Jagbir Singh concentrated all his attention to the taxonomic research on moths, commencing with the North-Eastern region of India. This humble, yet careful, beginning of science learning under tutelage of dedicated scientists like Dr H.R. Pajni laid the foundation of a lifelong interest in taxonomy for Lepidoptera and love for their conservation in nature by Dr Jagbir Singh which has greatly advanced the knowledge about these most elegant and useful, yet endangered, creatures.

Dr Jagbir Singh, who always thought of the rural masses who suffered the most, on one hand, the recurrent onslaught of the vector-borne diseases and health hazards of different kinds due to obnoxious insect bites and stings and, on the other hand, the annual stupendous damages to the crop and stored food products, understood much early in his career the need for translating the intriguing and arduous entomological researches done in the comforts of laboratories for an everlasting benefit to the end user – the common rural man! He, thus, popularized the entomological science to the masses including schools, colleges and rural populations. Using his vast knowledge on vectors of public health importance, he organized and addressed endless meetings and confabulated with people about the deadly mosquitoes transmitting diseases in their environments and as to how their concerted efforts could abate the onset of diseases by practicing simple preventive measures such as personal protection, hygiene and change in behaviour. To connect the masses across the country, he delivered many popular talks on Radio/TV top rimarily sensitize general public on health issues. This was a highly complicated and meandering task to perform since it demanded a lot of time and resources. To do this, often Dr Jagbir Singh had had to be away from home and his professional teaching assignments which he would not mind even on his personal cost of time and money since the first and the last word in the back of his mind was always ringing, “help”, to those who are deprived of knowledge the most and who deserve to have this first!! For the sake of bringing all the stakeholders pertaining to health issues on one platform, he regularly organizes national symposia on environmental issues almost every year. Besides, he designed many motivational programmes by selecting higher secondary students from Punjab for a week long program in Punjabi University campus and made them aware about the benefits of various different disciplines of sciences, igniting early in the budding mind the twinkle of hope and devotion of a scientist who mirrors the future growth and development of the nation in true sense. This program scripted by Dr Jagbir Singh is now so successful that +1 and +2 students are spontaneously attracted to opt for basic sciences once again in Punjab.

A major turning point in the professional career of Dr Jagbir Singh occurred when he providentially met Dr Teckwyn Johnes of Bionet International at London, in 1988, during an advance course on taxonomy. The output of this course was the establishment of a regional taxonomists’ network, SACNET, devoted to promote taxonomy in the South Asian Countries including India where Dr Jagbir Singh represented the country. It could not be a mere coincidence that a new discovery of an insect in the Namibian desert in southern Africa, leading to the erection of the latest and the 34th Order of insects, Mantophasmotoidea – a hybrid from mantids and phasmids, which not only revolutionized the entire insect classification, but also revitalized the slowly dampening science of insect taxonomy worldwide. The species, on the basis of which the Order Mantophasmotoidea (the Gladiator insects) was founded, was displayed to the participants of the Bionet International Meeting at Pretoria in 2002 where Dr Jagbir Singh witnessed a newly evolving science of systematics and phylogeny. Like a 'Ball of Fire', as fondly referred to him by Prof. K.S. Rai, a very renowned vector biologist from the Nortedam University, U.S.A, he is continuing to spread the light and removing the darkness in the realms of taxonomic researches.

Fig. 8.9. (a) Dr Jagbir Singh Kirti, (b) Dr B. Reddya Naik, (c) Dr Ashwani Kumar and (d) Dr Sajal Bhattacharya. (*Source:* Free Wikipedia/Dr B.K. Tyagi's personal archive)

Dr Jagbir Singh is an expert advisory member or chairman of many prestigious scientific and philanthropic societies such as (i) Member Program Advisory committee (PAC), Dept. of Sci. & Tech., Govt. of India, New Delhi, (ii) Chairman, North West India's Chapter of National academy of Vector borne Diseases, (iii) Convener, Punjabi University Environment Society, and (iv) President, Association of Indian Entomologists. He enjoys membership and/or fellowship of a large number scientific bodies, like (i) Indian Academy of Environmental Sciences, (ii) Indian Academy of Vectors & Vector-Borne Diseases, (iii) Indian Academy of Entomology, Chennai, (iv) Society for Advancement of Insect Sciences, (v) Society for Environmental Scientists, (vi) Punjab Academy of Sciences, Patiala, (vii) Society for Vector Ecology, USA, (viii) Founder member SACNET, Bionet - International, (U.K). He is also a recipient of many coveted prizes and awards prominent amongst which being : (i) Award of Honour by the Japanese Society of Electron Microscopy, in Amagaski (Osaka), Japan in May, 1997, (ii) Award for his outstanding work in vector biology presented by the National Academy of Vector-Borne Diseases at Patiala on Feb. 18, 2005, (iii) A Roll of Honour was awarded by Punjab University Alumni Association Toronto (Canada) for significant contributions in the field of entomology in October 2005 at Toronto, (iv) GOLD MEDAL presented by Indian Academy of Environment Sciences, Hardwar for the year 2011 on Feb. 12, 2011, (v) Biotech International Award 2011 for outstanding contributions in the field of vector biology and vector management by National Academy of Vector Borne Diseases on 15th Oct., 2011, and (vi) Global Vighyanik Award -2012 presented by Global Punjab Foundation on 13th September for outstanding contributions in the field of Environment. Dr Jagbir Singh has more than 132 research publications to his credit, three books and many in-house reports. He has also produced several PhD students many of them are well established academicians and/or scientists working in India or abroad.

Whether rural or urban environment the vectors were mostly, rather almost always as a rule, controlled both locally and nationally through application of chemicals. Since these chemicals, whether larvicides, adulticides or repellents etc. had an adverse impact on the environment. The challenge to tackle this problem was taken up by **Dr B. Reddya Naik** (*b.* 3rd April 1971). His early education was done at the Reddipally village school, at a distance of app. 2 km which he daily commuted afoot. Shri Sangameshwar was the first teacher to acquaint him with the English alphabets. The young Reddya Naik showed early in life signs of brilliance in studies. After the 3rd Class, he moved to another village called Pedda Chintakunta for further studies up to 4th Class which was 4 km from home. From the 5th to 10th Class he was put up in a social welfare hostel at Kowdipally, where, in the Zilla Parishad High School, Shri Chalamaiah, the Head Master an English teacher, made a remarkable impression on him about values of sincerity, puncuality and honesty. For his further studies young Reddya Naik moved first to Govt. Jr. College, Naraspur and then to Tara Govt. Degree College in Sangareddy Medak District. While studing graduation, he joined as a NSS volunteer and took a cue from Dr Veerasham, Lecturer in Zoology (NSS Programme Officer), to pursue further studies in Zoology. Since then the youthful Reddya Naik developed a keen interest in the disciplines of science particularly biology. Justifiably enough, he passed all his school and college examinations with flying colours. Due to his intense liking for animal science, he enrolled himself in the postgraduate subject of Zoology in Osmania University, Hyderabad which he passed with first division in 1993. By this time his interests in teaching and research had greatly intensified and he developed an insatiable desire for educating students in the finest tradition. Soon he registered for Ph.D. in Osmania University and conducted research on his thesis entitled, "*Usage of certain larvivorous fish for biological control of mosquito larvae in urban sewage water - A case study of Hyderabad*" in 2003. His thesis later became a source of inspiration to many other doctoral dissertations in the country.

Dr Reddya Naik had a highly distinguished and chequered career during which he earned both name and fame for his excellence in biological research. For his continued quality research work he had received many awards prominent amongst which are: (1) UGC Post-Doctoral Research Award from Selection and Award Bureau UGC, New Delhi, for a period of three years, i.e., from 1st May, 2004 to 30th April, 2007, and (ii) Mahatma Gandhi Peace Award-2014 presented by Subedar Amir Ali Khan Global Peace Foundation & Trust, Hyderabad. He maintains, nevertheless, that all undertakings must be carried out without even thinking about any kind of favourable return, "*Work within heart, God over head.*" His greatest virtue is 'self-belief' and would go any farther to accomplish his commitments sheer on the basis of his trust. He is credited with the exceptional quality of leadership. He organized many national/international conferences/ symposia/seminars/workshops as well as training programmes, most significant of which have been the following:

In 2008 Dr Reddya Naik organized a unique conference on the 150th Birthday Centenary of the world famous malariologist and Nobel Laureate (1902; conferred in 1904) Major Dr Ronald Ross at the recently rechristened Sir Ronald Ross Institute of Parasitology in Secunderabad-Hyderabad of which he was then the Director. Dr Tyagi was among many dignitaries who attended and delivered lectures during the colloquium. The conference was unique and pathbreaking mainly in two ways; (i) the Sir Ronald Ross

Institute of Parasitology (SRRIP) was spotlighted prominently to disseminate the spirit of focused, almost maddeningly, perseverance and application in scientific research, and (ii) many scientists, otherwise unknown or less known, who had significant memorial as well as documentary information associated with both the SRRIP and Dr Ross with which the world outside India has so far remained unknown completely, had confabulated about the chain of events that took place in course of Dr Ross's epochial and path-breaking discovery of association of the mosquito with the occurrence of human malaria, in spite of his abysmal entomological knowledge!

The State of Bengal, or for that matter the metropolis of Kolkata (old Calcutta), has more or always been the seat of knowledge pertaining to any major outbreak of a vector-borne disease. Consequently, several brilliant medical entomologists emerged from the region such as Professor A.K. Hati. To the same extraordinary league of dedicated medical entomologists belong to **Dr Sajal Bhjattacharya** (*b.* 3rd March, 1958). He completed his entire education under great financial and personal hardships till he notched up the enviable post of Professor in the field of biological sciences in Calcutta. Dr Sajal Bhattacharya's scientific career has been forged in three of the most renowned and oldest institutions of Calcutta: Asutosh College, Calcutta University, and Calcutta School of Tropical Medicine. He passed BSc from Asutosh College in 1976 and MSc with Zoology from Calcutta University (Ballygunge Science College) in 1978; the latter proving a turning point in the life of Sajal when he was also selected for a West Bengal Science and Technology's *Enquiry Research Scholarship* for undertaking a research project entitled, "Studies on *Culex vishnui* and other related vectors of Japanese encephalitis in rural West Bengal" at the famous Calcutta School of Tropical Medicine. During this period he was also selected for two other fellowships for accomplishing research on forensic science and wildlife conservation at Calcutta Medical College and Zoological Survey of India, Calcutta, respectively. But his heart was for research in biomedical science and so headed straight to enrol himself as a research scholar at the prestigious Calcutta School Tropical Medicine and received PhD in 1989 under the guidance of Prof. A.K. Hati, a renowned medical entomologist. This was the beginning of a never ending bonding between Dr Sajal Bhattacharya and the science of medical entomology, especially ecological investigations. All his life he had focused research on vector mosquitoes and diseases transmitted by them. His numerous students bear testimony to this rare development and endeavour hard to shine it further under his watchful solicitude. He has collaborated and interacted with a large number of scientists in Australia, Kenya, Hungary,

Italy, England, Portugal and Indonesia. He went to Portugal for his Post-Doctoral assignment at the Institute de Hygiene e Medicina Tropical (IHMT), Lisbon to work with Prof. Henrique Rebeiro in 2000, followed by another fruitful collaboration under the tutelage of legendary Medical entomologist Prof. C.F. Curtis at London School of Hygiene and Tropical Medicine, UK in 2004. His humility and great sense of camaraderie earned him lifelong friends and colleagues such as Prof. M. S. Chakraborty, Prof. A.P. Dash, Prof. Romeo Bellini, Dr B.K. Tyagi and Prof. Sugeng Mardihusodo, to mention a few. Prof. A.K. Hati had once said of him, "*Dr Sajal Bhattacharya is a contemporary outstanding scientist in the field of Medical Entomology. His research, especially on JE has added new dimension in the epidemiology of the transmission of the disease.*"

Dr Sajal contributed significantly in the field of vector biology particularly with reference to *Culex vishnui* but also rediscovered potential threat of malaria transmission by *Anopheles sundaicus* in the eastern peninsular India. He also worked extensively on arbovirus-vector inter-relationship. With respect to the JE virus maintenance he reinforced in West Bengal the thesis of principal and complementary vertebrate reservoir hosts keeping the virus circulating uninterruptedly in nature and also epizootic transmission occurs at the time of higher than average level of virus circulation. He is an elected Member/Fellow of the Zoological Society of Calcutta (1998) and the Royal society of Tropical Medicine and Hygiene, London (2005). Additionally he holds lifelong membership for half a dozen scientific bodies such as the Indian Science Congress Association, National Academy of Vector-borne Diseases, Association for Entomologists, International Society of Applied Biology etc. He has been also bestowed with many awards and distinctions prominent amongst which include Prof. J.B. Srivastava Memorial Award (1985) of the Indian Association of Pathologists and Microbiologists (WB), National Science Day Award (2004) of the Science Association of Bengal, Prof. G.C. Bhattacharya Memorial Award (2009) of the Science Association of Bengal, Mother Teresa International Award (2014) of the Mother Teresa International Award Committee. He acts as a reviewer for several national and international scientific journals. Dr Sajal has been the Regional Chairman of National Academy of Vector-borne Diseases (NAVBD) since 2008 besides being its executive member.

Vector Control Research Centre (ICMR), Pondicherry, ranks among the premier vector control institutes not only in the country but quite proudly in the whole world, thanks to the unstinting patronage by its founding stalwarts. Three of its scientists **Dr S. Sabesan, Dr P. Jambulingam** and **Krishnamoorthy** deserve special mention in the country's endeavours to mitigate the impact of burden from a string of vector-borne diseases, but most particularly the human lymphatic filariasis and malaria. When at Ramaeswaram island the trio worked under the charge of Dr Sabesan and had laudably worked by defining the factors of influx of malaria cases in the island through massive seasonal human migration from mainland and the neighbouring Sri Lanka island nation, highlighting at the same time the role of *Anopheles culicifacies*, the primary vector of malaria. Rameswaram (Population: 44,856 as per 2011 census), a holy town in the south of peninsular India, is wold famous for Ramanathaswamy Temple which is visited by devotees from all over the country. It was realized that the island site of pilgrimage suffered with malaria mostly due to population movement both intrinsically by its fishing human population and millions of pilgrims every year. Dr S. Sabesan, Dr P. Jambulingam and Dr K. Krishnamoorthy studied *de novo* the role of population movement on the persistent transmission of malaria in Rameswaram Island during early 1980s. While the island supports a population of 56,000, mostly fishermen, there is also a periodic, back and forth migration of fishermen between mainland villages and Rameswaram. This population movement greatly contributes to the high prevalence of malaria in both areas, since fishermen can be either donor or recipient of malaria in either place. The VCRC monitored and recorded the movement of fishermen in various seasonal camps by questioning them and by the VCRC staff accompanying them when possible. In 9 fishing camps 412 of 1098 families had migrated from mainland villages; 686 families had migrated from villages within Rameswaram Island. A mass blood survey found 138 of 4073 individuals examined positive for malaria; 107 of 680 fever cases examined were positive for malaria. Mosquito collections, the lack of permanent treatment facilities for the transient population, and ecological factors indicate a high receptivity for malaria on Rameswaram Island. With the island attracting between 1000-4000 tourists daily and over 200,000 travelers annually between India and Sri Lanka, evidence exists for considerable danger from the importation of chloroquinine-resistant malaria strains into Rameswaram. Adequate attention to human ecology will be needed for malaria control in this area.

Later, when at Koraput (Odisha) around 1985-86, Dr Jambulingam, who subsequently also held the reigns of the Centre in his hands, led the project on tribal malaria and established many unknown or less known facts about tribal malaria; discovery of *Plasmodium ovale*, incrimination of many vectors among which *An. culicifacies* was predominant, followed by *An, fluviatilis*; increased proportion of splenomegaly in tribal population etc. It was a great experience for me, too, as I was designated as "Co-ordinator" for the whole project, and explored the entire region along with Dr P.K. Das, Dr S.P. Pani and Dr P. Jambulingam.

Right from the time of Dr P.K. Rajagopalan research on lymphatic filariasis had been the Centre's main forte which Dr P.K. Das furthered effectively, and three of its scientists, namely, Dr Krishnamoorthy, Dr Subramaiam and Dr Vanamail eventually added substantial new knowledge by devising practically applicable mathematical models etc. in the elimination of filariasis. Dr S.L. Hoti and Dr Vasuki Venkatesan contributed enormously by inventing a xenomonitoring methodology for executing surveillance post-elimination of filariasis. To the later methodology's validation I was working as one of the evaluators of the technology which proved very effective. In the same line of extraordinary research on lymphatic filariasis the VCRC had during 1980s conducted a massive LF (*Brugia malayi*) control through environment management and community participation under the respective charge of Dr K.N. Panicker and Dr S. Sabesan, and in the process offered to the world a grand model of control of the disease by simply vector eradication through biological and environmental intervention with the community participation.

One of India's most beautiful places and the smallest State, Goa (3702 sq. km, with a coast line of 104 km and a population of 1.45 million), was endemic for malaria for decades. However, during eradication phase, malaria incidence was greatly reduced till mid- 1980s when only a few cases were reported. A serious epidemic of malaria occurred from 1986-88 in Panaji where after upon the request of Government of Goa, a field unit of National Institute of Malaria Research was established at Panaji on 6th Sept. 1989 with objectives to (i) Ascertain factors responsible for malaria outbreak and its social, geographical and health impact, (ii) Study bio-ecology of malaria vector/s and devise suitable control strategy, (iii) Demonstrate malaria control with bio-environmental measures engaging with cross section of the society, (iv) Transfer malaria control technology to the state, (v) Provide regular training and research support to the state NVBDCP and strengthen malaria control activities in the state, and (vi) Provide need based consultancy and advise to NGOs working in health sector and establishments. **Dr Ashwani Kumar**, who is currently the Director of Vector Control Research Centre, Puduchery, was earlier heading the field unit of National Institute of Malaria Research in Goa since its inception when malaria virtually affected the entire region. He is a medical entomologist with more than 25 years of experience in the field of vector-borne diseases. His main research interests include vector biology and control, epidemiology, disease burden estimation and malaria control/elimination. It was his meticulous and sturdy planning to control malaria in Goa through integrated vector control strategy using largely fishes as biocontrol agents that had not only brought the cases to near zero but offered a new hope to repeat the same example in many other sites in the country to eliminate malaria. Besides a member of Vector Control Advisory Group, Dr Ashwani is serving as an Adviser to WHO Global Malaria Programme, Malaria Policy Advisory Committee-Evidence Review Group on Global Malaria Burden Estimation. Honoured with as many as thirteen awards at National and International levels he is also head of regional office in India for the Society of Vector Ecology since 2018.

Likewise, malaria hit very badly the overpopulating 1980's Bengaluru metropolis, once a salubrious town with hardly any major outbreak of disease in old Bangalore of post-independence period, besides other parts in the State. Karnataka with the vast semi-urban areas, rich irrigated lands and good monsoon, facilitates prolific growth of mosquitoes and transmission of malaria. Malaria in the state has been due to *P. vivax* and *P. falciparum* infections with a high proportion of *Pv* compared to *Pf*. The major vectors of malaria in the state are *A. culicifacies* and *A. stephensi*, latter is an urban vector. The third vector *A. fluviatilis* also reported from foot hill regions. Previously Karnataka was categorized as one of the high-transmission zones in India. A total of 3650 cases reported in 1961, was increased to 37,725 in 1971 and thereafter the state has witnessed the upward trend with a maximum of 6,29,977 cases in the year 1976 and this high malarious situation continued till 2003 when a total of 1,00,220 cases were reported. However, from this point onwards malaria incidence started decreasing and a steady and continuous decrease with a reduction of 89% was recorded as compared to 6 the incidence reported in 2000. Malaria situation in the state is never a *Pf* predominant. Out of the total cases reported during 1961 to 2015, *Pf* cases range from a minimum of 2% to maximum of 59%. More than 30% *Pf* incidence was recorded largely during 60s. The API of the state ranges from 0.1 to a maximum of 22.3 in 1976 and the SPR ranges from 0.1% to a maximum of 20% again in 1976. **Dr S.K. Ghosh,** heading the Field Unit of National Institute of Malaria Research in Bengaluru (old Bangalore) since its inception, offered a great and practically replicable model for malaria control by simple yet meticulous biological and environmental tools (Ghosh *et al.*, 2012). He reinforced innovatively deployment of different fish species for controlling vector population copiously breeding in a large number of ponds and lakes in and around Bengaluru. His model was so successful that the entire state followed it in combating malaria. To begin with, in 1993 one highly malaria problematic primary health centre (PHC) Kamasamudram (population over 37,000) was selected as a demonstration site to evolve a non-insecticide method of malaria control. This PHC is situated 100 km east of Bangalore city. Entomological surveillance of vector species showed the presence of *An. culicifacies* and *An. fluviatilis*. After long deliberations, it was decided to undergo LSM against these vectors. Geographical reconnaissance was carried out to identify and locate the potential vector breeding habitats. Among various breeding sources, wells, ponds, and streams were the main breeding habitats of the two vector species. It was found that malaria was comparatively numerous in six sub-centres of this PHC. Interestingly, there were very less malaria in the remaining five sub-centres. This area is connected with the Kolar Gold Field which was operational from 1896. Further analysis of malaria distribution in the six sub-centres showed that villages having more number of malaria cases are surrounded by wells and ponds. On the contrary, villages located on streams had less number of malaria. Based on this information, genetic analysis of the vector species was carried out. Two sibling species A and B of *An. culicifacies* and, species T of *An. fluviatilis* were present. Species A was more numerous in villages having wells and ponds, whereas species B in villages located on streams. Based on laboratory experiments, species A is an efficient carrier of malaria parasites, while species B exhibited partial refractoriness to *P. falciparum* and complete refractoriness to *P. vivax* infections. All *An. fluviatilis* belonged to species T, and were zoophagic indicating non-participation in transmission. This information was very important that helped in designing the LSM of the vector species. A pre-plan for channelization of the streams with a huge cost was thus abandoned. Subsequent surveys in the non-malarious sub-centres detected the presence of Poecilia in most of the breeding habitats. Entomological surveillance also indicated the absence of the vector species. This was a great breakthrough in our search for planning a suitable vector control strategy. Accordingly, Poecilia was introduced in all the breeding habitats especially in ponds and wells in 1994. This resulted in almost 50% reduction on malaria incidence in 1995 from an average annual parasite index (API; cases per 1000 population per year) of over 40. Entomological surveillance and Poecilia monitoring indicated the ineffectiveness of this fish in ponds. In late 1995, approximately 500 Gambusia were released in a ditch connected with the main channel of Kamasamudram tank. In 1996 a heavy flood had occurred, and these fish were dispersed in the entire area along with the flood waters. Subsequently, this resulted in a total elimination of malaria in this area in the subsequent years for over 15 years. Entomological monitoring also showed very level of vector abundance that did not support malaria transmission. In 2001, the trial was extended in a highly malaria-endemic area having a population of 1.3 million. Here the average API was over 130 and malaria has reached to a near elimination phase. Now, the program is being extended to the entire state.

Lastly, but not with a cul-de-sac in our story-telling, **Dr Brij Kishore Tyagi**• has given a new meaning to the sciences of medical entomology and vector-borne diseases, since right from the beginning of his training as a scientist he considered scientific research as one dealing strictly and mainly with discoveries, inventions and innovations! Born on 23rd July, 1951, Dr Tyagi's formal education took place in Christian Intermediate College, Bareilly (1962-64), Parker Intermediate College, Moradabad (1964-67), Government Intermediate College, Meerut (1967-69) and later graduated from D.N. Degree College, Meerut (1969-72) and postgraduated from J.V. College, Baraut in Meerut district (1972-74). It was during his MSc that he developed interest towards research and completed a highly commendable project on freshwater fishes, thereby also securing highest marks in the practicals as well as first division in MSc. This sparked a turning point in his life and he yearned to undertake higher education by registering for the PhD degree under the guidance of Prof. S.K. Sangal, a noted odonatologist, at Dayanand Anglo Vedic (Post Graduate) College, Dehradun. Inspired by Dr Sangal's work on dragonflies, particularly, *Crocothemis servilia* Drury, Dr Tyagi came in contact with some of the greatest entomologists and cytogeneticists, such as Prof. P.K. Sen-Sarma, an internationally acclaimed termite expert and the Director, Biology Research, Forest Research Institute & Colleges, Dehradun, Mr. Pratap Singh, Forest Entomologist, *ibid.*, Prof. T.N. Ananthakrishnan, an unparalleled thysanopterist and the Director, Zoological Survey of India, Prof. G.P. Sharma, an internationally reputed cytogeneticist and Head, Zoology Department, Punjab University, Chandigarh, Prof. J.S. Yadav, a celebrated geneticist on Coleoptera, in particular, and Principal, Choturam College, Jind in Haryana, and of course Dr S.K. Kulshrestha, a colleague of Dr Sangal in the same department, Dr Santokh Singh, noted high altitude entomology expert and Head, School of Entomology, St. John's College, Agra, Dr H.N. Baijal, a famous collembologist and Head of Zoology Department, Agra College, Agra, Prof. J.M. van Brink, Editor of *Genetica* and *Genen en Phenen*, and a senior professor at the Utrecht University's Institute of Genetics, and many more...! But, one person who made the

• The material about Dr B.K. Tyagi has been adopted through his courtesy from Vijay Veer (2012).

most significant impression on his persona is Prof. Bastiaan Kiauta, Department of Animal Cytotaxonomy & Cytogenetics, Institute of Genetics, Utrecht University, The Netherlands and also the Executive Editor, ODONATOLOGICA as well as the Head, *Societas Internationalis Odonatologica* (SIO) and the Director, Khumbu Himal Research Expedition of the Netherlands Centre for Alpine Biological Research (N.C.A.B.R.), Utrecht in collaboration with Nepal Research Centre, Kathmandu. It was the early teachings of Prof. Kiauta who had not only guided Dr Tyagi towards accomplishing successfully his doctoral dissertation, but also taught him the lessons in performing science in the best tradition! After completion of his doctoral thesis in 1978 Prof. Kiauta editorially abstracted his thesis in the international journal ODONATOLOGICA and expertly commented on Dr Tyagi's PhD thesis (Dec., 1978; (*cf.* OA No. 2683) titled *"Studies on the chromosomes of Odonata of Dun Valley (Dehradun, India)"* as: "*The comprehensive volume is based on original observations on 45 spp. (the main cytotaxonomic data on 13 of which were subsequently published elsewhere (cf. OA No. 2665), leaving as new Anisopleura lestoides [n=13, m] only on a thorough literature study. It is organized into thirteen chapters, incl. a comprehensive bibliography and a synopsis of chromosome numbers recorded in the order up to 1977, and presents several novelties for the regional and/or Indian fauna. The book was not issued commercially, but a substantial abstract is available from the author."* – **(Abstractor's note: The author should be congratulated on the production of this work, which has been carried out under often hard personal circumstances and under a great financial stress).** At the same time Prof. G.P. Sharma, his first PhD thesis examiner, had commented ".... a phenomenal research work" while another evaluator, Dr P.K. Sen-Sarma wrote ".... such a massive research work must be published as a priority".

In 1980 Dr Tyagi, a CSIR senior research fellow then at the famous Forest Research Institute & Colleges, Dehra Dun, received his CSIR-postdoctoral fellowship, and soon later in January 1981, the Research Associateship, while undergoing training from the legendary termite systematics and evolutionary biologist Dr P.K. Sen-Sarma. There he investigated ecology and control of termites using certain fungi. Here he came briefly in contact with Prof. R.N. Roonwal, one of the greatest zoologists India has ever had!

In 1981 Dr Tyagi joined as Senior Research Officer on a WHO-TDR funded research project, "A laboratory and field study of malathion resistance in *Anopheles culicifacies*", under Malaria Research Centre (ICMR) in collaboration with the NMEP, Delhi. After obtaining necessary training and exposing himself to a variety of laboratory and field related malariology in the field of vector ecology, biology, insecticide resistance, cytogenetics, mosquito colonization and malaria epidemiology for three months, both in Delhi and Haldwani, UP, he was transferred to the project site at Ukai in Songarh tehsil in Surat district, south Gujarat State, as its Senior Research Officer in-Charge. During course of this project (1981-84) he mastered in ecology of vector species, *An. culicifacies* complex, as well as other malaria vector species, in the heavily forested and mountainous terrain of southern Gujarat, simulated hut experiments with malathion-resistant strains of *A. culicifacies*, larval breeding of various anopheline mosquitoes, and susceptibility status of major malaria vectors against insecticides in vogue then. He was principally responsible in organizing for sampling the newly described 'Species B' of *An. culicifacies* sibling species complex, from the Ukai and surrounding areas. In 1984, Dr Tyagi got the scientific cadre in the Indian Council of Medical Research at its institute of Vector Control Research Centre, Pondicherry. Here he researched on a variety of subjects like evaluating different types and insecticides, both conven-tional and candidate compounds, as larvicides, adulticides, JHs/IGRs/ IDIs, repellents/attractants etc. (important being the evaluation of a new amide group larvicide, VCRC/INS/A-23, and determination of mosquitocidal properties of ARTILIN paint, fenozycarb, arosurf, bendiocarb and alphamethrin etc.). Dr Tyagi was also in-Charge of various important outstation field projects such as (i) the Koraput project on control of malaria and ecology of vectors (1985-1987) whereupon he co-ordinated research and administrative activities of the Koraput Field Station at Jeypore (Orissa) with VCRC Directorate. His field work in Jeypore, Orissa, a tribal belt, was a very ambitious programme since the region had a history of hyper endemic malaria for decades, with *P. falciparum* contributing as high as 90% in certain areas. During course of malariological investigations in the Koraput district, he also worked on the tribal population in the seclusive Bonda hills and others in Malkangiri district, (ii) Bangalore Mosquito Control Project, where he was chiefly involved in planning, organization and implementation (by personally joining with the staff at the field station at Bangalore), and finally authoring country's one of the first and significant reports on the "Master plan of mosquito control in Bangalore metropolitan city (1987)", which proved a landmark for other metropolises as well. In 1988 Dr Tyagi was transferred to Desert Medicine Research Centre in Jodhpur and got promoted to the post of Assistant Director in July 1990 and to the post of Deputy Director in July 2000. Most of these years, i.e., between 1991 and 1999, he also looked after the work of DMRC as Local in-Charge. In The Thar Desert he conducted many novel research activities, while also winning for the Centre its first WHO (2001: Identification and classification of larval breeding habitats for determining adult population estimates, US$ 55,300.-; 2001: Distribution of Thar malaria mosquitoes belonging to the *Anopheles culicifacies* and *Anopheles stephensi* complexes; app. US$ 50,000.-) and IWMI (Control of vectors of malaria in relation to domestic/canal water supply strategies in Jaisalmer district, the Thar Desert, India; US$ 9,000.-) funded extramural research projects on malaria, in addition to ICMR Task Force (extra-mural) project (2001, Evaluation of disease burden due to leptospirosis, Rs. 1.60 Lakh). He thoroughly investigated ecology, biology and blood feeding behaviour of the principal malaria vectors, *An. stephensi* and *An. culicifacies* in the Thar Desert particularly under the impact of the extensive canalization by the Indira Gandhi Canal (640 km, with over 8000 km network of distributaries), and travelled all over the Thar Desert to understand the impact of anthropocentric behaviour over malaria exacerbation or conflagration. The culmination of his thirteen years long research work was the production of several internationally acclaimed research papers in high Impact Factor journals, such as *The Lancet*, topped with the unique monograph, "Malaria in the Thar Desert", published in 2002. Prof. P. Amerasinghe, an internationally acclaimed mosquito ecologist and malariologist of the International water Management Institute, Colombo, Sri Lanka, has once commended Dr Tyagi as, "*.... Tyagi is best known internationally among malaria and medical entomology circles for his long years of research into mosquitoes, malaria and irrigated agriculture in the Thar Desert of Rajasthan, India.*" (*cf.* Dr Felix P. Amerasinghe's *Foreword* dated 7th April, 2004 in Dr B. K. Tyagi's *The Invincible Deadly Mosquitoes: India's health and economy enemy # 1.* Scientific Publishers (India), Year 2004, pp. 276). He was the first to have offered a classification of the Thar Desert's physiography on the basis of malaria vectors distribution, linking this with that of the type of dominant malaria parasite in the Thar. Innovatively, he designed a *"Tanka"* lid to prevent the breeding of vector, *Anopheles stephensi*, the species predominant in the Thar's xeric environments and best adapted inherently in the local, ageold underground reservoir, the Tanka. He was also the first to have linked the traditional man-made well - like water reservoirs in the belly of ponds, called *"Beri"*, with the malaria vector breeding (Tyagi and Yadav, 1996). It was during this period of his research that two eminent scientists visited him and worked in his collaboration: (i) Prof. Peter Miller from Zoology Department, Oxford University, United Kingdom, who came to India on a fellowship facilitated jointly by the Indian National Science Academy, The Trustees of the Aneurin Bevan Foundation and the Indian

Council for Cultural Relations to work on Odonata in Madurai Kamaraj University, Madurai and at Jodhpur under the aegis of SIO-National office in India, and (ii) Dr Manu Thomas, Zoology Department, Madras Christian College, Tambaram, Chennai, who received a Visting Fellowship from the Indian National Science Academy, Delhi under Dr Tyagi's guidance at the DMRC, Jodhpur. With Prof. Miller, he even published a research paper (Tyagi, B.K. & Peter Miller, 1991. A note on the Odonata collected in south western Rajasthan, India. *Notulae odonatologicae* 3: 134-135).

The Thar Desert offered many challenges since some of the major vector-borne diseases were specific to this hostile environment, such as the guinea worm disease or dracunculiasis and cutaneous leishmaniasis, while some others were on the rise in incidence and intensity. He studied these diseases in great detail and reviewed them for the Thar (Tyagi, B.K., 2001. Dracunculiasis eradication in India, with special reference to the Thar Desert (Rajasthan). (In: *Helminthology in India.* (Ed.) M.L. Sood], International Books Distributors, Dehradun. pp. 543-555; Tyagi, B.K., 2003. Cutaneous leishmaniasis in the Thar Desert: a critical appraisal of the fast emerging "sand-disfiguring disease". In: *Advancement in insect biodiversity*. Ed. Prof. Rajiv Gupta, JNV Univ., Jodhpur, pp. 313-323). He served as the full-fledged Deputy Director & Officer in-Charge, Desert Medicine Research Centre (ICMR), Jodhpur during 1999-2001. He received his first international prize, the First WHO-TDR Prize in providing a panoramic view to the achievements, as well as future course of action, of the TDR, in 1995.

Having been recognized as a PhD Guide by many Universities including Madurai Kamaraj University, Madurai; Central University, Pondicherry; Mohan Lal Sukhadia University, Udaipur; Punjabi University, Patiala etc. he has examined several doctoral dissertations, notably of Dr Somasundaram (a student of Prof. S. Chokalingam), Dr R.J. Andrew (a student of Prof. D.B. Tembhare) and Dr Manu Thomas (a student of Prof. Gladstone), and many a Masters Dissertations.

At CRME (ICMR) in Madurai, his real *karmakchetra,* where Dr Tyagi joined in 2001, he became its Director in-Charge in May 2005, Dr Tyagi has brought in phenomenal changes in its research capabilities and infrastructural development. His leadership qualities are unmatched, being absolutely transparent and accessible in all his dealings. His dedication and intensity for high quality research was visible through a large number of research programmes funded by international bodies which he had initiated at the Centre during past nearly one decade. No wonder, therefore, that he soon became a scientist of many "firsts", in context with research in the science of medico-entomology, but transformed CRME to a formidable force of knowledge in medical entomology globally, as Prof. Dr Graham B. White, an international authority in medical entomology in the Department of Entomology and Nematology, University of Florida, Gainesville, USA, while recommending Dr Tyagi's nomination for the WHO's STAC membership wrote on 14th February, 2011 in his appreciation:

> *"... For the past three decades, Dr Tyagi has contributed greatly to research and leadership on vector-borne diseases across the spectra of biological science and organizational systems. During his career with the Indian Council of Medical Research (ICMR) his effectiveness and productivity transcend the norms of ICMR responsibilities: while zealously performing his formal research management duties, Dr Tyagi emphasizes the pastoral side of institutional development, and somehow finds the time and energy to produce inspirational books on biological themes. His personal research began with insecticidal product evaluation and development, collaborating with many agencies and industries for innovative technology, progressing to implementation research and cost-benefit analysis of components of integrated management and prevention of vector-borne diseases. One cannot exaggerate the versatile skill that Dr Tyagi brings from PI management of grants from TDR and other agencies, for projects spanning many States of India, various disciplines and groups of TDR client countries, on wide ranging questions regarding arboviruses, malaria, filariasis, genetic control, and VBDs generally. His well balanced enthusiastic stewardship of everything he takes on, the depth and breadth of his biological knowledge, multi-disciplinary and visionary approach, administrative and international experience in fields currently acquired by STAC, make Dr Tyagi highly suited and willing to serve TDR in his capacity."*

This truly sums up the leaps forward Dr Tyagi has made in both his research and professional arenas, taking all along the CRME's own unprecedented development and prosperity in research capability and infrastructure as his *numero uno* priority and responsibility. He always looked after the scientific development of all the employees of the Centre, be it scientific, technical and administrative. That is why, setting a rare example in the whole Council, he sent a large number of technical staff for overseas training. As if this was not sufficient, he even arranged training and/or conference participation abroad on several occasions for the scientists working in temporary and ad-hoc projects. At CRME, where he became its Director in-Charge in May 2005, Dr Tyagi has brought in phenomenal changes in its research capabilities and infrastructural development. A careful visionary and schemer of all his activities, he sees accomplishment of each and every assignment, however small and trifle looking, once he is committed. His leadership qualities are unmatched, being absolutely transparent and accessible in all his dealings. He rewarded the benefits of the institute to its employees, especially scientists, in the forefront of all his functions. He strengthened diverse collaborative research projects with a great number of institutions overseas like Institute of Hygiene and Epidemiology, Hanoi (Vietnam), Brandeis University, Waltham (USA)U, Bamako University (Mali, Africa), Mahidol University at Salaya (Thailand), Institute of Medical Research, Kuala Lumpur (Malaysia), University of Oxford /Oxitec (UK), and a large number of institutes within the country such as National Institute of Mental Health and Neurosciences, Bangalore; Osmania University, Hyderabad, Bharthair University, Coimbatore, Bharathidasan University and Anna University, Tiruchirapally; Vellore Institute of Technology, Vellore; Madurai Kamaraj University, Madurai; Mepko Schlenk Engineering College, Sivakasi; International Clinical Epidemiology Network, New Delhi, just to name a few, in addition to several of ICMR institutes.

A recipient of all major research fellowships, *viz.*, Junior Research Fellowship (Z.S.I., Calcutta; 1977), Senior Research Fellowship, Post Doctoral Fellowship and Research Associateship, all of CSIR, Delhi at the FRI & Colleges, Dehradun; 1979-1981), Dr Tyagi is a Fellow of Royal Entomological Society, Fellow of Indian Society of Malaria & Communicable Diseases, Fellow of Applied Zoologists Research Association, Cuttack, Fellow of the Academy of Zoology, Fellow of International Biotechnological Society, and Member of over a dozen national and international scientific societies, *viz.*, International Odonatological Society (SIO), Holland, Indian Sci. Congr. Assn., Calcutta, Zool. Soc. India, Gwalior, Odonata Specialist Group of the *Species Survival Commission* (IUCN), Gland, Switzerland, Z.O.O. Outreach Orgn., Coimbatore, Indian Society of Parasitology, Lucknow, Invertebrate Conservation & Information Network of South Asia (ICINSA) and Founding Fellow of the Entomology Academy of India, Chennai. Dr Tyagi, who himself had pioneered in producing a dragonfly journal, *Indian Odonatology* (1988), has been on the editorial/referee boards of about three dozen scientific periodicals published both nationally and internally, such as *Odonatological Abstracts* (Holland), *Notulate odonatologicae* (Holland), *Fraseria, Geobios New Reports,* Jodhpur, *Journal of Applied Zoological Research, Journal of Indian Society of Health Education and Environmental Research, Annals of Medical Entomology, Indian Biologia, Indian Journal of Environment and Eco-planning, Zoo Outreach Journal, Entomon, Indian Journal of Medical Research, Journal of Vector Borne Diseases, Journal of Environmental Biology, Journal of Parasitic Diseases, Advances in Medical and Dental Sciences, BioTechniques, Gene Research, Acta Tropica, BioMed Central Public Health, Tropical Biomedicine, Caspian Journal of Environmental Sciences, Polish Journal of Microbiology* etc. He has been a member of Executive Council of the International Odonatological Society (Societas Internationalis Odonatologica, SIO) 1981-95 (Holland); National Representative in-Charge, National Office of the International Odonatological Society (SIO) in the Indian Subcontinent, 1981-88; Liaison Officer of International Odonatological Society (SIO) in the Indian subcontinent1983-93; Head, Regional office in Southern Asia of the International Odonatological Society (SIO: Central Office in Bilthoven, Holland), 1988-95; Member in-Charge, Odonata Specialist Group of the species Survival Commission 1985-95; Member, Executive Council of Nature conservators (NATCON), Muzaffarnagar, India 1988; Member, Executive Council of Applied Zoologists' Research Association (AZRA), Cuttack, India; Scientific Secretary, 15th National Congress of Parasitology; Oct. 1-3, 2001, Jodhpur; President & Org. Secretary, 8th Int. Symp. of Vectors & Vector-borne Diseases (Oct. 13-15, 2006), Madurai; Member, Advisory Board, UMAA (Uttarakhand Mosquito Abatement Association); Member, Academic Council, Lady Doak College, Madurai, 2008.

Dr Tyagi has widely travelled across the globe (Australia, Bangladesh, China, Canada, Colombia, Czech Republic, Denmark, France, Malaysia, Mali, Mexico, Nepal, Singapore, South Africa, Sri Lanka, Switzerland, Thailand, The Reunion Republic, USA, The Netherlands etc. in connection with either presenting invited lectures in international conferences, chairing scientific session in symposia, representing as an expert on behalf of Indian Council of Medical Research or as advisor to the WHO. He had several occasions to discuss or work with a large number of legendary entomologists such as Prof. Minter J. Westfall Jr. (with Prof. James G. Needham co-authored *Dragonflies of North America*), Prof. Philip S. Corbet (*A Biology of Dragonflies*), Prof. J.M. van Brink (Editor, *Genetica* and *Genen en Phaenen*), and Prof. B. Kiauta (Executive Editor, *Odonatologica*) which greatly influenced his life. Dr Tyagi won his first international award/prize from the World Health Organization-Special Programme for Research &

Training in Tropical Diseases (TDR) for a worldwide competition in 2005-06. Recently, he was awarded with the ICMR's most coveted Dr M.O.T. Iyengar Memorial Award (2007) 'For Outstanding work in the field of Biomedical Research'. He received AZRA (Applied Zoologists Research Association) Award in 2002. He is the Founder President of Society of Medical Arthropodology (2016) and Indian Dragonfly Society (2017). Dr Tyagi is a scientist with many 'firsts':

(i) He is possibly the first medical entomologist to have filed a patent application under the ICMR in as early as 1992 (NRDC, New Delhi; Patent Grant No. 191635 in the year 2005).

(ii) He was the first to have given a classification based on the distribution of malaria vectors, *Anopheles stephensi* and *An. culicifacies,* corresponding to that of prevalence of the malaria parasites, *Plasmodium vivax* and *P. falciparum* in the Thar Desert region, in 1995.

(iii) He has provided for the first time a correlation between age old water conserving underground storage wells called '*beri*' and '*tanka*' in the Thar Desert with that of the existence of *An. stephensi* the truly desert species, in 1996.

(iv) In the Thar Desert, he designed for the first time a 'Tanka Lid' to prevent breeding of *An. stephensi* in the tanka, its most common breeding habitat in the xeric ecosystem, in 2000.

(v) He first demonstrated that *Aedes albopictus* in the Kerala's Western Ghat region is able to transmit dengue in nature without the support of the principal vector of dengue in the country, *Aedes aegypti,* and emphasizing role of the former in Kerala as a primary vector, in 2002.

(vi) He ventured to describe a new anopheline species, *Anopheles pseudosundaicus* Tyagi *et al.*, 2009, after a gap of nearly six decades of the last Anopheles species described in India.

(vii) He for the first time laid the foundation of the science of Medical Arthropodology in India through regularly held annual conferences since 2007.

(viii) He is possibly the only scientist in the country who brought out regularly biobibliography of prominent medical entomologists from across the country – an invaluable treasure, to be remembered and emulated by the future generations of entomologists. So far he has compiled 16 biobibliographies (*e.g.,*

Prof. T.N. Ananthakrishnan, Prof. C.F. Curtis, Dr Rachel Reuben, Dr M.O.T. Iyengar, Dr Vijay Veer, Dr R. Rajendran).

(ix) Dr Tyagi has also additionally published original biobibliographies of several nationally and internationally zoologists/ entomologists, both from India and overseas, such as Prof. Bastiaan Kiauta, Prof. T.N. Ananthakrishnan, Prof. P.K. Sen-Sarma, and Prof. I. Prakash.

(x) He was instrumental in executing the First Transfer of Technology on "Detection of antigen for Japanese encephalitis virus (JEV) from the wild-caught vector mosquitoes as a disease surveillance tool" by the CRME (ICMR), Madurai to the Directorate, Public Health & Preventive Medicine, Tamil Nadu, on July 25, 2010.

Dr Tyagi has to his credit over 600 titles, many of which have been published in the international journals of high impact factor. He has authored/edited as many as 37 books, which include several monographs, for example, *Malaria in the Thar Desert* (2002), *Handbook of Medical Entomology* (2003), and *The invincible deadly mosquitoes* (2004). Other highly acclaimed books written by him include (i) *Entomology: Ecology and biodiversity*, (ii) *Odonata: Biology of dragonflies*, and (iii) *Pests of forest importance and their management*. He has even written a highly thought provoking book, 'Scientist', which is widely read by young and budding researchers and students alike.

Dr Tyagi is functioning as a discipline expert/research project reviewer of various different departments/organizations, such as Department of Biotechnology, Ministry of Forests & Environment, Indian Council of Medical Research, Department of Science & Technology (West Bengal, Kerala, Delhi, Tamil Nadu) and Indo-US Science & Technology Forum (IUSSTF), University Grants Commission etc. He is an expert member of the DBT-RCGM's Subcommittee for inspection of the infrastructural facility and for monitoring the R&D work of Asian outcrossed RIDL OX-513A Strain of the GMO *Aedes Aegypti* mosquito" at both the M/S Gangubhushan Bhikulal Investment & Trading Ltd (GBIT), Jalana, Maharashtra and the International Institute of Biotechnology & Toxicology (IIBAT), Chengalpet, Tamil Nadu. The DBT also appointed him as their representative on the Institutional Biosafety Committees of (i) Department of Biotechnology, Lady Doak College, Madurai, and (ii) Mepko Schlenk Engineering College, Sivakasi, Tamil Nadu. He is also a member of the Expert Subcommittee of the DBT-RCGM for formulating Biosafety Guidelines to conduct and monitor Confined Research Trials (CRTs) on the genetically engineered (GE) insects. He is Chairman of DBT-RCGM's re-constituted Subcommittee for formulation of biosafety guidelines to conduct and monitor research trials (CRTS) on Genetically Engineered (GE) mosquitoes.

Having worked on a large number of insect orders and families, he is also one of very few entomologists who have an equal grip on more than one group of insects. Accordingly his publication units numbering over 600 are based on a vast variety of subjects and insect orders such as mosquitoes, dragonflies, house flies, sandflies, cyclops, springtails, thrips, termites, fungi, malaria, filariasis, dengue, chikungunya, and Japanese encephalitis etc. In addition, he has authored/edited books on as varied subjects as medical entomology, mosquitoes, malaria, biology of dragonflies, forest pests, desert ecology and biodiversity and vector control. In pursuance of diverse knowledge he has worked in almost all parts of the country. He has been always ready to move on to a new posting, and never considered one as matter of relegation; instead he considered such an opportunity to learn more and lead from the front!

Dr Tyagi described his first insect species, *Calicnemia doonensis* Sangal & Tyagi, 1984 from the Dehradun Valley and his latest, *Anopheles pseudosundaicus* Tyagi *et al.* (2009) from the post-tsunami Kerala. His colleagues from the CRME have recently dedicated a new species in his honour, *Toxorhynchites tyagii* Krishnamoorthy *et al.* (2013).

Dr Tyagi is a highly systematic planner and a careful visionary. A workaholic scientist, he arduously works every day for 14-16 hours, besides being an astute administrator. Prof. T.N. Ananthakrishnan, the Founding Director of Entomology Research Institute, Chennai and Ex-Director, Zoological Survey of India, Kolkata, and unquestionably India's most outstanding and extraordinary entomologist in modern times, who was also the Chairman of the Centre's Scientific Advisory Committee in the CRME's formative days in mid-1980s, has chronicled Dr Tyagi's growth in science since his PhD scholarship days and lauded his achievements (14th March, 2009) as:

> *"... I am privileged to be able to testify to the uncanny abilities of Dr B.K. Tyagi, Director, CRME, as a medical entomologist of high calibre and whose dedication and perseverance has earned for CRME the status of internationally acclaimed institution in the field of Medical Entomology. This is justified by the diversity of his numerous publications in journals of high standing. The solid*

foundation laid by the first Director, Rachel Reuben, has been further strengthened by his personal efforts, not to mention the interactions with his colleagues... . Friendly, alert, interested in diverse subjects concerned with Medical Entomology, Dr Tyagi has been considered as an outstanding Medical Entomologist whose efforts to tide over many diseases like Dengue have been well known. The Annual Reports prepared by him reflect the diverse activities of CRME under his care and needless to say that in Dr Tyagi we have a scientist with an uncanny ability to sustain the growth of an institution, which has adequately demonstrated the significance of the use of vector control technologies, some of which have had sufficient potentialities to end users. It is certainly a privilege to be associated with Dr Tyagi whose sense of dedication is unparalleled ... ."

Fig. 8.10. Dr B.K. Tyagi organized the 4th Conference of Medical Arthropodology (CRME, Madurai; 20-21 December, 2010) (*Source:* Dr B.K. Tyagi's personal archive)

Dr Tyagi is a completely indefatigable scientist with rare quality of leadership, as he professes "Leadership is in promoting team spirit and taking lead from the front, and not in domination." There is no Saturday or

Sunday or holiday for him. There is no tomorrow for him since he invariably lays emphasis on completing all the work on the day, leaving for tomorrow no work of today! His quest for knowledge continues, just as the famous maxim goes:

"The woods are lovely, dark and deep
But I have promises to keep,
And, miles to go before I sleep,
And, miles to go before I sleep."

Chapter 9

NOTEWORTHY INDIAN DISCOVERIES AND INVENTIONS ON MOSQUITOES

9.1. Introduction

Discoveries, innovations and inventions are the core constituents of science. Without these science leads to nowhere, like the heart without arteries and veins. These are offbeat researches often directed leeward. By nature a discovery or an invention is a daring effort of deviation from the conventional principles in work. Consequently, they are not easily accepted in the first sight and are, in fact, more often repudiated and junked! Peoples particularly those who are in the same business of scientific research, if not honest to research ethics or not sensitive enough to patiently hear new ideas, lack the sense of appreciation to others' novel thoughts and thus have a tendency to remain firmly bonded with obsolete and obscure dogmas of any act. They find a discovery or an invention quite hard to appreciate. It is most unfortunate, therefore, that while a discovery is applauded by some people, most others consider it a personal affront. Some people, more or less always within the known scientific community, consider a discovery by a scientist as a signal of insecurity to their own survival or professional supremacy. In other instances, some comrades, both within and outside the scientific fraternity, feel being lagged behind in the race to stardom resulting into raising intense jealousy and thereby lobbying against the discoverer when the time comes to reward the originator. In extreme cases, a discoverer or an inventor is inhumanly treated, ridiculed and even expatriated because the novel ideas underlying the discovery are far beyond the comprehension of the rude, arrogant and idiotic conventional minds. Nonetheless, if the discoverer or inventor has based his novelty on sound and demonstrable principles, sticks to his guns through all thick and thin and is resolved to prove his worth then he irrevocably braces the success with all in acceptance of his discovery finally. The road to success, however, is not easy and is like treading on a double-edged sword. Often the road is full of thorns (i.e., all kinds of professional hurdles, criticisms, inhuman social treatment etc. etc.) and the journey has to be completed solitarily by the inventor. In this often long and difficult peregrination his unflinching confidence, firm determination and sound knowledge of the subject matter of discovery are his/her only accomplices. The discoverer wins eventually, but not without passing through a fire-test. Thus, when analyzed from a mosquito's eye view, while the efforts of a genuine scientist are smitten with so many a counter-productive bickering, on one hand, races of the common enemy, the mosquito, are united to obfuscate successful survival of the former, on the other, and also enjoy the last laugh at the same time!

No one however great by social position or professional attainments, escaped the criticism, though it varied from person to person. While the early guessers of 'mosquito-malaria connection' like Lancisi, Beauperthuy and Rasori seemed to have had a low-keyed repudiation for their offbeat ideas during the 18th century, scientists like Laveran and Koch faced stern opposition and criticism pertaining to their initial discoveries along the same line. Crawford was unfortunate being maltreated and ostracized, while Finlay's 'mosquito-yellow fever theory' was almost hijacked by another genius in the field. Manson was a rather exception, though not without a trace of fault-finding, if not direct criticism, in both his 'mosquito-filariasis theory' and the undemonstrated 'mosquito-malaria theory'. The Grassi-Ross classical bickering, full of mud-slinging and work assassination for each other's contributions in deciphering man-mosquito-malaria mystery, is well known worldwide. A hundred years later, the scenario has not changed a bit, and the originator of a discovery is senselessly made to face the oft-repeated worthless criticism even today. Winston Churchill, the famed British Prime Minister during the World War II often used to say, *"You have enemy? Good. That means you have stood up for something, sometime in your life."* In fact, the intensity of criticism is a direct proof of its importance; that is, fiercer the criticism, the bigger in importance the discovery is.

9.2. Dr Ronald Ross's discovery of oocysts in the midgut of an Anopheles mosquito

The most famous discovery in mosquito science ever has crowned Dr Ronald Ross for his demonstration of the inextricable link between mosquito and malaria, paving way for a rather easier and direct way of eradication of the disease through vector control. This revelation was, however, not so simple a task as it appears and Ross had had to endure all kinds of hardships in achieving this goal. His courage, displayed itself in silent effort and endeavour, was more truly heroic than the achievements of physical valour which are rewarded by honours, titles and laurels. The time-tested maxim "*Courage is the thing, all is gone when this is gone*" applies letter-by-letter to Dr Ross whose combating spirit can be put forward than that of his life.

Dr Ross, a medical officer in British Indian Army since 1881, had routinely joined *inter alia* his regiment 19th Madras Infantry at Begumpet (now Begampet) Lines, situated near the outflow of the Husein Saugar (now Hussain Sagar) Lake on 18th June, 1897, after arranging an accommodation for his wife and children at Ootacamund, where, to his shattering dismay, the latter contracted contorting whooping cough after a while. At Hyderabad it was exceptionally dry and he initially whiled away in playing cricket and bicycling, and, otherwise, inditing long letters to his mentor and good guide Dr Manson: "**Company of wise men always offers good knowledge.**" He followed Manson's advice: "Follow the flagellum" and was almost convinced that he had seen the spectacular entity in his numerous malaria-infected human blood slides' preparations. The message about his malariological obsession had gone all around in the unit. For next few weeks or so nothing very notable happened in his monotonous life except that, on one particular occasion when he was challenged by one of his colleagues to give evidence for presence of malaria parasites in human blood, he miserably failed to demonstrate prevalence of the malaria parasite in the known malaria patient's blood under the microscope. He was crestfallen and ridiculed in silence. The humiliation penetrated deep in his heart but he valiantly fought to douse the anger within him. Next few days he, in a silent attempt to overcome depression, spent the time doing almost nothing in malariology and found some solace in writing poetry. Nonetheless. the continuous feeling of being drifted away from his objective forced him to find his track to tread, lest the Italians, French and Americans might take away all the hard-earned discovery! He became restless and pondered pensively, "One more effort and the thing will be done."

Ross religiously ensued Dr Manson's thesis to "follow the flagellum" and demonstrated as early as 1895 that the sexual bodies found in the human blood had unsuspectingly transformed into sphericals – distinctly different form, shape and size - and in many the flagilla were formed. Beyond that point he could not build a correct pathway of the flagellum in the stomach of the mosquito. Ironically, this latter part of Dr Manson's thesis that the parasite somehow get away into the ground water which, on drinking, make way into the human system was never demonstrated in any of Dr Ross's experiments since 1895. So, he thought to review afresh the whole issue of parasite biology in mosquito. He also understood that to prove his viewpoint he needed to have, in addition to his 'A' (grey) and 'B' (brindled)-type mosquitoes the third, ‘C’ type of mosquito (the ‘dappled wing’) which he had earlier encountered in Sigur Ghat. A major shortcoming in his seeking mosquito-related wisdom was that in spite of all efforts he could neither get to study any book or literature nor was there any person he knew he could contact for solving doubts. He had had to trust his own confidence, and by the mid-July, precisely on 16th July when monsoon had commenced, he decided to go forward, with the assistance of three natives. Before he could have encountered the ‘dappled wings’ mosquitoes once again, he however continued working with the mosquitoes in his hot office-cum-laboratory; he did not use ‘*punkha*’ fearing his dissected specimens of mosquitoes might be blown away! The result was invariably failure after failure, and he felt dizziness before his eyes and was even hallucinated due to overwork; leaving for his quarter almost exasperated. At dining table as well as at sleep he continuously pondered over the issue whether he were using the right mosquito group for malaria result? His microscope, too, was showing signs of ‘fatigue’ since its screws had got rust and the eye-piece cracked. But Dr Ross refused to stop and rather pushed forward with a reinvigorated charge to nail the bite once more. On 15th August, 1897, one of his three naive native attendants brought to him a bottle containing water full of wriggling mosquito larvae some of which to Dr Ross were quite different from those of other two types, **i.e.**, grey and brindled types. Again, on 16th August, his hospital assistant, an Indian native, showed him a ‘dappled winged mosquito’ (*Anopheles stephensi* ?) resting on the wall. He bottled it carefully and, after quickly incapacitating with tobacco smoke, dissected the mosquito hoping to see parasites under the microscope. Alas! He could see nothing of that sort. Failure, once again. The hospital assistant looked at his face which painted a dark shade of dismay and dejection.

“*Where there is a will, there is a way*” goes the saying as one of his attendants almost yelled with pleasure, “ Sir, I know a place where many such critters can be seen.” They were the same as those emerged in the bottle, Ross quickly found out. Leaving everything as it was in the office he dashed to the place he was guided to by his hospital assistant. He was amazed with elation to see about a dozen ‘dappled winged mosquitoes’, his ‘C’ type (*Anopheles stephensi* ?), trying hopelessly to pass through a gauze. He collected them; his first face-off with the ‘dappled winged’ mosquitoes ever in Secunderabad. Hussein Khan, his hospital patient with copious crescents in blood was immediately brought to a secluded room, and made to lie down on a cot under the mosquito net. Before putting Husein on cot he was body-bandaged to restrict mobility. At 12.20 p.m. he squeezed from the inside the bed net 10 fully blood-fed, and put one each in a separate test tube. Between 12.40 and 12.50 pm, first he subjected two of his catch to systematic operation or surgery to see under the microscope any crescent-sphere flagella metamorphosis in them. He succeeded little and for a disappointingly ephemeral time. His disappointment doubled up next day morning when he found two of his gorged mosquitoes had perished. For a moment he got panicked and hurriedly dissected two of the remaining mosquitoes, but in vain! In the raze of excitement he dissected, or rather wasted, all the remainders sans any sign of the crescent-sphere flagella metamorphosis in them. The man was so obsessed with dissection that he left no stone unturned in the process and even began dissecting fleas caught from a rat for trypanosomes. He was emerging infinitely fearless and his mind was growing stronger. Collecting all his energies he sent for his hospital assistant and the three native attendants, and asked them to take him to the place where one of the latter had originally found the 'dappled wing' mosquito. To his grave misfortune and sheer disappointment they miserably failed to spot the site again! Ross sank in his chair, deeply pensive. Next moment he jumped out of his comfort zone and began to jolt down details of his work to Dr Manson.

On the morning of 19th August, 1897 Dr Ross found out that the mosquito larvae earlier fetched to him by one of his native attendants had begun to hatch in the original bottle and they were his "dapple wings" type (*Anopheles stephensi* ?). He fed them on Hussein Khan again and carefully selected only two of them for dissection to see crescent-sphere flagella metamorphosis in them. Failure, once more. He became stressed for the time was running out of his hands and he desperately wanted results. From his already shrinking reserve, he picked up a third mosquito on the third day of blood feeding (so that the blood in the stomach was by that time completely digested), and dissected with utmost care. His dissection was not reasonably neat, yet he found there some "peculiar vacuolated cells in stomach about 10 μmm in diameter." It was certainly a "eureka" moment for him but he, like a good scientist, exercised complete restraint before reconfirming the presence of the peculiar bodies. Dr Ross was still subscribing to Dr Manson's thesis that mosquito feed only once in their lifetime, and then die after laying eggs on water. He hoped for better results with next dissections on the remaining samples – only three mosquitoes left from that batch. That night he could hardly sleep. Next day, 20th August, Dr Ross was informed the worst news of his life that all the three ready-for dissection mosquitoes had died last night. More so also because his native Indian attendants could not succeed in finding out a place to collect more larvae of 'dapple wings' mosquitoes. For a moment Ross felt having been evanesced but not a man of clay but of steel he determined to see the parasites in the cadaver, but failed as expected. He then took the seventh mosquito of his 16th August blood-fed ones (**i.e.**, on 4th day after blood feeding), with just one last left behind for the test! Already, due to bad sleep last night, his eyes sored and he yawned continuously. Nevertheless, with a phlebotomist's precision he was perfectly flawless this time; on cursory examination he did not find anything unusual and felt betrayed by his belief that the 'dappled winged mosquitoes' were the key channel to regulate parasite biology before re-entering a fresh human being. The lunchtime had passed and he continued empty stomach. Then just as he was once again dipping in distress he casually opened up wide the body wall covering the stomach. "Eureka", he cried spontaneously, at the top of his voice since this time lay before him in no uncertain way the oval-shaped spheres with black pigment on the stomach wall which represented one advanced stage in the development of the malarial parasite. He, as per army training, immediately jolted down and drew in his diary what he in his moment of triumph had just gone through – the oocysts!

Of the "Oocyst", totally twelve in count and each measuring about 12 μmm in diameter, embedded in the stomach wall, and scrupulously identifiable due to the black pigment exactly similar to the ones in the parasite crescents in the blood. He questioned himself as to how could parasite flagella move to the stomach wall retaining the black pigment (haemoglobin) numbering about 12 to 16? He was as sure as death about those cysts on stomach wall representing the parasite, but there was a remotely sounding miniscule doubt, too. To answer this irrefutably he needed to demonstrate growth in these cysts, with the passing time, in his last specimen – the eighth mosquito. Still, as far as his concept of malaria transmission was concerned, his 'Great Problem' had been thus solved, *i.e.*, the mosquito-malaria connection was established. He did not stop at this and went ahead to sacrifice on 21st August, on 5th day after feeding, the 8th and last survivor of the mosquito's batch fed on the 16th August and witnessed the presence of more pronounced 'cells' (oocysts; 12-14 μmm; the largest being 20 μmm) numbering twenty one in all each of which bearing the characteristic black pigments. The 'cells' (oocysts) were therefore parasites, and as they contained the characteristic black pigment, they were almost certainly the malaria parasites growing in the mosquito tissue – the stomach wall. The coordination of his mind's eyes with hands scripted an immortal verse:

This day designing God
Hath put into my hand
A wondrous thing. And God

Be praised. At his command,
I have found thy secret deeds
Oh million-murdering Death.
I know that this little thing
A million men will save –
Oh death where is thy sting ?
Thy Victory, oh grave?

Dr Ross prepared a manuscript on his discovery, titled, "On some peculiar pigmented cells found in two mosquitoes fed on malaria blood" which was published in **The British Medical Journal** on 18th December, 1897. Later he confirmed the finding in another type of 'dappled Winged Mosquito' (**Anopheles culicifacies**?).

9.3. JSB Stain for malarial parasites

Romanowsky-type stains are used to differentiate cells for microscopic examination in pathologic specimens, especially blood and bone marrow films, and to detect parasites such as malaria within the blood. Before Romanowsky staining, also known as Romanowsky–Giemsa staining, the blood constituents (hematology) including malaria parasite-infected red blood corpuscles (cytopathology) were largely monochromed and distinction between the malarial parasite's cytoplasm and the nuclear part remained indistinguishable, jeopardising the correct identification of the parasitic stage. But this staining required a bit longer time to complete the process. (Fig. 9.1)

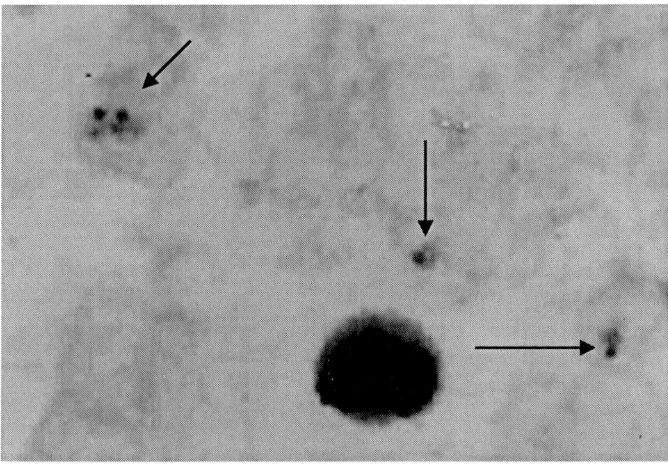

Fig. 9.1. JSB stain used for colouring malaria parasite, *P. falciparum,* in a thick blood smear. (*Source:* B.K. Tyagi personal archive).

To overcome this drawback Jaswant Singh & Bhattacharji (1944) invented a stain, commonly referred to as JSB stain, which not only rapidly stained the malaria parasites but also coloured chromatin in contrast to the cytoplasm of the parasitic body. Thus this method became most preferable for detection of malaria. Characteristically the stain proved useful for the diagnosis of malaria in thick smear samples of blood. The JSB stain has since been commonly used throughout India. The JSB stain consists of two solutions which are used in sequence to stain various parts of the sample. The first solution consists of methylene blue, potassium dichromate, and sulfuric acid diluted in water. This solution is heated for several hours to oxidize the methylene blue. The second solution is eosin dissolved in water. The pH of the buffer used to prepare the working stain is critical to the quality of staining. The preparation of JSB stain is done in two steps: (i) *JSB-1 solution*: For preparation of JSB-1 solution 0.5 mg of methylene blue was dissolved thoroughly in 500 ml distilled water in flask. 3 ml of one percent sulphuric acid was added drop by drop to the solution and then mixed with 0.5 gm of potassium dichromate. Addition of potassium dichromate led to the formation of some precipitate and at this step 3.5 gm of disodium hydrogen phosphate dihydrate was added. The mixture was boiled for half an hour. The stain was allowed to cool at room temperature and was ready for use as JSB-1 solution. This solution when used after maturation of 2-3 weeks gave better results; and (ii) *JSB-II solution*: It was prepared by dissolving 1 gm of water-soluble eosine In 500 ml of distilled water. A freshly prepared eosine solution, 15 although can be used immediately, may not yield as satisfactory result as the one which has turned deep red after some storage. Additionally buffer water is prepared by dissolving disodium hydrogen phosphate dihydrate (0.22 gm) and potassium acid phosphate (0.74 gm) in one litre distilled water. To maintain initial pH buffered water was kept in glass bottles. The slide staining process includes following steps: (a) 2-3 dips in JSB-II, (b) Washing ten times with buffer water, (c) A dip of 45 seconds in JSB-1, (d) Washing ten times with buffer water, and (e) Drying thoroughly before examination. Finally the stained slides are examined under the microscope with 10x and 40x objectives to see the positivity for malaria parasite. Positive slides were then thoroughly examined under an oil immersion lens (100 x) to ascertain the type of infection (i.e. *Plasmodium vivax, Plasmodium falciparum* or mixed type).

9.4. Pupae separator

Dr V.P. Sharma has been credited with many scientific and research achievements especially in the field of malaria research (*cf.* Chapter 8). His studies on the chemo-sterilisation of male mosquitoes are well documented. He is credited with developing a new technique for sex separation of mosquitoes at pupal stage for which he was awarded the Young Scientist Award. He has also worked extensively on the bioenvironmental malaria control which led to many innovations in the subject. He is also acknowledged for his contributions to vector biology.

9.5. Mosquito Sampler

Sampling is amongst the most important research activities warranted to determine a lot of various different bio-ecological characteristics about the mosquitoes, especially ones acting as vector species for communicable disease pathogens. In mosquito science, adult sampling, represented by per-man hour density has for long been in use, and in India it is still practiced through the use of a mouth-operated aspirator, though outdated, unethical and not practiced in other countries. This aspirator-device, comprising a plastic or glass tube connected to one end with a rubber tube, and separated with latter by a thin mosquito plastic or metallic netting at the joint (so that mosquitoes sucked are held there at the end of the tube), was in the earlier days recommended by even the World Health Organization for all countries combating malaria scourge. Even though this device has in course undergone enormous modification and, in some developed countries, like USA and UK, has also been successfully replaced by certain motorized sampling devices, however, in India the same mouth-operated device had more or less been exclusively employed in the exercise (except in some institutes where, too, the motorized samplers were obtained by purchase from abroad!) - totally unaware of the fact that the aspirator was a health hazard, besides of course not fool-proof to basic entomological vagaries generally encountered during adult sampling in field. This mouth-operated aspirator, though simple in appearance, could be operated up to nearly two feet above the erected hand (meaning a sampling person could collect adult mosquitoes everywhere but the roof, which in most urban set up is about 9-10ft. above the ground) but with the sampler held aloft he could barely access 8ft., considering average human height of 5′6″. Consequently only a part of the adults actually present indoors could be sampled during a specified period of time (meaning insufficient and misleading sampling), and, above all, the risk of contracting throat infection by the sampling human (the inhaled mosquito carries a lot of dust fouled with fungi, bacteria and or similar unwanted things). The entire sampling procedure involving mouth-operated aspirator seemed a farce in the name of scientific investigation.

When I joined malaria research in 1981 on a Tropical Disease Research (WHO) project under the then ICMR-Malaria Research Centre/National Malaria Eradication Programme, in the tribal belt of Ukai/Songarh project in Surat district, southern Gujarat, the mouth-operated aspirator was an integral companion of an entomologist and a replacement could not be even thought of in India. The tool was obsolete, cumbersome and hazardous but still an inseparable part of an entomologist's sampling kit as the alternative was hardly available to an insect collector. One incident, however, forced me to dwell on the subject; one of my staff members, Govind Ram, a laboratory technician, who regularly joined me on routine entomological outings, showed first signs of severe and uncontrollable coughing whenever he resorted to sampling with the mouth-operated sampler. His deteriorating health soon made him totally incompatible to adult sampling with the aspirator, and he finally left the job too, but before he resigned in late 1983 he had confided in me that he was diagnosed for tuberculosis. I felt very sad but clearly understood a need to find a replacement to the aspirator in vogue.

In 1984 I joined ICMR-Vector Control Research Centre (VCRC) at Pondicherry, and participated in numerous field-oriented entomological programmes in south and eastern Indian states, where the mouth-operated aspirator was in exclusive use for adult mosquito sampling indoors. I witnessed several of my accomplices resorting to spitting and coughing soon after each inhalation to capture a mosquito. Some even suffered with occasional heavy breathing dyspnoea. This spectacle fired in me the thought the seed of which was first sown on my mantle in 1981 of finding out a solution to the problem. I drafted my first drawing of a substitutive sampler, in my mind, though. Come summer of 1988 and I was moved on transfer to another ICMR institute (DMRC, Jodhpur) in the hostile Thar desert. The mouth-operated aspirator continued to be witnessed in all adult sampling ventures here, too. On March 22, 1990 I developed my invention of a mechanical mosquito sampler, which the ICMR subsequently named as the "Tyagi Sampler" or "Mechanical Sampler" and filed application for seeking patent with the National Research Development Corporation, Delhi, in 1996 (vide infra). This possibly was the first invention by an in-service medical entomologist/scientist in the ICMR's over eighty years history of existence that was of direct significance in field-based entomological investigations.

I have had numerous opportunities to read about the all-time great inventors and discoverers, right from Aristotle's time (500 BC), and was quite aware of the various difficult moments each of these scientists had

had to go through to prove the worth of his or her research. They were criticized, ridiculed and had faced fears of ostracism, yet, as they were, braved through all thick and thin by the dint of their ceaseless perseverance, undaunted determination and focused application of wisdom. Interestingly, often repudiations and innuendoes were hurled on these innovative scientists by their own contemporaries. Within research community due either to scientific rivalry for claiming credit on the discovery or, at times, out of jealousy, simply to create a foul atmosphere to discredit the inventor from the recognition due to him/her or, as occurred in case of my mosquito sampler, as a negative and destructive scientific approach or attitude, even procrastinate the process of realization of the significance and public utilization of the invention. The journey to seek patent certainly did not prove a cake-walk and I faced all the imaginable constraints which need be told here for the benefit of all the young discoverers, innovators and inventors who must not be distracted from their objective come what may!

Soon after I had joined at the Desert Medicine Research Centre in 1988 I began working on my hypothesis of a mechanical mosquito sampler. I developed the design on paper and began to assemble its parts from different sources; prehensible tubes from the bangle-making factory, a hair-blower, rear side mosquito holding jar, rechargeable 1.5V × 6 dry cell batteries, connecting wire etc. Then I befriended a certain Mr Hemant who ran an electric shop handling mostly the motor-wire winding or general electric goods' repair work in the city. Everyday for next six month during late 1989 I worked nonstop in Hemant's shop from 6 to 9 PM. I will actually drive everyday directly from my Centre to Mr Hemant's shop without going to my home first. Except my wife none knew where I would go after the office hours, though a couple of colleagues had become curious to find out reasons behind my disappearance without a word. So, one day soon after the office-hours my Officer in-Charge then, Dr K.A.V.R. Krishnamachari, along with some of my scientist-colleagues (Mr T. Ramnath, Mr J. Laxminarayan and Mr Mahapatra) appeared at my house in Milkman Colony, when I was obviously away on my unfinished work, but they were welcomed by my wife and children. Dr Krishnamachari apparently worrisomely inquired about my whereabouts and they were suitably answered that I was fine though not available at home at the moment! They had waited patiently till I arrived around 8.45 PM. I was much amused seeing him and others at home, and it did not take long for me to understand the purpose of their arrival. I too pacified his inquiry, though he never seemed satisfied with my answer. Next day I submitted a research proposal to the office about developing a new, indigenous and more effective "mechanical mosquito sampler", to be defended in the next Scientific Advisory Committee (1990) meeting.

Dr Krishnamachari had a quaint nature; he could hardly bear any scientist under him to come up with anything better! On the contrary he would become restless. He therefore clandestinely called upon a discussion upon my mechanical mosquito sampler and for this he chose Dr Namita Mahapatra, a rather new incumbent to the Centre, and was an entomologist. She wrote to her references in the USA and sought the drawings and other details for certain mosquito samplers available in the market on cost. Apparently Krishnamachari was very happy with all this information in his hand, to cut my project off from the list of future proposals of the scientists meant for the SAC meeting. Equipped with such shenanigans he called one day a meeting of all the scientists in his room, to discuss about the projects being recommended for review at the next SAC. During the meeting he omitted my project proposal to which I objected and emphasized upon its inclusion in the list merely on the basis of its being original, indigenous, highly effective and cost-effective. A pin-drop silence persisted in his room, and he, instead of directly facing me for question-answer, directed her query to Dr Namita. I objected to this as well and stressed that for my research proposal only I could respond. Dr Krishnamachari, as if left with no blood, looked blankly at me and said only one word, "OK." After this meeting, I could never be myself too sure about the inclusion of my research proposal in the list for SAC, but this had created tremendous pressure on me to complete my instrument before the meeting later that year. By early March 1990 I had completed all my laboratory and field tests on outdoor resting adult mosquitoes in different habitat settings, and generated all the possible physical and biological data related with my first invention, a mechanical mosquito sampler. The instrument was ready on 22nd March, 1990, much before the date of the SAC meeting.

The Scientific Advisory Committee of Desert Medicine Research Centre met during mid-1990, with a powerful Chairman - a retired Joint Director, Health Services of Maharashtra State. After I had presented on Mechanical Mosquito Sampler, the Chairman encouragingly asked me, "Dr Tyagi, can you sell your prototype to me at any cost since I will like the Maharashtra Health Services to manufacture this for their own use?" He further added immediately, "Please apply for the Patent at once." This kind of support from the Chairman came as a great encouragement and support to me and possibly like a bomb-shell, to Dr Krishnamachari, who had no answer but to agree with the Chairman. After the meeting I did not get any help from the Officer in-Charge to file my application for patent. Therefore, I decided to be myself privately going to ICMR HQrs in Delhi and take necessary approval. Prof. A.S. Paintal was the Director General of ICMR at that time, and I had heard a lot great remembrances about him as to how straightforward and fearless he was, as well as how helpful he was during my earlier encounters! I met the DG, ICMR in his chamber and explained to him the reason behind my coming to him. He called for Dr S.P. Tripathi, Additional Director General to be also a witness to my demonstration with the mechanical mosquito sampler, my invention. Dr Paintal was extremely delighted and asked Dr Tripathi to kindly provide all necessary help to me in filing papers for the patent. Dr Tripathi, to be doubly sure that my demonstration was really good, asked for same again and satisfied himself with a volley of inquiries regarding its originality, feasibility in field, usefulness in sampling adult mosquitoes, cost effectiveness, reproducibility of results etc. He directed me to see Dr C.R. Ramachandran, Senior Deputy Director General and Head, Non-Communicable Diseases under whose aegis my Centre was classified. There I demonstrated the function of my instrument for the third time, in a row on the same day, and his indifferent comment was, "I have my own instrument awaiting approval and pointed out, with almost a sigh, to a far-off corner of his room where a glassy distillation plant-like apparatus bearing heavy dust was resting on a wooden frame." He continued, "What is so new with your invention then?" I was crestfallen but not given up. I requested him to approve my application as the DG, ICMR has recommended. He approved my application for the formal and official consideration of the ICMR. Some months later I received a communication from the ICMR that the Council had favourably considered my invention and was desirous to file a patent for the same. Unfortunately in all those institutions present in Jodhpur none could guide as to where to get an application or where to apply for patent? So, once again I had had to privately make a visit to the ICMR HQrs with a view to seek necessary background information. Strangely, there too nobody was coming forth to deliver the desired information. Then, as if providentially I stumbled upon Mr R.K. Sharma, then Section Officer in NCD, and he candidly informed two things; (1) Nobody ever sought a patent in the ICMR in past and therefore nobody knew as to where to deal with on the subject, and (2) however, there is an instance of a young researcher-odontologist in the All India Institute of Medical Sciences (AIIMS) who got funding from the ICMR for his project. This young dentist had unfortunately logged on horns with the Council because he wanted the patent in his name whereas, as governed by the rules of ICMR funding, the Council wanted it to their name. Mr Sharma went through some files and let me know the address of National Research Development Corporation (NRDC) at 20-22, Zamroodpur Community Centre, Kailash Colony Extension, New Delhi, Delhi 110048. I approached the NRDC and sought the Patent Application, besides other guidance. I filed Patent Application for the "Mechanical Mosquito Sampler', or "Tyagi Sampler", in 1992 which was accepted by the NRDC in 1996 only. The patent application went through a rigorous process of scientific, legal and public acceptance and validation that itself took 10 years to complete!

Opposition to my invention had never ended – all by my own dear colleagues at Desert Medicine Research Centre even when I always went up regularly the ladder of promotions and all my detractors invariably stumbled time and again!! I maintained good peace and camaraderie with them, however. During 1999 I was appointed as the Officer in-Charge of DMRC and I coordinated organization of a ICMR-NRDC's joint "Workshop on Intellectual Property Rights (IPR) and Technology Transfer (TT)" which took place in the Auditorium of Defence Research Laboratory (DRDO), Jodhpur. Along with many scientist-colleagues from various sister organizations in jodhpur such as Central Arid Zone Research Institute (ICAR-CAZRI), Arid Forest Research Institute (ICFRE-AFRI), Zoological Survey of India, my colleagues at the DMRC also joined at the Workshop. At the conclusion of the workshop after a panel of experts from the NRDC and ICMR was established on the dias to face to queries of the participants, interestingly enough, one of the participants, Dr A.K. Dixit, from DMRC, treacherously asked a very irrelevant, preposterous and incriminating question, "Sir, a patent application on mosquito sampler was filed by a scientist from the DMRC. Some of us at the Centre consider that this was not an original invention and must be rejected forthwith." Everybody in the auditorium looked at the venomous jealousy spewed by his assertion to which Mr C.M. Gaind, Legal Expert at NRDC, responded that, as per the well-laid rules, the whole exercise will go through a extensive public acceptance process and only the genuine applications will meet the merit of award of patent. My "Mechanical Mosquito Sampler" won the Patent (# 191635) in 2005, at a time when I had been already promoted to the level of Scientist "G" (Director) & Director in-Charge, Centre for Research in Medical Entomology (ICMR) at Madurai. I have never regarded such criticism as an affront and kept moving forward. In next few years I was destined to transfer two technologies to the Health Department of Tamil Nadu - an accomplishment never ever executed by the ICMR before - which filled my heart with a feeling of true satisfaction.

9.6. Transfer of Technologies

During my tenure as Scientist "G" (Director) & Director in-Charge, Centre for Research in Medical Entomology, Madurai, I executed two Transfer of Technologies (TTs), namely, (i) JE detection in Wild-caught mosquitoes using TOXO-IFA System for JEV surveillance, and (ii) Multi-Angular Viewer for identification of mosquito species.

9.6.1. JE detection in wild-caught mosquitoes using TOXO-IFA System as a surveillance tool

The true originator and exponent for this methodology, accomplished in two stages, is Dr V. Thenmozhi (*cf.* Thenmozhi *et al.*, 2016), even though Dr A. Gajanana could be credited to have invented it (Gajanana *et al.*, 1997). Wild caught mosquitoes are separated species-wise. While Unfed mosquitoes were pooled (1 - 50) on the same day of collection, engorged female mosquitoes were held for 48 h for digestion of blood meals before pooling. Mosquito (only females) abundance was calculated as density (number collected per man-hour). Mosquitoes were pooled (minimum 5 specimens per pool) and stored at –80°C until processed for virus detection and isolation. Two systems were used :

(i) *Antigen capture ELISA*: Monoclonal antibody 6B4A-10 (reactive against all viruses in JE/WN/SLE/MVE complex) was used as capture antibody and monoclonal antibody peroxidase conjugate SLE MAB 6B6C-1 (reactive against all flaviviruses) as detector antibodies (supplied by Dr T.F. Tsai, Centres for Disease Control and Prevention, Fort Collins Co., USA). A mosquito pool was considered ELISA positive if its optical density value was ≥ mean + 4 standard deviation of the six normal pools.

(ii) *Insect bioassay* (Toxo-IFA): *Toxorhynchites splendens* mosquito larvae were inoculated with ELISA positive pools intracerebrally and incubated for 7-10 days at 29 °C and then tested by the indirect immunofluorescence assay (IFA) on head squeeze preparations (Toxo-IFA). Smears were tested with JEV-specific monoclonal antibody, MAB 112 (supplied by Dr Kimura Kuroda, Tokyo Metropolitan Institute of Neurosciences, Japan) and detected by Fluorescein isothiocyanate (FITC) conjugated anti-mouse immunoglobulin (Dakoppats, Denmark).

Virus infection rate in mosquitoes was expressed as minimum infection rate (MIR, i.e., Number of positive pools/Total number of mosquitoes tested × 1000).

CENTRE FOR RESEARCH IN MEDICAL ENTOMOLOGY
(Indian Council of Medical Research)
MADURAI

Transfer of Technology

"Detection of antigen for Japanese encephalitis virus (JEV) from the wild-caught vector mosquitoes as a disease surveillance tool"

to

The Directorate, Public Health & Preventive Medicine, Tamil Nadu
on this day the 25th July, 2010

Received by or on behalf of
Director, Public Health & PM, Tamil Nadu

Director-General, ICMR

Handed over by
Director in-Charge, CRME

Fig. 9.2. (a) Event of Transfer of Technology by Dr V.M. Katoch, Director General, ICMR & Secretary Department of Health Research, MOH&FW, GOI to Dr K. Durairaj, Addl. Director, the Representative from the Health Directorate, Tamil Nadu, in the presence of (from left to right) Dr D.V.P. Raja, Dr V.M. Katoch, Dr K. Durairaj, Dr B.K. Tyagi, Dr V. Thenmozhi, Mr Sanjeev Dutta, and Dr R. Paramasivan. (*Source:* Dr B.K. Tyagi, pers. archive)

The Technology Transfer (TT) was made to Dr K. Durairaj, Addl. Director, representing the Health Directorate, Tamil Nadu government on 25th July, 2010, in the presence of Dr V.M. Katoch, Director General and Mr Sanjeev, Dutta. Senior Financial Advisor, ICMR. This was for the first time the ICMR was transferring a technology directly to a State health directorate - for their and people's direct benefit (Fig. 9.2).

9.6.2. Multi-Angular Viewer for identification of mosquito species.

Since most field microscopes do not have an inbuilt system to view the object from all angles, identification of vector species generally offered many challenges. To overcome this constraint, a multi-angular viewer was developed to facilitate studying *in vivo* the taxonomic characters in the field-sampled specimens at 360°. This helped draw a clear and conclusive decision on the status of taxon and identify it properly. The body of the multi-angular viewer is small enough to be conveniently transported. It can be safely adjusted under the object lens of the microscope and the base stand. Viewing pinned specimens is characteristically free from any handicap, thereby facilitating drawing of any part of the pinned specimen of mosquito or fly without any lapse. The instrument was developed indigenously with locally available material, hence economically viable, too.

9.7. 'Tanka' lid for preventing *Anopheles stephensi* breeding

The Thar Desert is believed to have come into existence between 4000 and 1000 BC, although palynological evidence indicates that the development of this desert set in during late Cretaceous. Prior to this a well-knit river system existed. However, in the ensuing passage of time a series of events took place, such as the deterioration of climate and land physiography, the rise of the Himalayas, lowering of the Aravallis, the diversion and disappearance of three classical rivers from their original courses (the Yamuna, the Saraswati and the Ghaggar), and the lowering of the underground water-table, which possibly caused increased aridity in the area. Presumably agriculture was introduced in the region from the Middle East through Baluchistan and Sind. The Thar Desert of Rajasthan, comprising 12 districts, is spread over a 28,600 km^2 area, which is 12% of the mainland of the country and 62% of the total area of the State. It harbours a population of 13·4 million, with an average density of 64 persons per km^2, making the Thar Desert region one of the most populated deserts in the world. The climate is characterised by extremes of temperature, varying between –4 °C in winter and above 50 °C in summer. The rainfall is poor and erratic, ranging from 400 mm in the eastern part of the desert to less than 100 mm in the western fringe (aver. 310 mm). Since the late twenties, however, three major canals (*viz.,* Gang Canal, Bhakra Canal and Indira Gandhi Canal), with over a 10,000 km network of distributaries, have been providing water for drinking as well as irrigation in an area of 1·3 million ha. Although the introduction of canals in the region has extensively changed the physiography of the desert, nonetheless the rural areas continue to subsist on sources like 'nadi' (ponds), and 'tanka' and 'beri' (both man-made well-like structures never charged with underground water on their own). In the desert, the primary malaria vector *Anopheles stephensi* has been reported to rest and breed primarily in the latter two structures – a fact that was highlighted for the first time by Tyagi (1995) and Tyagi & Yadav (1996).

Both '*tanka*' and '*beri*' are integral parts of the desert ecosystem and found in all the 12 notified desert districts of Rajasthan State. The need for a '*tanka*' was forced upon the desert dwellers by such compelling conditions as low rainfall and scarcity of ground-water. By definition, a '*tanka*' is an earthern pit made either indoors for household use or outdoors for community use. It is exclusively used as a reservoir for storing drinking water. Potable water is fetched from the far-off villages generally 20–30 km apart, which have perennial ponds, or from the canals traversing the area. Camel carts, which are a typically desert-based means of transport, equipped with huge metallic drums or bins, are the normal mode of transportation of drinking water in the desert villages. Consequently, the '*tanka*' is never charged up on its own from the underground water. A small portion of *'tanka'* always protrudes above the ground. The household '*tanka*' is generally larger in size (diameter 1·5–2·5 m and depth 2–5 m) than the community based one, which is again of two types. When built in the outskirts of the village the community '*tanka*' is generally smaller in size (diameter < 2 m and depth < 4 m), but, when built inside the village to meet the frequent requirement of the villagers the '*tanka*' is huge (diameter 4–5 m and depth 5–7 m). Each '*tanka*' has a mouth at the top (45–60 cm diameter) which can be round or squarish in shape. It is generally covered with a metallic lid, wooden plank, loose cloth pieces or a collection of twigs.

Additionally, a small orifice (< 15 cm in diameter) is provided at the junction with the ground of the protruding part of the '*tanka*'. Potable water, fetched from distant places through camel carts, is poured into these reservoirs through the apical hole. The orifice below is essentially meant for the runoff water of the monsoon rains to drain in and is generally left uncovered (Fig. 9.3). The temperature is about 25 ± 2°C in the '*beri*'. Relative humidity in '*tanka*' is maintained between 50–70%, while in '*beri*' it varies between 70–80%. Compared with the water of '*tanka*', the water of '*beri*' is cleaner, less odorous and colourless. In '*tanka*', when the water is stored for long periods of time, some algal growth along with plankton is also observed. A '*beri*' is essentially a community-based well-like subterranean pit made in the bed of a seasonal pond only. It is normally 1m wide and loosely lined by stones on the mouth. Several such structures are built in close proximity. Sometimes they may become fused underground due to erosion of soft soil, making a huge tank underneath. The mouth of a '*beri*' is loosely covered mostly using twigs or wooden planks. When the monsoon rains hail the water gushes into '*beri*' and fills them. The number of '*tanka*' and '*beri*' is less in those areas where adequate canal water and/or bore-well water has been made plentiful, for instance, in the Sri Ganganagar district in northern Rajasthan. In such cases, however, *An. stephensi* breeding in 'tanka' gave way to *Aedes aegypti,* the vector of dengue, chikungunya and zika virus (Tyagi & Hiriyan, 2004).

Fig, 9.3. An example of a '*tanka*' (above) and a '*beri*' (below) (each mound of stones in the middle ground marks the entrance to a '*beri*' in the Thar Desert). (*Source:* Tyagi personal archive)

Notwithstanding tremendous variations in the ambient temperature in the Thar Desert, both '*tanka*' and '*beri*' maintain more or less the same temperature throughout the year. On average, the temperature of an adequately protected and covered '*tanka*' stays about 26 ± 2°C, Among all the vector species collected from the desert so far, *An. stephensi* is the most significant vector (Tyagi, 1992; Tyagi, 1994a-f; Tyagi & Yadav, 2001; Tyagi *et al.*, 1994, 1995). It prefers to breed in '*tanka*' and '*beri*' and is the only anopheline species that is prevalent throughout the year in extreme xeric conditions, and also invariably breeding in these two habitats. However, when the '*tanka*' is either unused or discarded altogether after becoming dilapidated or polluted, *Aedes aegypti*, the vector for dengue (including dengue haemorrhagic fever), is also observed to breed in it. *Anopheles stephensi* uses these structures mainly for two purposes, *viz.*, for resting both after a blood meal (normally Cella II–III stage) as well as soon after emergence. Besides, *An. stephensi* seems to use these resting sites to tide over the extreme ambient temperatures both during summer and winter. Chorogeographically, *An. stephensi* is a desert species which appears to have made headway to India centuries ago from the Middle-East deserts (Zahar, 1990). Therefore, the distribution and preponderance of *An. stephensi* in The Thar Desert are parallel to the occurrence of these breeding sites. It has been observed experimentally that vector breeding in unused and discarded '*tanka*' can be controlled effectively by applying expanded polystyrene beads, though the technique also renders the water unusable for drinking.

Conclusively, the control of vector breeding in potable '*tanka*' water became very difficult since villagers prefer not to spray Abate (temephos) which is employed in urban situations to control breeding in wells, etc. Minor engineering work on the lid of a '*tanka*' makes it possible to cover the hole and prevent the entry of the mosquitoes to some extent but never completely. Although both '*tanka*' and '*beri*' occur in the entire desert region, they are more scarce in those places where canalized water has become abundantly available. Therefore, in the Shri Ganganagar district, which is extensively canalised, *An. stephensi* is the third most dominant species after *An. culicifacies* and *An. subpictus* (Tyagi & Verma, 1991). Here it breeds both in the '*tanka*' and wells as well as in the seepage water from canals. Consequently, its breeding index (BI) in tanka (0.01) is comparatively far less than in either the seepage water collections (0.23) or the flood waters (0.56). Interestingly, however, even in irrigated areas, *An. stephensi* is the only species breeding in wells or '*tanka*' (Tyagi & Verma, 1991). In irrigated areas, the most dominant vector species is *An. culicifacies* which breeds profusely in irrigation channels (BI of 6.28) and flood waters (BI of 6.32). In contrast to this, *An. stephensi* is the most dominant species among all the anophelines collected hitherto in the truly desert surroundings (Tyagi *et al.*, 1994a). It could, therefore, be safely surmised that 'tanka', besides 'beri', is the most important breeding habitat for *An. stephensi,* and, thus, play a very significant role in maintaining vector density high enough to enable the vector species to transmit malaria in the desert community, and for this reason, in the first place, it is inevitably indispensable to device and deploy a vector breeding preventer in these '*tanka*' on community basis so as to mitigate the density of the vector mosquito in the rural Thar Desert.

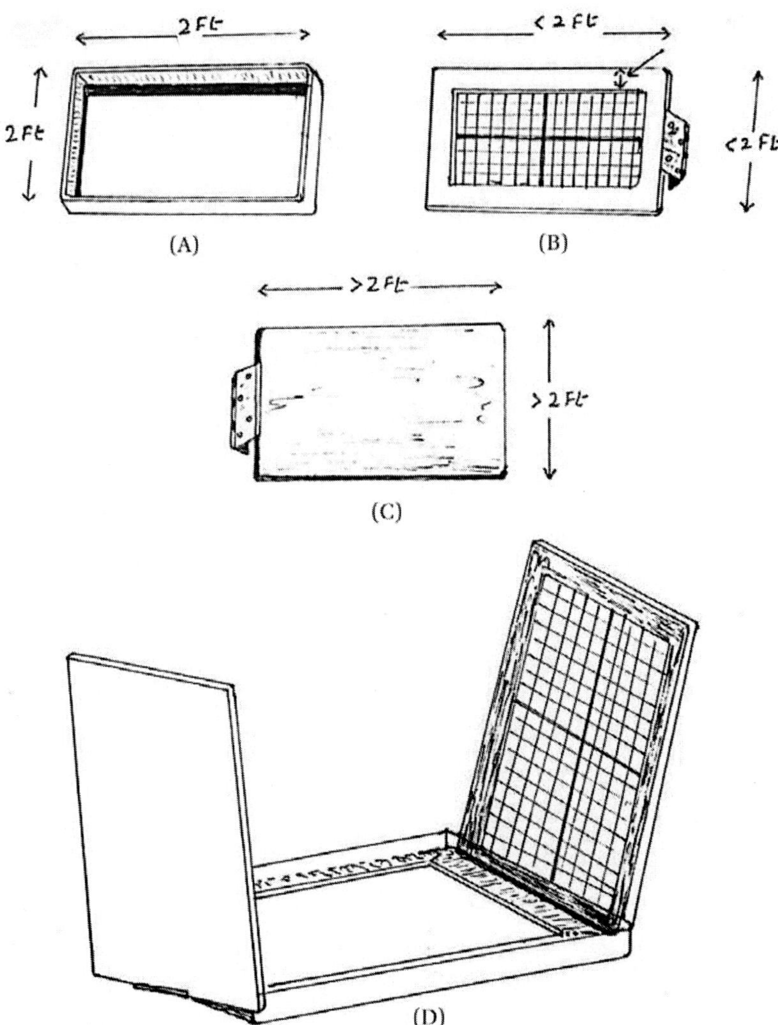

Fig. 9.4. A prototype design of the 'tanka lid' (*Resource:* Dr B.K. Tyagi pers. Archive).

During early 1998, inspired by my conclusion about 'tanka' being a major source of breeding of malaria vector *Anopheles stephensi*, I designed and fabricated as well as assembled an "*Anopheles stephensi* breeding preventer", which was also limitedly tested in field to ascertain its fool-proof *modus operandi* (Fig. 9.4). In 2000 when a malaria outbreak struck in desert districts, I had suddenly received a telephone call from the Jaisalmer district health authorities to help control mosquito breeding in Tanka (*cf.* Tyagi, 1995). I immediately informed the district authority, Mr Rajat Mishra, District Collector, Jaisalmer, of my invention and sent them a copy of the drawings of the prototype so that they could replicate the same and get it deployed *en masse* in desert villages. (Fig. 9.5).

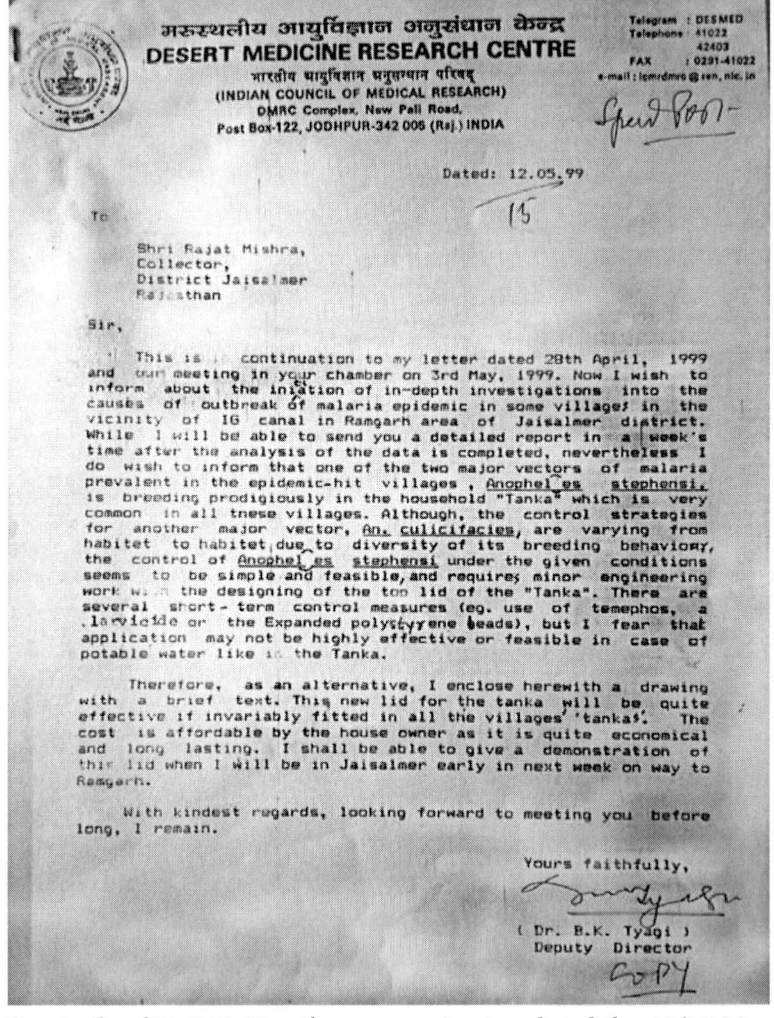

मरुस्थलीय आयुर्विज्ञान अनुसंधान केन्द्र
DESERT MEDICINE RESEARCH CENTRE
भारतीय आयुर्विज्ञान अनुसंधान परिषद्
(INDIAN COUNCIL OF MEDICAL RESEARCH)
DMRC Complex, New Pali Road,
Post Box-122, JODHPUR-342 005 (Raj.) INDIA

Telegram : DESMED
Telephone : 41022
42403
FAX : 0291-41022
e-mail : icmrdmrc@sen.nic.in

Speed Post

Dated: 12.05.99 15

To

Shri Rajat Mishra,
Collector,
District Jaisalmer
Rajasthan

Sir,

This is in continuation to my letter dated 28th April, 1999 and our meeting in your chamber on 3rd May, 1999. Now I wish to inform about the initiation of in-depth investigations into the causes of outbreak of malaria epidemic in some villages in the vicinity of IG canal in Ramgarh area of Jaisalmer district. While I will be able to send you a detailed report in a week's time after the analysis of the data is completed, nevertheless I do wish to inform that one of the two major vectors of malaria prevalent in the epidemic-hit villages, Anopheles stephensi, is breeding prodigiously in the household "Tanka" which is very common in all these villages. Although, the control strategies for another major vector, An. culicifacies, are varying from habitet to habitet due to diversity of its breeding behaviour, the control of Anopheles stephensi under the given conditions seems to be simple and feasible, and requires minor engineering work with the designing of the top lid of the "Tanka". There are several short-term control measures (eg. use of temephos, a larvicide or the Expanded polystyrene beads), but I fear that application may not be highly effective or feasible in case of potable water like in the Tanka.

Therefore, as an alternative, I enclose herewith a drawing with a brief text. This new lid for the tanka will be quite effective if invariably fitted in all the villages' 'tankas'. The cost is affordable by the house owner as it is quite economical and long lasting. I shall be able to give a demonstration of this lid when I will be in Jaisalmer early in next week on way to Ramgarh.

With kindest regards, looking forward to meeting you before long, I remain.

Yours faithfully,

(Dr. B.K. Tyagi)
Deputy Director

COPY

Fig. 9.5. Facsimile of Dr B.K. Tyagi's communication dated the 12/15 May, 1999 to the Collector, Jaisalmer re: the '*tanka lid*'. (*Source:* Dr B.K. Tyagi, pers. comm.)

Much later, first in 2014 (during the Scientific Advisory Committee meeting of the ICMR-National Institute of Malaria Research, Delhi) and second time in during the 12th Joint Annual Conference of Indian Society for Malaria and Other Communicable Diseases & Indian Association of Epidemiologists, held on 1-3 September, 2017 in Armed Forces Medical College, Pune, Dr A.C. Dhariwal, the then Director, National Vector Borne Diseases Control, gave me a very pleasant surprise by informing that how much useful the “Mosquito Vector Breeding Preventer” had proved to mitigate the breeding in ‘*tanka*’ almost immediately, and consequently also the malaria cases, in Jaisalmer district. This apparently delighted me more than any award - the most appropriate *summa cum laude* compliment!

9.8. N,N-diethylphenylacetamide (DEPA):

Mosquito-borne diseases, including malaria, filariasis, dengue, and encephalitis viruses continue to cause morbidity and mortality in humans, domestic animals, and wildlife. Arthropod "eradication" measures envisaged five decades ago failed to make any significant impact. A change to the philosophy of insect "control" or "management" has continued to evolve. The use of insect repellents seems to be one of the best available means forpersonal protection against annoyance and diseases associated with hematophagous organisms, especially because repellents protect subjects without having a major adverse impact on the environment. Amides, imides, esters, ethers, and poly-functional compounds are known to be good repellents. Selected petroleum oil fractions and essential oils repel *Aedes aegypti* (L.). Dimethylphthalate (DMP) was the first synthetic chemical that had considerable repellent efficacy and has been used widely as a mosquito repellent. N,N-Diethyl m-toluamide (DEET) emerged in 1954 as a superior arthropod repellent. Cost considerations arising from the unavailability of m-toluic acid for the production of DEET in India prompted investigation into an array of alternative compounds bearing high repellency, safety and cost-effectiveness. Consequently, this resulted in the development of a new insect repellent, N,N-diethylphenylacetamide (DEPA) (Kalyanasundaram, 1982). N,N-diethylphenylacetamide, popularly called as DEPA, has repellent activity against hematophagous insects including mosquitoes, black flies, horse flies, muscoid flies, rat fleas, and ticks, as well as land leeches and cockroaches. The efficacy of DEPA is comparable to that of N,N-diethyl-3-methylbenzamide (DEET) in both laboratory and field tests. Toxicological studies indicate that DEPA is safe for human use. Different formulations as well as fabrics impregnated with DEPA are effective for personal protection from insect bites. Dr Vijay Veer, former Director, DRDO, during his deputation in the DRDO’s Defence

Research Development Establishment (DRDE) at Gwalior, sweated hard to swing the use of this great repellent in favour of Indian armed personnel serving in frontiers, saving the nation in the process crores of rupees otherwise going out in the purchase of DEET etc.

9.9. Discovery of *Anopheles pseudosundaicus* Tyagi *et al.*, 2009

Sixty three species of Anopheles have so far been recorded in India (Knight & Stone 1977; Christophers, 1933; Rao, 1984; Nagapal & Sharma, 1995; Tyagi *et al.*, 2015) and as many as ten species are considered primary or secondary vectors of malaria (Nagpal & Sharma, 1995; Tyagi & Yadav; 2001, Tyagi 2002, 2004); the last endemic species, *An. pinjaurensis* Barraud, 1932 having been described from India more than seven decades ago. In India only a few species, *e.g.*, *An. subpictus* Grassi and *An. sundaicus* (Rodenwaldt), were reported breeding in the fresh and/or brackish water of coastal areas. The coastline of the Indian peninsular region is a rich, yet little explored habitat for mosquitoes. Studies initiated after the tsunami on December 26, 2004 revealed a new mosquito species, *Anopheles pseudosundaicus*, belonging to the Pyretophorus Series of Anopheles (Harbach 2004). This species was encountered in Kerala State, southwestern India. Although, the new taxon seems to be close to *An. subpictus* and *An. sundaicus*, nevertheless, it differs distinctly from either of them both in morphological (adult and immature) and molecular characteristics. Larval specimens were collected from the tsunami-hit coastal localities of Kuzhithura, Kollam District and Valiazheekal, Alleppey District, Kerala State in southwestern India, from January to July, 2005. Samples were individually reared to adult stage for species identification. To plot the genetic variation, individual adult mosquito specimens of *An. subpictus* and *An. pseudosundaicus* (CRME Museum Ref #A605) were used for isolating mtDNA, while comparative molecular characteristics for *An. sundaicus* were obtained from GenBank. DNA was isolated using "DNA Isolation Solution" kit (Genei, Bangalore, India) following manufacturer's instructions, with slight modification.

(i) *Molecular investigations*: A polymerase chain reaction (PCR) was set to amplify the subunit 1 of cytochrome oxidase gene of the mtDNA using GeNeiTM PCR Master Mix (Genei, Bangalore, India) following manufacturer's instruction. The universal CO1 primers C1-J-1718 (5' - GGA GGA TTT GGA AAT TGA TTA GTT C - 3') and C1-N-2191 (5' - CCC GGT AAA ATT AAA ATA TAA ACT TC - 3') were used following the thermal cycling of initial denaturation at 94 °C for 10 min., followed by 40 cycles of denaturation at 94 °C for 30 sec., annealing at 50 °C for 30 sec., and elongation at 72 °C for 1 min. followed by final

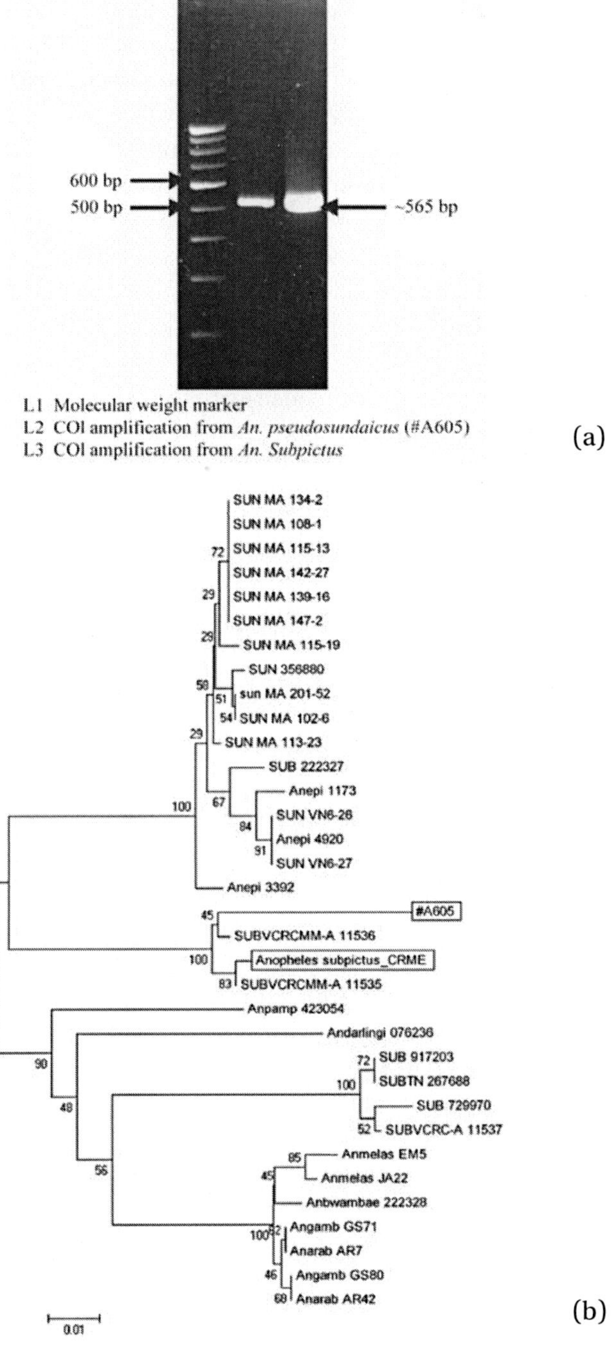

Fig. 9.6. *Anopheles (Cellia) pseudosundaicus* Tyagi *et al.*: (a) molecular characterization, and (ii) phyletic relationships between *An. pseudosundaicus* and other related anophelines. (*Source:* Tyagi *et al.*, 2009)

elongation at 72 °C for 10 min. The thermal cycling was carried out in Master cycler (Eppendorf, Germany). The amplicons were visualized in a 1.2% agarose gel stained with ethidium bromide. The PCR amplicon was purified by ultrafiltration (Millipore, India). The purified PCR product was sent for custom sequencing service (MWG Germany, India Office, Bangalore city, India). DNA sequencing was performed using Big Dye terminator reaction Kit in ABI, Foster City, CA, USA. To determine the sequence identity, the sequences were subjected to BLAST CLUSTAL W analysis was used for multiple sequence alignment. The evolutionary history was inferred using the Neighbor-Joining method. The bootstrap consensus tree inferred from 1000 replicates is taken to represent the evolutionary history of the taxa analyzed. The tree is drawn to scale, with branch lengths in the same units as those of the evolutionary distances used to infer the phylogenetic tree. The evolutionary distances were computed using the Kimura 2-parameter method and are in the units of the number of base substitutions per site. All positions containing gaps and missing data were eliminated from the dataset (complete deletion option). There were a total of 363 positions in the final dataset. Phylogenetic analyses were conducted in MEGA4.

(ii) Morphological Description of Female (Figs.1 a, b)

Head: antenna dark brown about 0.7 length of proboscis (mean 0.7; range 0.58–0.75; n=13); proboscis dark brown, about 0.9 the length of maxillary palpi (range 0.75–0.99; n=13); palpus with three pale bands; apical pale band little shorter than the subapical dark band; subapical pale band very narrow, about 1/5 the length of subapical dark band. Vertex covered with broad dark brown scales and erect white scales, frontal setae white, large and bifurcate. *Thorax*: antepronotum and postpronotum without scales; propleuron with 2 or 3 setae; pleura with broad pale scale patches on mesepimeron and mesokatepisternum; coxae with a patch of broad pale scales. Abdomen: abdominal segments covered with numerous golden setae, the last two segments intermixed with few narrow yellow scales also; terga I–VIII each covered with broad pale scales and few dark brown ones posteriorly; sterna I–VIII each covered with broad dark brown scales and pale scaled patch basally. *Legs*: femora I, II mainly covered with dark brown scales with few scattered pale scales; femur III largely covered with pale scales anteriorly; tibae I–III dark brown scaled with a narrow stripe of pale scales; tarsi I–III mainly with dark brown scales; fore tarsi with tarsomere 1 with a narrow apical pale

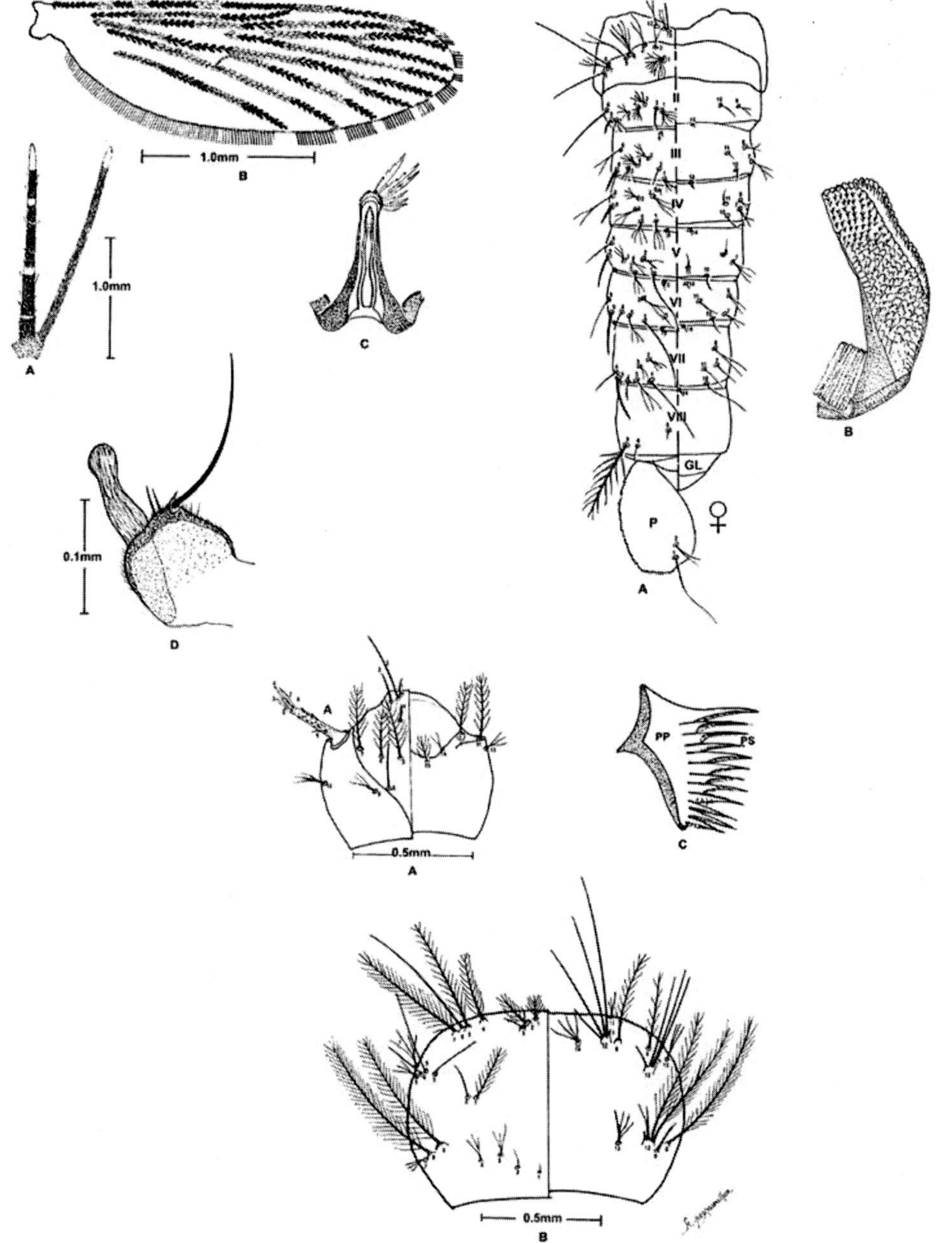

Fig. 9.7. Some of the adult and immature characters of *Anopheles pseudosundaicus.* (*Source:* Tyagi et *al.*, 2009)

band, tarsomeres 2–4 with both basal and apical pale bands, tarsomere 5 completely dark; mid-tarsi with tarsomeres 1–4 with narrow basal and apical pale bands; tarsomere 5 dark; hind tarsi with tarsomeres 1–4 with narrow apical bands, tarsomere 5 completely dark. Wings: mainly dark scaled, about 3mm in length; prehumeral usually completely dark scaled, occasionally interrupted with few pale scales, costa with the following pale spots: humeral, presector, sector, subcostal, and preapical; dark spots other than humeral and preapical are large; pale fringe spots at the end of all veins and rarely a feebly pale fringe spot between 1A and base of wing.

Pupa (Figs. 2 a, b; Table 1). *Cephalothorax*: moderately pigmented; *Trumpet*: heavily pigmented with scattered spicules, index 2.34–2.83, mean 2.68 (sd ± 0.23), meatus about 0.40 of the trumpet length. *Abdomen*; Seta 7-I 2-5(3) branched, shorter than 6,9-I *Paddle*: width about 3/4 of its length, index 0.7–0.8; refractile border about 4/5 of paddle length (index 0.8) with small clearly defined teeth, fringed hairs on posterior border with hooked tips; 1-P long and hooked; 2-P bifid.

Larva (Fig. 3 a, b, c; Table 2). Head: setae 1-4-C slender and single, 2-C about twice the length of 3-C; 4-C about equal length of 3-C. Thorax: setae 1-P, 7-15 branched; setae 9-12-P all single; seta 1-M with 25–30 branches, 4-M single, rarely bifid ; setae 9-12-M all single; seta 3-T poorly developed with 3-5 branches. Abdomen: seta 1-I (palmate hair) weakly developed with 4–9 leaflets, seta 1-II-VII (palmate hair) well developed, evenly pigmented, filaments sharply differentiated, about twice the length of the blade; tergal plates small to moderate sized, smallest on abdominal segment II; seta 1-VIII single to 7 branched; pecten plate with 12–15 (mean 12) uneven teeth.

(iii) Type material

Holotype female (coded A #538) with associated larval (Le #254) and pupal (Pe #256) exuviae mounted on slides with following collection data:- India: Kerala State, Kollam District, Kuzhithura, 22 January 2005, collected as larva from brackish water pool, G. Baskaran collector, deposited in the Museum of Centre for Research in Medical Entomology (MCRME), Indian Council of Medical Research, Madurai, India. The allotype male (A #543) with associated larval (Le #259) and pupal (Pe #261) exuviae with collection data same as holotype also deposited in MCRME, Madurai. Paratypes (36 males, 72 females, pinned and 78 Le and 78 Pe mounted on slides) with the same collection data as holotype are also deposited in CRME Mosquito Museum.

9.10. Discovery of *Aedes albopictus* as the primary and all important vector of dengue in Kerala State, India

In 1999, I declared at the annual Scientific Advisory Committee meeting of my institute (Centre for Research in Medical Entomology, Madurai) that *Aedes albopictus* carried dengue virus in Kerala State and that the mosquito was predominantly prevalent in the diseased area. This information about *Ae. albopictus*, which was later supported by a large number of investigations both in India and other severely dengue-affected endemic countries in Southeast Asia, such as Malaysia, Thailand and Sri Lanka and the WHO too took notice of this research, was a new biological revelation. This discovery made a great impact on the understanding of virus transmission biology and disease control strategy that was hitherto entirely based on the vector biology of *Ae. aegypti*. Significantly, soon thereafter many countries in Southeast Asia began reporting similar observations, substantiating strongly my viewpoint about both the unprecedented 'primary' role of *Aedes albopictus* in disease transmission and control. The discovery turned out to be far more gigantic than I had possibly imagined!

The real proof of 'my discovery on *Aedes albopictus*-dengue transmission in Kerala' flared inferno by not less than a thunder of the bombshell as an 'iconic' vector-borne disease control specialist, Dr N.L. Kalra, then employed as a Consultant at the WHO SEARO, New Delhi complained in writing to my employer and put up his own claim for the discovery instead during or soon after publication of my ICMR Bulletin paper (Tyagi *et al.*, 2006). In his entirely irrelevant, preposterous and false complaint this unscientific mind had not only criticised my research gains but also challenged and abused scientific prowess! In my explanation, which I formed after a deep research into the past research and administrative experience on VBDs by this charlatan who was also doing an editorial work for a journal on dengue, I exposed him entirely. About an year before his repudiation of my research work and putting instead his own claim I had actually submitted a manuscript formulated on the same concept to the journal on which he was employed as a consultant to review all the manuscripts submitted to the journal. This manuscript was rejected and returned to me. First, about an year or so later I found his own paper (co-authored with the Editor of the journal) appearing in the same journal and characterized by marked plagiarism for picking word-by-word and sentence-by-sentence from my own 'rejected' manuscript submitted earlier to his journal. Its important to remember here that "Plagiarism is defined as the use, without giving reasonable and appropriate credit to or acknowledging the author or source, of another person's original work, whether such work is made up of code, formulas, ideas, language, research, strategies, writing or other form". A plagiarist is a cheap human being and not a scientist at all. Secondly, he had not worked on dengue, or its vectors, ever in his career spanning over four decades. And thirdly, he had never even visited Kerela officially or privately on any dengue related scientific investigation. My answer to his treacherously ill-claim was thus simple, straight forward and objectively eyed; I proved him a plagiarist, a charlatan and a non-scientist. My job, which was once threatened, was not only intact, but also my employer was too happy that, with due consideration to many of my achievements in vector-borne disease control science, they had honoured me with the prestigious ICMR-MOT Iyengar Memorial Award. The citation for the award read out that the Award was being given to *"honour a scientist for carrying out research in the field of malaria, filariasis, plague or medical entomology. The award for the year 2007 is being presented to Dr Brij Kishore Tyagi, Scientist 'G' & Director in-Charge, ICMR's Centre for Research in Medical Entomology, Madurai for his research work on 'Medical Entomology'. Dr Tyagi has done extensive research work on malaria in The Thar Desert and coined the term 'desert malaria' defining classification of the Thar's physiography based on the distribution of malaria vectors, besides developing a tanka lid for preventing the breeding of the local vector species, Anopheles stephensi. His work on the dengue vector, Aedes albopictus, in Kerala has redefined dynamics of the disease and its control aspects. Dr Tyagi, along with his team, promoted the thesis of advantageous use of a single annual dose of DEC - albendazole, compared to only DEC, in eliminating human lymphatic filariasis in the affected endemic population in Tamil Nadu, particularly when the vector control was used as an adjunct to the MDA programme. Dr Tyagi has extensively promoted the science of mosquito taxonomy by undertaking revision of the Fauna of British India. Dr Tyagi has authored over 400 articles, many of which were published in international scientific journals. He has also authored - edited 15 books, in addition to a patent to his credit. He has been felicitated with the WHO - TDR's First Prize for the best conceptual drawing for future decadal achievement in 2006."*

9.11. *AlboTyag* – An *Aedes albopictus* breeding preventer in rubber plantation latex-collecting cup in Kerala

During the course my research on *Aedes albopictus* in upper reaches of Western Ghat where it transmitted dengue it was proved that it bred predominantly in the rubber collecting cups tethered along with the main trunk of the tree. To overcome this nuisance I developed on 2nd July, 2008 an *Aedes albopictus* breeding preventer as depicted in the figure given below. (Fig. 9.8).

Fig. 9.8. *AlboTyag* - An *Aedes albopictus* breeding preventer. (*Source:* Dr Tyagi, pers. archive).

9.12. Coining the term "Desert Malaria"

All over world malaria has entered in diverse ecosystems and paradigms, including also the periphery of deserts where it is referred to as 'Desert Fringe malaria' (Zahar, 1985). In modern India, mosquito abundance and diseases transmitted by them have been accentuated by several mega hydroelectrical or hydro-agricultural projects, and one glaring example is that of Indira Gandhi Nahar Pariyojana (IGNP) in The Thar Desert region of northwestern Rajasthan State, which is also the world's one of the largest such projects in a desertic environment. One usually does not associate malaria with a desert environment. Common sense dictates that mosquitoes, that breed only in stagnant water, would give arid, water-scarce spaces such as a desert a wide go-by. But that is not the case in The Thar Desert (Rajasthan), India, which of late has transformed into a land of persistent "desert malaria", a term coined by Tyagi (1995a,b, 1997) after his landmark investigations into malaria in the desertic environments (Tyagi, 2002, 2004).

Chapter 10

THE MOSQUITO THEORIES

10.1. Introduction

Although for almost 2,500 years man had known mosquitoes being associated with disturbance in sleep, public nuisance and loss of general tranquillity by virtue of their painful and irritating biting habits, yet it was only in the third quarter of the 19th century that a serious attention was paid to their possible association with transmission of certain pathogens like protozoans, viruses, worms etc. (Table 10.1). The idea that mosquito could have been involved in some way in spreading certain communicable diseases had been born out even a century ago but these were mostly ill-defined conjectures and assumptions without any material display. It eventually occurred in the early 1880s when Dr Patrick Manson, a surgeon officer with the British army in China, first provided an irrefutable and scientifically demonstrable evidence in support of his 'mosquito theory of human lymphatic filariasis', thus also sowing the seeds for the birth of that fascinating and adventurous discipline called 'Medical Entomology'. The subsequent mosquito theories, particularly 'the mosquito theory of malaria' propounded by Dr Ronald Ross, in 1897, in the footsteps of Patrick Manson's only, have been rife with much claims and counterclaims, and often with a pinch of bitterness!

Table 10.1. Facts and figures about mosquitoes and diseases they transmit (*Source:* Tyagi, 2004)

Human disease	Pathogens involved	Vector(s) involved	Date of vector incrimi-nation	Country	Authority
Filariasis	*Wuchereria* and *Brugia* spp.	Mosquitoes	1886	Taiwan, China	Dr Parick Manson
Malaria	*Plasmodium* spp.	Anopheline mosquitoes	1897	India	Ronald Ross
Yellow Fever	Togavirus	Mosquitoes (Primarily Aedes aegypti)	1890	Cuba	Finlay

Human disease	Pathogens involved	Vector(s) involved	Date of vector incrimi-nation	Country	Authority
Dengue	Flavivirus	Mosquitoes (Primarily Aedes)	1906	Africa	Graham
Chikungunya	Togavirus	Mosquitoes (Primarily Aedes)	1932	Africa	-
Zika	Flavivirus	Mosquitoes (Primarily Aedes)	1957	Africa	-
Encephalitides - JE, WEE, EEE, SLE etc. - Lyme, borreliosis & Ehrlichiosis	Virus/ Bacteria	Mosquitoes, Ticks	1930-1990	Various countries	Various scientists

10.2. The mosquito theories

Like many other hair-raising stories in the annals of medical science by many great scientists of yore which were not without a tinge of battle for supremacy or primacy (*e.g.*, Edward Jenner's vaccine for small-pox in 1796 and Luc Montaigner and Françoise Barré-Sinoussi's discovery of HIV in 1983 were embittered with verbal and other treacherous socio-political moves by certain charlatans), the so called mosquito theories, too, were full of dramas that saw much ups and downs, making the science ever so interesting. But, before reaching a conclusion on who claimed a given mosquito theory with respect to a particular infection, it is inevitably indispensable to comprehend beforehand, in truly lexiconic sense, as to how the meaning of 'theory' differs from a mere conjecture or 'hypothesis'. According to The New Pocket Oxford Dictionary (2002), the word 'Theory' is defined as '*an idea or set of ideas that is intended to explain something or a set of principles on which an activity is based*' whereas the word 'Hypothesis' is spelt out as '*a proposed explanation based on limited evidence, used as a starting point for further investigation*'. Thus, it is comprehensible that a theory is differentiable from a hypothesis in its inexorability of demonstration of an idea or act just as those being enumerated below.

10.2.1. The mosquito theory of filariasis

Fewer tropical diseases appear to frighten as fiercely as possibly the lymphatic filariasis. Some of the chronically affected grotesque cases of elephantiasis, a form of filariasis, leave an indelible imprint on mind of the people who have discerned a case to such a great extent that they simply can not avoid being affected in some way by its intensity of severity and incapacitation to the affected. During the British Raj, a self-acclaimed period of global governance by the British, particularly in the South-east Asia, one of the major concerns of the British government was to save its people from the deadly and/or debilitating mosquito-borne diseases amongst which, after malaria, lymphatic filariasis threatened them most, exacting enormous costs, even though the disease is hardly ever fatal. Dr Patrick Manson, then in his late twenties, and working as a British medical officer in Taiwan (at that time known as Formosa, annexed with China), saw a challenge in deciphering the cause responsible to carry transmission from one human to another. He witnessed a large number of cases of elephantiasis, on several of which he could work to drain and surgically remove the bloated flesh resultant to this disease as no antidote was invented to treat elephantiasis. For over a decade, he worked assiduously with an unparalleled devotion, often secretly performing autopsies, to encounter the animal associated with the disease and to track down its life cycle outside the human body.

Manson drew the opinion that the elephantiasis disease in patients had been caused by the victim's immune system reacting to the presence of the copulating worms (Spielman and D'Antonio, 2001), All along this venture, Manson understood that filarial worm could not lay trapped to their dead end in the body's lymph nodes and appreciated nature's ingenuity in devising means and ways to unlock and set free the elephantiasis-causing worms from their human abode so as to be able to pass on to another human through possibly an organism that fed on the blood of the infected one. Mosquito, for several reasons, was the first probable culprit to be thought of, and he began to work meticulously on this hypothesis which was already aired more than nearly two decades ago in cases of malaria and yellow fever, though, by Beauperthuy and Nott in the Americas. Still more, the idea had actually trickled down over about 100 years when Lancisi, an Italian physician, conjectured a vague association between insect-infested swamps and the resultant human illnesses. Finally, pursuing rigorous and painstaking path of systematic human and mosquito dissections, Manson, in early 1880s, was able to demonstrate that the filarial worm lived and developed in both man and mosquito, although his part of mosquito theory was not totally flawless. But, even these follies only show how monumental was Manson's single-handed accomplishment in opening an altogether new chapter in furthering our understanding of this socially and physically the most devastating infection.

10.2.2. The mosquito theory of malaria

Even though Charles Laveran, a French Army doctor on an assignment to Algeria to check epidemics that plagued the French outposts, had discovered the first of the four human malaria parasite species on November 6, 1880, and many assumptions implicating the mosquito for playing some role in the transmission of this and other diseases were already making rounds, yet it was arguably an epoch-making finding of Dr Patrick Manson, a British medical officer, who, following his posting to Taiwan (then a part of China under the name Formosa) demonstrated with certitude in 1880s the involvement of the mosquito, *Culex pipiens*, in playing the role of an intermediate host, the vector, in executing the transfer of the lymphatic filarial worms to another, possibly, uninfected human. A little more time ago, the Cuban scientist Dr Carlos Finlay had practically proved the inexorable role of the mosquito, *Stegomyia fasciata* (which we now know as *Aedes aegypti*) in transmitting yellow fever in Havana, on October 13, 1880, albeit strong repudiation for about two decades till the mosquito's prowess as a vector of malaria was finally confirmed by Ross and the Italians. Later, on October 23, 1900, a USA giant of a scientist Dr Walter Reed had proclaimed having deciphered the yellow fever enigma by charging *Ae. aegypti* for carrying the disease pathogens. These prolonged scientific feats had had to leave an indelible imprint on the young mind of Ronald Ross, who, though was an unlikely, if not reluctant, scientist, later rose to write the history by solving the malaria puzzle almost single-handedly.

Ronald Ross, eldest of a British general's ten children, was born in the foothills of the Himalaya, at Almora, India, on May 13, 1857. A mediocre in the school, he showed nothing much remarkable, save for his strong inclination and a distinct talent for subjects such as art, music and literature. However, since his father was determined that he should become a doctor, young Ross reluctantly joined a medical school in London where he managed to pass. For joining the Indian Medical Service, he took the examination a second time, and was selected for serving in the Madras Presidency. Ross served in the Indian Army from 1881 to 1899, before he returned to England. During this period of his highly troubled service in India he studied malaria almost by chance and due to his unending desire to do something great. Amidst the pressure of his frequent movements due to transfers, Ross began studying malaria without even having seen a parasite until much later time of Laveran's discovery of *Plasmodium falciparum/malariae*, in 1880, due mostly to his initial contretemps to this new finding. In addition to this, he was also unable to identify mosquitoes which compounded his problems manifold (ironically,

Ross could not have identified the mosquito he called "dappled wings", possibly *Anopheles stephensi*, which he had been successful to get infected in his attempt to prove Anopheles mosquito's role in malaria transmission and was for long confused about only a few anopheline species' power to carry the parasite).

Ronald Ross became a medical scientist almost by accident. Although, always yearning to do something remarkable, he was utterly unsuccessful in sighting malaria parasites, despite his obtaining further training in public health, bacteriology and microscopy during his first furlough to England in 1888. Soon back to India, this time married, he, a truly polymath, had plenty of time to spend playing golf and fulfilling his deep interests in poetry and music. He published a romance novel in London and composed poetry, but even with this initial success he was bored and yearned intensely for achievement. During his second furlough in 1894, he returned to London and met Patrick Manson, who had by then already demonstrated mosquitoes role in transmitting lymphatic filariasis and, rightly so, earned the reputation of the Father of Tropical Medicine. At the Seaman's Hospital, where Manson, a true giant of medical science, had taken a position after leaving China, he showed to Ross malaria parasites for the first time, and demonstrated to him how to find the parasites in blood smears. Manson had inferred, based on his earlier experiences in China in respect of filarial worm, that mosquitoes could also be responsible for transmitting malaria, a theory then circulating, although Ross felt sceptical about this. Yet, Ross took note of his advice quite seriously, and returned to India fully determined to prove Manson's "Grand Induction" by observing malaria's growth and development in mosquitoes. Ross looked in Manson a father figure and a mentor - a bond endorsed by the 173 letters, rich in both scientific and human terms as well as remain one of the celebrated scientific correspondences, that the two exchanged between 1895 and 1899. These letters record, in great detail, Ross's research in India elucidating the role of mosquitoes in the transmission of malaria and worked to firmly cement their intimacy. These are a testimony to not only Ross's excitement in his pursuit to make the discovery, but also his immense frustration and vexation he had had to endure on many fronts - his army bosses' annoyance, day-to-day cumbersome duties, frequent outstation postings and constant failures in his endeavours to incriminate a mosquito with a malaria parasite. This massive correspondence is, ironically, also a testimony to the rise and fall of a relationship, as Ross's preoccupation with his place in the history of malariology led to a breach between the two men. That Ross was completely obsessed with garnering every possible laurel in the wake of his "mosquito malaria theory", it ultimately spoiled his relationship with his mentor. Themes of priority, nationalism, and personal vanity punctuate this latter correspondence, which also reveals new insights about the golden years of tropical medicine. This happened primarily because Manson thought to credit Grassi, who was a Professor of Zoology with extensive mosquito/malaria-related research experience and was considered, rightly so, by many scientists to be the pre-eminent medical zoologist of his time, for his commendable contributions, but Ross turned cool toward his former teacher, mentor and father figure. Ross continued to critically despise Manson, ignoring latter's immense assistance and guidance in his pursuing malaria research as evidenced from scores of affectionate letters that passed between them. This bitterness culminated to its zenith when Ross, recuperating from a heart stroke by the end of his life, wrote a highly unpleasant book charging that Manson even hindered, rather than helped, his malaria work.

Ross worked relentlessly about finding the mosquito's role in malaria transmission. Though unaware of mosquito identification and dissection techniques, he burnt oil all through night dissecting all the mosquitoes thought of carrying the pathogen. Alas! Success still eluded him. Ross all the while burning with that unquenchable thirst for achieving something very great, continued to search for the parasite in the mosquito body. The most significant step forward occurred on 20th August, 1897, when he got the first sight of 'pigmented cysts' in the stomach of a mosquito. Ross celebrated this historical event as "Mosquito Day", and the jubilation showed up in his memorable poem that he wrote soon after the discovery was made (see Chapter 8).

10.2.3. The mosquito theory of yellow fever

Yellow fever had been the scourge of the Caribbean and the coastal and southern United States for centuries. In the 18th and 19th centuries, devastating epidemics of yellow fever had claimed thousands of lives. In 1792, yellow fever claimed 10 percent of the population of Philadelphia (4,000 out of 40,000). New Orleans was swept by two epidemics: in 1853 almost 8,000 died and 4,000 more in 1878. That same year over 5,000 died in Memphis. Many thousands more died in other American cities and numerous Caribbean islands, The sheer terror that the epidemics brought was often overwhelming. Despite years of efforts by the best medical experts, no one knew how it was spread, why some were stricken and other family, friends or neighbours simply passed over. Equally frightening, there seemed to be no way to protect yourself or your loved ones. It appeared that no defence, no barrier, no precaution could stop its deadly progression. The epidemic in Memphis was so volcanic that it overpowered normal human behaviour: the frightened men, women and children poured out of the city by every avenue of escape, by every possible conveyance such as the hacks, carriages, buggies, wagons, furniture vans, street drays, bateaux, or by anything that could float on the river, and by the railroads. The ordinary courtesies of life were ignored, politeness gave way to selfishness, and the desire for personal safety broke through all social amenities.

"How could Yellow Fever, the dreaded killer, be stopped?", was the daunting task of the Yellow Fever Commission that was established in Cuba some 100 years ago. It consisted of Walter Reed, James Carroll, A. Agramonte and Jesse Lazear. In the first few years, unsuccessful attempts were made to prove a relationship between yellow fever and the bacteria, *Bacillus icteriodes*. It failed, and the Commission members focused their efforts determining how yellow fever was spread. They had found mounting evidence against the commonly held fomite theory of transmission. Among the several other theories, Dr Jesse Lazear was most intrigued by the possibility of a mosquito vector. The mosquito theory of transmission of yellow fever had been around for many years.

Possibly the first to discuss it was Dr Josiah Nott of Mobile, Ala., in 1848. It had been proposed again by Dr Carlos Finlay of Havana in 1881. But try as he might, Finlay had been unable to prove his theory in almost 100 experiments on human subjects. Finlay's mosquito theory remained in abeyance until attention was again drawn to it by the demonstration in recent years of the part played by these insects, the mosquitoes, in the causation of other tropical diseases. Many in the medical field belittled him because of his passion for his mosquito theory, but the discovery in 1897 by Dr Ronald Ross of the British Indian Medical Service that malaria was transmitted by the mosquito (*vide supra*) had given needed weight to this possibility. Lazear, who had previous experience with mosquitoes during his malaria research, was most interested in the theory. Finlay was convinced that only the female *Culex fasciatus* mosquito (now called *Aedes aegypti*) transmitted the disease. Lazear carefully hatched the mosquitoes and allowed himself, along with another Army doctor, A.S. Pinto, to be bitten by mosquitoes that had fed on yellow fever patients on the fifth day of illness which had been 12 days earlier. They remained well. Seven other volunteers were bitten, they also remained well. The researchers were beginning to doubt there was anything to this mosquito theory.

On Aug. 27, 1900, Lt. James Carroll and Lazear were discussing their lack of results when Carroll suggested that they try one on the mosquitoes on him. Carroll`s mosquito had bitten a yellow fever patient on the second day of illness, 12 days before. Two days after the bite, Carroll was not feeling well and on the third day he was clearly ill with fever. A blood test for malaria was negative and as Carroll became sicker, it was obvious he had yellow fever. Lazear and Aristides Argamonte were very concerned about Carroll's health as he was somewhat frail and well past 40 years of age. However, they were excited that the mosquito bite could have caused his yellow fever. This would be impossible to prove conclusively, as Carroll had been in contact with other yellow fever patients at the hospital in Havana and in the autopsy room. He could have gotten it in any of these places; no one could know for sure. On Aug. 31, the day Carroll was diagnosed with yellow fever, Lazear was in the post hospital at Columbia Barracks. Lazear and Argamonte were much relieved several days later when it became apparent that Carroll was going to survive.

There had been no cases of yellow fever on the post or in the hospital. Pvt. William H. Dean of the Seventh Cavalry was about to be discharged from the hospital for a minor condition when Lazear asked him if he would be willing to submit to an experiment involving his mosquitoes. Dean was newly arrived in Cuba and had never been off post. He agreed and was experimentally bitten that same day. Being careful not revel in someone's illness, they must have been elated when Dean became ill and was diagnosed with yellow fever. Dean's was the first experimental case of yellow fever transmitted by the bite of a mosquito that was irrefutable as to its origin.

This discovery was, however, subjected to much harsh treatment and was always in the centre of stormy debates. Even celebrated men of science became locked in fierce disagreement. These differences in opinion gave rise to wildly different approaches to prevention and treatment. Around 1793, in Washington, D.C., Dr Albert King hypothesized that mosquitoes were the real culprits, but he was ridiculed and repudiated openly for his ideas to protect his nation from such infectious mosquito bites. Another hundred years would pass before scientists realized that King was on the right track. In 1881, a Cuban physician, Dr Carlos Finlay, made a discovery that would revolutionize disease control and ultimately make possible the construction of the Panama Canal. He identified the Aedes aegypti mosquito as the carrier of yellow fever. Finlay's breakthrough had glitches, however. He correctly had surmised that the mosquito was spreading the disease by biting people previously infected with the disease and then transmitting the disease agent to other people through its bite. But when he attempted to prove his theory by having infected mosquitoes inoculate healthy volunteers with the disease, he failed. What Finlay didn't know was that the virus needed to incubate for approximately 12 days inside the insect in order to develop. He was attempting to inoculate his volunteers with the virus using mosquitoes that had only been infected a few hours. More years would pass before these missing pieces were added to complete the puzzle. Finlay continued his quest to identify the agent causing yellow fever and reveal how the mosquito transmitted it. At the same time, building on the work of Sir Ronald Ross, a British physician and Nobel laureate (1902) who identified a malaria plasmodium in the stomach of an Anopheles mosquito like the one Laveran had found in human blood, Finlay formulated a detailed plan for combating both yellow fever and malaria.

Initially, Reed did not put much stock in Finlay's mosquito theory, efforts focused initially on ridding the Cuban capital of the filth and contamination presumed to cause the disease, But when the yellow fever death toll rose despite a massive clean-up effort, Reed decided to test Finlay's theory. Reed constructed two small buildings through which air could circulate freely to first show that the disease was not airborne. In one building he had volunteers sleep for 20 nights in the foul bed-sheets and garments of yellow fever patients. In the other building, volunteers slept next to infected mosquitoes, but were separated by screens. In this experiment, Reed did not use recently infected mosquitoes like Finlay had, but rather insects that had been infected for some time. Then one volunteer was exposed to infected mosquitoes for three successive nights. This volunteer was the only one to develop yellow fever. Proof at last was conclusive: *Aedes aegypti* was the culprit. Reed had yielded the knowledge that would free Havana from yellow fever, although even now certain geniuses of the time, like Gorgas, remained unconvinced of Finlay's theory but resorted to eliminate the mosquito and its breeding places, besides quarantining the patients to prevent them from the remaining infective mosquitoes. The results were spectacular as within less than a year yellow fever had been virtually eradicated. Final conclusion certainly is that it was Dr Carlos Finlay of Havana, who in 1881, propounded the theory that mosquitoes were the carriers of Yellow Fever.

10.2.4. The mosquito theory of dengue fever

Incidentally, dengue is amongst the main mosquito-borne infections to be causing great concern to both the people who are at risk of infection with its deadly forms, the dengue haemorrhagic fever (DHF) and dengue shook syndrome (DSS), and the health planners all over the world, but particularly in the tropics, due to disease's tendency to spread to non-endemic areas in all the world's five continents. Owing to the growing concern associated with its causing enormous morbidity and mortality, and the subsequent sequelae to health retardation at various levels, the Tropical Disease Research of the World Health Organization has in recent past incorporated dengue within its fold of priority research.

As it does even today, the disease was often mistaken for either yellow fever or other similar illnesses in Asia, Africa and America. Dengue came to be recognized as a separate entity only during first part of the nineteenth century when a severe epidemic occurred in the Caribbean country. Finally, the mosquito *Aedes aegypti* was firmly confirmed to be playing a vector's role in transmitting dengue fever virus in 1906. Subsequently, other species were also identified to carry the transmission, but much less seriously, and the most notable such a species is the Asian tiger mosquito, *Ae. albopictus.* The latter species, from whom dengue virus was detected in India as late as 1988, has in recent years gained tremendous significance due to its implication in disease transmission in Kerala State, hitherto largely a *terra incognito* for epidemic dengue (Tyagi *et al.*, 2006). In Malaysia a simian zoonosis has been identified which has rendered an altogether new dimension to the dengue epidemiology.

10.2.5. The mosquito theory of encephalitis

Almost each of the encephalitis-mosquito theories, all of which have been floored during the 20th century, is as full of scintillating and stimulating scientific zeal, charm and obsession as were characterized any of their predecessors with respect to other mosquito-borne infections. Yet, nearly all these certainly went on to be accepted in the concerned research community without an unreasonable confabulation or consternation. The Japanese encephalitis' mosquito theory dates back to 1930, while that of the infamous St. Louis encephalitis is credited to 1942.

10.2.6. The mosquito theory of Zika

Zika, a flavivirus transmitted mainly by mosquitos in the genus Aedes, was discovered in 1947 in Uganda. From the 1960s to 1980s, human infections were found across Africa and Asia, typically accompanied by mild illness. The first large outbreak of disease caused by Zika infection was reported from the island of Yap (Federated States of Micronesia) in 2007, as the virus moved from south-east Asia across the Pacific. During an outbreak in French Polynesia in 2013–2014, Guillain-Barré syndrome was linked to Zika infection and cases of microcephaly in new-born children were also retrospectively linked to this outbreak. The World Health Organization (WHO) received the first reports of locally-transmitted infection from Brazil in May 2015. In July 2015, health ministry officials from Brazil reported an association between Zika virus infection and Guillain-Barré syndrome in adults. In October 2015, WHO received reports from Brazil of microcephaly in babies whose mothers had been exposed to Zika during pregnancy. At this time, there was no proof of a causal link between Zika infection and these neurological complications. In February 2016, as infection moved rapidly through the range occupied by Aedes mosquitos in the Americas, WHO declared that Zika infection associated with microcephaly and other neurological disorders constituted a Public Health Emergency of International Concern (PHEIC). By the start of February 2016, local transmission of Zika infection had been reported from more than 20 countries and territories in the Americas and an outbreak numbering thousands of cases was under way in Cabo Verde in western Africa. Beyond the range of its mosquito vectors, Zika virus infections are expected to be carried worldwide by people as they travel and be transmitted by travellers to sexual partners who have not been to places where the virus is endemic.

In 2015 Brazil saw a dramatic increase in the number of babies born with microcephaly, a condition in which a baby's head is much smaller than normal due to abnormal brain development while in the womb, often resulting in developmental delays later in life. Normally microcephaly is a relatively rare occurrence usually caused by exposure of a pregnant woman to drugs, alcohol or environmental toxins, but nearly 3,000 cases were reported in Brazil, about 20 times the number of cases reported in the prior year 2014. Initial epidemiological investigations implicated an infection during pregnancy with the Zika virus as the cause. Prior to the microcephaly epidemic, Zika virus was a relatively unknown pathogen. Transmitted through the bite of an infected female mosquito *Aedes aegypti*, Zika virus normally causes few or no symptoms in children and adults, although in rare instances it can result in an autoimmune disease, Guillain-Barre syndrome, in which the body's immune system attacks part of the nervous system leading to muscle weakness and paralysis.

The past several decades have seen an alarming increase in numbers of *Aedes aegypti*, which also carries the viruses for yellow fever, dengue, and chikungunya, affecting millions of people worldwide. This enormous increase in numbers of mosquitoes most likely reflects the explosion of human-created trash that litters the world. Needing only small amounts of water that puddle in abandon tires, non-biodegradable containers, and plastic bags to lay their eggs, the mosquito has evolved to live in close quarters with humans, especially in poorer regions of the world where trash removal, mosquito netting, air conditioning, and insect control are limited or non-existent. Zika virus has been observed by virus hunters making slow progress from the Zika Forest in Uganda to Indonesia to Micronesia and the Polynesian Islands mirroring the movement of populations across the Pacific. Until 2013, Zika virus seemed mostly innocuous, and not worth worrying about. Then in 2015 it was clear that Zika virus was becoming a very serious concern (Fig. 10.1). The virus is believed to have arrived in the blood of infected travellers attending the 2014 World Soccer Cup. Because most Zika infections show few or no symptoms, these travellers may not even have realized they were infected. The combination of many individuals carrying the virus in their blood coming in contact with and being bitten by the mosquito vector present in great numbers resulted in an unprecedented rise in Zika infections and the terrible consequences for unborn babies. It is not been determined whether Zika has a non-human animal reservoir.

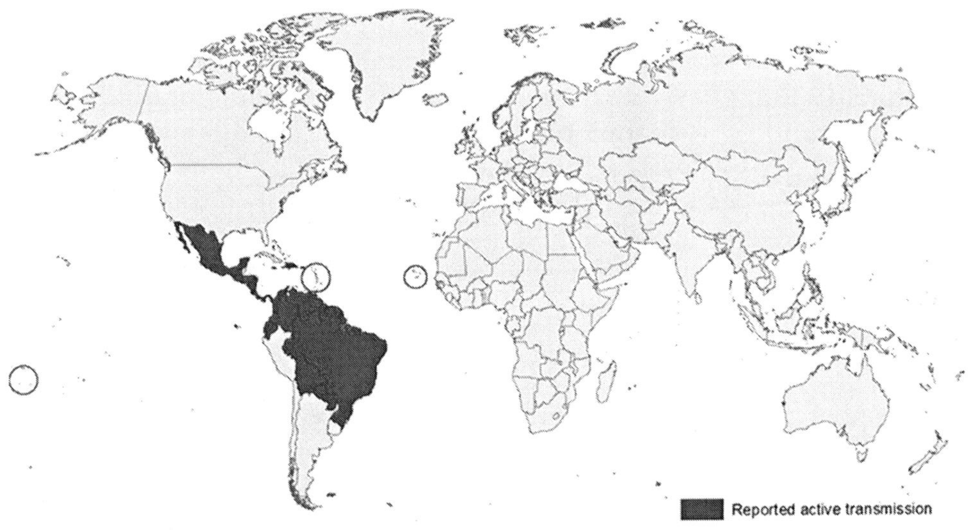

Fig. 10.1. Areas of major Zika infections as of February 2016. (*Source:* Free Wikipedia)

The rapid spread of Zika throughout Central and South America may also be the result of a change in the virus itself. Mutations may have altered the virus enabling it to infect humans more easily or increasing the number of Zika virus particles in human blood. Either of these changes could increase the rate of Zika spread by increasing the chances of a mosquito picking up the virus from an infected individual and transmitting it to another person.

Chapter 11

MAN-MOSQUITO: *FACE OFF*

11.1. Introduction

It was a world war with a difference: the enemy, a tiny and feeble-bodied insect - the mosquito, was still not quite known for its role in bringing deadly infections to man but was always looked upon with draconian suspicion and great scientific interest alike. For this tiny, feeble-bodied mosquito was soon going to occupy the central position of all the World's strongest minds' attention and discourses unmatched in the annals of science of any other field. There have been many a great mosquito-related battle but none was fought like the battle of 'malaria mosquito', in the last quarter of nineteenth century! The tiny mosquito assumed such a high importance that the conflict engulfed more countries into its bitter folds than probably the real World War One nearly a quarter century later! But, even this 'great malaria-mosquito war' was fought in a military style. This was mankind's first war against another species. This too was an intra-specific human war, with mosquito playing the pivotal role in the entire tumultuous saga of research primacy and priority over the discovery of malaria- mosquito connection.

11.2. Prelude to the 'Great Malaria-Mosquito War'

The 'Great Malaria Mosquito War' (which l have discussed here as the "First Mosquito World War") erupted with the first glimmer of proclamation on 20th August, 1897, by Dr Ronald Ross, whom many claim as an unlikely, if not reluctant, scientist or malariologist, that mosquitoes transmit malaria (amazingly, and still interestingly, Ross did not yet know identifying one type of mosquito from another, leave aside one genus against another). However, based on colour patterns, he classified them for his convenience as grey or bar-backed mosquitoes (one of which perhaps was *Culex quinquefasciatus,* the common house mosquito), brindled mosquitoes (which he perhaps concluded as *Aedes aegypti,* the yellow fever mosquito) and the big brown mosquito with striped (actually spotted) wings which he later called as the dapple-wing mosquito (which was in fact an anopheline mosquito belonging to the genus Anopheles, members of which are capable to transmit human malaria parasites and pass them to people; whether it was *An. stephensi* can only be conjectured). For this reason only, Dr Grassi, considered amongst medical zoology fraternity a preeminent scientist with enormous experience in malaria/mosquito-related subjects, regarded Ross, who was completely unknown to mosquito taxonomy, as an amateur and an interloper. In the face of the background of the announcement of establishment of the prestigious Nobel Prize an year before in 1896, endowing in prize money for scientific discoveries the largest monetary payments ever, the first seeds of the great 'malaria mosquito war or the war of giants were seemed to have been sown, which mushroomed in a more vicious way in years to come, for the rest of their lives, both Ross and Grassi, the former in particular, were horn-locked hurling insinuations of all types on each other with a view to secure priority or supremacy in several discoveries!

11.3. The Nobel Prize factor

During the late 1870s - early 1880s, mosquito had begun to attract much attention of tropical disease researchers and Manson's assertion of mosquito's role in transmitting human lymphatic filariasis worked to further ignite subject-ebullience at a fevered pace all over the world. All the three serious diseases of the day, *i.e.*, malaria, filariasis and yellow fever, were being looked into with a microscopic precision. Still, the attraction which malaria commissioned was unparalleled in the history of vector-borne tropical diseases around the globe. Malaria always asserted some claim to be special. Widespread and debilitating, malaria - the most common of all tropical diseases - was of all-time great interest, as scientists in France, Italy, Germany, United States, England and India vied in their independent and isolated pursuit with a rather nationalist fervour for research excellence and discovery claims. The struggle was no less intense than the sprint to discover the New World or 'the golden bird that India had once been; and it foreshadowed future scientific contests over the polio or small-pox vaccine and the atom (Spielman and D'Antonio, 2001). Most of the names in circulation were those of much acclaimed and seasoned scientists; the Italian genius Professor Giovanni Battista Grassi, commanding a team of as dedicated and qualified researchers as Dr Amico Bignami, Dr Marchiafava and Dr Camillo Golgi, was already a hugely respected scientist with a unique knowledge of both medicine and Zoology, and a background of several years of mosquito-malaria research; Dr Patrick Manson, truly a genius and creative mind always perceiving connections and resemblance between disparate phenomena through his sound and logical imaginations, was an undisputable authority on tropical diseases;

Dr Charles Luis Alphonse Laveran, though unimportantly dispatched to Algeria, in 1878, on an official assignment to help save the lives of French troops from some kind of arcane infection scientifically unknown to the mankind till then and providentially pitched into investigating malaria, was later credited for discovering the first human malaria parasite *Plasmodium malariae*. In contrast to all these and many others who worked on malaria and mosquitoes, Dr Ronald Ross, though egoistical and ambitious, like Laveran, and often regarded as an unlikely, if not reluctant, scientist, on one hand, and an inexhaustible and obsessed malariologist and a polymath, on the other, was in fact a genius par excellence who was ultimately credited for unravelling the long-time baffling secrets of malaria parasite's journey via mosquito! All these scientists regardless of their academic and professional background had one thing in common; they were endowed with the seemingly limitless energy, patience and optimism that are indispensable traits of discoverers.

In this different kind of war game the enemy was unknown and the actors from different countries played their part of roles in all directions with a phlebotomist's cautious precision, hoping to strike the bull's eye first amongst all. While Manson's pursuit of the filarial worm between late 1870s and early 1880s cleared all doubts and dark points pertaining to mosquito's role in effecting transmission of the filarial worm and ignited equally subtle minds around the globe, scientists particularly in Europe and North America competed with nostalgic resolution and pace to win the ultimate race to scientific discovery, which was certainly like hitting the jackpot. All kinds of means were adopted to press in the claim of primacy; advance reports of 'successful' experimentations were regularly advertised in the premier dailies, scientists indulged in verbal and written duels to forward first their 'original' claim to secure credit, some scientists, if not all, lobbied not only in the world of science but in the public realm as well to muster in support to win the all-important victorious vote in the event of a contest for priority over discovery or when it mattered most, and, worst of all, a few scientists clandestinely spied over others' progress in work! Scepticism and doubt loomed large even though scientists, in order to understand the enigmatic matters of malaria a bit more better, logically involved in exchanging written, published or unpublished records, including microscopic slides. Patriotic appeals for increased funding in research became intense and nationalistic sense of competition took the uppermost tone in expressing solidarity to the home economy and health of the countrymen. Notwithstanding all this precarious environment, the scientific fraternity, comradeship and cooperation amongst scientists of various different countries too continued, though not always unscathed, as illustrated through the scores of correspondence among Manson, Grassi and Ross in the midst of the year of discovery -1897.

Destined, as if it were, to effervesce an already ebullient quest for malaria mystery through mosquito, the announcement of setting up of the Nobel Prize in Medicine and Physiology (as a part of a series of world prizes instituted in 1896 following the death will of the dynamite king Alfred Nobel, a Swede who had a great benevolent heart and appreciation for scientists' remarkable contribution in making a better world to live in), with the largest monetary payments ever for scientific discoveries, apparently ignited every mind around genuinely engaged in deciphering malaria enigma and to proclaim first to reach the ever awaited conclusion! The contest eventually narrowed down to two scientists of contrasting characters, Grassi and Ross. Professor Grassi was a zoologist of international repute and he had been seriously toiling to prove a malaria-mosquito connection for several years. In contrast, Dr Ross, in the words of Grassi, was an interloper and an amateur who could not distinguish one mosquito from another. For the 1902 Nobel Prize, the prize committee had initially recommended a joint award for both Grassi and Ross, in recognition of their independent and highly valuable malaria research. But, one of the members of the jury, Dr Robert Koch, a German who also wielded tremendous scientific influence and had still not forgotten the ill treatment he had received at the hands of Grassi several years ago during an invited visit to Italy, got an opportunity to settle a personal score and, using his influence, maneuvered to vote against Grassi in the final round. Professor Grassi Was most unfortunately dropped, and Dr Ross emerged a victor.

The 'Mosquito World War' which had already begun around early 1890s now reached he climax - a point of no return. Henceforth both Grassi and Ross battled on virtually every issue of priority in several discoveries in malaria research they encountered. Influenced by the widespread British Empire and, in pert, Ross's own genius in effecting timely scientific publications and/or exposition of his achievements in established scientific forums, mostly through generosity of the ever helping Dr Manson, in Great Britain, Ross was destined to be more widely recognized and rewarded. Ross was said to be so insolently obsessed with squeezing all the possible honours that he eventually marred his highly affectionate and long-time relationship with Dr Manson whom Ross had been calling his mentor and teacher. When Manson wished to highlight the malaria research contributions by Grassi, Ross harshly discounted Manson's science and even suggested that Manson hindered rather than helped his malaria

Work. Thus, the Great Malaria Mosquito War had set in with full-blooded attacks and counterattacks over the primacy of mother of all discoveries, that is an Anopheles mosquito obligatorily transmits the malaria parasite from man to man.

11.4. The Italian folly

Scientists in Italy had a long tradition of working on insect-borne illnesses, including 'intermittent fevers' allegedly resulting from the widespread 'mal-air emanating from swamps, dangerously prolific with mosquito productions. The scientific community in Italy was blessed with a string of geniuses like Grassi, Celli, Golgi, Marchiafava, Guarnieri, Dionisi, Bastianelli and many others. Yet, in spite of all the treasure of data generated through long years of dedicated perseverance, they would best be known to have won in the last lap of the medley by pinpointing irrefutably the role of particular Anopheles species in mediating malaria transmission as well as the malaria cycle, in human. In a parallel to what they earlier did following Laveran's discoveries, they repeated history by completing, rounding out and refining Ross's discoveries. And, therefore, even with a massive scientific background they lagged behind a solitary medical scientist in India.

In working out a connection between malaria and mosquitoes, at least two follies were significantly discernible in their endeavours. First, they 'procrastinated' publication of their findings. If only intensity, dedication and volume of work done are considered then Prof. Grassi, the most eminent living Zoologist then, would certainly have finished first in the race to pin claims to the discovery of the transmission of malaria by the mosquito. But no, he reserved his well-studied judgement. He busied himself in arguing that the parasite, if it was a parasite, had to have a nucleus and none had yet been found. He did not cease at it and further argued that it had to eat and no one had definitely shown that it did. So, little realizing about Ross's amazing writing prowess and an insatiable hunger for seeing his manuscripts published at the earliest, Grassi waited for the last tittle of evidence which was accomplished only in 1889 after several scientists' researches. Later on, picking up the old threads, the Italians began work in the middle of July, 1898. By that time Ross's proof was complete and partly published. By the end of summer, Ross was able to see his results made public both by circulating his official report to all those working on the subject of 'mosquito malaria connection', including the Italians, and the lecture that Manson presented in Edinburg, printed in two reputed journals, the British Medical Journal and the Journal of Tropical Medicine, in 1898.

Secondly, none, even the preeminent Grassi, stuck to the 'mosquito malaria connection' on sustained and long term basis, and there were many gaps in resuming researches. The kind of unfazed zeal, focused dedication and resolution, and maddening obsession that Ross exhibited in every action, verbal or written, was nowhere seen at any stage of 'mosquito malaria connection' research by the Italians. Black sheep and scientific charlatans, who existed everywhere looking for suitable opportunities, lost no time to impugn scientific credibility of those toiling hard to serve the humanity by unravelling malaria mystery. Consequently, the duel Grassi-Ross war turned more ugly with the introduction of great German bacteriologist, Dr Robert Koch, who too was tempted to have claimed as the first on the 'mosquito-malaria connection' discovery as early as 1883, without putting forth much proof, though! Finally, the rivalry took a triangular tangle, where Grassi was sandwiched between the giant Koch and the flamboyant Ross. This equation had its own very unfortunate and eventful bearing on a later date. Subsequently, when the Nobel Prize jury met to select the winner for the 1902 Prize in Medicine and Physiology, it was Koch, who was himself locked in animus with Grassi and was filled jealousy ever since his Italy visit and Grassi's hurried publication of Plasmodium cycles in Anopheles mosquito and human in 1890 without any reference or acknowledgement to Koch's own hypothesis, and railed against the idea of conferment of the medal jointly to Grassi and Ross. The result: Grassi was dropped!

11.5. Mosquitoes under suspicion

Truly speaking mosquitoes were suspected for long to have connection with certain illnesses, particularly intermittent fevers. Malaria was already known widely and adequately characterized clinically amongst the folks in various parts of the world. It indeed occurred as a special malignity of the environment. Europeans did know through expeditions to South America from 17th century onwards that cinchona bark extraction was able to cure intermittent fevers but they had classified all fevers as one phenomenon. Europeans, whose dominance around the world then was phenomenal, associated malarial sickness with wetlands, hence the name '*paludisme*' [which in French means from swamp; *palus* (Latin) = swamp]. Rome was particularly a malarial inferno owing to Pontine swamps in vicinity, hence the local name 'Roman fever? Clearly malarial fevers related to swamps, as unhealthy and notorious fever breeders, along with disturbed lands, forests under denudation and lands under ploughing, particularly the last one, hence the emergence and acceptance of the "telluric" and 'miasmic' theories for centuries. Yet, though traditional epidemiologists stuck to their unhealthy soil (telluric) and malicious air (miasmas) theories, some noted 18th century Italian physicians, like Laricisi, tried to find out aetiology of the disease by explaining the fevers were caused due to inhalation of a living contagium or animalcules, called 'serafici' invisible to naked eyes, which multiplied in human body under unmeasurable conditions. A hypothesis subsequently advanced by Rasori noted that different kinds of fevers were caused by different types of species - an extraordinary thought so close to the true interpretation of malaria which would still take several decades until discovery of advanced microscopes good enough to spot microorganisms. The relationship between swamp-emanated inhaled bad air, comprising whatever toxin of Lancisi and others, with the fever was so obvious that the Italians, possibly in the 18th century, naturally saw the cause of fever as mal' aria (bad air), which was later refined as malaria in Great Britain. Subsequently in late 19th century when Laveran, and others, discovered causative parasites - Plasmodium species, the word 'malaria' was readily accepted unchanged for the disease itself. Both the telluric and miasmic explanations circulated for a long period were added by another explanation of mixing of the toxin or poison from the soil into water which resulted in fever after being drunk. This hypothesis changed the course of thinking about malaria which was now classified along with other important gut diseases as cholera and typhoid. This meant that, as typhoid and cholera - which were associated with unhealthy aquatic, land and living environments - had to be taken care of through improved sanitary measures, malaria too - considered as a disease of rural decay, spawn of an ill-kempt, savage and uncultivated land, yielding to good husbandry and civilization - fit the pattern and invited similar countermeasures. This thinking of control of disease through good sanitation is still largely practiced. So, water was understood beyond doubt the key factor to malaria prevalence. Interestingly, it was discovered inadvertently that malaria distribution virtually overlapped the distribution of anopheline mosquitoes (for details on Anopheles mosquitoes see Chapter 1). Why the miasmas were associated with these mosquitoes is due to the fact that most of the malaria-carrying species bite following the night is set in, a pointer to the fears surrounding night miasmas. Another interpretation is the feeble-bodied adult females, in particular, face the danger of desiccation if exposed to the heat of the day, hence they venture out and about more often when the air is damp than when it is dry which is very much likely during the daytime. While a school of thought of the day prudently tried to build the connection between intermittent fevers and the mosquito prevalence, another school silently advocated the presence of the pathogenic microorganisms. Unfortunately, the fewer doctors who ventured out to bridge observed associations between mosquitoes and intermittent fevers were actually scientifically greatly impoverished due to lack of demonstrable facts. Some approached the solution in a different way; Dr John Crawford, an American physician, while working on 'yellow fever mosquito theory' opined in 1807 that these are the eggs 'insinuated into our bodies' which caused the malarial fever. He was faced with all round criticism, but little did the world know that he was more nearly right than anyone else would be for ninety years in future.

In 1854, Louis Beauperthuy, a French physician, better known as the prototypical author of the mosquito theory, boldly connected mosquitoes with intermittent fever - in warm weather but not in the cold. He was however firm in narrating that mosquitoes injected toxins, picked up from the putrefied vegetation in swamps, into humans. Rasori, an Italian physician, thought mosquitoes might carry the contagium or animalcules. All these early thinkers had no idea or a scrap of evidence of what exactly that imaginative contagium was or how and where it came from, albeit the basic notion that parasites might shuttle between different animal species having been already proposed in 1790 by Peter Christian Abilgaard!

11.6. Mosquitoes under scrutiny

Mosquito has been conceived to cast evils on man for centuries and by 1870-80s some concrete evidences also began to emerge manifesting mosquito's role in transmitting certain diseases. The first important work on the role of insects was done by Dr Patrick Manson in China during the 1870s When he showed that mosquitoes might be vectors of the disease colloquially known as elephantiasis from the gross swellings, particularly of the legs and scrotum, that it often produces (Harrison, 1978). The hypothesis underlying mosquito's role in transmitting filariae precisely dawned on Manson in 1875 when on a furlough to England. Reasons which he forwarded were absolutely irrefutable: he found out the young microfilariae, which unlike their procreator moved freely in blood stream and were always enclosed in a delicate membrane, could neither feed nor move of their own volition, and therefore must complete their life cycles somewhere else - the mosquito? But, Manson whose inquisition into the ingenious ways of nature was ebullient and effervescent conjectured by working meticulously and arduously for seven long years on his 'filariasis mosquito theory', all along with his fulltime and highly successful medical practice, and despite excruciating illness due to general gout which he contracted at the age of twenty six and suffered from it throughout his life - often totally crippled for days or weeks. Yet, finally in early 1880s, Manson succeeded in demonstrating that mosquito, *Culex pipiens quinquefasciatus*, was responsible to effect man to man transfer of the not so deadly but fiercely dreaded and debilitating lymphatic filariasis worm. This discovery was by all standards a great victory in furthering human knowledge about mosquito's placement in the man-mosquito-disease triangle. However, despite the partly true nature of his discovery, Dr Manson, a surgeon-physician, had made two major mistakes while proclaiming mosquito's intermediate role in disease transmission: (i) First, he thought that these worms completed their development in mosquitoes by five days after feeding - which we today know is more than 10 days - which implied that worms had still not matured, and (ii) Secondly, believing that the mosquito fed only once in her lifetime, he postulated that the parasites infected people only when they consumed drinking water having disintegrating body of an infected mosquito. It bore no adverse significance, considering Manson's phenomenal recognition as the tropical medicine genius and his firm demonstration that mosquitoes performed an essential role in generating human diseases, particularly his guidance to Dr Ronald Ross about malaria and mosquito - a subject that would soon reveal the mosquito's role in producing explosive outbreaks of the disease, both in terms of human maladies of morbidity and mortality exacted by the Plasmodium-Anopheles nexus, as well as the unprecedented scientific rivalry between the geniuses of the day. The filariasis-mosquito discovery by Manson went largely unopposed.

Taking cue from his 'filariasis-mosquito connection' theory, Manson began to speculate on malaria. Like Laveran, he too was baffled with the encounter of 'exflagellation' phenomenon of the parasite, Following Laveran's assertion that these flagellae were the mature stage of parasite, as well as the speculation that mosquitoes picked them up in that stage for transfer to a new host, Manson, in l894, began to navigate his thinking the filarae way, though quite faultily. Already, for the benefit of Manson, the emergence of flagella in only half a minute, or more, old drawn blood was highlighted by the Russian Sakharov in 1889 and confirmed by Grassi in 1892. By this time, too, Manson - a warm, patient and avuncular doctor of 50 years of mature yet youthful age - had emerged without question the outstanding expert in tropical diseases in England, and it had become widely known in medical circles that he, following his remarkable 'mosquito filariasis theory', was seriously embarking to explain the malaria tangle by way of mosquito's role in transmission. However, Manson's newly found filaria-related mosquito theory weighed heavily on him and he drew the conclusion that the red blood cells which enclosed the malaria parasite were comparable to the filarial sheaths. Consequently, he landed with a totally misplaced explanation. Even so, Manson's mosquito theory (first presented preliminarily in 1894 and later in the fully developed form in lectures of 1896) was a brilliant imaginative construction to pave the way for future research. Ross picked up the evidence where Manson had left!

Almost around the same period, precisely in 1880 in Havana, Dr Carlos Finlay aired his finding of *Aedes aegypti* (then known as *Culex fasciatus* or *Stegomyia fasciatus*) transmitting the yellow fever disease. He arrived this adrenaline-charged conclusion by pursuing a scientific methodology, otherwise next to impossible at present due ethical reasons. He made yellow fever patients to be bitten by mosquitoes, only to be let loose to interact in course of time with the uninfected healthy persons. He did succeed partially in effecting yellow fever in some healthy subjects, but failed to realize the minimum time period the infected mosquitoes needed to be fully infective (actually, Finlay only conjectured about the disease based on its jaundice-implicated nature impaling the patient, without any knowledge about the pathogen, a virus, which was not yet known to the science). But, unlike Manson in case of filariasis mosquito, Finlay rightly discovered that mosquitoes bite more than once - a trait that truly pointed out towards mosquitoes role as a biological transmitter of the disease. However, as it would normally go with every exciting new discovery, Finlay was not only widely ignored, he was ferociously repudiated for this observation of his - a Cuban, particularly by the then dominating English and German workers, but also by his contemporaries in Latin America. Finlay's yellow fever mosquito theory meandered a long journey through all thick and thin to be finally accepted, almost in the name of another person, a truly famous mosquito-man himself though, the great Dr Walter Reed, when in 1990 a universal acceptance was accorded to this great piece of scientific inquiry. Such was the deep impact of world-wide recognition of the scientific genius of Dr Reed, a U.S. Army officer, that the 'discovery of yellow fever mosquito theory' was without any further innuendoes accepted globally. However, unlike malaria where the pathogen, a protozoan parasite belonging to genus Plasmodium, was discovered much earlier than its mosquito vector, the world had had to wait for some thirty years before the yellow fever virus could be identified, along with another *au fait* discovery regarding feral primate populations in the American and African forests harbouring a reservoir of the disease. Luckily to Finlay, his notion on mosquito theory for the yellow fever was propped up by the ultimate discovery of mosquito's prowess as a vector of malaria by Ross and the Italians, amongst which Grassi was most prominent, in 1897.

11.7. The international competition

The three major mosquito-related researches in connection with demystifying transmission cycles of some of the most deadly and debilitating tropical diseases included well-placed and rather globally important 'mosquito yellow fever theory', 'mosquito filariasis theory' and 'mosquito malaria theory' - all during the fourth quarter of the 19th century. Of these three milestones, the yellow fever and filariasis connections to mosquito went rather peaceful or unchallenged, though low-voiced disagreements or oppositions could always be audible. Both these mosquito-borne revelations were made much earlier than the one for malaria or, more importantly than anything else, the event of establishment of the highest prize money in the name of Nobel Prize in 1896, to be commenced for awarding from the year 1901. Malaria thus asserted some special claim, as already was paralleled - with still greater attention due to its prowess to stay on year after year - with the population-decimating deadly plague, a reputation no other tropical disease was bestowed with! Enemies not in clear vision then included Plasmodium parasites and the Anopheles vector mosquito species, and, true to the demand of Nature, they were to be found out by none other than the two military men, Doctors Alphonse Laveran of French Army and Ronald Ross of British Indian Medical Service, who made basic discoveries that unlocked the secrets of malaria. Both of these scientists had characteristic opposite basic human qualities (Harrison, 1978): Dr Ronald Ross, a complicated and chronically maladjusted man, was nevertheless filled with ceaseless energy, endurance, positive thinking and, above all, the desire to do something extraordinary - something that nobody had earlier done and which would single him out amongst the world's greatest men of honour. He took life as a struggle and a challenge, and himself as a soldier eternally pitched against the people and forces who sought, from stupidity or maliciousness, to frustrate him and hold back the course of human progress. Ross was exaggeratingly ambitious for progress and would not accept anyone as hindrance in his way. Dr Alphonse Laveran, on the other hand, was by nature a sober and noble human having a special penchant for scientific research and professional progress.

Initially, it all started as a pure international competition, with the noted German pathologists Dr Heinrich Meckel, in 1847 and Dr T. Frerichs, in 1858, along with Dr Rudolf Virchow, having described and characterized the black particles in the blood smears of malaria patients of being formed of degraded haemoglobin from the affected red blood corpuscles, respectively, and many a French doctor like F.C. Mallot, Leon Colin, P.J. Keiner and Acheille Kelsch, particularly the last one, who had enormous experience in malaria pathology, showing presence of innumerable tiny black particles, especially conglomerated in spleen, liver and, sometimes, brain to give the affected organ a tinge of slate grey.

Many explanations were floored as to the origin of these malaria specific black particles, and the science of understanding moved ahead without much hassles. Dr Kelsch, who worked in Algeria in early 1870s, apparently missed the discovery of the first malarial parasite viewed in the human, which was some six years later done by Laveran in Algeria only. Professor Kelsch, true to the nature of his profession of a pathologist, looked at everything malarial with a classical pathologist's eye only, even having probably routinely killed the contents of the freshly drawn blood from the malaria patients - in which there existed every likelihood of finding what Laveran had subsequently discovered, the 'crescents'. Though, Kelsch had passed by the brink of discovery, he nevertheless had demonstrated that the black pigments emerged with the onset of fever. He obviously did not follow the question of uniqueness of black pigment in malaria, and focused on other important lines of malaria research pertaining to the disease's impacts on red and white blood cells. Thus, the central theme was left unattended and unsolved, for the man who had a keen eye for observation and a determination to find answers to most baffling queries - Charles Louis Alphonse Laveran.

Laveran, sent to Algeria in 1878 as an official necessity and certainly not on any mission to research on malaria, was nonetheless forced to go into studying malaria as military bosses were desperately keen to cease the outbreaks (of malaria) that constantly wrecked their soldiers there and exacted astronomical costs in replacing dead and the sick as well as in boosting up their soldiers' morale. He was familiar with the work of Kelsch and others and soon understood the value of studying alive the contents of the freshly drawn blood to which he had access in plenty. He met with success rather early on 5th November, 1880 when he clearly saw 'spheres' (the exflagellating male gametes). For the next four years he worked to know more about the malaria parasite which he described in extenso in a book titled '*La Nature parasitaire des accidents de l'impaludisme*', published in 1881, followed by two more elaborative books in 1884 and 1898, respectively. When he reported his findings to the Academy of Medicine in Paris he did not forget to underline his claim to have been the first to discover the malaria parasite. It was indeed the iirst glimpse of one form of the malarial parasite, *Plasmodium malariae*, a protozoan - still not known by that name. ironically, even such a spectacular discovery had to go unattended in appreciation. There were contemporary scientists who stooped low to belittle the significance of Laveran's grand discovery by giving one or another bizarre explanation. Others, more vocal, including the celebrated haematologist Dr William Osler, Baltimore (USA), tot/ally refused to believe his observations and an event such as the 'exflagellation' in blood.

Disagreement with Laveran's single-celled protozoan (which he referred to as the germ) led to divert the attention of several scientists of the day to consider an alternative possibility - the bacteria. The bacillus theory of *Bacillus malariae* prevalent in Italy's infamous malarious swamps, raised by two Italians, was as soon sidelined as it arose in spite of undercurrent of support from a newly emerging highly promising discipline of bacteriology directly under the guidance of as immensely famous bacteriologists as Louis Pasteur and Robert Koch. However, this dethronement of malarial bacillus hypothesis did not quite help ready acceptance of Laveran's germ theory but, on the contrary, Laveran had to take upon himself the responsibility to convince in person by practical demonstration of 'exflagellation' to the fellow scientists both in Algeria (Dr Richard] and in Italy (Drs Ettore Marchiafava and Angelo Celli), but largely in vain, since both the Italian experts had their opinion reserved in favour of *Bacillus malariae,* thought as the most likely agent of malaria. Not only this, Marchiafava and Celli mischievously hid from him their own research outcome in next two years conforming to Laveran's findings, which they published without any reference to Laveran's remarkable original work with a view point of making their claim as first for the discovery. Laveran was immensely exasperated by this fraudulent act of Italians who, in turn, believed what they experienced in fresh blood preparations was not seen by Laveran and was certainly different. In an attempt to put their stake for original claim of the discovery they named their form as *Plasmodium malariae.* Dr Camillo Golgi, whose primary interest was the nervous system, confirmed their findings, with some startling advancement in knowledge. With this the initial coolness of trust in comradeship and scientific fraternity seemed to have volatilized, and the first seeds of an embittered future world war of priorities and primacy had been sown sans healthy international competition even when new human malaria parasite species continued to be discovered with a large number of excellent scientific inputs by Golgi and others.

Laveran, as said earlier, was a never-say-die type of scientist and his consistently good relation with scientists worldwide paid dividends before long. Starting in 1885, Laveran's discovery soon spread to America, Great Britain and India, and by the end of 1890 it was said that acceptance had been universally granted. The technique with a lot of malaria-related knowledge would in time pass on to Dr Patrick Manson in Great Britain who, in return, descended the wealth of facts to his protege, Dr Ronald Ross. Both Grassi and Ross would later take their own time to accept or reject these wonderful discoveries, but never to sit quite even a bit moment when it came to their own research findings. The stage for gruesome "First Mosquito World War" had just been set in.

11.8. Mosquito-malaria world war of priority and primacy

Dr Ronald Ross, who had been ushered into the medical profession against his intention to become a writer or a musician, did not even imagine ever to pursue research in his early life in British Indian Medical Service, leave aside the exemplary Nobel Prize winning mosquito-malaria research which of course he did beating everybody else with his unmatched discipline and focused attention. The transformation was not subtle, In fact, Ross seemed to have never come to terms with the extreme tropical climate of India, rife with blood-sucking mosquitoes everywhere almost round the year. It was after all his father's plan, not his, to serve in the British Medical Services in India where the senior Ross was lucky to have risen himself to the rank of a full General! Ronald Ross, a good son, obeyed his father to this extent, but felt eternally distant from medicine and research, and quite closer to matters deeply imprinted on his mind - the music, poem and novel writing (he did succeed in publishing a romance novel, *The Spirit in the Storm*, in London in 1896) and, of course, mathematics. In his initial days Ross had felt tormented by both the dusty heat of Indian plains (he once even got a brief posting to Nasirabad close to the eastern edge of the Thar Desert in the erstwhile Rajaputana Estate) and the callous behaviour of his seniors in sending him quite frequently on postings to less cosy and recreational, and monetarily least attractive places, still mostly as a forced bachelor having had to leave the family behind either in the native Great Britain or the previous posting. Consequently, he silently confronted his military bosses on one or another pretext, and, out of agony and exasperation, even threatened to quit the military job and return to England. Frustration enveloped Ross - *hook, line and sinker*! He had plenty of time to golf away and compose music, but the impact of continued humiliation in service kept constantly blasting off his ever alert mind to do something extraordinary that nobody in the world had earlier done and doing what he could rise to the top honours of the world. He was eternally restless but did not know where to go, or navigate his life to reach his objective.

Greatly lost in the labyrinthine medical disciplines and researches, he decided to coolly think meticulously within the precincts of his medical assignments and carefully narrowed down to malaria research which indeed was the hottest topic of the day! Ross immediately jumped with a burning desire in his eyes. He thought he had got the objective, but still did not know how to strike the bull's eye! The problem was, however, that Dr Ross, with almost total ignorance of the literature, knew very little about malaria, had never seen a parasite, did not know the necessary techniques to examine malaria parasites, despite the fact that Laveran and, subsequently, the Italians and Americans have demonstrated microscopically the presence of various stages of Plasmodium species throughout the second half of the 1880s, and could not at all distinguish one mosquito from another. Overcame he did yet most of these one by one as the inexorable laws of time and destiny had in store for him something the whole humanity will be later proud of. As if providentially designed to reach his objective, he first met the great Dr Patrick Manson under whose careful tutelage Ross rose to enormous heights of success, winning finally the race to pinpoint Anopheles mosquito's role in malaria transmission during 1897 and, more rewardingly for Ross, the Nobel Prize in 1902 (actually awarded in 2004). His discovery of the role of Anopheles mosquito (which he ignorantly referred to as 'dapple wing') in malaria transmission and later a complete life cycle of the bird malaria parasite in an avian using a natural culicine vector mosquito species are a classical saga of exemplary determination, perseverance, courage, patience and ambitiously focused application.

At a time when Ronald Ross was groping in dark and ceaselessly endeavouring to grapple with the necessary microbiological and microscopic techniques, the Italians, particularly the genius of Giovanni Battista Grassi and his team of as celebrated scientists as Drs Marchiafava, Celli and Golgi, were already at the problem of solving the malaria mystery through implication of mosquito. It appears that both Ross and Grassi knew about each other's moves in demystifying the hidden secrets of malaria cycle, but none really showed any signs of this awareness. Ross, true to his military training was meticulous in keeping his day-to-day account of mosquito-malaria experiments as well as bringing the findings to the notice of all through Manson in England and quick dispatches to the scientific journals of international fame. Thus, not only his phenomenal express reporting of the pigmented cells in two mosquitoes fed on malaria patient (Ross, 1897), even publication of his proof of mosquito's involvement in an avian model did not take long to get public as within a few months he ensured its publication following the revelation of pigmented bodies in Anopheles mosquito (Ross, 1898a,b). He carefully planned his bracing the tape first by every possible, logical means in the international race for staking the claim for the unique discovery that coincided knowingly or unknowingly with the news of institution of the Nobel Prizes, including one in Physiology and Medicine, following the death of their willer, the dynamite king Alfred Nobel, in 1896, which, if won, meant unimaginably colossal prospects ahead both monetarily and from honours point-of-view.

Around the same time, and possibly a bit earlier to Ross's revelation of August 20, 1897, Professor Grassi, who had been working relentlessly on the mosquito-malaria connection for several months, then immediately unaware of Ross's discovery, true to his reputation and genius patiently waited for the last iota of truth to emerge to fool-proof his own 'first' discovery of the same matter as Ross's. Grassi's first publication from his observations taken since mid-July, 1898 through summer 1899 came to see the Sun in October, 1899 and then in December, 1899, wherein he announced, referring to his associate colleagues (Celli, Bignami, Dionisi and Bastianelli), having arrived at a definitive solution of the question of mosquito-malaria riddle and that the victory was completely theirs. The rush to publish the results and air the news was unusual and inclined to beat Robert Koch who too had been advocating his own mosquito-malaria theory since as early as 1883. Ironically, it was Ross, and not Koch, who was hit. Ross felt treacherously ignored by the Italians for his 'first and original' discovery. Italians' superior techniques and work infrastructure were always highly revered in scientific circles. Even Ross was full of praise for both Grassi's and Bignami's ingenuity in medical research and in identifying the vector, and confirming the mosquito cycle of the Plasmodium. In return, Grassi returned courtesies full of applause to Ross for his efforts.

These pleasantries between Grassi and Ross were only ephemeral and all the sense of mutual respectability effervesced with the initial news aired about Ross's having brought first his claim for the primacy of the discovery of Anopheles mosquito being invariably responsible to transmit malaria. The competition readily reached its lowest and ugliest point with the war for credit of priority and primacy of the discovery. Ross charged Grassi's work of several years as childish which he could have accomplished in a few weeks. Grassi, who was rightly considered by many to be the preeminent medical zoologist of his time, shot back calling Ross as an amateur and an interloper. Ross was stung to an instant wrath. He was a person once launched in anger was never to pull back. Dr Ronald Ross and

Professor Giovanni Battista Grassi envied each other for the rest of their lives, broaching over priority in various different mosquito-malaria related researches. Ross, feeling eternally insecure for his place on various discoveries, was more vociferous and hostile, having spared no opportunity to attack Grassi and inflating his own achievements, Ross engaged in a lifelong polemic to quash all Italian pretensions and reassert with ever-diminishing qualifications that he and only he had demonstrated the transmission of malaria by mosquitoes.

Grassi, who in course of the continuing controversy on primacy, lost association of two of his best colleagues Bignami and Batianelli, the chief architects of Grassi's mosquito-malaria work, due mostly to his growing arrogant and egoistical contentions, battled Ross ceaselessly with matching intensity. Ross's obsession with garnering every possible laurel was so intense that he did not even hesitate to hurl insinuations on his mentor, Dr Patrick Manson, his science and scientific integrity, just because Manson, truly a generous, impartial and honest grand man of tropical disease sciences, indicated to credit Grassi for his outstanding contributions in malaria research.

Ross, in spite of having lost association of many friends in course of the never-ending battle with Grassi, did have many others to support his cause and value. One such well-wisher was Dr Robert Koch. Haunted by Grassi's ignominious treatment to his Italian visit in late 1890s and later the surreptitious correspondence, Koch had turned into an eternally implacable foe of Grassi. Koch's persistently vehement attacks on Grassi's scientific integrity, climaxing into his violent opposition for the Italian genius against a consideration for awarding the 1902 Nobel Prize jointly to both Grassi and Ross, in all likelihood cost the former the fortune. Ross won the Nobel Prize, besides the prestigious knighthood a few years later. For whatever happened between these two geniuses the humanity was greatly benefited from their tireless inquiries in the mystique mosquito-malaria relationship and what benefits we reap today in saving precious lives to the savage attacks of the insidious malarial disease is due entirely to their, and other similarly devoted scientists' tireless and focused hard work.

Chapter 12

MAN'S HEROICS AGAINST THE MOSQUITO - *EXAMPLES OF SUCCESS AND FAILURE*

12.1. Introduction

Mosquitoes, despised for their blood-sucking habits and, in the process, also transmitting several dreaded and/or debilitating infections, are a relatively much greater source of nuisance and disturbance amongst arthropods all over the World, but particularly in the tropical and subtropical countries (Table 12.1). Many significant attempts have been executed in Americas, Africa and Asia to wipe out the mosquito which, barring a fewer examples, remained invincible (Ross, 1903; Speilman & D'Antonio, 2001; Tyagi, 2004a,b; Tyagi *et al.*, 2020). Actually, any attempt to exterminate forms of life, such as the mosquito, which compete too successfully with us may be as mistaken as it has generally proved in practice. While successes were restrictive in glory, failures, so universal and apparently ineluctable in their exacerbation, were instructive to remind that we, having basically misconstrued the mosquito problem *per se*, have proved incompetent warriors.

India during the British regime was for obvious and comprehensible reasons of widespread colonial prowess and unquestionable administrative governance in most parts of Asia, in particular, on one hand, and the ever increasing impoverished state of health and economy, on the other, glorified as a perfect experimental model ground for a variety of research investigations and demonstrations of control programmes especially about malaria - the greatest scourge of humans at that time, killing more than a million people every year across the country. This philosophy of "Indianness of all vector-borne diseases experimentations" providentially also brought in light to witness some of the most startling and bold researches ever conducted on mosquitoes, such as Ronald Ross's epoch-making discovery of an Anopheles mosquito transmitting malaria, in the late 1890s. It is highly interesting and stimulating to know about these pioneers and their brave moves against the mosquito, which at times appear to take the form of a real battlefield between two enemy forces hellbent to try against each other every possible arsenal of knowledge, energy and experience through all thick and thin. From man's point of view 'he' overpowered 'her' in some encounters, but not without tasting a pinch of defeat or frustation in several others. Notwithstanding, victory or defeat, each of these ventures was a saga of man's enormous courage and determination to wipe off the villainous mosquito to inherit a life without any bite or infection.

Mosquitoes' prowess to transmit disease pathogens is matchless in the whole animal kingdom, as they are involved either solely (*e.g.*, malaria, yellow fever, dengue) or mainly (*e.g.*, lymphatic filariasis and several types of encephalitides) in effecting these infections in humans. At the same time, of no less significance are the discoveries of other non-mosquito arthropod vectors involved in transmission of the diseases prevailing outside India, in Africa and the Americas, in particular, although the arthropod/insect group sans the concerned transmitting vector species along with the pathogen involved do exist in the Indian environment; in 1895, David Bruce, while working in South America, has incriminated triatomid bugs (*Triatoma* spp.) for transmitting trypanosomiasis; Bruce and David Nabarrow, while working in Africa, determined the role of tsetse flies (*Glossina* spp.) in transmitting African sleeping sickness in 1903, while around the same time Carlos Chagas found that triatomine bugs carried the *Trypanosoma cruzi* infection; in 1926, the Scottish physician, Donald Blacklock, while working in Sierra Leone, confirmed the role of blackflies (Simuliidae) in transmitting river blindness or onchocerciasis, and, between 1930 and 1990, various virus-borne encephalitides transmitted by mosquitoes (by *Culex tarsalis*), but also equally deadly infections effected by ticks, were correctly understood about their transmission. Arthropods, particularly mosquitoes, have developed several novel schemes to tide over difficult situations and the vertical mode of virus transmission is able to help the vector species maintain the virus in nature even if the same is eliminated from the human population. Various mosquitoes exercise this mode of transmission, but the first instance of vertical transmission by any arthropod was exemplified by *Boophilus annulatus*, a cattle tick, in 1891 in USA by Theobald Smith and F.L. Kilbourne.

Table 12.1. The world's most important mosquito-transmitted protozoa, helminths and viruses causing human diseases.

Vector	Pathogen	Disease	Geographical distribution	Distribution in India
Anopheles spp.	*Plasmodium vivax, P. falciparum, P. ovale, P. malariae, P. knowlesi*	malaria	Tropical and subtropical countries (worldwide)	Yes
Culex quinquefasciatus *Mansonia* spp.	*Wuchereria bancrofti, Brugia malayi*	Bancroftian filariasis Brugian filariasis	South East Asia and Africa	Yes (Only *W. bancrofti* and *B. malayi*)
Aedes aegypti, Ae. albopictus	Dengue virus 1-5	Dengue	Tropical region	Yes
Aedes aegypti, Ae. albopictus	Yellow Fever virus	Yellow Fever	Africa and South America	No.
Culex tritaeniorhynchus	Japanese encephalitis virus	Japanese encephalitis	Asia, pacific	Yes
Culex annulirostris	Murray Valley encephalitis virus	Murray valley encephalitis	Australia and Papua New Guinea	No.
Culex spp.	Rocio virus	Rocio	America and Brazil	No.
Culex pipiens and Culex quinquefasciatus	St. Louis encephalitis virus	St. Louis encephalitis	North and South America	No.
Culex spp.	West Nile Virus	West Nile	Africa, Asia, Europe and America	Yes
Culex spp.	Rift Valley virus	Rift Valley fever	Africa, Middle East	No.

Vector	Pathogen	Disease	Geographical distribution	Distribution in India
Aedes triseriatus	La Crosse virus	La Crosse encephalitis	North America	No.
Aedes triseriatus	California Encephalitis virus	California encephalitis	North America, Europe, Asia	No.
Aedes aegypti, Aedes albopictus	Chikungunya virus	Chikungunya	Africa and Asia region	Yes.
Culex annulirostris, Ae. vigilax, Ae. camptorhynchus	Ross River virus	Ross River	Australia, South Pacific	No.
Culicidae Haemagogus spp.	Mayaro Virus	Mayaro	South America	No.
Anopheles funestus, Anopheles gambiae	O'nyong-nyong virus	O'nyong-nyong fever	Africa	No.
Culex spp.	Sindbis virus	Sindbis	Africa, Egypt, Israel, Philippines and Australia	No.
Aedes notoscriptus	Barmah Forest virus	Barmah Forest	Australia	No.
Culiseta melanura and Cs. morsitans	Eastern equine encephalitis virus	Eastern equine encephalitis	North, Central and South America and the Caribbean	No.
Culex tarssalis	Western equine encephalitis virus	Western equine encephalitis	Americas	No.
Aedes taeniorhynchus	Venezuelan equine encephalitis virus	Venezuelan equine encephalitis	Americas	No.

12.2. The successes

Even though there are a large number of instances where man had clearly beaten both the mosquito and disease transmitted by her, I am, for reason of exemplification only, quoting here a most famous event from outside India and many indigenous efforts which invite their narration time and again to reinforce confidence in modern generation of scientists that it's possible!

12.2.1. Yellow Fever at Panama Canal:

This was the daunting task that the Yellow Fever Commission scientists (Reed, Caroll, Agramonte and Lazear) faced in Cuba more than 100 years ago. How could these four men possibly stop this dreaded killer? They had spent their first full month, July 1900, in unsuccessful attempts to prove a relationship between yellow fever and the bacteria, *Bacillus icteriodes.* Because there was little hope of identifying the etiologic agent, the commission members decided to focus their efforts determining how yellow fever was spread. They had found mounting evidence against the commonly held fomite theory of transmission. Among the several other theories, Dr Jesse Lazear was most intrigued by the possibility of a mosquito vector. The mosquito theory of transmission of yellow fever had been around for many years. Possibly the first to discuss it was Dr Josiah Nott of Mobile, Ala,, in 1848. It had been proposed again by Dr Carlos Finlay of Havana in 1881. But try as he might, Finlay had been unable to prove his theory in almost 100 experiments on human subjects. Many in the medical field belittled him because of his passion for his mosquito theory, but the discovery in 1897 by Dr Ronald Ross of the British Indian Medical Service that malaria was transmitted by the mosquito gave added weight to this possibility.

The tale of the Panama Canal, or "the big ditch", as it was then locally called, is a passionate one, embodying within its innumerable folds dramas of every kind - the joy, the sorrow, the disappointment, rewards and honours, and even comeuppance and death (Fig. 12. 1). It is the story of an impassioned and persistent struggle against the forces of nature, marked with the blood, sweat, and tears of tens of thousands of labourers. engineering challenges, political turmoil, and bureaucratic headaches each added its own seasoning to an already spicy and exciting chapter in history exemplifying man's indefatigable ventures to command victory over the seemingly invincible looking land-mass, and, of course, the real villain - the yellow fever disease and its vampish carrier, *Aedes aegypti.*

Fig. 12.1. The Panama Canal. (*Source:* Free Wikipedia).

Cutting a nearly 50-mile swath through mangrove swamp and mountains, the Panama Canal was considered the most daring and ambitious engineering feat of its time. But before completion of "the big ditch" in 1914, tens of thousands would die, international tempers would flare, and the biggest villain had to be held at bay. There were many starts and stops before scientists, sanitary engineers, and canal builders would come to agree that the biggest enemy was actually so small as to pass unnoticed, yet it was so powerful that it very nearly destroyed the centuries-old dream of a passage connecting the world's two greatest oceans ~ the Atlantic and the Pacific. Before the canal was finally built, more than 30,000 people would succumb to this tiny twin-enemy, Anopheles spp. and its distant ally *Aedes aegypti,* the mosquitos that carry malaria and yellow fever respectively. The dream of opening a waterway across the isthmus of Panama dates back to 1524, when Emperor Charles V commissioned a study for constructing a canal to link the Chagres River, which flowed into the Atlantic Ocean, with the Pacific Ocean. Charles' successor, Philip II, quashed the project, justifying that the act would be against God's wish. Interest in the idea waned until the mid-nineteenth century, when gold was discovered in California, igniting man's rapacious desire to rush to the newfound golden land. This penchant for gold inevitably also engendered a rejuvenated interest in finding a quicker east-to-west route than the tedious trip around Cape Hom at the tip of South America. France, Great Britain, and the United States - the three commerce giants at the time - drew up different plans for cutting the distance. One resulted in a 50-mile railway built across the Panamanian isthmus in the 1850s. Here, railroad workers had their first run-in with what canal laborers would later face: malaria and yellow fever waged war on all comers to their tropical soil. The non-immune labourer population was particularly sickened. The French were the first to tackle canal-building in Panama. Ferdinand de Lesseps, a great engineer who had just successfully completed the Suez Canal, linking the Mediterranean and the Red Sea, was basking in international glory of bringing the Far East nearer to the West. He was naturally the first choice to undertake the job and, looking at the enormous financial prospects, he was only too eager to repeat the feat, without having made a good homework. But, he soon realized that Panama was not Egypt. He was met with thick jungle vegetation, torrential rains and floods, and oppressive heat and humidity, compounded by vastly different soil conditions than those of the arid desert. The Culebra and the Chagres, the mountain and river that barred his path, could not be overcome in any other way. Disease and death were taking toll of those brought from France to work at the Panama canal with such a frenzy that the site infamously earned the ghost name of a pest house. Once the digging began, the mosquitoes were driven out of their habitats, bringing them in direct contact with canal workers. Yellow fever and malaria killed an estimated 3,500 people a year during the 1881-1889, forcing de Lesseps to terminate the attempt precociously, at huge costs of character assassination, charges of financial bungling and amassing public wealth, and of course international repudiation about his engineering skills. The malaria and dengue mosquitoes were obviously the main villains in driving out the genius of de Lessips, who, following felonious charges of fund-swindling and failure to complete the task of building the canal, was abominably tried out, with comeuppance settled - heavy fines and imprisonment - by the French Government. De Lesseps died on December 7, 1894, in his ninetieth year. Mosquito-triggered malaria and yellow fever had beaten man in their first major encounter!

At the time, widespread disagreement existed about what caused these epidemics. One popular belief was the "marsh miasma" theory, which held that the invisible mists and vapours given off by swamps and decaying organic matter produced yellow fever and malaria. Even men of science became locked in fierce disagreement. Debate raged between those who believed that these pestilences were communicable diseases caused by some infectious agent and spread by materials contaminated by patients, and those who clung passionately to the idea that they were air- or waterborne. These differences in opinion gave rise to wildly different approaches to prevention and treatment. For instance, during a yellow fever epidemic in Philadelphia in 1793, Dr Benjamin Rush, a well-known professor and physician, attempted to cure his patients by bleeding them to get rid of the environmental contaminants that he believed had invaded their bodies. Around the same time, in Washington, D.C., Dr Albert King hypothesized that mosquitos were the real culprits, but he became the object of much ridicule when he proposed that a gigantic mesh screen be erected around the U.S. capital to keep the insects out. Another hundred years would pass before scientists realized that King was on the right track. In 1880, as de Lesseps was *en route* to Panama to begin work on the canal, Dr Charles Alphonse Laveran, a French physician working in the then French territory of Algeria, discovered Plasmodium, the organism that causes malaria, in the blood of a patient. The news came too late for de Lesseps. On the other hand, it might not have mattered because de Lesseps did not believe in the mosquito theory. The next year, a Cuban physician, Dr Carlos Finlay, made a discovery that would revolutionize disease control and ultimately make possible the construction of the Panama Canal. He identified the *Aedes aegypti* mosquito as the carrier of yellow fever. Finlay's breakthrough had glitches, however. He correctly had surmised that the mosquito was spreading the disease by biting people previously infected with the disease and then transmitting the disease agent to other people through its bite. But when he attempted to prove his theory by having infected mosquitos inoculate healthy volunteers with the disease, he failed. What Finlay didn't know was that the virus needed to incubate for approximately 12 days inside the insect in order to develop. He was attempting to inoculate his volunteers with the virus using mosquitos that had only been infected a few hours before. More years would pass before these missing pieces were added to decipher the enigma. His discovery got a pat when Sir Ronald Ross, a British physician in India, identified a malaria parasite in the stomach of an Anopheles mosquito. Finlay formulated a detailed plan for combating both yellow fever and malaria.

In 1898, the United States declared a war with Spain colonized in Cuba. During the hostilities, it soon became apparent to the U.S. military that yellow fever was claiming more lives than the actual war itself. True to its name, the diseased person turned yellow and, in most instances, readied to follow a nearly certain death. It had taken its toll on thousands of U.S. Army recruits during the brief Spanish-American War, and it was thought that the death due to yellow fever among American troops was much more common than that from bullets. Yellow fever had been the scourge of the Caribbean and the coastal and southern United States for centuries. Despite years of efforts by the best medical experts, no one knew how it was spread, why some were stricken and other family, friends or neighbours simply passed over. It appeared that no defence, no barrier, no precaution could stop its deadly progression. In response, the U.S. Government dispatched a medical commission, whose members comprised Walter Reed, James Carroll, A. Agramonte and Lazear, to find a way of halting the epidemic. The commission mounted a vigorous campaign to rout the disease, but since Reed did not put much stock in Finlay's mosquito theory, efforts focused initially on ridding the Cuban capital of the filth and contamination presumed to cause the disease. But when the yellow fever death toll rose despite a massive clean up effort, Reed decided to test Finlay's theory. Reed constructed two small buildings through which air could circulate freely (to first show that the disease was not airborne). In one building he had volunteers sleep for 20 nights in the foul bedsheets and garments of yellow fever patients. In the other building, volunteers slept next to infected mosquitoes, but were separated by screens. In this experiment, Reed did not use recently infected mosquitoes like Finlay had, but rather those mosquitoes that had been infected for some time. Then one volunteer was exposed to infected mosquitoes for three successive nights. This volunteer was the only one to develop yellow fever. Proof at last was conclusive: *Aedes aegypti* was the culprit. Reed had yielded the knowledge that would free Havana from yellow fever. Despite this proof, many in the medical community, including General Gorgas, remained unconvinced of Finlay's theory. But he set about eliminating the mosquito and its breeding places in Cuba where he succeeded in virtually eradicating yellow fever within one year, forcing him to agree with Finlay's mosquito theory. Gorgas then proceeded to destroy Anopheles species' habitats and scored similarly dramatic success against malaria. The Havana anti-mosquito campaigns, the first large-scale systematic public health effort of its kind anywhere in the world by that time, not only provided a formula for controlling epidemics elsewhere, they served as an early model for effective health promotion and community involvement. This victory over yellow fever laid the foundation of yet another significant and mammoth task of cleaning up the area of the future Panama Canal. However, there were many serious political constraints. At the time, Panama still belonged to Columbia. A bitter civil war ensued from 1900 until 1903 when, with the U.S. support, it finally won its independence.

Meanwhile, a second French company had bought the canal franchise in 1894, largely to keep the project alive. But, they sold their holdings to the U.S. for $40 million in 1904, where after plans to resume construction began. Taking stock of the huge numbers of workers the French had lost to malaria, yellow fever, and other ills related to Panama's swampy environment, the newly formed Panama Canal Commission decided that the first step was to tackle the sanitary situation in the Canal Zone. President Theodore Roosevelt sent a group of experts that included Gorgas, Reed, and others who had taken part in the Havana triumph. Since they knew now who the enemy was, they made plans to wage an all-out mosquito war, not without being shown a contretemps by the bureaucrats who dismissed the mosquito theory - *hook, line and sinker!*

In the meantime, a veritable health crisis was brewing. As hundreds of non-immune workers reached Panama, the number of yellow fever cases and deaths began to grow steadily, raising the spectre of an uncontrolled epidemic. By June 1905, fear of the disease had grown so intense that three-quarters of the canal work force, including the chief engineer, had fled Panama. It appeared that, once again, *Aedes aegypti* would foil attempts to build a canal. Faced with the prospect of failure, President Roosevelt fired all the commissioners, appointed new ones, and named John F. Stevens the new chief engineer. Stevens believed in Gorgas and gave him free reign. Without the bureaucracy, Gorgas was able to get the isthmus rid of the mosquitos and yellow fever. His group began by eliminating *Aedes aegypti* breeding places. At the same time, they carried out a sweeping fumigation campaign. In Havana, which had 200,000 inhabitants, only the houses of yellow fever patients and neighbouring houses had been fumigated. Panama City, which had only 20,000 inhabitants, and Colon, which had a much smaller population, were fumigated house by house. For the first time in the history of the isthmus, yellow fever was stamped out and the victory was truly fantastic. Yet, it was eclipsed by an even bigger threat that remained - Panama's most serious problem was malaria, which most believed to have killed even more people than yellow fever.

The two diseases differed in several ways. Yellow fever confers immunity, but malaria does not. Repeated attacks are frequent, and people who have been infected remain capable for years of transmitting the disease to others through the bite of an anopheline mosquito. Gorgas estimated that most of the 12,000 people living along the Canal Zone carried the infection. The mosquito carriers also had very different habits. Whereas *Aedes aegypti* laid its eggs in fresh, still water, Anopheles mosquitoes bred in swamps which were prodigious in Panama. So Gorgas and his men researched and, noting that Anopheles mosquitoes could fly only short distances, ordered the drainage or destruction of all potential breeding places located close to human dwellings which resulted in their elimination. Obligatorily doors and windows were screened to keep the adult mosquitos out, and underbrush that might shelter them was cut. In addition, the antimalarial drug quinine was dispensed to all inhabitants as a preventive measure. The results were simply spectacular: the number of malaria deaths fell to zero and the number of cases among canal workers also dropped dramatically. The isthmus had been freed from the grip of yellow fever and malaria. At last, the canal could be completed after a long peregrination of time spanning over 34 years, on August 15, 1914!

Thus, making of the Panama canal is also a saga of the fierce battle between man and mosquito in which the former finally edged out the later by freeing the isthmus from both malaria and yellow fever, largely due to the invaluable public health contributions, directly or indirectly, of indefatigable men like Laveran, Ross, Finlay, Gorgas, and Reed, but never to forget in the least men of great courage like Carroll, Agramonte and Lazear, the last of whom even sacrificed his life for a noble cause of deciphering a centuries-old puzzle - the Cause of transmission of yellow fever. The malaria and yellow fever campaigns carried out in Panama provided a working blueprint for the control of these diseases that was subsequently applied around the world.

12.2.2. *Mosquito/malaria control in Kheda district (Gujarat State), India*

In the beginning of the first half of the 20th Century there were an estimated 100,000,000 cases of malaria each year in India - roughly 25% of the population. The insecticide DDT, introduced in health programmes in 1950s, was dramatically successful in controlling malaria vectors and the incidence plummeted to 100,000 cases per year in the mid-1960s. The momentum of the then National Malaria Eradication Programme (NMEP), which was first rechristened as National Anti-Malaria Programme, in late 1990s and more recently subsumed along with other similar national programmes into a newly erected integrated scheme called, `National Vector Borne Disease Control Programme' (NVBDCP) by a government decree dated 2nd December, 2003, could not be, however, sustained due to multiple reasons, such as the vector resistance to insecticides, decreasing community acceptance of house-hold spraying and high costs of alternative insecticides, compounded by various financial and administrative constraints. At the same time malaria incidence rose rapidly to get around 2 million mark per year, with *Plasmodium falciparum* proportion rising steadily, notwithstanding.

Recognizing plausible drawbacks in controlling malaria vectors with insecticide spraying, the Indian Government through the Indian Council of Medical Research (ICMR/Malaria Research Centre (MRC) launched, under the able stewardship of Dr V.P. Sharma, Director of Malaria Research

Centre in Delhi, a malaria control programme with bio-environmental approaches to vector control (including engineering modifications, environmental manipulation involving local communities and the use of larvivorous fish and other predators to reduce mosquito abundance) at a dozen sites throughout India in a variety of urban and rural locations, of which Kheda district was most significant.

Kheda is basically a rural site, typical of much of the high malaria risk areas, located close to the industrial capital of Gujarat State in western India. The project was started in mid-1983 and soon various details were gathered about mapping, surveys, malaria stratification, site selection, fish production, requirement and procurement of equipment. The project carried out work in seven villages with a population of 25,000 but expanded yearly to cover a population of 700,000 by 1987. The project undertook several field level activities, like (a) environmental management to reduce mosquito breeding sites by minor engineering works, drainage, soak pits and application of polystyrene beads to cover water sources, (b) biological control of mosquito larvae through the introduction of larvivorous fish, (c) education of the community concerning malaria, its transmission and control, (d) surveillance and chemotherapy to reduce transmission of malaria, and (e) other environmental improvements and income generating activities (*e.g.*, social forestry, fish breeding, the introduction of smokeless stoves). In addition, the project also undertook considerable amount of research, management, supervision and training activities.

A substantial case reduction was achieved in environmental control area villages than those which were conventionally insecticide-sprayed by the NMEP, although the annual parasite incidence (API) remained more or less same in both types of localities (*cf.* Tyagi, 2004).

It was effectively demonstrated that environmental control measures have an edge over the insecticide-spraying methodology. The main vector of malaria in Kheda district is *Anopheles culicifacies* which has developed resistance against DDT, dieldrin/HCH and malathion. Therefore, entire emphasis was laid on reducing populations of this vector through various different environmental measures.

There are obviously a variety of associated benefits from the environmental intervention. A few examples are worth quoting here; (i) the introduction of fish to reduce larvae and mosquito densities has provided a lucrative additional benefit in the form of fish for consumption. The gross income from the ponds during 1986-1987 was Rs. 78,000 and was retained within the villages for their development fund, (ii) tree plantations in wasteland and marshy areas to reduce malaria breeding sites have encouraged the development of village nurseries which are expected to generate income either during their lifetime or on felling (in one village alone the tree yield is estimated to be worth Rs. 200,000). The net cost per head for environmental control is thus clearly less than Rs. 4.5.

This project was obviously a great success, igniting a fresh hope to abate both mosquito population and the malarial disease they transmit, in a true 'Gandhian' way - by practicing safe environmental measures. Dr V.P. Sharma, the dynamic crusader of this unique experiment, was a clear victor! Later, during late 1980s/early 1990s, Dr V.P. Sharma had done a similar wonder in Chennai City where *An. stephensi* was the main vector for urban malaria.

12.2.3. Mosquito/bancroftian filariasis control in Pondicherry, india

Most part of the Coromandel coastal region along the south-east peninsular India is endemic for lymphatic filariasis caused by *Wuchereria bancrofti,* a nematode belonging to family Filariidae, and transmitted to man by the infective bites of haematophagous mosquitoes. *Culex quinquefasciatus* is the major vector responsible for transmission of bancroftian filariasis in India. As such, the areas in and around Pondicherry Union Territory have also been affected badly with both the disease and mosquito infestation, and in order to develop a model for their containment, the Indian Council of Medical Research through its institute known as Vector Control Research Centre, Pondicherry had undertaken a filariasis control demonstration project, better known as the Pondicherry Project, through an integrated disease vector control approach, under the pioneering leadership of the Centre's Director, Dr P.K. Rajagopalan, ably assisted by Dr P.K. Das. Project's Assistant Director (and later the Director, VCRC) and a team of dedicated scientists and technicians. I was associated with this project between 1984 and 1986, and made many a field-oriented explorations including high-level meetings with Chief Secretary and Health Secretary, Pondicherry.

Pondicherry, a former French dominion with a total area of 59,38 sq.km and population of 272,250 (at 1981 census level), lies between 11.45° to 12.15° latitude north and 79.35° to 80.00° longitude east on the Coromandel coast, about 165 km south of Chennai (old name Madras). its climate is dissymmetric type, with high humidity, rainy days and a relatively cool season. Despite general hygiene in urban agglomeration in Pondicherry being of relatively high order, it nevertheless bore enormous mosquitogenic conditions largely due to unplanned urban growth, choking or obliterating in the process the existing drainage storm-water canal system of the township. The major contributory breeding sites, of course were cesspits, cesspools and choked drains etc. Over the years Pondicherry presented a grave situation of the disease prevalence. In 1957, of about 12% people surveyed, out of a total population of 59,835 resident in the town, for various epidemiologic parameters the *mf* rate was found as high as 10.3%, with the disease rate, infection rate and infectivity rate as 27%, 3.1% and 1.4%, respectively. In 1980, according to the National Filariasis Control Project, the *mf* rate, disease rate, infection rate and infectivity rate obviously declined to 4.89%, 0.62%-, 1.20% and 0.24%, With both *mf* (aver. 8.59%) and disease rate (45.81%) being highest in the core urban area. In this core urban area the vector mosquito, *Cx. quinquefasciatus*, also was found in the highest per-man hour density (3224). The control of disease was brought about mainly by reducing mosquito population through integrated vector management strategy comprising (i) inter-sectoral cooperation, (ii) community participation, and (iii) integrated vector control by source reduction and allied methods.

A detailed plan of activities was drawn which was subjected to evaluation and re-evaluation periodically to assess the vector preponderance in each classified area. Amidst a control programme utilizing extensively only environmental and biological methods, insecticides were also deployed where inevitably required. At the conclusion of the project, in 1985, the *mf* rate dropped in Pondicherry from 8.41% in 1981 to 6.34% in 1985, and to 6.35 in 1986 (showing a reduction by 24.51%) from 8.93 in 1981 in the VCRC control demonstration area, which was comparable to the NFCP zone figures of 6.33 in 1986 and 8.98 in 1981. Of great interest was the observation that *mf* rate in the 0-5 yrs. of age group wherein it was 2.22% in 1981 and 1.76% in 1986 in the NFCP area, but got dropped to 0.35% in 1986 in the VCRC control demonstration area. The disease rate in the VCRC area was found to be 2.26% in females and 13.67% in males, with older age group people manifesting more of disease symptoms. Both the diseases and *mf* rate had clearly been shown to decline. The ingenious leadership of Dr P.K. Rajagopalan, ably assisted by his team especially Dr P.K. Das, emphatically demonstrated that the filariasis vector mosquito, *Cx. quinquefasciatus*, could be substantially controlled through known, simple and yet strictly implemented environmental measures and with minimum use of insecticides. I, being a part of the project team, witnessed the major part of development of this project between 1984 and 1986. Victory in the end was surely theirs - Dr Rajagopalan and his team.

12.2.4. Control of lymphatic filariasis in Tamil Nadu, south India

India has been suffering with the scourge of lymphatic filariasis for the time immemorial and till 1990*s* was one of the leading countries in reporting the highest incidence of the infection (*Wuchereria bancrofti*) throughout the world (Tyagi, 2018). The World Health Assembly in 1997 has targeted the elimination of lymphatic filariasis (LF) by 2020, the goal for India was originally set 2015 to be achieved by annual single dose mass drug administration (MDA). The country needed a model which could be replicated all over the affected parts in the country. The Centre for Research in Medical Entomology where I was at the helm of the affairs between 2005 and 2015 came out a very successful model (Tyagi, 2014). Since the control/elimination of lymphatic filariasis was, under policy decision, based on anti-filariasis drug, diethyl carbamazine citrate (DEC), and the community acceptance to the drug as such was not encouraging, it was necessitated to demonstrate if the community acceptance and use of this drug, with or without another anti-helminthic, albendazole, could be escalated or not. The Centre carried out a long-duration experiment in six villages of Trikoilur and proved the point that annual single dose of DEC + albendazole could reduce the microfilaria (*mf*) rate to less than 1% and the corresponding antigenaemia to less than 0.1% (Rajendran *et al.*, 2004) (Fig. 12.2).

Fig.12.2. Map showing six filaria endemic study villages in Tirukoilur, south India. (*Source:* Sunish *et al.*, 2007)

Still 100% control/elimination was not being achieved merely with the anti-filarial drug (+albendazole), therefore, due emphasis was laid on the vector control to demonstrate for the first time in the country if the vector control (VC) could be used as an adjunct to DEC+albendazole in our fight against lymphatic filariasis Table 12.2).

Table 12.2. Entomological parameters estimated from the indoor resting *Culex quinquefasciatus* in Tirukoilur (India). (*Source:* Sunish *et al.*, 2007)

Treatmen t in village	Period	Mosquitoes caught per worker hour	Number dissected	Number infected (%) with 95% CI[a]	Number infective (%) with 95% CI	Mean number of third-stage larvae (L^3)	Trans-mission intensity index[b]
Mass drug adminis-tration (MDA) alone	July 1999-March 2002	23.70	2108	44 (2.09)[c] [1.6-2.8]	9 (0.43)[d] [0.02-0.8]	3.44	0.3486
	April 2002-August 2005	39.15	3073	43 (1.40)[c] [1.0-1.9]	6 (0.20)[d] [0.1-1.9]	4.33	0.3312
MDA with vector control	July 1999-March 2002	1.49	373	6 (1.61)[c] [0.7-3.5]	2 (0.54)[d] [0.1-1.9]	1.00	0.0080
	April 2002-August 2005	0.83	354	3 (0.85)[c] [0.3-2.5]	0 (0)[d] [1.1]	0	0

a 95% binomial confidence interval.

b Transmission intensity index = mosquitoes caught per man hour x proportion infective × mean L_3.

c No significant difference among % infected (based on ^{2} or Fisher's exact test).

d No significant difference among % infective (based on ^{2} or Fisher's exact test).

Sunish *et al.* (2007) offered a new insight that the vector control could keep the incidence under check in case the mass drug administration could not be implemented on time, since vector control was able to reduce the Transmission Intensity Index (TII) in vectors to zero. This (VC) was also additionally able to impact the antigenemia rate substantially which indeed was a good sign. The Trikoilur Project (Tyagi, 2014) was a highly successful and path-breaking masterpiece and a model in the field of lymphatic filariasis elimination through vector control.

12.2.5. The Klang and Port Swettenham Project

Almost parallel to the infamous Mian Mir experiment of mosquito and malaria control in India in the beginning of the early 20th century, malaria control activities were also initiated in Malaysia in 1901, aided with the new information on the role of mosquitoes in the transmission of malaria (*cf.* Konradsen *et al.*, 2005). The town of Klang, harbouring 3,500 people, was located in hilly terrain on the Klang river in one of the most important districts of rubber and coffee cultivation. The town experienced enormous malaria problem afflicting both civilian and service personnel as well as labourers who were largely incapacitated and prevented to attend to work. Anophelines were found breeding copiously in the swamps in and around the town as well as large number of open well and pools along the road. The foothill lands being rich in high water table often created flooded conditions following rains.

Control measures aiming at bringing down the mosquito population and reducing malaria burden included:

(i) Dense forest vegetation and undergrowth from within and without the town was removed.

(ii) As many as possible swamps present in the township were filled or drained.

(iii) Large-scale drainage systems were put into use, based on construction of concrete turf-lined inverts. This type of drain was found to be relatively inexpensive and the turf sides facilitated water inflow from subsoil.

(iv) Special constructions were designed to prevent tidal water from entering the drains.

(v) Open drains that followed the contour of the hill at its lowest point were constructed along the foothills. The idea behind this was to capture the Water from the springs and surface runoff before it could accumulate on the plains outside the town.

Within a few years the entire town perimeter was drained with the cost primarily covered by the State and, to a lesser extent, by private landowners and enterprises. After some years experiments with subsoil drainage systems were introduced with a good result in terms of improved drainage and reduced mosquito populations. Mosquito menace and malaria were also rampant further down the Klang river at Port Swettenham surrounded by mangrove swamps. Though not very old, the port immediately began to report malaria cases, with virtually 100% labour force falling sick. The infection later spread even to crews from the labourers returning to work. To abate vector mosquito breeding, embankments were raised at the edge of port township to prevent inflow of the sea water, aided with a network of surface drains to channel any tidal water. An iron pipe was attached with each drain in such a way that the intruded water would be fluxed out during low tide but the pipe was closed during high tide. However, the mosquito menace could be partially checked because of a railway line on a raised bund which prevented clearing of deposited water across the bund albeit frequent culvert constructions. Mosquito breeding was prevented, therefore, by both chemical and environmental methods and the cases were treated with quinine. Basically it was a joint effort of several organizations working in coordination with a meticulously drawn plan. The result was spectacular both on Klang township and the Port Swettenham as both malaria cases and mosquito menace were reduced substantially. This experiment proved to be highly successful and replicated in many parts of Malaysia, Singapore and elsewhere in Asia, as well as in United States of America.

12.2.6. Control of Urban Malaria in Bangalore City

General impression one carries about Bangalore is that the salubrious city was free from malaria during a time when it earned the reputation as, "A retired army officer's first choice to settle for the rest of life" during early part of the 20th Century. Malaria situation was alarming with spleen rate, 23.2%, and parasite rates ranging from 6.5 to 21.2% (Sweet & Rao, 1934). *Anopheles stephensi,* breeding in several hundreds of wells, was regarded as the vector for disease. Malaria in the City was controlled by the use of fishes like, *Gambusia affinis,* in wells, reducing the spleen rate below 1% and parasite rates to about 5%. It is noteworthy that after a brief and benign quiescence malaria built up gradually through seventies and eighties unless the infection virtually engulfed the entire and fast developing metropolis when not only the original vector *An. stephensi* but also the newly inveigled vector *An. culicifacies* helped transmission happen there (Ghosh *et al.*, 2012).

12.2.7. Control of Urban Malaria in Pune City

Pune City has been notoriously known for its malariousness for a long time (Barber & Rice, 1938). The main vector was *An. culicifacies* and the secondary vector was *An. stephensi* var. *mysorensis.* Following extensive surveys and larval control in 1944-45, both the spleen and parasite rates fell down by 33% and 20% of what they were before. Later during 1950s DDT was brought in use to further the impact of overall control measures.

12.2.8. Control of Urban Malaria in Bombay City

Bombay (now Mumbai) was nearly always a malariated town, especially in the beginning of the 20th Century (Bentley, 1911). For example, in 1903, about 74% were infected with malaria parasite. *Anopheles stephensi* was the vector, breeding in all kinds of stored and potable water sources, particularly wells. Thorough laws under Municipal Act were deployed and all the wells were managed in a way to reduce vector breeding completely, resulting in absolute elimination of malaria from Bombay (Covell, 1928).

12.2.9. Control of Urban Malaria in Delhi

Delhi city was notorious for the prevalence of malaria so much so that at the outset of the 20th Century the army garrison had to be shifted to the Ridge about 3 km to north and west during the worst months (Covell, 1934). Closely at the heels of the inevitable Second World War II a planned anti-malaria operation was carried out and *An. culicifacies* was confirmed as the main vector assisted by *An. stephensi* in the disease transmission. Malaria and mosquito control was effected by use of fishes (*Gambusia affinis*), engineering works with drains and canals of all kinds, including straightening and lining of all nullahs, and filling of wells and ditches. After DDT became available it was used in peripheral semi-urban areas, along with the anti-larval works. The malaria problem of Delhi City could really never be overcome completely but it had surely dipped a little after the advent of DDT and better vector management.

12.2.10. Control of Irrigation Malaria in Sarda Canal Project

The paradox of irrigation and malaria has been well known and, in India, the Sarda Canal Project, built on Sarda river in Banbasa near Indo-Nepal border, is a glaring example (Clyde, 1931). The canal construction, aiming mainly to provide water for irrigation in Uttar Pradesh with a network of 6,500 km over a command area 7 million acres, was accomplished between 1920 and 1929. Using environmental and insecticidal methods malaria was significantly controlled even though there was an outbreak again in 1928 under exceptional weather conditions.

12.2.11. Control of irrigation malaria in Irwin Canal Project (now Visveswarya Canals)

In this highly noteworthy irrigation project in Mandya District, Karnataka the main vectors were *An. culicifacies* and *An. fluviatilis* which specialized transmission of disease in irrigated lands. Control measures exercised included management of irrigation waters, rotation of crops, and use of Paris Green in addition to therapeutic intervention (Rao, 1945). It appeared that the disease cases were greatly reduced particularly with the deployment of antilarval methods. This was further improved subsequently by the pyrethrum spraying during 1940s and the DDT in 1947-48.

12.2.12. Control of irrigation malaria in Cauvery-Mettur Project

One of the most encouraging examples of controlling *An. culicifacies*-induced malaria comes from the irrigated area in Pattukkottai taluk, Tanjore District of the Madras Presidency, once the taluk being free from malaria before the introduction of irrigation in 1933 (Knipe & Russell,

1942). Malaria outbreak occurred in the project area in the mid-1930s. An exemplary four year-long work in the control of *An. culicifacies*-driven malaria was directed against the breeding of the vector in a group of villages with a population of 3,390 in an area of 7 sq.m. In this area there were 3.3 miles of main canal, 4.6 miles of branch canal, and many miles of poorly designed and ill-kept field channels. There were 27 tanks varying in size from a few hundred to several thousand square feet. There were 1,300 wells used for irrigation of ground-nuts between January and June; the canal water was used for the irrigation of rice from June to January. There were 1,900 acres of irrigated land, including 175 acres of swamp and 25 acres of seepage. Countless borrow pits produced *An. culicifacies.* The malaria control measures employed in undertaking a formidable task included filling and draining, tidying of channels, the use of Gambusia, organic pollution of breeding places, the use of Paris green by dustless and automatic methods, and intermittent irrigation. Some 55,000 yards of filling were used. The measures were applied step by step and with an additional aim of raising the standard of living, by securing better agriculture, better roads, better washing, and bathing facilities. Noteworthy results were achieved. In three villages the spleen rates, in the malaria season of 1937 were 48, 54 and 56; in the malaria season of 1941 the spleen rates of these villages were 4, 19 and 11. In a comparable untreated village the corresponding spleen rates were 53 and 57. The fall in the parasite rates during the same period was from 42, 50 and 49 to 0, 10 and 3; in the untreated village it rose from 43 to 48. The total cost of the four-years programme was Rs. 4.60 which is equivalent to annual cost of Rs. 1.15 per capita. This cost was considered pretty low.

12.2.13. Control of Rural Malaria in North Kanara District, Karnataka

It's a fable of control of malaria by eliminating *An. fluviatilis* in the difficult terrain of Western Ghat extending in the North Kanara District (Viswanathan, 1946, 1950). The anti-larval measures, along with pyrethrum spraying, was continued for two years and moderately good results were obtained; Spleen rate reduced to 32% from 62% and parasite rates to 14% from 22% in Haliyal town within one year of operation.

12.2.14. Control of rural malaria in Bengal-Nagpur Railway

A good example of malaria control hails from the erstwhile Bengal-Nagpur Railway (now South East Railways), where *An. fluviatilis,* along with secondary vectors like *An. varuna,* and *An. annularis* flourished, by use of antilarval measures such as filling of borrow pits, canalization, and use of Paris Green as well as oil (Venkat Rao, 1949). *Anopheles sundaicus* was also an important vector in certain areas such as coastal Orissa but its control via anti-larval measures was not satisfactory. It was DDT eventually which helped eliminate malaria from the area and complete the project.

12.2.15. Control of malaria in coal fields

Malaria in coal field offered a very challenging situation and needed a special attention in coal field areas such as Raniganj Coal Field (now in west Bengal) in the year 1931. Though not highly effective yet the early intervention measures included antilarval consisting of application of Paris Green, oiling, de-weeding, canalization and minor engineering works. Later these were reinforced with pyrethrum spraying to achieve moderate success. However, with the advent of DDT for adult killing it was introduced against vectors from 1948 onward and the rest is mere history – a great saga of victory over malaria in coal fields (Sen & Azeez, 1963).

12.2.16. Bangalore Mosquito Control - Master Plan.

The once famous as a retired military officers' preferred abode the Bangalore city with its everlasting salubrious environment and verdant fields, throughout dotted with numerous lakes and ponds, making it a virtual paradise till the mid-1950s, slowly began to turn into a mosquito haven - thanks to the senseless and unplanned development of urbanization, pollution of water bodies, choked drains and population boom (from less than 10 lakhs in 1950 to more than 25 lakhs by 1981). Though the major nuisance was caused by *Culex quinquefasciatus* (97%), nevertheless other mosquitoes such as for example *Cx. Gelidus, Cx vishnui* group mosquitoes, *Cx* (Lutzia) *fuscanus, Mansonia annulifera, Aedes aegypti, Ae. albopictus, Armegeres subalbatus, Anopheles subpictus* and *An. vagus* escalated the loss of tranquility of the inhabitants. Incidentally two of these mosquitoes, Aedes aegypti and Ae. albopictus, unleashed a spate of outbreaks of dengue and chikungunaya fevers in Karnataka State, including Bangalore (now Bangaluru) city. The problem of nuisance due to mosquitoes became so grave in Bangalore that Mr Ramakrishna Hegde, Chief Minister of Karnataka State had had to call a high level meeting on 23rd August, 1983, with the ICMR-Vector Control Research Centre to study the problem of mosquito nuisance and prepare a Master Plan for Control of Mosquitoes in the Bangalore City. A two-year long strenuous study in the 151.16 sq. km Bangalore City Corporation area plus an additional 496.64 sq. km area under the Urban Malaria Scheme was detailed out in the document "Bangalore Mosquito Control: Master Plan" (Rajagopalan *et al.*, 1987), the first ever such a guidance book in the history of Indian Metropolices. This document was a corroboration of Dr Ross's well-judged dogma of public health responsibility, commitment and execution with absolute resolution: "A genuine campaign does not consist merely in the formation of expert committees, the passing of ordinances ...; and the issue of wide advice to the public... it must always be permanent concern of the state, requiring careful measurement of the amount of the sickness present, a nice appreciation of measures most suitable for the locality, exact estimates of their cost compared with the cost of the disease, a well-considered organization, and, above all, a fixed determination to succeed."

12.3. The Failures

As for the successes, failures are rather prodigious, but instead of feeling disappointed and dejected they had largely taught and equipped scientists for future face-off on a number of occasions. Any failure, if taken in the right spirit, teaches a good lesson, and makes the first rung of a long ladder to success!

12.3.1. Freetown, Accara and Lagos in West Africa

Freetown, Sierra Leone, the was first of Britain's West African colonies. The Freetown malaria control programme had failed because Dr Ronald Ross, who was posted there to fight malaria, underestimated the number of breeding pools and the sheer number of vectors that he was trying to control. Ross had very limited funding and the best available technology was to pour oil on the numerous breeding sites around Freetown. The problem was not only that Ross severely misjudged the severity of the number of breeding sites, but he didn't appreciate that as soon as the oil treatments stopped, breeding would begin again. Ross redoubled his efforts with increased funding from private sources and ensured the removal of all potential breeding sites, including rubbish, broken bottles and other potential water containers. Despite these concerted efforts, the programme did not succeed and was remembered more for its impact on the Freetown's rubbish than with malaria control. Ironically, despite the fact that Stephens and Christophers had seen the programme fail in Freetown, however, they both believed that malaria could be dramatically reduced at malaria-ridden Mian Mir way back in India.

12.3.2 Mosquito control in Mian Mir

One of the earliest and highly challenging mosquito control campaigns was undertaken in India, or for that matter in the entire Asia, at the beginning of the 20th century. Then Mian Mir was a part of undivided Punjab in the pre-independent India, and is now a suburb in Lahore in Pakistan. Malaria was particularly severe in Mian Mir and was therefore taken up as an experimental site and model for intervention programmes elsewhere with equally high malaria endemicity. Mian Mir comprised a huge and sprawling army garrison, which continued to be severely hit by malaria despite extensive sanitation activities, and traditional farming areas punctuated with scattered residential localities.

The physiography of Mian Mir, though alluding toward an arid environment, included very flat terrain allowing limited natural drainage (*cf.* Konradsen *et al.*, 2005). Soil of the region was impervious to water after 1-2 hr of rains which would result into formation of indiscriminate puddles inveigling mosquito breeding if persisted for longer time. Since the gradient in land was very poor, even the storm-water drains did not keep constant flow, leading to form a perennial breeding ground for certain mosquitoes. Another significant feature of Mian Mir was its extensive irrigation network of shallow channels from which peasants drew water by means of Persian wheels to irrigate field crops and home gardens.

Christophers (1904) has described in detail the baseline investigations carried out before initiating vector control activities, which included assessment of malaria endemicity in the area, seasonal transmission dynamics of malaria and stratification of malariogenic conditions particularly the vector breeding sites. Besides malaria being rampant amongst the civilians, the infection also came down heavily on the army personnel. Thus, a special focus of the investigation was the burden of disease among the services men and the vector breeding in the vicinity of army barracks. Soon, it was revealed that *Anopheles culicifacies*, which bred profusely in the smaller irrigation channels growing some kind of vegetation along the edges, was the vector of endemic malaria in Mian Mir, signifying importance of special attention on types of irrigation network in the region.

Sometimes, these water courses would overflow, or even get breached, creating breeding opportunities along the channels. The poorly designed and maintained small earthen drainage canals were other important breeding sites. The practice of providing excess irrigation water on daily basis to small gardens immediately surrounding homesteads created breeding opportunities, too. The army quarters being very close to the irrigation canals, numbering ten, were the worst affected. The Mian Mir mosquito-malaria menace was conceived as a problem of environment degradation, therefore, most of the methods adopted to reduce vector breeding comprised repairing of the banks of irrigation channels, clearing silt from the irrigation channels to maintain good flow without allowing overflow;

(i) reclamation of low-lying areas otherwise tending to collect water with the smallest of rains,

(ii) removal of detritus and/or debris from the irrigation canals near to the army barracks so as to reduce chances of pool formation,

(iii) drying up of seasonal water pools, as much as possible,

(iv) emptying the domestic water containers undesired for any good domestic purpose and which unnecessarily collected rain water or otherwise for unduly long periods,

(v) creating superficial small drainage lines from large temporary ground pools in order to dry them up by thinning under the sun, and

(vi) during the most troublesome rainy period of June and July, sustained efforts were made to reduce the breeding potential of large pools formed in drainage canal either by using buckets to empty the pools or by applying kerosene oil. Simultaneously, malaria cases were treated with through mass distribution of quinine.

The Mian Mir case was one of the first documented cases resulting in heated debate over the success achieved in reducing vector population and the malaria incidence through mainly the environment based methods. Since, however, very little reduction in adult vector population was achieved, along with nearly indiscernible reduction in malaria cases particularly in the army residential areas, the prestigious Mian Mir campaign was branded a failure. Ross and others, were critical from the outset itself about the improper design of the operation. Ross (1910) claimed that not much efforts were actually put in to environmentally sanitate the region, besides insignificant investment into the campaign. If only the costs on sick soldiers at Mian Mir as made by the government was to be considered then the overall expenditure for malaria prevention made up only a $1/400^{th}$ proportion of this cost.

The Mian Mir failure offered opportunities to many raise their eye-brows and question the *modus operandi* as well as the expertise of the authorities in reference. It was argued that the area was never sufficiently drained and proper outlets should have been constructed from the drainage canals, possibly with pumping devices, to assist the drying of canals. The Mian Mir case was also one of the first cases resulting in a debate over the role of irrigation in the spread of malaria and how to combine the necessity of food production and livelihood with public health. It also gave rise to some of the fundamental questions for effective malaria vector control, *i.e.*, how large an area should he covered to have an effect on the protection of human settlements and how much reduction in adult population will have to be achieved to measure an impact of malaria on human beings? The subject of cost effectiveness in environment-based control methodology became the central point of subsequent discussions and enquiry - a damnifying and marooning experience for both Dr Christophers and Dr Stephens, the most experienced and energetic malariologists of their time!

Chapter 13

VECTOR CONTROL AND FUTURE NOVEL TECHNOLOGIES

13.1. Introduction

Vector mosquito control is an integral part of the disease management. There are several different methods available both at individual and community levels. Mosquito control can be divided into two areas of responsibility: individual and public. Most often it's performed following the Integrated Mosquito Management (IMM) concept. IMM is based on ecological, economic and social criteria and integrates multidisciplinary methodologies into pest management strategies that are practical and effective to protect public health and the environment and improve the quality of life. IMM strategies are employed in concert with insecticide. These include source reduction, which incorporates physical control (digging ditches and ponds in the target marsh) and biological control [placing live mosquito fish (Gambusia) in the ditches and ponds to eat mosquito larvae]. Other non-chemical control methods include invertebrate predators, parasites and pathogens to control mosquito larvae. Adult mosquito biological control by means of birds, bats, dragonflies and frogs has been employed by various agencies. However, supportive data are anecdotal and there is no documented study to show that bats, purple martins, or other predators consume enough adult mosquitoes to be effective control agents.

Pesticides continue to hold the main plank against vector-borne diseases especially during epidemics. Unfortunately many mosquito vector species have lately developed resistance against most of conventional insecticides, and likelihood of availability of newer insecticides is highly precarious, cost challenging, tedious and long exercise, as indicated in Fig. 13.1. They may be applied to control larvae (larvicides) or adults (adulticides). Applications of adulticides or larvicides are made after the presence of mosquitoes has been demonstrated by surveillance procedures. Application is made by prescribed standards. All insecticides must have the name and amount of active ingredient (AI) appearing on the label; examples are DEET and pyrethroids. No pesticide is 100 percent safe and care must be exercised in the use of any pesticide. Material Safety Data Sheets (MSDS) contain basic information about a product intended to help you work safely with the material.

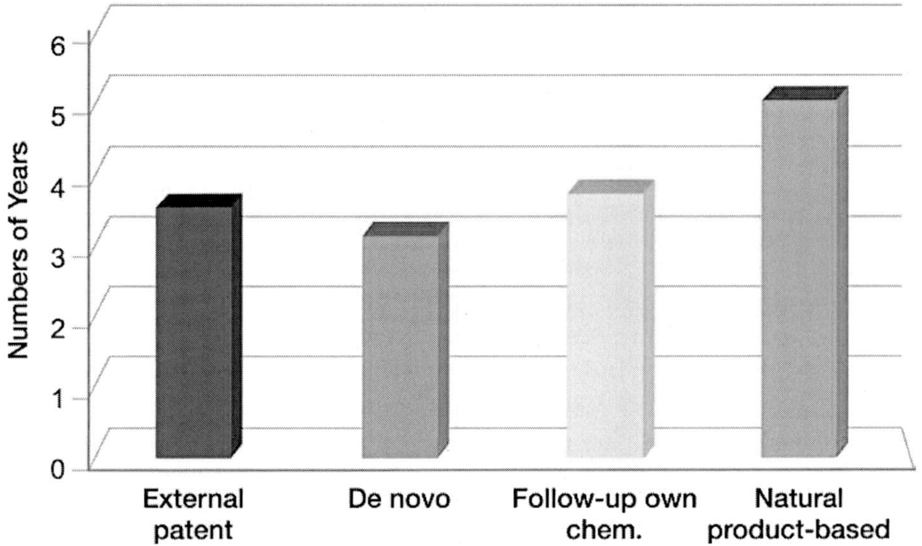

Fig. 13.1. Approaches for insecticide discovery: 1990-present Mean Time to Discovery (measured from the time an effort starts to when the compound (that ultimately becomes a product) is first made - n=40. *(Source*: Sparks, 2013)

Mosquitoes mediate transmission of some of the world's deadliest and most debilitating vector-borne diseases that are prevalent in entire tropics and subtropics, but are also gradually extending their distribution to temperate regions (Fig. 13.1). Currently dengue, transmitted by two vicious mosquito biters, *Aedes aegypti* and *Ae. albopictus*, has replaced malaria in both the intensity of dissemination and severity of the infection including annual mortality and new cases. While in the pre-DDT era malaria alone was responsible for about one lakh deaths and over 100 million cases in India each year, resulting into unlimited human misery including economy as well as physical and intellectual deprivation (Sinton, 1935; Sharma & Mehrotra, 1986). With the discovery of insecticidal property of DDT in 1939, and its availability for health programmes by the late 1940s, it became the mainstay for controlling vector mosquitoes and reducing the disease burden. However, soon it was discovered that the main malaria vector, *An. culicifacies*, which effected almost two-third of the total malaria cases in the country, had developed resistance against DDT, rendering it ineffective even when deployed at the maximum permissible limit of 2gm (ai)/sq. m. This situation warranted immediate deployment of alternative insecticides, notwithstanding some of them were highly toxic to human being, birds and fishes (*e.g.*, dieldrin), on the one hand, and others less effective and long-lasting but highly expensive, on the other. The control of vector mosquitoes appeared a near possibility when less than 0.1 million cases without single death in early 1960s were reported under the command of National Malaria Eradication Programme (NMEP). The euphoria was however short-lived and malaria resurged with a vengeance within next few years, by 1975. In the ensuing years it became extremely difficult to control vector mosquito populations with insecticides like DDT, BHC, malathion (as adulticides) and Abate and Baytex (as larvicides) due to several technical and administrative reasons, although these insecticides continue to be used selectively. Lately use of these and chemicals in public health has been continuously opposed due to their multifarious impacts on human health and environment. Insecticides are becoming increasingly unacceptable to the community.

Table 13.1. Mosquito distribution and cases worldwide.

Vector	VBD	Parasite/ Pathogen	Global Distribution	Population at risk	Number of people infected
Anopheles spp.	Malaria	*Plasmodium* spp.	Tropical and subtro pical regions- Sub-Saharan Africa, Asia and the Americas. 60 % of cases occurring in Africa south of the Sahara	3.6 billion	219 million
Aedes aegypti, Aedes albopictus	Dengue	Flavivirus	Tropical and subtropical countries. 75% of people at risk (1.8 billion) live in Asia-Pacific region	2.5 billion	50-100 million
Culex and *Anopheles* spp.	Lymphatic Filariasis	*Wuchereria bancrofti*	65% of people at risk reside in South-East Asia Region, 30% in the African Region and the remainder in other parts of the tropical world	1.43 billion	120 million
Aedes aegypti, Aedes albopictus	Chikungunya	Alphavirus	Africa, Asia and European region	5.17 million	1.67 million
Aedes aegypti, Aedes albopictus	Yellow Fever*	Flavivirus	Tropical and sub-tropical areas in Africa and South America, but not in Asia. 90% of infection occur in African continent.	900 million	0.0026 million$

Vector	VBD	Parasite/ Pathogen	Global Distribution	Population at risk	Number of people infected
Culex tritaeniorhynchus	Japanese encephalitis	Flavivirus	Countries of Asian Pacific region	3 billion	0.068 million[$]
Culex pipiens and *Culex quinquefasciatus*	St. Louis encephalitis	Flavivirus	North and South America	900 million	0.0001 million
Culex spp.	Rift Valley fever	Phlebovirus	Sub-Sahara Africa	No data	No data
Culex tarssalis	WEE	Alphavirus	United States		
Culiseta melanura	Eastern equnine encephalitis	Alphavirus	United States		
Culex pipiens, Culex tarsalis and *Culex quinquefasciatus*	West Nile Virus	Flavivirus	Africa, Central Asia, South-West Asia, Europe, the Middle East and North America	No data	No data

13.2. Need for alternative strategies

In spite of the well-recognized fact that the large usage of insecticides can no longer suppress the vector population, the national disease vector control agencies are even today practically solely dependent on these chemicals for instant results. Three main reasons can be cited here to discourage the deployment of conventional insecticides in future. First, many vector species have developed resistance against most of the insecticides in vogue. All over the world, 392 species under Arthropoda have developed resistance, of which 136 (35%) species are vectors of diseases of public health importance. These include 47 anophelines resistant to dieldrin (and BHC), 34 species to DDT, 10 species to organophosphate compounds, 4 species to carbamates and 7 species to synthetic pyrethroids. There are also 42 species of culicine mosquitoes that have developed resistance against these insecticides. In India, the most important malaria vector, *Anopheles culicifacies*, has become resistant to DDT in most parts of the country, and to BHC in some areas. In Gujarat and Maharashtra, it has become resistant to DDT, BHC and malathion.

Secondly, costs involved in the insecticidal operations are exorbitantly high. It has been estimated in 1990s that to produce one effective insecticide nearly Rs 4000 million (US$ 20 million) were needed, whereas the chance for discovering a successful insecticide was only 1/1500 compounds screened (0.06%). It can thus be understood that the astronomical costs involved in the production of insecticides and their subsequent deployment in vector control operations are highly challenging and discouraging for any government, particularly in a developing nation like India. Yet, between 1960 and 1980 only, a total of 77.75 mt DDT, 123 mt

BHC and 68.7 mt malathion had been applied under various health programmes in India. During that period the Government of India was spending approximately Rs 2000-3000 millions annually on control of malaria alone, and ever since there has been persistent demand for higher allocation each subsequent year.

Thirdly, the persistent and judicious use of insecticides has brought about drastic changes in our environment, DDT, which has been particularly overused due to its low mammalian toxicity, low costs and high degree of residuality on sprayed surfaces, has been often detected in the human blood, milk and tissue. According to one estimate in the mid-1980s, DDT occurred 20 times the permissible limit in the vegetables in Delhi. In Maharashtra, the bottled milk was found to contain 4.8-6.3 ppm of DDT which is 7.2-9.5 times the permissible limit. The daily intake of DDT in an average Indian diet was estimated to be 0.27 mg or around twice the dose accepted by WHO, while the DDT residues in human milk was estimated around 0.51 ppm which is quite alarming. As the DDT continues to be the cheapest insecticide available its injudicious use in health programmes can be judged from the fact that 110 mt DDT had been deployed for vector control against only 38.23 mt DDT in agriculture in India till 1978. All insecticides have close interaction with the environment and their use in any way is bound to affect man and his associates severely.

In view of these facts intensive search for innovative and alternative measures had been constantly made out. These measures or tools were comparatively cost effective, less hazardous, community-friendly and did not easily pose risks of development of resistance in the target vector species. Some important methodologies are discussed below which have been tried not only in the field but also proved quite promising.

13.3. Chemical-based measures

Some of the promising-looking methodologies are listed below:

13.3.1. Insect growth regulators (IGRs)

These are also referred to as the third generation insecticides. The IGRs interfere with the hormonal mechanisms of target organisms and results into various different kinds of morphological, anatomical and physiological abnormalities so that the target species does not reach the final, viable adult stage of development. There is no likelihood of development of resistance against these IGRs. However in case the phenomenon somehow occurs (just as in case of diflubenzuron) it would naturally take a long duration of time to develop, and, secondly, it can be easily overcome by the addition of synergists like piperonyl butoxide and S' S' S' tributyl phosphorothioate. A large number of IGRs, both juvenoids and chitin synthesis inhibitors, have been evaluated for the vector control, but only very few of these were found effective and viable commercially in varied water conditions, *e.g.*, diflubenzuron, Methoprene, fenoxycarb (Tyagi, 1985; Tyagi *et al.*, 1987; Amalraj *et al.*, 1988a,b). The IGRs have an added advantage of being used at a relatively very low dose compared to the conventional insecticides.

13.3.2. Surface active agents

These are non-ionic, biologically degradable chemicals whch on application to the aquatic media (mosquito breeding habitats) spread spontaneously to form a monomolecular film on the water surface, *e.g.*, Arosurf ISA 20E [ployoxyethylene (2) isostearyl ether (2) mole ethoxylate of isostearyl alcohol] (Das *et al.*, 1986). These films bring about mortality in the immatures by a "physical" rather a chemical factor as observed with the chemical larvicides. The monomolecular films form a barrier on the surface and the larvae/pupae of the target species die off due to anoxia. The films also lower the surface tension of the treated water body (25-29 cm/dynes) causing suffocation in the tracheal system, on one hand, and disallowing inflation of the wings of the emerging adults, on the other. At the constant rate of 11.2 l/hr Arosurf has been actively used to control *Culex quinquefasciatus, Anopheles stephensi* and *Aedes aegypti* in different types of breeding habitats such as cesspits, cesspools, drains and wells. Due to their self-spreading property it has been constantly advocated that Arosurf and other surface active films can be deployed in combination with a fast acting and residual larvicide like fenthion for a better coverage of the breeding habitats and long effective life of the film in such inaccessible habitats like marshes, bogs and lagoons.

13.3.3. Insecticide-impregnated paint

This technology has perhaps been never realized on large scale in India, but these paints are in use for decades in some developed nations. Das *et al.* (1986) demonstrated one such an insecticide-impregnated paint, Vernacide, against both the mosquito, *Culex quinquefasciatus,* and the cockroach, *Periplanata americana,* in urban conditions. In both water- and oil-based formulations this paint was found 100% effective for over one year. Its ineffectiveness was regained after a cold water wash to the painted surface. The long-lasting effect of the paint could be possible because the insecticide is embodied in an acrylic medium with small openings which help to prevent its fall-out from the surface. Consequently, these paints are far more safer than the conventional water-dispersible powder (WDP) formulation of the adulticides, and can be deployed in public places such as airports, hospitals, hotels, cinema theatres, ships etc where pest-free conditions are desirable on permanent basis.

Apart from these paints, insecticide-impregnated fabrics have also been developed and used primarily to protect from the mosquito bites. While Jambulingam *et al.* (1989) demonstrated the permethrin-impregnated bed nets (@ 0.5 gm (ai)/sq. m.) in reducing the vector density of *An. culicifacies* in the tribal villages of Orissa (now Odisha), which were highly endemic for malaria, for more than three months. Sharma & Yadav (1995), and Sahu *et al.* (2003) demonstrated the practical significance of cyfluthrin- and lambdacyhalothin-impregnated bed nets and deltamethrin-impregnated ropes in human dwellings in various parts of the country. Additionally fabric-knitted hoods, long sleeves, gloves and trousers are also privately used in many western and European countries for personal protection.

13.3.4. Controlled slow-release formulations

These formulations are now commercially available in different forms. They have invariably a biodegradable base carrier which is imbibed with a relatively high concentration dose of a larvicide so that when placed in a mosquito breeding habitat the base carrier spontaneously biodegrades releasing in the process the larvicide slowly in the desired, low, yet toxic, concentration for the target species. Such a methodology is highly desirable in highly polluted and expansive water bodies where repeated spraying of the conventional larvicide (*e.g.*, fenthion or Baytex) is very much cost prohibitive and time consuming. *Culex quinquefasciatus* and *Cx. tritaniorhynchus* can be significantly brought under control by this methodology. These formulations are required to be sprayed on weekly basis to bring about good results.

13.4. Non-chemical measures

Expanded polystyrene (EPS) beads are a synthetic product of a very light thermocol-like material widely available commercially at a price of Rs. 80/kg. These beads are spherical products, measuring about 0.2- 0.8 mm in diameter. These are non-biodegradable and can remain effective for months in the treated water bodies such as wells and latrine pits. The EPS has an inherent property to keep buoyant irrespective of the level of the breeding habitat's water at any time. The EPS beads are light in weight, inert, nontoxic, non-wettable and resistant (do not interact) to sea water, salt solutions, soap and wetting agents. Direct exposure to intense sunlight can turn them yellowish and brittle due to ultra-violet radiation. These are not acted upon by any micro-organisms, are non-biodegradable and remain for years on the surface of water on a single application. Being light in weight, the EPS beads when applied @ 500 g to 1 kg/m2 in different habitats, float on water surface in several layers. Since mosquitoes lay eggs only on the water surface, the physical barrier formed by the floating blanket of EPS beads prevents them in doing so. Further, the immature stages of mosquito trapped under the layer of the beads die of suffocation and mosquitoes do not emerge from the treated habitats. Curtis and Minjas (1985), Dua *et al.* (1989) and Sivagnaname *et al.* (2005) have demonstrated control of malaria vector, *Anopheles stephensi,* in live wells with good drinking water and filariasis vector, *Culex quinquefascaiatus,* in unused wells with turbid and/or polluted water and latrine pits. Soltani *et al.* (2008) demonstrated another form of EPS called Shredded Waste Polystyrene (SWAP) Beads for Control of Mosquitoes. Recently, Sunish *et al.* (2018) used EPS as a component of the Integrated Vector Control approach in controlling filariasis vector, *Cx quinquefasciatus,* in the endemic villages of Tamil Nadu, south India. Considering its long life, non-hazardous nature and one-time application, the EPS is considered cost effective and feasible in certain important situations.

13.5. Biological control agents

Mosquitoes represent the major arthropod vectors of human disease worldwide transmitting malaria, lymphatic filariasis, and arboviruses such as dengue virus and Zika virus. Unfortunately, no treatment (in the form of vaccines or drugs) is available for most of these diseases and vector control is still the main form of prevention. The limitations of traditional insecticide-based strategies, particularly the development of insecticide resistance, have resulted in significant efforts to develop alternative eco-friendly methods. Biocontrol strategies aim to be sustainable and target a range of different mosquito species to reduce the current reliance on insecticide-based mosquito control (Fig. 13.2).

13.5.1. Bacterial agents

Bacillus thuringiensis H-14 and *B. sphaericus,* commonly called *Bt* an *Bs,* respectively, have shown great promise to control a variety of vector and pest mosquitoes, particularly those breeding in extensive marshes and polluted water bodies. While *Bt* is commercially available as Deltor, Bactimos and Tekner in different formulations (*e.g.,* liquid concentrate, and water dispersible powder), *Bs* is relatively less persistent in checking the vector populations. Both *Bt* and *Bs* are highly effective against all stages of anopheline and culicine mosquitoes. Deltor and Bactimos at the rate of 1 gm and Tekner at the rate of 3 ml per sq. m. of surface area were found to reduce the larval density drastically. Also, these bacterial species, albeit being pathogenic to the target vector mosquito species, are non-hazardous to beneficial and non-target organisms.

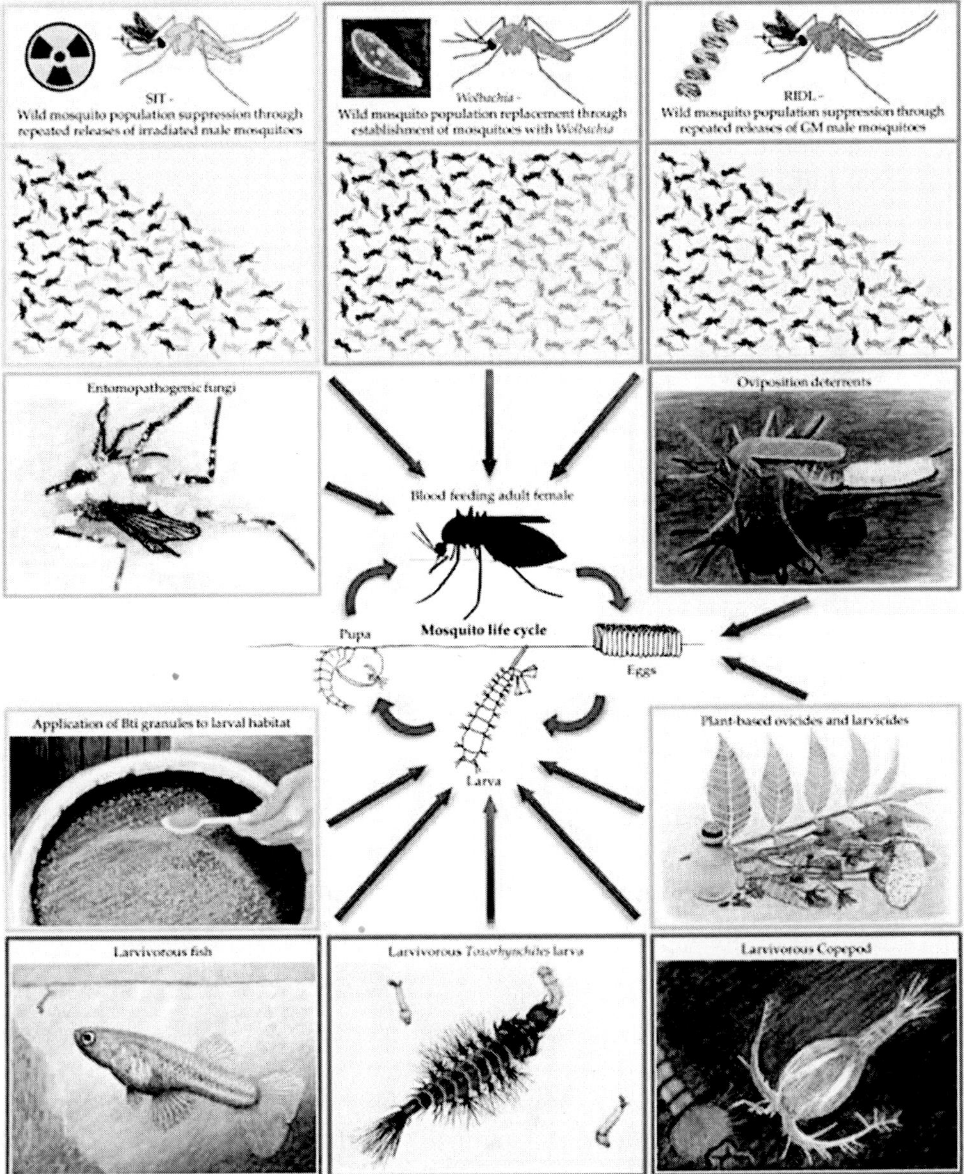

Fig. 13.2. Mosquito biocontrol strategies targeting different development stages in the mosquito life cycle. (*Source*: Benelli *et al.*, 2016).

13.5.2. Fungal agents

Because of their property to destroy larval populations under natural conditions, several fungi have been for long considered to be highly suitable for checking vector population, on one hand, and developing fungi as a potential biological tool against the vector mosquitoes, on the other. Field testing has, therefore, been made with several kinds of fungi, such as, for example *Culicinomyces clavosporum* and *Tolypocladium cylindrosporum* (fungi imperfecti), *Lagenidium gigenteum* (Oomycetes), and several species of *Coelomomyces* (*Chytridiomycetes*). In India extensive work was done on *Culicinomyces, Lagenidium* and *Coelomomyces* species isolated from natural sources. Conidia of *Coelomomyces* spp. can be suspended in water and applied with manually operated pressure sprayers or motor-driven knapsack sprayers at the application rate of 10^{13} conidia/ha. *Lagenidium* spp. were introduced (seeded) into breeding habitats at a rate of 10^{11} zoospores/ha or 10^{10} zoospore/ha during early hours of the breeding season, although repeated applications were required subsequently. Coelomomyces was introduced by placing in medium sporangia or moribund larvae that contained resistant sporangia. While in the breeding season, single introduction of 1000 infected moribund larvae/ha is good enough to check the larval breeding effectively in the early stage, yet repeated applications will be desired subsequently for achieving a long term control.

13.5.3. Nematodes against mosquitoes

There are certain mermethids that attack and kill vector larvae. *Romanomermis culicivorax* and *R. iyengari* were for long consistently cultured in Indian laboratories. It is the infective preparasite stage (nema) which penetrates the mosquito larvae. After 6-8 days when the parasite emerges out as a postparasite, this invariably results into the death of the immature mosquito form. Since, however, both these mermethids tolerate very low levels of salinity, alkalinity and pollution, these can be safely released only in fairly freshwater bodies, such as the rice-fields.

13.5.4. Larvivorous fishes and predaceous arthropods

Larvivorous fishes have been considered for decades an important vector control tool because they are effective predators and are prevalent over a wide range of ecological habitats, as well as because the operational costs involved in the integrated control programmes are low. Fish like *Cyprinus carpio, Ctenopharyngodon idella, Catla catla* and *Gambusia affinis* are found to have a special liking for the Anopheles mosquitoes, and to some extent also Culex, Aedes and Armigeres species.

Among predaceous insects, *Anisops bouvieri* (Notonectidae) and *Toxorhynchites splendens* (A non-biting mosquito) have been found to control vector larval population a great deal under specific conditions. Both these insect species can be mass-cultured and deployed once in a year to the breeding habitats. Single fourth instar of *Tx. splendens* can consume 114 larvae of the vector species. Recently, some dragonfly species too have been successfully used to control the yellow fever/dengue vector, *Aedes aegypti.* Sebastian *et al.* (1980) deployed larvae of dragonfly *Crocothemis servilia* in suppressing *Aedes aegypti* breeding in ornamental vases and other water holding pots, and demonstrated how the dragonflies had an advantage over other predaceous insects. Labellulid nymphs were found to predate on *Ae. aegypti* larvae and pupae readily. The rate of consumption was found to be 133 +/- 2 all stages of larvae per medium size nymph per 24 hours. In container habitats complete elimination of all larvae and pupae were achieved between day 4 and 9 depending on density of aquatic stages. Corbet (1986) has also advocated the application of dragonflies in eliminating vector larval populations in certain specific conditions., such as cement tanks, pools and ornamental pots etc. as they have effective preying mouthparts and relatively longer larval life. Venkatesh & Tyagi (2015) used the dragonfly *Bradinopyga geminata* for controlling dengue/chikungunya vector *Aedes aegypti,* in the laboratory. The feeding rate of 8th instar *B. geminata* on *Ae. aegypti* showed maximum predation on 1st instar larvae (86%), followed by 2nd , 3rd and 4th instars (72%, 66% and 48%), respectively. In the first hour, the consumption rate was maximal for all instars and a low intake (about 5%) was observed in subsequent hours. In 12th instar *B. geminata* larvae maximum predation was observed for the 1st and 2nd instar larvae (98%) of *Ae. aegypti,* followed by 3rd and 4th instars (92% and 78%), respectively. The feeding rate of 12th instar *C. coromandelianum* larvae on *Ae. aegypti* larvae showed that the maximum predation was of the 1st instar larvae (82%), followed by 3rd and 4th instars (92% and 78%), respectively. The feeding rate of 12th instar *C. coromandelianum* larvae on *Ae. aegypti* larvae showed that the maximum predation was of the 1st instar larvae (82%), followed by 2nd, 3rd and 4th instars (51%, 35% and 24%) respectively. The first hour consumption rate was maximum for all instars and no significant intake was seen in the following hours. The predation of Aedes larvae by the 2 spp. of odonate larvae was compared for the 4 mosquito larval instars by using one way ANOVA. No significant difference was found between them for 1st instar larvae of *Ae. aegypti* but there was a significant difference ($P<0.05$) in predation on the other 3 instars, with *B. geminata* consuming more mosquito larvae. A single anisopteran larva is sufficient for eliminating the huge mass of larval mosquitoes breeding in a cement tank or a cement cistern. Therefore, this biological control agent could be released to control Aedes larval production in areas of dengue epidemics.

13.5.5. Cyclops as predator of Aedes aegypti

Dengue is the most spread mosquito-borne viral disease in the world, transmitted by Aedes mosquitoes. Since dengue is still not having any antidote to treat a patient nor a vaccine is commercially available to prevent the onset of infection, vector control of the vectors, *Aedes aegypti* and *Ae. albopictus,* is the only way to stem out the disease. These vectors breed in potable waters, therefore, larvicides are not generally recommended, save for fortnightly/weekly treatment in certain sites by temephos @ 0.5-1 ppm. A noticeable means for Aedes control is the use of copepod predators. Since the discovery of copepod predation on mosquito larvae, they have been used to control Aedes larvae in temperate, subtropical and tropical regions (Tuno *et al.,* 2020). So far, copepod application has been proven to suppress mosquito populations in some southeast Asian countries such as Vietnam, but copepod application alone was not effective in southern Vietnam, although it had been argued if copepods alone can suppress the vector populations. Thus, it is proposed that the effectiveness of copepod application varies according to the differences in community structures and environmental conditions.

Copepods predate on younger (mainly first instar) mosquito larvae more efficiently in comparison with later-stage larvae. Therefore, the effectiveness of copepods to suppress mosquito increase may be reduced at high temperatures or under rich resource conditions, because mosquito larvae develop faster under such conditions and have shorter vulnerable periods. The effectiveness of copepods would also be reduced if food resources other than mosquito larvae are available. In addition, their predation efficiency is reduced in the presence of detrital substrates, possibly because such substrates provide refuge for mosquito larvae. Furthermore, mosquito larvae are able to compete with copepods for food when they grow larger. Competitive interactions would lower survivorship and fecundity and prolong developmental time both in copepods and mosquitoes, and the outcome of their competition would differ if environmental conditions change.

13.6. Repellents

Repellents are made based on chemical products with an offensive smell or taste to mosquitoes. Plant-derived compounds with repellent properties are most likely chemicals that are produced in defence against insects that pose a threat to the plant itself. These chemicals can be grouped into different categories, based on the functional groups present. They include nitrogen-containing compounds, terpenoids, phenolics, proteinase inhibitors and growth regulators. These compounds are generally produced to fight off a broad spectrum of insects including mosquitoes. Plants with better repellent properties fall into distinct families, with the Poaceae family (citronella based especially Cymbopogon spp.) being the pre-eminent one. Species of Lamiaceae, Fabaceae and Asteraceae also show promising results. Prior to the extensive use of synthetically produced repellents, aromatic/essential oils were commonly used. The military was a significant consumer of these oils, creams containing citronella, camphor and paraffin. Numerous essential oil-producing plants from the Lamiaceae, Poaceae, Rutaceae and Myrtaceae families have very well-known repellent activity. Synthetic products that have been used as repellents include indalone, dimethylphthalate 2-ethyl-1, 3-hexane diol (Rutgers 612) and N,N-diethyl-m-toluamide (DEET). DEET is by far the most effective and widely used repellent. The Indian counterpart of the DEET is N,N-diethyl phenyl acetamide (DEPA) which is now commercially available in different formulations and has been claimed to have the protection rate of about 8h, against not only the mosquito pests and vectors but also other hematophagous arthropods as well as ticks, mites and even leeches.

13.7. Insect Traps

The ovitrap or oviposition trap was mostly used and invented for the surveillance of Aedes vectors and then modified to render it lethal to adults or larvae of *Ae. aegypti*. In principle, ovitraps could kill adult mosquitoes if the ovistrip was treated with insecticide or destroy progeny by using fine nylon netting for trapping the larvae. Lethal ovitraps (which incorporate an insecticide on the oviposition substrate) with deltamethrin-treated ovistrips killed 89% of *Ae. aegypti* adults and produced more than 99% larval mortality during 1-month field trials in Brazil. The autocidal ovitraps (which allow oviposition but prevent adult emergence) and sticky ovitraps (which trap the mosquito when it lands) have been used on a limited basis. The autocidal ovitrap was used in Singapore for the control of Aedes vectors in urban areas with a high density of Aedes. The advantages of lethal ovitraps include their simplicity, their specificity for and effectiveness against container breeders like *Ae. aegypti* and its potential for integration with other chemical or biological control methodologies. A newer approach to killing mosquitoes in a nontoxic way is to use a device that burns propane, thus generating carbon dioxide, warmth and water vapour which draws the mosquitoes towards the propane flame, where they are then sucked into a net or holder where they are collected. Some newer mosquito traps or known mosquito attractants are also available which are based on the principle of disabling its odour receptor as host-seeking female mosquitoes are guided by attractant odours released by their target. *Anopheles gambiae* is attracted to ammonia, lactic acid and other carboxylic acids naturally present in the body odour and sweat produced by warm-blooded animals. There are traps like black hole mosquito and midge trap and Jakmax insect trap which were developed based on this concept which release heat and CO_2 and captures mosquitoes, sand flies, midges, termites and other flies that are attracted by black light.

13.7. Strategies based on genetic manipulation

Genetic modification (GM) of vector mosquitoes is one of such technologies, which may be mainly used either to suppress or replace the wild populations of a vector so as to decrease vector populations or reduce the vector's ability to transmit. These applications include the release of reared mosquitoes (GM) in the environment to introduce modified genetic traits in wild population. GM encompasses multiple approaches, which are broadly categorized into two types. The first category includes sterilized insect technique (SIT) for population suppression and the second category is the gene drive systems for population replacement or manipulation. The sterilized insect technique (SIT) includes the dominant lethal gene systems (RIDL), Wolbachia-mediated cytoplasmic incompatibility (CI) and classical radiation-induced male sterility, while population manipulative technologies include Medea-based gene drive, under-dominance gene drive, homing endonuclease, Wolbachia-mediated heritable biocontrol or genetically modified midgut bacteria and transposable element like Piggy Bac. These techniques have been put into field application in many countries (Cayman ils., Florida, Brazil, Mexico, Malaysia etc.). However in India most of them are still in laboratory development and/or contained stage. Two private institutions, namely, International Institute of Biotechnology and Toxicology (IIBAT) in Tamil Nadu and Gangabishan Bhikulal Investment and Trading Limited (GBIT)'s Hybrid Seeds Co. (MAHYCO), Dawalwadi in Maharashtra State began laboratory based investigations on the RIDL *Aedes aegypti* during mid-2000. While IIBAT discontinued after a few of years of investigation, MAHYCO progressed satisfactorily from laboratory to containment studies with splendid results (Patil *et al.*, 2014). Further, these investigation require different region-based field testing under independent monitoring in order to prove the result to gain the consent of society to implement. These techniques require large volume of mosquitoes to be released into the environment at different period of interval to either suppress or replace mosquito population. The Wolbachia-based research for controlling Aedes aegypti and dengue has been recently taken up at the ICMR-Vector Control Research Centre, Puducherry in collaboration with Monash University, Australia, whereby naturally occurring Wolbachia bacterium will be introduced into dengue virus-carrying mosquitoes to "inhibit" viral transmission. The efficacy of the disease-control method will be tested at the VCRC before undertaking any field trial (*cf.* The Economics Times, Feb. 07, 2017).

13.8. Integrated vector control strategy

This is defined as "the utilization of all technological and management advancements to bring about an effective degree of vector population's suppression in a cost-effective manner." The significance of such a strategy was already demonstrated in Kheda district by controlling malaria through vector suppression and in the urban agglomeration in Pondicherry for controlling filariasis (Sharma *et al,* 1986; Rajagopalan & Das, 1988). Integrated vector management (IVM) is based on sound ecological principles and integrates multi-disciplinary methods in developing ecosystem management strategies that are practical, effective, economical and protective to both public health and environment. However, it is important to know that this approach is specific to local needs and problems, and the operational strategy, therefore, changes from area to area accordingly. This method offers a desired degree of control at a lower cost or with a greater long-term benefit. Health education, participation of community and supportive legislation are three major factors that make the integrated vector management successful. Thus, all the above multiple ways of vector control measures are all independent yet inter-linked constituents of an IVM only.

REFERENCES CITED

Adie, J.R. and A. Alcock, 1905. On the Occurrence of Anopheles (Myzomyia) listoni in Calcutta. *Proceedings of the Royal Society of London.* Series B, Containing Papers of a Biological Character 76(510): 319-321.

Alfaro-Murillo, Jorge A., Alyssa S. Parpia, Meagan C. Fitzpatrick, Jules A. Tamagnan, Jan Medlock, Martial L. Ndeffo-Mbah, Durland Fish, María L. Ávila-Agüero, Rodrigo Marín, Albert I. Koand Alison P. Galvani, 2016. A Cost-Effectiveness Tool for Informing Policies on Zika Virus Control. *PLoS Negl Trop Dis.* 10(5): e0004743. doi: 10.1371/journal.pntd.0004743

Alnwick, D., 2000. Roll Back Malaria - What are the prospects? *Bulletin of the World Health Organization* 78:1377.

Amalraj, D., V. Vasuki, C. Sadanandane, M. Kalyanasundaram, B.K. Tyagi and P.K. Das, 1988a. Evaluation of two new juvenile hormone compounds against mosquito vectors. *Indian Journal of Medical Research* 87: 19-23.

Amalraj, D., V. Vasuki, M. Kalyanasundaram and B.K. Tyagi, 1988b. Laboratory and field evaluation of three insect growth regulators against mosquito vectors. *Indian Journal of Medical Research* 87: 24-31.

Barber, M.A. & J.B. Rice, 1938. Malaria in Poona and its vicinity. *J. Mal. Inst. India* 1: 37-55.

Barraud, P.J.,1934. *The fauna of British India, including Ceylon and Burma. Diptera: Family Culicidae; Tribes Megarhinini and Culicini.* Vol. 5, Taylor & Francis, London, 463pp.

Beklemeshev, W.N., T.S. Detinova and V.P. Polovodova, 1959. Determination of physiological age in anophelines and age of distribution in anopheline populations. *Bull. Wld. Hlth. Org.* 21 :223-232.

Benelli, G., Claire L. Jeffries and Tomas Walker, 2016. Biological control of mosquito vector: past, present and future. *Insects* 7 :52 ; doi :10.3390/ insects7040052

Bentley, C.A., 1911. *Report on malaria in Bombay.* Government Press, Bombay.

Bhattacharya D.R., Anil Prakash P.K. Mohapatra and D.K. Sarma, 2009. *Armigeres (Armigeres) mahantai,* a new mosquito species from India. *Mosq Syst.* 25(1):1-5.

Bhattacharyya, D. R., A. R. Rajavel, P. K. Mohapatra, P. Jambulingam, J. Mahanta, A. Prakash, 2014. Faunal richness and the checklist of Indian mosquitoes (Diptera: Culicidae). *Check List* 10(6): 1342–1358.

Bhattacharyya, D.R., Anil Prakash, P.K. Mohapatra and J. Mahanta, 2004. *Uranotaenia dibrugarhensis* a new species in subgenus *Pseudoficalbia* from Assam, India. *Mosq Syst* 20(1):1-5.

Bhattacharyya, D.R., S.C. Tewari, Anil Prakash, P.K. Mohapatra, 2004b. *Verrallina (Neomacleaya) assamensis,* a new species from Assam, India *Mosq Syst.* 20(2):115-120.

Christophers, S.R., 1904. Second report of the anti malarial operations at Mian Mir, 1901–1903. Scientific Memoirs by the Office of the Medical and Sanitary Departments of the Government of India, Calcutta, pp. 1–37.

Christophers, S.R., 1933. *The Fauna of British India, including Ceylon and Burma. Diptera: Family Culicidae; Tribe Anophelini.* Vol. 4, Taylor & Francis, London, 271pp.

Clements, A.N., 1992. *The Biology of Mosquitoes. Vol. 1. Development, Nutrition and Reproduction.* CABI Publishing, 571 pp.

Clements, A.N., 1999. *The Biology of Mosquitoes.* Vol. 2. *Sensory Reception and Behaviour.* Wallingford: CABI Publishing, 752 pp.

Clements, A.N., 2011. *The Biology of Mosquitoes. Vol. 3: Transmission of Viruses and Interactions with Bacteria.* CABI Publishing, 592 pp.

Clyde, D., 1931. Reports on the control of malaria during Sarda Canal construction (1920-1929). *Rec. Mal. Surv. India* 2: 49-110.

Cohuet, A., C. Harris, V. Robert and D. Fontenille, 2010. Evolutionary forces on Anopheles: what makes a malaria vector? *Trends Parasitol.* 2010;26:130–136.

Corbet, P.S., 1986. Using dragonflies to suppress mosquitoes in domestic water-storage containers. *Waterlines* 4, 10–11.

Covell, G., 1927. A critical review of the data recorded regarding the transmission of malaria by the different species of Anopheles, with notes on distribution, habits and breeding places. *Indian Medical Research Memoirs,* 7: 87–99.

Covell, G., 1928. *Malaria in Bombay.* Government Centre Press, Bombay, 113pp.

Covell, G., 1934. Notes on the control of mosquitoes and malaria in Delhi. *Rec. Mal. Surv. India* 4:273-289.

Curtis, C. F. and J. Minjas, 1985. Expanded polystyrene bead for mosquito control. *Parasitology Today.* 1: 36.

Das, B.P., R. Rajagopal and J Akiyama, 1990. Pictorial key to the species of Indian Anopheline mosquitoes. *J. Pure App. Zool.* 2, 131-162.

Das, P.K., B.K. Tyagi and M. Kalyasundaram, 1986. Vernacide, a new insecticide impregnated paint for controlling mosquito vector, *Culex quinquefasciatus* and cockroach, *Periplanata americana. Indian Journal of Medical Research* 83: 268-270.

Das, P.K., B.K. Tyagi, N. Somachari and V. Vasuki, 1986. Efficacy of Arosurf, a molecular surface film in controlling *Culex quinquefasciatus* Say, *Anopheles stephensi* Liston and *Aedes aegypti* (L.). *Indian Journal of Medical Research* 83: 271-276.

Dev, V. and V.P. Sharma, 2013. The dominant mosquito vectors of human malaria in India. *In:* Manguin, S., editor. *Anopheles mosquitoes - New insights into malaria vectors*. Rijeka: InTech Open Access.

Dhananjeyan, K.J., R. Paramasivan, V. Thenmozhi, R.Chandrasekar and B.K. Tyagi, 2014. Insect exuvium extracted DNA marker: a good complementary molecular taxonomic characteristics with special reference to mosquitoes. (*In: Short Views on Insect Biochemistry and Molecular Biology*. Vol. 1. Eds. Raman Chandrasekar, B.K.Tyagi, Zhong Zheng Gui and Gerald R. Reeck), pp.355-364.

Dossou-yovo, J., J. Doannio, , F. Riviere and J. Duval, 1994. Rice cultivation and malaria transmission in Bouake city (Cote d'Ivoire). *Acta Tropica* 57: 91–94.

Dua, V.K., S.K. Sharma and V.P. Sharma, 1989. Use of expanded polystyrene beads for the control of mosquitoes in an industrial. complex at Hardwar, India. *J. Am. Mosq. Contr. Assoc.* 5 (4): 614–615.

Dutta, P., S.A. Khan, A.M. Sharma, C.K. Hazarika and J. Mahanta, 2003. Survey of medically important mosquito fauna in Mizoram. *Entomon,* 28: 237-240.

Duvallet G, N. Boulanger and V. Robert, 2018. *Arthropods: definition and medical importance.* In: Boulanger N, editor. *Skin and Arthropod Vectors.* London: Academic Press. pp. 29–54.

Edman, J., J. Day and E. Walker, 1984. Vector-host interplay-factors affecting disease transmission. In: *Ecology of mosquitoes.* Proceedings of a workshop, Florida Medical Entomology Laboratory, Vero Beach, pp. 273-285.

Eldridge, B.F. and John Edman, 2012. *Medical Entomology*: A Textbook on Public Health and Veterinary Problems Caused by Arthropods. Springer Science & Business Media, , 672 pp.

Failloux, A.B., A. Bouattour, C. Faraj, F. Gunay, N. Haddad, Z. Harrat et al., 2017. Surveillance of arthropod-borne viruses and their vectors in the Mediterranean and Black Sea regions within the MediLabSecure network. *Curr. Trop. Med. Rep.* 4:27–39

Fontenille D, L. Almeras and C. Garros, 2009. Concepts et méthodes d'identification des espèces d'arthropodes. In: Duvalet G, Fontenille D, Robert V. [editors]. 2017. *Entomologie Médicale et Vétérinaire.* Marseille: IRD Editions & Quae Editions;. p. 61–78;

Gajanana, A., R. Rajendran, P.P. Samuel, V. Thenmozhi, T.F. Tsai, J. Kimura-Kuroda *et al.*, 1997. Japanese encephalitis in South Arcot district, Tamil Nadu, India: a three-year longitudinal study of vector abundance and infection frequency. *J. Med. Entomol.* 34 : 651-9.

Ghosh, Susanta, K., Satyanarayan, S.C. Tiwari, and Vijay P. Ojha, 2012. A Renewed way of Malaria Control in Karnataka, South India. *Front Physiol.* 3: 194.

Gupta, R.K., 1975. The landscape and land forms. *In:* Gupta, R.K. & Prakash, I. (Eds), *Environmental Analysis of the Thar Desert,* pp. 1–21. Dehradun: English Book Depot. 484 pp.

Gupta, Uma Das, ed., 2010. *Science and Modern India: An Institutional History,* c.1784-1947: Project of History of Science, Philosophy and Culture in Indian Civilization, Volume XV, Part 4. Pearson Education India. p. 587.

Harbach R., 2014. Mosquito taxonomic inventory. http://mosquito-taxonomic-inventory.info/node/11667 (assessed on 31st July 2014).

Harbach, R., 2007. The Culicidae (Diptera): a review of taxonomy, classification and phylogeny. *Zootaxa* 1668:591–638.

Harbach, R.E., 2004. The classification of genus Anopheles (Diptera: Culicidae): a working hypothesis of phylogenetic relationships. *Bull. Entomol. Res.* 94(6):537-53.

Harrison, Bruce A. and John E. Scanlon, 1975. *The subgenus Anopheles in Thailand (Diptera: Culicidae). Medical Entomology Studies – II. Contrib. Amer. Ent. Inst.* 12 (1):1-307.

Harrison, G., 1978. *Mosquitoes, malaria and man: a history of the hostilities since 1880.* London: Tohn Murray, 314 pp.

Hiriyan, J.1. and B.K. Tyagi, 2004. Cocoa pod (*Theobroma caco*) - a potential breeding habit of Aedes albopictus in dengue-sensitive Kerala State, India. *J Am. Mosq. Control Assoc.* 20(3):323-325.

Jagbir, S.K and K. Jagdish, 1999. Mosquitoes of Malwa region of Punjab. *Geobios new Reports* 18:75-76.

Jambulingam, P.K., K. Gunasekharan, S.S. Sahoo, P.K. Hota, B.K. Tyagi and M. Kalyanasundaram, 1989. Effect of permethrin-impregnated bednets in reducing population of malaria vector, Anopheles culicifacies in a tribal village of Orissa State (India). *Indian Journal of Medical Research* 89: 48-51.

James, M.T. and R.D. Harwood, 1969. *Herms's Medical Entomology.* 6th Edition, MacMillan, New York, pp. 286.

Jong, R. de & B.G.J. Knols, 1996. Selection of biting sites by mosquitoes. In: Cardew, G. (ed.), *Proceedings of CIBA foundation* No. 200: Olfaction in Mosquito-Host Interactions. Wiley, Chichester, pp. 89–99.

Jong, R. de and B.G.J. Knols, 1995. Olfactory responses of host-seeking *Anopheles gambiae* s.s. Giles (Diptera: Culicidae). *Acta Tropica* 59: 333-335.

Kalyanasundaram, M., 1982. A preliminary report on the synthesis and testing of mosquito repellents. *Indian J. Med. Res.* 76: 190-194.

Knight, K.L. and A. Stone, 1977. *A catalog of the mosquitoes of the world (Diptera: Culicidae).* II edition. 6, 1-611. College Park, MD: The Thomas Say Foundation, Entomological Society of America.

Knipe, Fred W. and Paul F. Russell, 1942. A Demonstration Project in the Control of Rural Irrigation Malaria by Antilarval Measures. *Journal of the Malaria Institute of India* 4(4):615-631.

Kogan, P.H., 1990. Substitute blood meal for investigating and maintaining *Aedes aegypti* (Diptera:Culicidae). *Journal of Medical Entomology* 27:709-712.

Konradsen, Flemming, Wim van der Hoek, Felix P. Amerasinghe, Clifford Mutero and Eline Boelee, 2005. Engineering and malaria control: learning from the past 100 years. Papers and Abstracts for the SIMA Special Seminar at the ICID 18th International Congress on Irrigation and Drainage, Montreal, 23 July 2005. *Malaria in Irrigated Agriculture. Working Paper* 47. SIMA Document 2. p.5.

Krishnamoorthy R., A. Munirathinam, K.J. Dhananjeyan, J. Hiriyan, T. Mariappan, Philip Samuel P. and A. Venkatesh, 2013. Description of a new species, *Toxorhynchites* (Toxorhynchites) *tyagii* (Diptera: Culicidae), from Nilgiri hills, Western Ghats, southern India. *Zootaxa.* 3701:447-459.

Maxwell, C.A., C.F. Curtis, H. Haji, S. Kisumku, A.I. Thalib, S.A. Yahya, 1990. Control of bancroftian filariasis by integrating therapy with vector control using polystyrene beads in wet pit latrines. *Trans Roy Soc Trop Med Hyg.* 84: 709–714.

Munirathinam A., K. Krishnamoorthi, A. Venkateshand and B.K. Tyagi, 2014. A note on spatial and temporal distribution of Anopheline mosquitoes in Western Ghats, Southern India. *International Journal of Mosquito Research* 1:14-18.

Nagpal B.N. and V.P. Sharma, 1995. *Indian Anophelines,* Oxford and IBH Publishing Co. Pvt. Ltd., New Delhi.

Najera, J.A., Kouznetzsov, R.L., Delacollette, C., 1998. Malaria epidemics: detection and control, forecasting and prevention. *WHO/MAL/98.1084,* pp. 1–81.

Natarajan, R., A.R. Rajavel and P. Jambulingam, 2016. Description of a new species of the genus Hulecoeteomyia (Diptera: Culicidae) from Meghalaya, India. *Zootaxa* 4137(3): 330–338.

Natarajan, R., and A.R. Rajavel, 2009. Description of a new species *Culex (Lophoceraomyia)* from Orissa, India. *Journal of American Mosquito Control Association* 25(4):403-408).

Nayduch D. 2009. Molecular tools used in medical and veterinary entomology. In: Mullen GR, Durden LA, editors. *Medical and Veterinary Entomology.* 2. London: Academic Press; pp. 543–555.

Paramasivan R., Philip Samuel P., V. Thenmozhi, R. Rajendran, Victor Jerald Leo S., K.J. Dhananjeyan, R. Krishnamoorthi, N. Arunachalam, and B.K. Tyagi, 2009. Chikungunya virus isolated in Lakshadweep islands in the Indian Ocean: evidence of the Central/East African genotype. *Japanese Journal of Infectious Diseases* 62, 67 - 69.

Patil, Prabhakar B., B.P. Niranjan Reddy, Kevin Gorman, K.V. Seshu Reddy, Shirish R. Barwale, Usha B. Zehr, Derric Nimmo, Neil Naish and Luke Alphey, 2014. Mating competitiveness and life-table comparisons between transgenic and Indian wild-type Aedes aegypti L. *Published online in Wiley Online Library*: DOI 10.1002/ps.3873

Prasad, Arti and Girima Nagda, 2018. Sporadic incidence of lymphatic filariasis in non-endemic state of Rajasthan and control of the vector (*Culex quinquefasciatus* Say, 1823) with innovative botanicals, with possible hypotheses on spread of 'endemism'. (In: *Lymphatic filariasis: epidemiology, treatment and prevention - the Indian perspective*. Springer, 314pp. Ed. B.K. Tyagi), pp. 171-183.

Puri, I.M., 1949. *Anophelines of the Oriental region.* Boyd's Malariology. W.B. Saunders 85 Co., New York, pp. 483-505.

Puri, I.M., 1955. *The Distribution of Anopheline Mosquitoes in India.* Health Bulletin No. 17; Malaria Bureau No. 7, Government of India. 104 pp.

Puri, I.M., 1960. *Synoptic tables for the identification of the full-grown larvae of the Indian anopheline mosquitoes.* Health Bulletin No. 16, Malaria Bureau No. 7., Government of India Press, Calcutta. 104 pp.

Rajagopalan, P.K. and P.K. Das, 1988. *Fileriasis control Demonstration Project.* Vector Control Research Centre, Pondicherry.

Rajagopalan, P.K., P.K. Das, M. Kalyanasundaram, B.K. Tyagi and N. Arunachalam, N. Somachary, C.B.S. Reddy and C.M.R. Reddy. 1987. *Bangalore mosquito control project: Master plan.* Vector control Research Centre, Pondicherry, pp. 309.

Rajavel, A.R. and R. Natarajan. 2006. Mosquitoes of the mangrove forests of India: Part 3-Andaman and Nicobar Islands, including an update on the mosquito fauna of the islands. *Journal of the American Mosquito Control Association* 22: 366–377. (doi: 10.2987/8756-971X(2006)22[366:MOTMFO]2.0.CO;2).

Rajavel, A.R. and R. Natarajan. 2011. First record of genus Ayurakitia in India: occurrence of Ayurakitia peytoni in Meghalaya state. *Journal of the American Mosquito Control Association* 27: 79–80 (doi: 10.2987/10-6062.1).

Rajavel, A.R., R. Natarajan and A. Munirathinam. 1998. Occurrence of *Culex (Culiciomyia) spathifurca* (Edwards) (Diptera: Culicidae). *Journal of the American Mosquito Control Association* 14: 346–347

Rajavel, A.R., R. Natarajan and K, Vaidyanathan, 2005a. Mosquito of mangrove forests of India: Part I- Bhitarkanika, Orissa. *J. Am, Mosq. Control Assoc.* 21; 131-135.

Rajavel, A.R., R. Natarajan and K, Vaidyanathan, 2005c. Mosquito of mangrove forests of India: Part II- Sundarbans, West Bengal . *J. Am, Mosq. Control Assoc.* 21; 136-138.

Rajavel, A.R., R. Natarajan and K. Vaidyanathan. 2004. *Aedes fumidus* and *Aedes amesii* in mangrove forests of India – two new country records. *Journal of the American Mosquito Control Association* 20: 315–316. (http://www.mosquitocatalog.org/files/pdfs/MQ0153.pdf).

Rajavel, A.R., R. Natarajan and K. Vaidyanathan. 2005b. Mosquito collection in the Jeypore hill tracts of Orissa, with notes on three new country records, *Culex (Lophoceraomyia) pilifemoralis, Culex (Lophoceraomyia) wilfredi* and *Heizmannia (Heizmannia) chengi. Journal of the American Mosquito Control Association* 21: 121–127

Rajavel, A.R., R. Natarajan and K. Vaidyanathan. 2005d. Mosquitoes of the mangrove forests of India: Part 1-Bhitarkanika, Orissa. *Journal of the American Mosquito Control Association* 21: 131–135 (doi: 10.2987/8756-971X(2005)21[131:MOTMFO]2.0.CO;2).

Rajavel, A.R., R. Natarajan, K, Vaidyanathan and N.P. Saniya, 2005e. Mosquito collections in the Jeypore hill tracts of Orissa, India, with notes on three new country records. *Culex (Lophoceraomyia) pilifemoralis. Culex (Lophoceraomyia) wilfredi, and Heizmannia (Heizmannia) Chengi. J. Am, Mosq. Control Assoc.* 21; 121-127.

Rajavel, A.R., R. Natarajan, K. Vaidyanathan and N.P. Saniya, 2005f. A list of the mosquitoes housed in the mosquito museum at the Vector Control Research Centre, Pondicherry, India. *J. Am. Mosq. Control Assoc.* 21 (3):243-51

Rajavel, A.R., R. Natarajan, K. Vaidyanathan and P. Jambulingam. 2011. Systematic list of the species added to the mosquito museum at the Vector Control Research Centre, Pondicherry, India. *Journal of the American Mosquito Control Association* 27: 8–14. (doi: 10.2987/8756-971X-27.1.8)

Rajendran, R., I.P. Sunish, T.R. Mani, *et al.*, 2004. Impact of two annual single-dose mass drug administrations with diethylcarbamazine alone or in combination with albendazole on *Wuchereria bancrofti* microfilaraemia and antigenaemia in south India. *Trans. R. Soc. Trop. Med. Hyg.*, 98: 174-181.

Ramalingam, 1987. On the restriction of *Armigeres durhami* Edwards and the description of *Armigeres kessli* n.sp. (Diptera: Culicidae). *Tropical Biomedicine* 4:55-65.

Rao, B.A., 1945. Malaria in the Irwin Canal area in Mysore State. Part I. *J. Mal. Inst. India* 6:101-108.

Rao, T.R., 1984. *The Anopheline of India* (Revised Edition.), Malaria Research Centre, Indian Council of Medical Research, New Delhi, 518 pp.

Rao, T.R., V, Dhanda, H.R/ Bhat and S/M/ Kulkarni, 1973. A survey of Haematophagous arthropods in western Himalays Sikkim and hill districts of West Bengal: A general account. *Indian J. Med. Res.* 61, 1421-1461.

Rattanarithikul, Rampa, Prachong Panthusiri and Suvanee Supavej, 2005. Illustrated keys to the mosquitoes of Thailand: *Southeastern Asian Journal of Tropical Medicine and Public Health* 36: 1-80.

Rebeiro, J.M.C., 1992. Characterization of a vasodilator from the salivary glands of the yellow fever mosquito, *Aedes aegypti. Journal of Experimental Biology* 165:61-71.

Rebeiro, J.M.C., 1995. Blood-feeding arthropods: live syringes or invertebrate pharmacologists? *Infect. Agents and Dis.* 4 (3): 143-152.

Reeves, L.E., C.J. Holderman, E.M. Blosser, J.L. Gillett-Kaufman, A.Y. Kawahara, P.E., Kaufman, N.D. Burkett-Cadena, 2018. Identification of Uranotaenia sapphirina as a specialist of annelids broadens known mosquito host use patterns. *Commun Biol* 1(1): 1-8.

Reinert, J.F., 2006. Tewarius Reinert, a new genus of Aedini (Diptera: Culicidae). *Proc. Entomol. Soc. Wash.* 108(3): 639-645.

Reuben, R., H. Kaul and Soman, S., 1988. Mosquitoes of arboviral importance in India. *Mosquito-borne Dis Bull.* 5: 48.

Reuben, R., S.C. Tewari and J. Hiriyan, 1993. Studies of the mosquito fauna of South India. In: *Proceeding of Entomologist Extraordinary: A festschrift in honour of Botha De Meillon*, 47-50.

Reuben, R., S.C. Tewari, J. Hiriyan and J Akiyama,1994. Illustrated key to species of *Culex (Culex)* associated with Japanese encephalitis in southeast Asia (Diptera: Culicidae*). Mosquito Systematics* 26:75-96.

Ross, R., 1897a. On some peculiar pigmented cells found in two mosquitoes fed on malaria blood. *British Medical Journal* 2: 1786-1788.

Ross, R., 1898a. Further observations on the transformation of crescents. *Indian Medical Gazette* (July, 1898).

Ross, R., 1898b. Preliminary report on the infection of birds with Proteosoma by the bites of mosquitoes. Government Press Calcutta (11 Oct, 1898); Indian Medical Gazette (Jan. 1899); *British Medical Journal* (18 Feb., 1899).

Ross, R., 1903. Malaria in India and the Colonies. Journ. R. Colonial Institute. (Dec, 1903).

Ross, R., 1910. Report on the measures taken against malaria in the Lahore (Mian Mir) cantonment. *Lancet* 2: 1371-3.

Russell, P.F. and Rao, H.R., 1940. The Anopheles of rice fields in South Eastern Madras. *Journal of Malaria Institute of India* 3: 427–446.

Russell, P.F., 1938. Malaria due to defective and untidy irrigation. A preliminary discussion. *Journal of Malaria Institute of India* 1: 339–349.

Sahu, S., P. Jambulingam, T. Vijayakumar and M. Kalyanasundaram, 2003. Impact of alphacypermethrin treated bed nets on malaria in villages of Malkangiri district, Orissa, India. *Acta Trop* 89(1):55-66.

Sebastian, A., M.M. Thu, M. Kyaw, M.M. Sein, 1980. The use of dragonfly nymphs in the control of Aedes aegypti. *Southeast Asian J Trop Med Public Health* 11(1):104-7.

Sen, R.N. and S.A. Azeez, 1963. History of malaria and its control in different coal fields of India. A review of 15 years in retrospect. *Ind. J. Malariol.* 17:157-178.

Service, M.W., 2008. *Medical Entomology for Students.* 4th edition. Cambridge University Press, 306pp.

Sharma, G.K., 1986a. *Malaria and its Control in India.* I. Delhi: Directorate of National Malaria Eradication Programme, 347 pp.

Sharma, G.K., 1986b. *Malaria and its Control in India.* III. Delhi: Directorate of National Malaria Eradication Programme. 618 pp.

Sharma, Ravendra K., H.G. Thakor, K.B. Saha, G.S. Sonal, A.C. Dhariwal, and Neeru Singh, 2015. Malaria situation in India with special reference to tribal areas. *Indian J Med Res.* 141(5): 537–545; doi: 10.4103/0971-5916.159510

Sharma, V.P., 1999. Current scenario of malaria in India. *Parassitologia* 41(1-3):349-353.

Sharma, V.P. and Mehrotra, K.N., 1986. Malaria resurgence in India: A critical study. *Social Science & Medicine* 22: 835–845.

Sharma, V.P. and R.S. Yadav, 1995. Impregnating mosquito nets with cyfluthrin. Study in the mining settlements of Orissa, India to control malaria. *Public Health* 12:8–17.

Sharma, V.P., Srivastava, A. and Nagpal, B.N., 1994. A study of the relationship of rice cultivation and annual parasite incidence of malaria in India. *Social Science & Medicine* 38:165–178.

Sharma. V.P., R.C. Sharma and A.S. Gautam, 1986. Bioenvironmental control of malaria in Nadiad, Kheda district, Gujarat. *Indian Journal of Malariology* 23: 95-117.

Singh, G and S. Prakash, 2008. A new species of *Lutzia* (Diptera: Culicidae) comparable with *halifaxii* Theobald 1903 in a semi-arid zone of India. (In: *Vector-borne diseases; Epidemiology and Control.* Ed. B.K. Tyagi), pp.195-204.

Singh, Himmat, Om P. Singh, Nasreen Akhtar, Gunjan Sharma, Ankita Sindhania, Nivedita Gupta and Neena Valecha, 2019. First report on the transmission of

Zika virus by *Aedes (Stegomyia) aegypti* (L.) (Diptera: Culicidae) during the 2018 Zika outbreak in India. *Acta Topica* 199: https://doi.org/10.1016/j.actatropica.2019.105114

Sinton, J. A., 1935. *What Malaria Costs India, Nationally, Socially and Economically*. Thacker's Press & Directories, 77 pp.

Sirivanakarn, S., 1976. Contributions of American Entomological Institute-III (A revision of the subgenus Culex in the Oriental region (Diptera: Culicidae). *Cont. Amer. Ent. Inst.* 12: 1-272.

Sivagnaname, N., D. Dominic Amalaraj and T. Mariappan, 2005. Utility of expanded polystyrene (EPS) beads in the control of vector-borne diseases. *Indian J Med Res* 122: 291-296.

Soltani A., H. Vatandoost, H. Jabbari, A.R. Mesdaghinia, A.H. Mahvi and M. Youne, 2008. Use of Expanded Polystyrene (EPS) and Shredded Waste Polystyrene (SWAP) Beads for Control of Mosquitoes. *Iranian J. Arthropod-Borne Dis,* 2(2): 12-20.

Spielman, A. and M. D'Antonio, 2001. *Mosquito: the story of man's deadliest foe.* London: Faber & Faber, vi+247 pp.

Stone, A. and M.D. Delfinado, 1973. *Family Culicidae: Catalog of the Diptera of Oriental Region.* Vol. I. Suborder Nematocera. University Press. Hawaii, Honolulu 9, 1-618.

Stone, A., 1961. A synoptic catalogue of the mosquitoes of the world. Supplement 1 (Diptera: Culicidae*). Proc: Entom. Soc. Washington* 63, 97-152.

Subbarao, S.K., K. Vasantha, T. Adak and V.P. Sharma, 1983. *Anopheles culicifacies* Complex: Evidence for a new sibling species, Species C. *Annals of the Entomological Society of America* 76(6): 985-988.

Subbarao, S.K., N. Nanda and Raghavendra, K.,1999. Malariogenic stratification of India using *Anopheles culicifacies* sibling species prevalence. *ICMR Bulletin* 29(7):75-80.

Sunish, I.P., M. Kalimuthu, Ashok Kumar, R. Rajendran, A. Munirathinam, N. Nagraj, N. Arunachalam, and B.K. Tyagi, 2018. Integrated vector control for the elimination of bancroftian filariasis in the villages of Tirukoilur, south India. (In: *Lymphatic filariasis: epidemiology, treatment and prevention – the Indian perspective.* Springer Nature Pvt. Ltd., 314pp. Ed. B.K. Tyagi), pp. 185-192.

Sunish, I.P., R. Rajendan, T.R. Mani, A. Munirathinam, A.P. Dash and B.K. Tyagi, 2007. Vector control complements the effects of mass drug administration against bancroftian filariasis in Trikoilur, south India. *Bulletin of World Health Organization* 85:138-145.

Sweet, W.C. and T.R. Rao, 1934. Notes on malaria in Mysore State. Part V. The control of anopheline breeding in Bangalore City and its cost in Mysore State. *Rec. Mal. Surv. India* 2:655-657.

Tewari, S.C., Hiriyan, J., Reuben, R., 1992. Description of two new species of *Aedes* (Diceromyia) from south India (Diptera: Culicidae). *Mosquito Systematics* 24:154-175.

Tewari, S.C., J. Hiriyan and R. Reuben, 1991. Description of a new species of *Aedes* (*Rhinoskusea*) from south India. *Mosquito Systematics* 23:123-131.

Tewari, S.C., J. Hiriyan and R. Reuben. 1995. Description of *Aedes (Finlaya)* niveus (Diptera: Culicidae) from south India. *Mosquito Systematics* 27:167-176.

Tewari, S.C., J. Hiriyan, K. Ayanar, A. Munirathinam, A. Venkatesh, R. Reuben Tyagi B.K., 2007. *CRME Mosquito Museum: An Annotated Checklist of Indian Mosquito Species.* Contributions of the Centre for Research in Medical Entomology, Madurai, , pp. 175.

Thenmozhi, Velayutham, Thiruppathi Balaji, Kasiviswanathan Venkatasubramani, Kutty Jagadeeswaran Dhananjeyan, Alagarsamy Selvam, Veeramanoharan Rajamannar, Brij Kishore Tyagi. 2016. Role of *Anopheles subpictus* Grassi in Japanese encephalitis virus transmission in Tirunelveli, South India. *Indian Journal of Medical research* 144 (3): 477-481.

Theobald, F.V. 1910b. Second report on the collection of Culicidae in the Indian Museum, Calcutta, with descriptions of new genera and species. *Rec. Indian Mus.* 4:1-33, 3 pls.

Theobald, F.V., 1910a. A monograph on Culicidae mosquitoes. *Brit. Mus. (Nat. Hist., London,* , 1-646.

Tuno Nobuko, Tran Vu Phong and Masahiro Takagi, 2020. Climate Change May Restrict the Predation Efficiency of Mesocyclops aspericornis (Copepoda: Cyclopidae) on Aedes aegypti (Diptera: Culicidae) Larvae. *Insects* 11(5): 307. doi: 10.3390/insects11050307

Tyagi, B.K., 1992. Control of malaria vectors in India. *Indian Review of Life Sciences,* 12: 211–238.

Tyagi, B.K., 1994a. Distribution of arthropod vector-borne communicable diseases and control of their vectors in India. *Indian Review of Life Sciences* 14: 223-243.

Tyagi, B.K., 1994b. Thanatosis as an adaptive defence mechanism in Anopheles stephensi larvae breeding in “Tanka” and “Beri” in the desert villages of Rajasthan. *Geobios New Reports* 13: 132-135.

Tyagi, B.K., 1994c. A note on the breeding of vector mosquitoes in cement tanks and pit latrines. *Journal of Applied Zoological Researches* 5: 149-151.

Tyagi, B.K., 1994d. Distribution of malaria transmitting mosquitoes (Culicidae: Anophelinae; Anophelini) in the Thar desert. *Abstract Paper, Meeting of Experts on Gaps in Research on the Faunal Diversity in the Thar Desert,* Jodhpur, p. 37.

Tyagi, B.K., 1994e. Need for imparting training to personnels undertaking insecticide spraying for vector control with special reference to xeric conditions in western Rajasthan. *Abstract Paper, IV Symposium of Vectors & Vector-borne Diseases*, Bhubaneshwar, p. 82.

Tyagi, B.K., 1994f. Alternative control strategies for mosquito vectors of human diseases in India, pp.359-372 (In: *Perspectives in Entomological Research*) Scientific Publishers (India), Jodhpur, pp.406.

Tyagi, B.K., 1995. Malaria in the Thar Desert. ICMR Bulletin 25: 85-91.

Tyagi, B.K., 1997. Malaria epidemics in the Great Indian Thar Desert and the validity of the Bauma-van der Kaay's El Nino Southern Oscillation theory as an early warning system for future epidemics. *Annals of Medical Entomology* 6: 19-24.

Tyagi, BK., 2001. Major vector-borne diseases in the Thar Desert, with special reference to the changing climate, pp. 103-112 (In: *Priorities of research for development of arid regions*. Eds.: Dr. S.K. Verma, Dr. N.M. Nahar, Dr. R. Raj-Bhansali, Dr. S.K. Jindal and Dr. Satya Vir) Scientific Publishers, (India), Jodhpur, pp. 238.

Tyagi, B.K., 2002. Malaria in the Thar Desert: Facts, Figures and Future. *Agrobios (India)*. pp. 165.

Tyagi, B.K., 2003. *Handbook of Medical entomology*. Scientific Publishers (India), Jodhpur. pp. 262.

Tyagi, B.K., 2004a. *The invincible deadly mosquitoes: India's Health and Economy Enemy No. 1*. Scientific Publishers, India, pp 266.

Tyagi, B.K. 2004b. A review of the emergence of *Plasmodium falciparum*-dominated malaria in irrigated areas of the Thar Desert, India. *Acta Tropica* (Special Issue: Malaria & agriculture) 89: 227–239.

Tyagi, B.K., 2014. *The Tirukoilur Project*. Centre for Research in Medical Entomology, Madurai, pp. 291.

Tyagi, B.K., 2018. *Lymphatic filariasis: epidemiology, treatment and prevention – the Indian perspective*. Springer, pp 314.

Tyagi, B.K. and R.C. Chaudhary, 1997. Outbreak of falciparum malaria in the Thar Desert with particular emphasis on physiographic changes brought about by extensive canalization and their impact on vector density and dissemination. *Journal of Arid Environments* (UK) 36: 541-555.

Tyagi, B.K., R.C. Chaudhary and S.P. Yadav 1995. Epidemic malaria in Thar desert, India. *The Lancet*, 346: 634–635.

Tyagi, B.K. and J. Hiriyan, 2004. Breeding of Dengue Vector *Aedes aegypti* (Linnaeus) in Rural Thar Desert, North-western Rajasthan, India. *Dengue Bulletin* 28: 220-222.

Tyagi, B.K. and S.P. Yadav, 1996. Malariological and sociological significance of 'tanka' and 'beri' in the Thar Desert, Western Rajasthan, India. *Journal of Arid Environments*, 33: 497–501.

Tyagi, B.K. and S.P. Yadav, 2001. Bionomics of malaria vectors in two physiographically difference areas of the epidemic-prone Thar Desert, north-western Rajasthan (India). *Journal of Arid Environments* 47: 161-172.

Tyagi, B.K., and K.V.S. Verma, (1991). Anopheline mosquitoes of Sri Ganganagar district (Rajasthan) transmitting malaria parasite. *Journal of Applied Zoological Researches* 2: 85-91.

Tyagi, B.K., S.K. Bansal and K.V.S. Verma, 1991. Intraspecific morphological variations in two ecotypes of Anopheles stephensi Liston in northwestern Rajasthan (Diptera: Culicidae). *Annals of Entomology*, 9: 85–89.

Tyagi, B.K., Sajal Bhattacharya and B. Reddyanaik, 2020. *Dr Ronald Ross: Mosquito, malaria, India and The Nobel Prize*. Scientific Publishers, India, pp 258.

Tyagi, B.K., M. Kalyanasundaram, P.K. Das and N. Somachary,1985. Evaluation of a new compound (VCRC/INSA/A23) with juvenile hormone activity against the mosquito vectors. *Indian Journal of Medical Research* 82: 9-13.

Tyagi, B.K., Hiriyan, J., Tewari, S.C. and Thenmozhi, V. 2002-2003. *Studies on dengue emergence in Kerala state, India*. Annual Report. Centre for Research in Medical Entomology, Madurai, p.49.

Tyagi, B.K., J. Hiriyan, , S.C. Tewari, V. Thenmozhi, and Philip Samuel, P., 2003-2004. *Studies on dengue emergence in Kerala state, India*. Annual Report. Centre for Research in Medical Entomology, Madurai, p.48.

Tyagi, B.K., J. Hiriyan, S.C. Tewari, K. Ayanar, Philip Samuel P, N. Arunachalam, R. Paramasivan, R. Krishnamoorthy, K.J. Dhananjeyan, S. Victor Jerald Leo, R. Rajendran., 2009. Description of a new species of Anopheles pseudosundicus (Diptera : Culicidae) from coastal Kerala. *Zootaxa* 2219: 49-60.

Tyagi, B.K., J. Hiriyan, P. Philip Samuel, S.C. Tewari and R. Paramasivan. 2006. Dengue in Kerala: A critical review. *ICMR Bulletin* 36 (4-5): 13-22.

Tyagi, B.K., A. Munirathinam and A. Venkatesh, 2015. A catalogue of Indian mosquitoes. *International Journal of Mosquito Research* 2 (2): 50-97.

Tyagi, B.K, A. Munirathinam, R. Krishnamoorthy and A. Venkatesh, 2012. *A field-based handbook of identification keys to mosquitoes of public health importance in India. Volume 1*. Taxonomy and Biodiversity Cell. Centre for Research in Medical Entomology. pp1-55.

Tyagi, B.K., A. Munirathinam, R. Krishnamoorthi, G. Baskaran, A Venkatesh, M. Kalimuthu and R. Govindarajan, 2014. *A field-based handbook of identification keys to 4th instar larvae of mosquitoes of public health importance in India. Volume 2*; Taxonomy and Biodiversity Cell. Centre for Research in Medical Entomology, pp 1-45.

Tyagi, B.K., K.V. Singh, S.K. Bansal and S.P. Yadav 1994. Malaria epidemic in some villages of north-western desert Rajasthan. *Journal of Applied Zoological Researches*, 5: 152–155.

Tyagi, B.K., N. Somachari, V. Vasuki and P.K. Das, 1987. Evaluation of three formulations of a chitin synthesis inhibitor (fenoxycarb) against mosquito vectors. *Indian Journal of Medical Research* 85: 161-175.

Venkat Rao, V., 1949. A critical review of malaria control measures in India. *Ind. J. Malariol.* 3:313-326.

Venkatesh A. and B.K. Tyagi, 2015. Predatory potential of *Bradinopyga geminata* and *Ceriagrion coromandelianum* larvae on dengue vector *Aedes aegypti* under controlled conditions (Anisoptera: Libellulidae; Zygoptera: Coenagrionidae; Diptera: Culicidae). *Odonatologica* 42(2):139-149.

Vijay Veer, 2012. Dr. B. K. Tyagi: Arthropodologist and polymath. 6th Conference of Medical Arthropodology, *CRME Bio-Bibliography Ser. No.* 8; pp.49.

Viswanathan, D.K., 1946. The seasonal prevalence of malaria transmitted by *Anopheles fluviatlis* James, 1902, in different regions of Bombay Presidency. *J. Mal. Inst. India* 6:253-257.

Viswanathan, D.K., 1950. *Malaria and its Control in Bombay State.* Aryabhushan Press, Pune.

Wattal, B.L., 1963. Keys to the *Anopheles* of the world. 5. Southern and northern Asia. *In*: *Practical malariology*. Second edition. Ed. , P.F. Russell, L.S. West, R.D. Manwell and G. MacDonald, Oxford University Press, London), pp. 680–696.

Watttal, B.L. and N.L. Kalra. 1961. Region wise keys to female Indian *Anopheles. Bulletin of Natural Society of India Malaria Mosquito Borne Diseases* 10:55-138.

WHO SEARO, 2007. *Anopheline Species Complexes in South and South-East Asia.* WHO Regional Office South East Asia, Technical Publication No. 57, pp. 1-102.

WHO, 2013. *World Malaria Report, 2013.* World Health Organization, Geneva, 1-284.

Zahar, A.R., 1985. Vector Bionomics in the Epidemiology and Control of malaria. Part I.III. East Africa. Geneva: *World Health Organization, WHO/VBC/85.3.*

Zahar, A.R., 1990. Vector bionomics in the epidemiology and control of malaria. II. The WHO European Region and The WHO Eastern Mediterranean Region. Vol. II. Applied Field Studies. *WHO/VBC/90·1, Mal/90·1* pp.90.

GLOSSARY

ABBRASION – Damaged area of skin.

ABIOTIC FACTORS – The climatic characteristics (temperature, humidity, saturation deficit, rainfall, light, wind) and edaphic characteristics (type of soil, soil texture) of a region or habitat.

AEDEAGUS – A finger-like evagination of the ventral body wall of male mosquito enclosing the terminal section of the ejaculatory duct.

ALLERGY – A state of hypersensitivity to a given material (called an allergen), acquired through exposure; re-exposure to the Allergen shows altered reactivity.

ANTENNAE – Paired appendage of head of insects and other arthropods. Usually with many segments. Antennae carry chemosensory organs.

ANTERIOR – The front end.

ANTERODORSAL – The front of the upper surface.

ANTEROVENTRAL – The front of the under surface.

ANTHROPOPHILIC – Man-biting (literally: liking man).

ANTICOAGULANT – Stops blood from clotting.

APICAL – At the tip or end of (APEX).

ARBOREAL – Living in trees.

ARTHROPOD – An animal belonging to the phylum Arthropoda in the animal kingdom, having a hard jointed exoskeleton and paired, jointed legs. The phylum includes, among other classes, the Arachnida and the Insecta, many of which are important medically as parasites or as vectors of organisms capable of causing disease in man.

ATTACK PHASE (Of a control programme) – the phase in which all houses are treated, regardless of whether or not the house was infested.

BIOECOLOGY – The study of the interrelationships between plants and animals and their abiotic environment.

BIOLOGICAL CONTROL – Control measures by means of living organisms.

BIOLOGY – The science that deals with the phenomena of life and with living organisms in general.

BIONOMICS – The (quantitative) relation of the development, reproduction and survival of an organism or population of organisms to factors in the environment.

BIOTIC FACTORS – Factors connected directly with the organisms living in a given environment, *e.g.*, nature and availability of hosts, water quality, plant environment, etc.

BIOTYPE – A small topographic unit, including the biotic community.

BRIQUET'l'ES – An insecticide mixed with clay, Plaster-of-Paris (gypsum) or other substance that will allow it to be moulded and hardened into brick-like blocks that will slowly dissolve in water, thereby releasing the insecticide over an extended period.

CHITIN – Hard skeletal material found in the majority of Arthropoda.

CHROMOSOME – A linear structure within the nucleus of a cell composed of proteins and DNA which holds the genetic material or genes of an organism, the arrangement of which is unique to each organism.

CIBARIAL PUMP – The anterior part of the pharynx, joined to powerful muscles inside the head of insect, which is used to suck up blood in the course of a blood-meal.

CONTAMINATION – The unintentional entrance, into a material (e.g., food) of foreign substance, matter, or organism.

CONTROL – Of insects (or other undesirable animals), the restriction of the population density of such insects to a level below that at which they can be harmful to the interests of man.

COPULATION – Physical union of male and female during mating.

COXA – The first basal segment of the leg of an insect.

CUTICLE – A non-living layer secreted by, and overlying the epidermis.

CUTICULAR HYDROCARBONS – Hydrocarbon chemicals in the surface layer of the cuticle which can be used to separate sibling species. Examination requires expensive and sophisticated apparatus, but has the advantage of using dried material.

CYTOSPECIES – Species which because of their morphological similarity, have been identified primarily by cytological characters, such as the banding patterns of the chromosomes.

CYTOTAXONOMY – The description of species based on a study of the banding patterns of the chromosomes.

DEC – Diethylcarbamazine, a microfilaricidal drug.

DENSITY (of flies) – Number of flies per given area or other unit.

DENSITY-DEPENDENT – A term used to describe factors that vary according to the population density, and provide a form of "negative feedback" to stabilize population numbers. All living populations must be limited by density -dependent factors operating on either their birth-rate or death-rate (or both), otherwise their populations would increase forever. In some cases, however, populations can be limited by density-independent factors, before density-dependent factors come into operation.

DERMATITIS – Inflammation of the skin.

DERMIS – The true skin or elastic vascular layer lying below the outer protective layer of the skin (the epidermis).

DIPTERA – An order of insects containing the two-winged flies which have only the anterior pair of wings well developed.

DISPERSION – The movements of flies in and around a locality or area in which they emerged or are found at a given time.

DIURNAL – Daytime - usually relating to a cycle of activity. The opposite of nocturnal.

DIVERTICULAM – A long sac-like pocket, closed at the far end.

DOMESTIC – Associated with houses.

DORSAL – Upper surface or back.

ECOLOGY – The study of the inter-relationship between organisms and their environment (including other organisms).

ECTOPARASITE – A parasite that lives on the external surface of an animal, e.g. fleas, lice, mites, ticks.

EMBRYO – The early or developing stage of any organism, especially the developing product of fertilization of an egg.

EMBRYOGENESIS – The formation or development of eggs or offspring.

EMIGRATION – The movement out from a given locality or area.

EMPODIUM – Thin claw-like structure arising medially between the claws of the tarsi.

ENDEMIC DISEASE – A disease that is caused by factors that are constantly present within a given geographical area or population group.

ENDOPHILIC INSECTS – Insects that regularly enter houses and spend at least part of their life indoors.

ENVIRONMENT – The sum total, at a given moment, of all external influences and conditions to which an organism or object is subjected, including all physical, chemical and biological factors.

ENZOOTIC – Pertaining to a disease that occurs in animals especially one that is restricted to a limited geographical area.

ENZYMATICALLY – Separation of isoenzymes to identify some sibling species of mosquito sibling species complexes.

EPIDEMIC – A period of increased prevalence of a given disease in a population (usually, but not necessarily, as a result of factors that are not constantly present in the populations).

EPIDEMIOLOGY – The study of the factors that determine the frequency and distribution of disease and of other health-related states in populations.

EPIDERMIS – The epithelium covering the body.

ERADICATION – Complete removal or destruction (e.g., of an infectious agent in a given geographical area).

EXOPHILIC INSECTS – Insects that normally do not enter buildings.

EXUVIUM – The cast outer skin which is shed when an insect moults to its next development stage.

FACETS – The hexagonal shaped divisions on the surface of the eyes.

FECUNDITY – Reproductive potential.

FEMUR – The third segment of an insect's leg, distal to the trochanter.

FILAMENTOUS – Thin and thread-like.

FUMIGATION – A process that uses gaseous agents to kill animals.

GENETIC – A hereditary factor controlled by genes of the previous generation.

GENITALIA – The external genital organs of the insect; the gonads and their ducts and all associated accessory organs.

GENUS – (plural, genera) a taxonomic category within a family, consisting of one or more generally similar species.

GONADS – Internal sex organs.

GONOTROPHIC CONCORDANCE – The production of an egg-batch immediately following a blood meal, so that blood-feeding and egg-laying follow each other in a regular sequence.

GONOTROPHIC CYCLE – The events between successive egg-laying, or the time interval between them.

GRAVID – Swollen and full of nature or nearly nature eggs.

HABITAT – The physical place (home) where an organism lives.

HAEMATOPHAGOUS – The term applied to an insect that feeds exclusively on blood; in mosquitoes only females blood-feed.

HAEMOCOEL – The physiological fluid in which the internal organs of insects are bathed.

HEAD – The anterior region of an insect, which bears the mouthparts, eyes, antennae and houses the brain.

HOST – The animal parasitized by a parasite or disease organism.

HUMIDITY OR RELATIVE HUMIDITY – The content of water vapour in the air measured as a percentage of the content of saturated air (100% RH).

HYPERENDEMIC – Heavily endemic.

HYPOPHARYNX – A very fine stylet joined to the unpaired salivary duct and used for the injection of saliva during biting.

IMMUNITY – Resistance to parasites or other noxious agents or organisms in a given time, e.g., a year.

INCUBATION PERIOD – The interval of time from the moment of entrance of the micro-organisms into the animal and the first appearance of symptoms of the disease.

INFECTION – The invasion of a host by pathogenic micro-organisms (establishing a host-parasite relationship) and the subsequent multiplication of the latter within the body of the host.

INFESTATION – l. The invasion of the surface of the body of the host by parasites, 2. The harbouring of parasites on the surface of the body.

INSECT DEVELOPMENT INHIBITORS (IDI) – Chemicals which inhibit the development of larvae or pupae.

INSECTICIDE – A chemical substance or a mixture of substances used to kill insects. It may be applied as a liquid, powder, fine spray or as a vapour. The term "larvicide" is usually applied to an insecticide that kills immature stages of insects, and the term "adulticide" to an insecticide that kills mature or adult forms.

INSECTICIDE RESISTANCE – A characteristic that renders a given population of an insect species resistant to a given insecticide although the species is normally susceptible to that being no longer controlled by the insecticide in the area concerned.

INSECTIVORE – Any insect-eating (insectivorous) group.

INSTAR – The stage of an insect's life between successive moults, for example the first instar is between hatching from the egg and the first moult.

INTEGRATED CONTROL – Applied pest control that contains and integrates chemical, physical and biological control measures (including the use of natural enemies of pests.

IVERMECTIN – A drug which has been used in the veterinary field for 1nany years, and has now been approved for limited use against human onchocerciasis. So far appears to act as a microfilaricide only.

LABIUM – The lower part of the proboscis forming, together with the labrum, a canal through which the blood is sucked up (food canal) and containing the hypopharynx.

LABRUM – The upper part of the proboscis of mosquito, sealing off the food canal.

LARVA – An immature form occurring in some animals after hatching, differing markedly from the adult form.

LARVICIDES – Chemicals which kill larvae directly or by preventing them from developing into adults.

LIFE CYCLE – Stages of development through which an organism passes between fertilization of the egg of one generation and the same stage in the subsequent generation. The term is sometimes used loosely to mean the number of days between egg- deposition and the attainment of sexual maturity.

MAMMAL – An individual belonging to the class Mammalia, a division of vertebrate animals, including all that possess hair and suckle their young.

MANDIBLE – toothed paired stylets in mouth used for sawing the skin.

MESONOTUM – The notum (tergum) of the mesothorax.

METAMORPHOSIS – In insects, the series of changes through which the insect passes in its development from the larval through the pupal to the adult form. If the pupa is inactive and does not feed, metamorphosis is said to be "complete"; if there is no pupa, or if the pupa is active and feeds, it is said to be "incomplete".

MICROFILARICIDE – A drug which kills microfilaria but not the adult worms. The most common drugs are DEC and ivermectin.

MICROORGANISM – Any microscopic or ultramicroscopic animal or plant, especially any of the bacteria, protozoa, or viruses.

MORTALITY – The action of death on populations; more specifically, the number of deaths in a given population in a given period of time.

NATURAL HOST – A host in which a given pathogenic microorganism (or parasite) is commonly found and in which the pathogen can complete its development. The term "natural host" implies that the host is the usual one; it has the same meaning as "typical host".

NOCTURNAL – Night time, usually applied to an activity cycle. Opposite of diurnal.

NOMENCLATURE – The scientific (Latin) names given to species, genera, families, etc. of animals and plants.

NULLIPAROUS - The term applied to a female mosquito that has not yet laid the eggs.

OBLIGATE - Organism that must live as a parasite of its host.

OLFACTORY - Associated with the sense of smell.

OVARIOLE - Each individual egg in its tube, including the ovarian follicle and oocyte, many of which make up an ovary.

OVIDUCT - An unpaired duct through which the eggs pass, opening into the uterus at its anterior extremity, and dividing into a pair of oviducts, each of which ends in one of the two ovaries of the female mosquito.

OVIPAROUS - Producing young by means of eggs expelled froln the body before they are hatched.

OVIPOSITION - The act of depositing eggs.

PARASITE - An organism which derives its nutrition by competing with its host, often with deleterious results to the host, or by feeding upon its tissues.

PAROUS - Having matured and laid eggs. Therefore by inference having mken at least one blood meal. UNIPAR - one gonotrophic cycle; MULTIPAR - more than one gonotrophic cycle.

PATHOGEN - An organism which causes disease.

PATHOGENIC - Causing or giving rise to disease.

PATHOGENS - Microorganisms (germs) such as viruses, bacteria, protozoa and eggs and cysts of worms, which can cause disease if man is infected.

PERIDOMESTIC - Associated with outhouses, storage houses, chicken houses or other animal enclosures, etc., near the house.

PEST - Any mosquito species that is harmful in any way, biting or otherwise, or present in sufficient numbers to be considered a nuisance by man.

PESTICIDE - An agent or a substance or a mixture of substances used to kill species regarded as pests by man.

PHYSIOLOGICAL AGE - The stage of oogenesis (development of the ova) in the female, from which the true age of the insect can be determined with a high degree of precision.

PHYTOPHAGOUS - Feeding on plants.

POLYTENE - Chromosome which have divided many times but have not separated, thereby forming thick bundles of up to a thousand parallel strands. After staining, their banding pattern can be seen under moderate magnification, and they are therefore convenient for examination. They are found in the nuclei of the cells of a few specialized tissues only.

POLYTENE CHROMOSOMES – Chromosomes which have divided many times but have not separated, thereby forming thick bundles of up to a thousand parallel strands. after staining, their banding pattern can be seen under moderate magnification, and they are therefore convenient for examination. they are found in the nuclei of the cells of a few specialized tissues only.

POPULATION DENSITY – The number of individuals of a given population per unit area or volume.

POTENTIAL – Existing and ready for action, but not yet active.

PRECIPITIN TESTING – A serological technique whereby the source of a blood meal can be identified by testing it with antibodies to blood from known host species.

PREDATOR – An animal which preys upon insects or other animals.

PREPARATORY PHASE (of a control programme) – The planning phase, when areas are mapped and surveyed for infestations, prior to the attack phase.

PROBOSCIS – The mouthparts of mosquito, which form a tube containing the stylets which penetrate the host tissues in order to suck up the host fluids.

PRURITUS – Severe itching.

PULVILLI – Fleshy pads at the extremity of the last joint of the tarsus of the legs of insects; when an insect settles on a surface covered with a contact insecticide, the latter enters the body of the insect through the pulvilli.

PUPA – The free living, highly mobile but not feeding stage of development of a mosquito that occurs between the larval and adult forms.

RESIDUAL TREATMENTS – Treatments with insecticide formulations having a long-term effect (weeks or months).

SIBLING – Very closely related. Literally means brothers or sisters. ln taxonomy means the species which make up a single species complex and usually cannot be distinguished by morphological features.

SILVATIC – Not associated with domestic or peridomestic environments, but literally associated with the forest.

SPECIES – The smallest unit of classification commonly used, i.e., the group whose members have the greatest mutual resemblance, are able to interbreed (if not separated in space or time), but not to breed with organisms of other groups (species). Species may be divided into subspecies which differ in certain genetic characters but are able to interbreed.

SPERMATHECA – An organ of the female reproductive system in which mature sperms are stored after copulation.

SPERMATOZOON – The male gamete.

SPIRACLES – External openings of the tracheae, the air tubes providing oxygen for the tissues of insects.

STERNITES – Hard chitinous plates.

SUBSPECIES – A geographical group of local populations, differing taxonomically from the other groups of the same type forming part of the species under consideration.

SURVEILLANCE – Observation of a thing, an animal, or a person. The term also implies supervision or inspection.

SURVEY – To examine for some specific purpose; to inspect or consider carefully; to review in detail.

SYNCHRONOUS – Simultaneous or taking place at the same time.

TARSUS – The last segment of the leg of insects: it consists of five joints, to the last of which are attached the claws and pulvilli.

TAXON (plural TAXA) – The basic taxonomical units. Often refers to species, but can refer to an element of the classification at any level.

TAXONOMY – Classification of animals and plants divided into species, genera, families, etc.

TERGITE – Hard chitinous plate forming part of the cuticle of a segment (dorsal surface).

TERMINALIA – The last two or three visible segments and their appendages at the tip of the abdomen which make up the externally visible sex organs of either sex.

TERRESTRIAL – Living on the ground.

THORAX – Flies and other insects are divided into three regions: head, thorax and abdomen. The thorax carries the legs and wings.

TIBIA – The fourth segment of an insect's leg, distal to the femur.

TROCHANTER – The second segment of an insect's leg, distal to the coxa.

ULV – Ultra-low-volume concentrate.

VECTOR – An arthropod or other animal that carries a parasite from one host to another host. The vector may or may not be essential for the completion of the life cycle or the parasite. If it is not essential, it is referred to as a "mechanical vector".

VENTER – Underneath surface of the abdomen.

VENTRAL – Pertaining to the front or lower surface.

VISCERA – Internal organs of the abdomen.

VOLATILITY – Capability to evaporate rapidly.

WETTABLE POWDER – A water dispersible but insoluble, formulation of an insecticide, in which the active ingredient is bound to an inert insoluble carrier material such as talcum powder. Wettable powder formulations are mixed with water to be sprayed, and usually have greater residual activity than liquid formulations of the same insecticide.

ZOONOSIS – A disease of animals that may be transmitted under natural conditions to man.

ZOOPHILIC – Feeding on animals other than man.

ZOOTIC – Pertaining to animals.